Cognitive Radio, Software Defined Radio, and Adaptive Wireless Systems

Edited by

HÜSEYIN ARSLAN
University of South Florida, Tampa, FL, USA

Springer

A C.I.P. Catalogue record for this book is available from the Library of Congress.

ISBN 978-1-4020-5541-6 (HB)
ISBN 978-1-4020-5542-3 (e-book)

Published by Springer,
P.O. Box 17, 3300 AA Dordrecht, The Netherlands.

www.springer.com

Printed on acid-free paper

Contents

Preface

Today's wireless services have come a long way since the roll out of the conventional voice-centric cellular systems. The demand for wireless access in voice and high rate data multi-media applications has been increasing. New generation wireless communication systems are aimed at accommodating this demand through better resource management and improved transmission technologies.

The interest in increasing Spectrum Access and improving Spectrum Efficiency combined with both the introduction of Software Defined Radios and the realization that machine learning can be applied to radios has created new intriguing possibilities for wireless radio researchers.

This book is aimed to discuss the cognitive radio, software defined radio (SDR), and adaptive radio concepts from several perspectives. Cognitive radio and cognitive networks are investigated from a broad aspect of wireless communication system enhancement while giving special emphasis on better spectrum utilization. Applications of cognitive radio, SDR and cognitive radio architectures, cognitive networks, spectrum efficiency and soft spectrum usage, adaptive wireless system design, measurements and awareness of various parameters including interference temperature and geo-location information, physical layer access technologies, and cross-layer adaptation related concepts are some of the important topics that are covered in this book.

Broad Topical Coverage

Following the Introduction (Chapter 1), the structure of the book consists of basically four parts:

- Fundamental concepts and architectures (Chapters 2–7)
- Awareness, measurements, and sensing (Chapters 8–10)
- Physical layer access technologies (Chapters 11 and 12)
- Applications of cognitive radio and SDR (Chapters 13 and 14)

In Chapter 1, a broad coverage of cognitive radio concept along with SDR is provided. A brief history and motivating factors for cognitive radio, including the economic aspects, are given. In Chapter 2, cognitive network concept is introduced and compared with cognitive radio and cross-layer adaptation. Cognitive network design and the related entities are also provided in this chapter. The architecture of an ideal cognitive radio, particularly with respect to the critical machine learning technologies, is described in Chapter 3. The components of the cognitive radio architecture and their functions, the cognition cycle, the inference hierarchy, and how to build cognitive radio architecture on the top of SDR architecture are all explained in detail in this chapter. The software defined radio architectures and relations with cognitive radio are further described in Chapter 4. Ideal and practical SDR architectures are given. The building blocks of SDR architectures are also described in detail. In Chapter 5, communications value-chain is discussed and some ways in which the value-chain can be altered by cognitive radios and cognitive networks are explored. Dynamic spectrum access in cognitive radio is extremely important concept as it enhances the spectrum efficiency greatly. Therefore, Chapter 6 is allocated for the discussion of dynamic spectrum access aspect of cognitive radio from game and coding theoretic perspectives. In Chapter 7, co-existence of different types of cognitive radios is covered, as well as modeling the efficiency of the cognitive radios from the Medium Access Control (MAC) perspectives.

Chapters 8 through 10 discuss the issues related to awareness, sensing, and measurement in cognitive radio systems. General concepts, various cognitive measurements and awareness functions are provided in Chapter 8. In Chapter 9, one of the most important awareness in cognitive radio, which is the spectrum awareness, is discussed in detail. Multi-dimensional spectrum space is explained, methods and challenges to sense the spectrum space are given. Another very important awareness, which is location of users, is given in Chapter 10. Location information management in cognitive wireless network is discussed from various dimensions.

The physical layer aspects of cognitive radio are discussed in Chapters 11 and 12. Orthogonal Frequency Division Multiplexing (OFDM) has proved itself to be the access technology for future generation mobile radio systems. In Chapter 11, the suitability of OFDM technology for cognitive radio is discussed. Another very interesting access technology that could potentially be very useful for cognitive radio for both communication and accurate positioning is ultra wideband (UWB). In Chapter 12, UWB and its suitability for cognitive radio are discussed.

Finally, Chapters 13 and 14 discuss the applications of cognitive radio and SDR. A generic overview and various application scenarios are given in Chapter 13. Chapter 14 focuses on the application of cognitive radio from cross-layer adaptation and optimization perspective.

Audience

This book is intended to be both an introductory technology survey/tutorial for beginners and an advanced technical overview intended for technical professionals in the communications industry, technical managers, and researchers in both academia and industry. We expect that a reader could skip the advanced technical treatments of the topics and still benefit greatly from the book.

A basic background of wireless communications is preferable for a full understanding of the topics covered by the book.

Course Use

The book provides an organic and harmonized coverage of cognitive radio, from radio hardware and digital baseband signal processing, all the way to applications. Within this framework, the book chapters are quite independent from one another. Therefore, different options are possible according to different course structures and lengths, as well as targeted audience background. The topics are covered in both descriptive and technical manners, and can therefore cater to different readers needs. For each chapter we expect that a reader may skip the advanced technical description and still greatly benefit from the book.

The objective of this book is to provide an introduction to the major research issues and challenges in cognitive radio that are currently occupying research attention worldwide. As such, the book is primarily intended to serve as a reference for comprehensive understanding of recent advances in both theory and practical design of cognitive radio and networks.

Acknowledgements

I wish to thank all my colleagues who are authors of this book for working with us and giving their time and effort in making and supporting this book project.

I would also like to particularly thank our editor Mark de Jongh, as well as all the entire editorial staff at Springer.

Finally, I would like to thank my immediate family (my wife, my son and daughters, and my graduate students) for coping with me and for their great support.

Hüseyin Arslan

1

Introducing Adaptive, Aware, and Cognitive Radios

Bruce Fette, PhD.

General Dynamics C4 Systems, Scottsdale Arizona, U.S.A.

Spectrum is the lifeblood of RF Communications[1]

In the late 1990s, nearly all telecommunications radios were built using digital signal processor (DSP) processors to implement modulation and signal processing functions, and a General Purpose Processor (GPP) to implement operator interface, network signaling, and system overhead functions. This architecture is attractive to a manufacturer because the same basic electronics can be used over and over for each new radio design, thereby reducing engineering development, enabling volume purchasing, and optimizing production of a common platform, while retaining the flexibility for sophisticated waveforms and protocols. A few manufacturers called their radios "Software Defined Radios" (SDRs), recognizing the power and market attractiveness of the customer community being able to add additional functionality that is highly tuned to market specific applications.

In the early 2000s, a few of these vendors made application layer software functionality available for additional value added functions to aid the user. Users could add music, games, or other applications, as long as the radio waveform functions remained untampered.

Similarly, in the late 2000s, many radios will make the software functionality available for various classes of adaptivity and significantly extend user support functionality. This will initiate the generation of "Cognitive Radios". Thus, with minimal additional hardware, additional software features will enable users, network operators, spectrum owners, and regulators to accomplish much more than with the fixed application radios of an earlier generation.

To understand this important design trend, we must first understand the background, history, and terminology.

[1] Quote originally from Cognitive Radio Technologies, Chapter 5 by Preston Marshall, B.A. Fette, editor, Newnes, 2006.

H. Arslan (Ed.), Cognitive Radio, Software Defined Radio, and Adaptive Wireless Systems, 1–16.

1.1 In the Beginning

The field of radio architecture has recently undergone a revolution of design, significantly enabled by Moore's law of computational evolution, where sufficient computational resources are available in DSPs and GPPs to implement the modulation and demodulation and all of the signaling protocols of a radio as a software function. This threshold was crossed in 1992 when the Department of Defense issued a contract called Speakeasy I and subsequently Speakeasy II to demonstrate that SDR was feasible. It was demonstrated by General Dynamics C4 Systems' Speakeasy II equipment in 1997, and gradually morphed from Speakeasy to Programmable Modular Communication System (PMCS), to Digital Modular Radio (DMR), to Joint Tactical Radio System (JTRS).

The SDR Forum industry group[2] was formed to guide the industry at large in standards to make the industry practical. Joe Mitola of Mitre Corporation and Wayne Bonser of Air Force Research Labs were the visionaries who started this organization and brought the industry together to work these hard design problems, and to begin establishing the new standards it would require. During this time, Joe Mitola was performing his Ph.D. research at the KTH Royal Institute of Technology, University of Stockholm,[3] and began his study of what would be possible once radios were software defined radios (SDRs). He recognized that it is feasible for a radio to become aware of its user, aware of its network (choices and features), and aware of its spectral environment. In fact, it could then be adaptive, and ultimately could have the software to learn various adaptations to its current environment that are desirable support to the user, network, operators, spectrum owners, and regulators. Dr. Mitola introduced the terms "aware", "adaptive", and "ideal Cognitive Radio" (iCR) to reflect the different levels of cognitive capability.

Paul Kolodzy of the Federal Communications Commission (FCC), and Preston Marshall of the Defense Advanced Research Programs Agency (DARPA) developed multiple programs to demonstrate the possibilities and the significance of these capabilities. One very significant demonstration is a demonstration of dynamic spectral access (which subsequently spawned an IEEE conference called DYSPAN) and Interference Avoidance (IA). The International Telecommunications Union (ITU), the IEEE P1900 study group, and the European Union End to End Reconfigurability Project (E2R) have subsequently launched several significant studies into the technical and economic issues of cognitive capability.

---

[2] SDR Forum was first called Modular Multifunctional Information Transfer System Forum (MMITS Forum) to reflect that these systems are capable of being far more than just a radio, and was initiated by Wayne Bonser of AFRL.

[3] Joseph Mitola III, *Cognitive Radio: An Integrated Agent Architecture for Software Defined Radio* (Stockholm: KTH, The Royal Institute of Technology) June 2000.

1.1.1 Support for the User

Ultimately, it is the user of the radio who must see incremental value in using his radio. In fact, there is a yearly upward spiral of utility in the industry. Each year new functions are recognized as "must-have killer applications" that nearly all customers will want. Unique musical ring tones were extremely popular, just so everyone could recognize their own phone from all the other cell phones. Camera phones, to send your smile to Grandma, came next. Then Bluetooth wireless and hands-free connections to the earphone became popular. Every year, these new user support features extend market demand and fuel growth.

Cognitive Radio envisions that the radio will have sufficient ability to recognize common user activities so that it begins to learn to assist the user and the network with common tasks. Mitola likens this to Radar O'Reilly,[4] who recognizes what tasks the Colonel must perform and is always ready to support those functions. A radio with sufficient flexibility can recognize that the selection of wireless access methods is a function of location and time, and can remember locations where cellular service, WiFi service, WiMAX service, Bluetooth, and other services are available and useful.

Awareness may also be able to help the user locate services when he travels in a new country (i.e., please help me find the: restroom, baggage, car rental, train, taxi, next flight to ..., restaurant, tickets to the opera, etc.). Concierge services are easy for a radio to provide, if the radio has location awareness. More sophisticated services are readily provided by database servers, each of which has a narrow but sufficient depth to be truly helpful to its subscribers (Mitola provides the example of airline reservation services). Language translation is another example of such as a service.

1.1.2 Support for the Network

Radios, just like computer networks, generally have a full protocol stack running from the physical layer to the application. The ISO standard defines seven layers for protocol. Other stack models combine some of these layers resulting in a four layer model. Of these layers, only the physical and MAC layers are intimately designed and specific to radio/wireless protocols. Higher layers frequently follow protocols that have been developed for the wired world including Internet Protocols (IP), or telephony protocols (SS7, for example). Many years of optimization have gone into these wired network layer protocols; re-optimizing them for wireless networks is a current research topic. Cognitive radios operate on the principle that by measuring the performance metrics of each layer, those layers and other layers can be optimized. However, the physical layer and MAC layer are very specific to radio/wireless applications, and

[4] Radar O'Reilly from the TV series Mash 1973–1982.

optimization of these layers as a real time activity based on measurements of local spectrum is the dominant part of Cognitive Radio research and design at this time.

Radio systems are often supported by a vast but unseen network. For example, "the array of towers synonymous with the cellular wireless telephony network will each be associated with base stations, which must be linked to a core network infrastructure containing the billing system and through which connectivity to the terrestrial wireline telephone network is made".[5] Even taxi and police radio functions usually include transmitter and receiver sites, interconnect, dispatch operators, terrestrial telephony connections, and a rich support network. On the university campus, it is common to have a network of WiFi access points and a wired network to aggregate traffic to the internet service point, to the university computer and database servers.

The network layer of the smart radio can be an important contributor to the stability and robustness of the network. The radio can select an access point which has less traffic, or where the traffic load presented consumes less total resource, at least in instances where there is more than one access point available. Similarly, the radio can shape its traffic load so that no user experiences unacceptable network performance variations.

The radio can also select the most appropriate network for the service requests of the user. For example, the radio can select a Bluetooth network access for a local printer, rather than using the WiFi access point, while using the WiFi access point for internet database searches. Similarly, voice or video traffic can be provided through the most appropriately resourced network.

1.1.3 Support for the Network Operator

Some network operators have recognized the benefit of convergence of wireless services delivered through multiple access networks. Thus, cellular, WiFi, WiMAX, Bluetooth, broadcast digital audio, and video are bundled so that the service provider selects which resource is most efficiently able to accommodate a user's current needs, and that the transition across different network access is both seamless and economical for the user and for the network operator. This topic is likely to become the differentiating feature of successful network operators over the next decade.

1.1.4 Support for the Regulatory Organization(s)

GSM cellular radio access has been harmonized across much of the globe. However, many other radio technologies have not. Each of the nearly 200 different countries has different regulatory rules and procedures for designing radios, defining spectral mask of each radio application waveform, manufacturing radios, and acquiring rights to use various parts of spectrum. Vendors

[5] Personal communication, Stephen Hope, 1/19/07.

who sell the radios would like to be able to manufacture a common design, and must have the ability to adapt the design as required for each country, where they have the right to utilize the design.

Cognitive radios include a policy engine. The policy engine's role is to be aware of the utilization rules for all countries where the equipment is licensed, and to assure that the radio only operates in those allowed modes in those geographic regions. The policy engine can also include additional rules that define network usage policy, network operator policy, and even manufacturer policy. For example, if software download is allowed into one of the processors in the radio, the policy may state the provisions and requirements to enable the downloaded software. Such a policy may be country specific, allowed by the regulators, and the software may need to be exhaustively validated by the manufacturer as compliant to regulations and non-deleterious under all network conditions, ranging from innocuous to abnormal.

We assume that the policy and downloaded software will follow security guidelines, assuring that the software cannot be modified by unauthorized sources, and that it is validated and un-tampered as required by all parties. Generally this involves exhaustive testing and validation, use of standardized protocols and software sources, cryptographic sealing, signatures, and certification authorities, as described in SDR Forum Security literature.

1.1.5 Support for the Spectrum Owner & Users

The physical and MAC layers of wireless are now undergoing significant research on real time optimization given the ability to measure the spectral environment. Through such measurements, it is possible to minimize interference while supporting very significant increases in traffic density. Measurements suggest that cognitive radio systems should be able to increase traffic density at ratios between 7:1 and over 20:1 (spectrum efficiency is often measured in number of spectrum users supported by measuring bits per second per square kilometer per MegaHertz of spectrum used). In addition to supporting significant increase in users, use of spectrum awareness and adaptation techniques can also indicate interference properties, multipath properties, signal strength, and in turn, these real time measurements can guide adaptation of the transmission waveforms to be more robust, providing the user improved quality of service in multiple dimensions.

Many organizations are now involved in the issue of dynamic spectrum allocation, which is unquestionably the field where cognitive radio is receiving the greatest attention. Dynamic spectrum allocation is attracting the same degree of attention that Internet Protocols and optimization received ten years ago. Activities range from analysis, to testbeds, to languages for specifying regulatory policy, and many new activities are now being funded by research organizations around the world.

All of this now converges in the commercial electronics world, as the next generation of Personal Digital Assistants (PDAs) assimilate cellular connectivity, WiFi, WiMax, Zigbee, and Bluetooth wireless network access and wireless protocols. Users will expect these devices to efficiently select amongst these networks to accomplish the user's most important objective of the moment, and to do it cost effectively. Cellular connectivity may swap with WiFi or WiMax connectivity as the user moves about his activities. Since the availability, range, throughput, and economics are vastly different models, users will come to expect the system to make cost effective and performance sensitive decisions. Eventually users will expect these decisions to be automatic.

1.2 Economics of Cognitive Radio

Many things are technically possible, but what justifies the time and effort to advance various degrees of cognitive functionality into the radio network, either in subscriber equipment or in the network infrastructure is the economics. In this section, we will scratch the surface of cognitive applications and the revenue stream.

Since adaptive spectrum has gained so much attention we will study it first.

1.2.1 Value of Spectrum

Without spectrum, no wireless telecommunications or wireless internet services would be possible. The telecommunications industry is now a 1 Trillion (10^{12}) dollar per year industry,[6] and the wireless part is growing very rapidly, while the wired telecommunication services are experiencing a relatively flat business. As of 2006, the wired and wireless businesses were approximately equal in revenue. Spectrum is required to support these wireless businesses. In the United States, the continuing increase in cellular telephony demand is supported by increasing density of cellular infrastructure.

However, in some regions of the globe, the cellular infrastructure is at peak capacity and increased infrastructure density is impractical. In these early warning hot spots, other means to support continued growth of demand is necessary to continue serving the market demand. Technologies that enable continued growth include adaptivity, smart antenna technology, and more efficient use of existing spectrum. More efficient use of spectrum in these high demand hot spots translates immediately into continuing to support the growing revenue stream for the network operators.

There are also regions which are under-served by the wired community, and the wireless community. In mainly rural regions, where fiber or cable

[6] Research and Markets, "2004 Telecommunications Market Review and Forcast – Trends, Analyses and Projections to 2007"; http://www.researchandmarkets.com/reportinfo.asp?report_id=226592.

is uneconomical to install, and where there is nearly no spectrum use (for example, in many rural areas, a single broadcast TV source may be nearly 200 miles away and there is little or no local TV service) there are significant opportunities to provide internet and telecommunication services using this under-utilized spectrum.

In each of these cases there are ways in which to utilize cognitive radio principles to make additional spectrum available. These methods are being thoroughly analyzed, and business cases are being examined.

1.2.2 Spectrum Adaptivity

Current research in selection of waveforms or protocols that provide higher throughput, and reduce interference to adjacent channels has focused on the exploitation of Orthogonal Frequency Division Multiplex (OFDM) types of waveforms. In these waveforms, high data rates are achieved through use of multiple carriers. Furthermore, the order of modulation may be adaptive so that the number of bits delivered on each carrier is adapted to the noise floor and the propagation performance at each carrier frequency. OFDM is already in use for Digital Audio Broadcast (DAB) and High Definition TV (HDTV) and is a research topic rich with additional opportunity.

1.2.3 Smart Antennas

One way to get enhanced spectral density and user density is with smart antennas. Smart antennas can provide narrow beam patterns for either transmitter or for receiver or both. In the Industrial Scientific and Medical (ISM) band frequencies of 2.4 GHz, practical narrow beam antennas can provide 9 dB of gain, and reduce interference to other communication activity off the sidelobes of the antenna beam patterns. Even more significantly, smart antennas can also be used for interference nulling, and can provide up to 30 dB of null depth. Either of these strategies significantly increases the feasible user density. Smart antenna techniques may be economically costly for certain applications, and may be impractical for certain form factors, but much research attention must be focused on this topic, as it provides the greatest opportunities for increased user density.

1.2.4 MIMO

Multi-Input Multi-Output (MIMO) is a communication technique in which the multipath properties of the channel are utilized to support greater data throughput. In a MIMO system, the transmitter transmits multiple channels of data traffic through multiple antennas; the receiver learns the channel behavior between the transmitter's multiple antennas and the receiver's multiple antennas, and uses signal processing to compute what waveform was transmitted by each transmitting antenna, and the corresponding data stream.

In this way, the same frequency is reused in the same geographic region to deliver greater amounts of data traffic than could be expected from a single transmitting and receiving antenna (SISO) system. Some MIMO systems also have the ability to learn to suppress interference from unrelated transmitters, further enhancing network performance. These MIMO techniques are practical when sufficient space for mounting antennas on the radio or platform are available.

1.2.5 Spectrum Subleasing, Sharing

In dense urban applications, there is opportunity to sublease spectrum from spectrum owners whose loading is temporarily light. Such owners can define access rules, and economics of such transactions, as well as rules to take back spectrum should demand or emergency arise. SDR Forum, IEEE, and others are working to establish standardized protocols to enable such transactions. For lightly loaded system, spectrum owners may be able to recover up to 60% of the cost of spectrum ownership in this fashion. In the public safety market, there have been discussions that some organizations might pay for new public safety infrastructure in exchange for shared spectrum access on a guaranteed non-interfering basis. This clearly shows just how valuable the spectrum is.

In rural applications, there has been discussion of using the relatively unused rural UHF TV broadcast bands for adhoc/mesh networks to deliver internet and other data services. Protocols to assure non-interference are currently under evaluation. However, concern from TV broadcasters that such techniques will eventually cause interference to broadcast TV, at the rural–suburban boundaries where such wireless networks will fail to recognize presence of the TV signal, or where local area networks may deliver spotty performance. Given that much of rural areas are served by S Band satellite TV, it is not clear what long range outcome of this debate will be.

Policy Assurance of Behavior to the Economic Stakeholders

Several significant studies have been performed on the economics of dynamically adaptive spectrum.[7] However, this work has also generated a great deal of concern, on the part of regulators, network operators, and spectrum owners. It is therefore important to understand the basic architecture of a cognitive radio. As shown in Fig. 1.1, a cognitive radio must include a variety of sensors, many of which support the user, a learning capability (possibly accomplished remotely), and a set of rules based on a policy engine which allow and disallow actions. It is the job of the policy engine, to define inviolable rules and behaviors of each stakeholder. Each stakeholder's rules will be expressed as

[7] "Software Radio: Implications for Wireless Services, Industry Structure, and Public Policy," W. Lehr, S. Gillett, F. Merino, *Communications and Strategies*, IDATE, Issue 49 (1st Quarter 2003) 15–42.

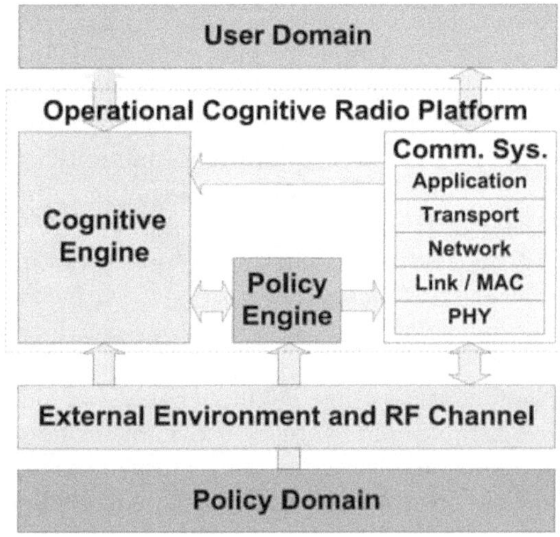

Fig. 1.1. Architecture of a cognitive radio. Figure 1.1 includes a cognitive function to analyze existing spectrum users, and measure properties of its own communication channels, and a set of rules expressed through a policy engine which define what the radio is allowed to do, and what it is not allowed to do.[8]

policies: policies for the regulator, for the spectrum owner, for the network operator, for the manufacturer, and for the user.

While policy engine languages are now a subject of R&D development throughout the industry, and in SDR Forum in particular, notionally, each policy engine must perform an analysis of actions proposed by the radio's cognitive engine, and must determine that the proposed action is allowed by at least one rule, and that the action is not disallowed by any rules. All actions which are allowed by the policy engine may then be compared and contrasted for spectrum efficiency, interference, quality of service and battery drain considerations, and the best choices applied. For example, if the cognitive engine were to propose that idle spectrum is available for a telephone call at 107.9 MHz, using a CDMA waveform, and a Universal Mobile Telecommunications System (UMTS) protocol, transmitting to close a link over 5 km, in Alamogordo, New Mexico, at 0.5 W, the policy engine will examine whether the user is entitled to perform that function at that frequency with that waveform, at that location and that power level. There must be one rule that allows it, and no rules that disallow it before the radio can implement the proposed adaptation. In this example, we assume that a regulatory policy would disallow use of an FM broadcast frequency within the continental U.S. for telecommunications applications, but it might be allowable in countries where there is no FM band broadcast activity. Thus the policy engines will disallow all

[8] Cognitive Radio Technologies, Chapter 7, Rondeau & Bostian, edited by B. Fette, Newnes 2006.

behaviors not in keeping with all stakeholders. Similarly, the network operator, spectrum owner, and user may all have various policy expressions, and each must approve before the radio would execute the mode.

A policy engine is an efficient encoding of these rules, but most importantly, an efficient method for managing radio functionality across the entire globe. If the location of the radio changes, then the radio can look up the rules for its new location. If the rules in a certain country change in order to allow a new capability, then an update of the policy database can allow the radio to perform new capabilities. Therefore, a cognitive radio will not learn or perform unacceptable behaviors, because those behaviors will be disallowed by the policy engine.

A very important consideration is that changes of policy must be prevalidated to assure reliable and stable performance under both reasonable and unreasonable conditions and must be administered by those trusted to distribute policies. Thus, the stakeholders must place their trust in the policy engines to enforce the rules of behavior, etiquette, and protocol to assure their participation in the revenue stream.

Value of Spectrum to the Stakeholders

"Spectrum is the lifeblood of communication systems."[9] Without spectrum, there is no electromagnetic communications. Many significant papers have addressed the value of spectrum. Each country has defined different mechanisms to assess and manage spectrum, radio production and use of spectrum, and the associated revenue. Some countries have recognized spectrum as publicly owned, and a value to be shared by all. In such a regulatory environment, spectrum should be used for the greatest public value. As such the primary issue is to prioritize current needs and grant access to spectrum best matched to most valued public need (higher frequencies to shorter range applications, etc.).

In the United States, and in many other countries, spectrum access is a source of federal government revenue. However, costs of spectrum auction are passed on to users as an increase in cost of services with incremental overheads of all participants in the value chain. In the United States it is assumed that the free market forces will assure that spectrum access, as a result of spectrum auction, will result in greatest cost effectiveness use of valuable spectrum.

Since July 1994, the FCC has conducted 33 spectrum auctions raising over $40 billion.[10] European auctions conducted in 2000 raised nearly $100 billion. Clearly the value of a spectrum license depends on many dimensions, including:

---

[9] Quote originally from Cognitive Radio Technologies, Chapter 5 by Preston Marshall, B.A. Fette, editor, Newnes, 2006.

[10] Peter Crampton, "Spectrum Auctions", *Handbook of Telecommunications Economics*, Elsevier Science, Chapter 14, 2002.

1. amount of bandwidth in MHz (directly related to the ability to service as many subscribers as possible, number of subscribers within the service region;
2. demand within the service region;
3. duration of the contract;
4. opportunity for growth of services within the service region and:
5. cost of installing and providing service. As a calibration, auctions of 2×10 MHz (paired spectrum) +5 MHz (unpaired Spectrum) range from \$2.60 per subscriber per year to \$107.20 per subscriber per year.

Network operators provide a rough guideline of cost of spectrum access versus cost of subscriber equipment. In the United States, one network operator reported that the subscriber pays approximately twice as much per year for the subscriber equipment as he pays for the spectrum. In Europe, one operator indicated that the subscriber pays approximately twice as much for the spectrum, as he pays for the subscriber equipment. While such numbers are not published, we can consider spectrum access to range between 50% and 200% of the cost of subscriber equipment as a guideline.

In Fig. 1.2, Forward Concepts presents the 2006 handset revenue by waveform/protocol type. It is clear that the global telecommunications market uses more than 11 protocols, and that the total subscriber hardware market is \$130B (1.3×10^{11}). So we can project that spectrum access ranges between \$65B and \$260B in yearly cost to subscribers. So very clearly, the subscriber and the network operator have much to gain in lowering spectrum access costs. We can also conclude that one radio design capable of supporting 11 protocols, knowing which country to use them in, and how to register for access would support global requirements.

This shows a clear and significant value of spectrum to enable valuable applications. It also reflects a cost to the user, which will be paid with interest over the lifetime of the license, and a future revenue stream opportunity to the operator that acquires the spectrum.

Spectrum owners consist of many communities, each with different interests such as AM, FM, and TV broadcast operators, Federal State and Local public services (police, fire, FAA, FBI, etc.), Telecommunications network operators (cellular phone providers), the Department of Defense (Army, Navy Air Force, Marines, Coast Guard, others), and dispatch service providers (taxi, UPS, FED X, etc.). There are also spectrum users which are authorized by a license passed through the manufacturer to private citizens to use or as a public commons (Citizens band, Family Radio Service, WiFi, UWB, etc.). Existing spectrum owners are understandably reluctant to give up spectrum regardless of whether it is used or unused for public good, while advocates of cognitive radio argue that unused spectrum can be put to productive use without impacting primary existing users.

Much has been published about the distribution of spectral activity. However, there is a very clear trend. Between 1% and 15% of the spectrum is busy,

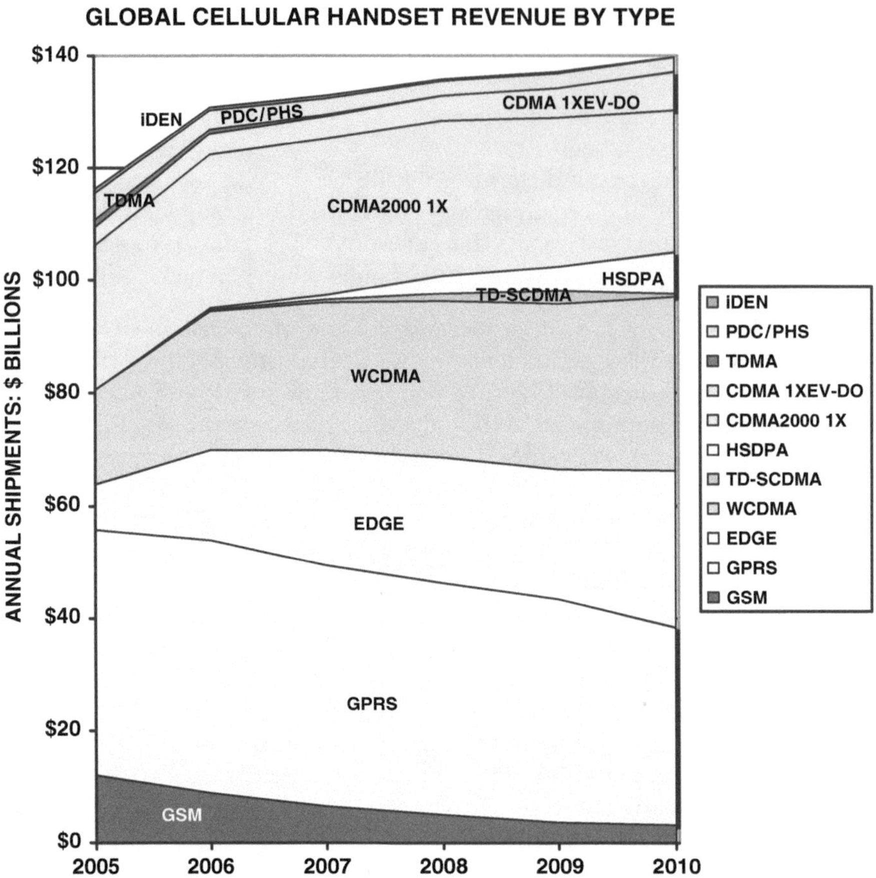

Fig. 1.2. Cellular handset market analysis, Forward Concepts DSP Market Analysis, 2006.

depending on what location the measurement is made in, what frequency the measurement is made at, and over how much time, frequency and geographic space the averaging is performed.[11] The implication is that while spectrum owners have paid dearly to own these time–frequency–space spectrum blocks, it is not put to 100% utilization. Nor would we necessarily want it to be fully utilized. In order to accommodate peak needs, and instant access, we must retain reserve capacity somehow. How much reserve capacity should be retained to accommodate peaking? This is a function of localized statistics, reserve access protocol, access latency, and urgency.

[11] Survey performed by Mark McHenry of Spectrum Signal Processing, as part of the Darpa XG program, and also funded by SDR Forum to survey spectrum use at the Republican National Convention September 2004.

1.2.6 Local Statistics

The Bell Telephone Company did extensive studies of call capacity and blocking probability on circuit switched telephony in the 1940s and 1950s, defining a unit of call bandwidth as an Erlang. In this work, the probability of dropped, lost, or blocked call decreases as the ratio of capacity to average demand increases. Their goal was to keep call blocking probability under 1% across each management region. In this case, the only resort beyond use of the full capacity of all existing trunk lines was to install more trunk lines – an unacceptably long access latency for unexpected peak capacity requirements. In these systems, normal daily peak utilization is about 30% of peak capacity (Mother's Day peak utilization usually exceeds capacity).

In today's modern age, cellular telephony operates with a somewhat higher ratio of average daily peak utilization to peak capacity. However, cellular systems have interesting local statistics. Airports have very high capacity requirements, and experience huge peaks as a consequence of delayed flights. Sports stadiums have extremes of peak to average traffic ratio, with the end of game resulting in very high peaks in traffic. Finally, traffic jams are also a source of high peak to average call ratios.

1.2.7 Peaking Support

Capacity to support a demand peak can be provided in many different ways. In the context of cellular telephony, additional demand can be supported by other nearby base stations until all channel capacity of all nearby base stations has also been consumed. In the context of laptop computers performing internet data access, additional demand is supported by reducing the available throughput support to all subscribers. In the case of FM broadcast, new capacity is being added as FM broadcasters add DAB subcarriers thereby offering new audio and data services. Where do police, fire, and other emergency departments get additional capacity? Within urban jurisdictions, they plan additional capacity for likely peak events such as emergencies. However, if the scope of an emergency exceeds the planned peak capacity, there is generally not any method to access additional spectrum. This occurred in both New York City and Washington D.C. on September 11, 2001. In remote regions, peaking capacity for emergencies (such as in rural Montana, rural Arizona, and rural Texas) is often supplied by cellular telephony if available, or by satellite transponder. It usually takes a few days to lease, deliver, and install the satellite transponder base station equipment.

1.2.8 Spectrum Leasing

Spectrum leasing is a way to support demand peaking, such that the ratio of the average utilization to peak capacity can be higher. In this scenario, multiple services either sublease from a spectrum owner for a time-frequency-spatial block, or owners cross license their current unused capacity to each other giving each the opportunity to support demand peak, or to prioritize access to spectrum depending on greatest public need.

At least one serious business proposal has been made to set up a spectrum leasing business. However, as of this time, no cellular subscriber equipment has the flexibility of frequency, waveform and protocol to accommodate varying spectral assignment, nor have industry accepted protocols been adopted to perform these transactions.

1.2.9 Spectrum Awareness Databases

Many radios can capture their local view of spectrum activity, position, and time. That information can be directly shared. It can also be aggregated into a regional database, that provides for awareness of local emitters, local policies, and knowledge of areas where signal dropouts are likely, resulting in radio performance predictions. One such data structure, developed by Zhao and Le is called a Radio Environment Map (REM). Such a database can be a significant aid to cognitive radios both in time to locate and allocate unused spectrum to a specific purpose, and time to acquire the policies of network access for local services.

1.2.10 Value of Concierge Services

Densely populated and popular urban areas lend themselves to concierge services. Tourists and business travellers can create significant demand because they are unfamiliar with local services. This is why major hotels have concierge desks to provide assistance with local arrangements. Concierge desks are often required to function around the clock. However, visitors may be unwilling or unable to use the local concierge service because of language barriers or unavailability of the identified need.

One company has been successfully selling concierge services as an internet service, and reports \$77M/year revenues.[12] Another reports combining concierge services with telematics, location, and vehicle tracking as a successful business.[13] A third company reports the global concierge Ecommerce business as \$12B/year with a 30% compound annual growth rate, and have established a monthly fee structure to test the market.[14] Frost and Sullivan has a market report focused specifically on telematics (concierge services focused specifically on luxury car owners). Their report, in short, indicates that of 250M cars on the road in the US, 30M are telematics capable, and 10M actually use the service. That service is reported as a \$1.3B/year market.

[12] MetroOne Telecommunications.
[13] Skynet Telecom.
[14] Agillion http://www.businessweek.com/ebiz/0011/ec1107.htm.

With attention to convergence of voice and data services, this could easily grow from 30% market acceptance to a far larger market.

From this we conclude that making the radio sufficiently location and position aware to be able to assist in concierge services for the radio user has a significant international value to the users, and to the value added service providers, and is a cognitive radio market immediately serviceable with existing technology.

1.2.11 Cognition

Learning is a significant part of cognitive radio research. Rondeau and Bostian have studied the use of Genetic Algorithm (GA) to learn how a radio might best respond to a spectral environment given a set of objective metrics. They have developed prototypes that show that a radio can successfully learn proper adaptions to a spectral environment that optimize for the objective metrics. Neel and Reed have studied how to apply game theory to a radio as a member of a cognitive network. With game theory they were able to analyze protocols to determine whether cognitive radio behavior could result in stable network behavior, and they conclude that it can. Mitola, Kokar, and Kovarik have independently studied knowledge representation (Ontology) that a cognitive radio would need to perform reasoning functions. Mitola proposes that a radio should be able to reason about the owner's voice and visual characteristics, should be able to take verbal commands, should be able to visually recognize locations and conditions. Kovarik describes the difficulty of building a sufficiently large information database that a radio would be usefully intelligent to be able to reason over any but the narrowest range of topics. Mitola concludes that narrow fields of human activity can be captured with sufficient depth of knowledge representation to be useful. Thus niche markets will develop by creating a deep representation of requirements for specific services. Where these can be brought forward as an economically successful business, we will see new products and services offered even before general cognition is practical in a radio.

Radios that have learned or adapted to local spectrum, channel, waveform, or protocol conditions can share their learning with other radios that have not yet learned these local optimizations. This learning can be infused to other radios via a network database which provides local optimization, or it can be shared directly from radio to radio. Since network operators correctly worry that the network behavior be stable, and predictable, and within FCC guidelines, it is most appropriate that learned behaviors be shared from a database, where they can be checked, and validated as producing a net benefit to the network before being used. It seems that the Radio Environment Map is an example of a method for providing such services.

Thus we conclude that cognition can be local to the radio, or can be served as a download of knowledge structures (or software) from other radios which have performed the learning and now make it available. Market studies have not yet characterized the value of cognitive radio learning technology. However, if learning to use spectrum wisely is an example, it will be very valuable.

1.3 Summary

In summary, we conclude that Cognitive Radio technology is a way in which one radio or even a network of radios are able to learn a useful degree of adaptivity, that aids the user, the network, and/or the spectrum owner. There are powerful economic incentives to provide new capabilities, through existing telecommunications infrastructure, and cognitive radios will provide those capabilities. As new services are offered, more spectrum will be needed and cognitive radios will provide the means for radios to communicate with greater spectrum efficiency.

2

Cognitive Networks[1]

Ryan W. Thomas, Daniel H. Friend, Luiz A. DaSilva, and Allen B. MacKenzie

Virginia Tech, Blacksburg VA 24060

2.1 Introduction

Current data networking technology limits a network's ability to adapt, often resulting in sub-optimal performance. Limited in state, scope, and response mechanisms, the network elements (consisting of nodes, protocol layers, policies, and behaviors) are unable to make intelligent adaptations. Communication of network state information is stifled by the layered protocol architecture, making individual elements unaware of the network status experienced by other elements. Any response that an element may make to network stimuli can only be made in the context of its limited scope. The adaptations that are performed are typically reactive, taking place after a problem has occurred. In this chapter, we advance the idea of cognitive networks, which have the promise to remove these limitations by allowing networks to observe, act, and learn in order to optimize their performance.

Cognitive networks are motivated by complexity. Particularly in wireless networks, there has been a trend towards increasingly complex, heterogeneous, and dynamic environments. While wired networks can also take on any of these characteristics (and are not excluded from potential cognitive network applications) wireless networks are a natural target because of their inter-node interactions and the size of their system state space. Previous research into cognitive radio and cross-layer design have addressed some of these issues but have shortcomings from the network perspective. Cognitive networks represent a new scope and approach to dealing with this complexity.

This chapter provides the reader with a primer on the cognitive network concept, as envisioned by the authors. It begins by explaining the need for cognitive networks, how they are defined, and possible applications for the technology. Then the chapter examines how cognitive networks are related

[1]Portions reprinted, with permission, from "Cognitive Networks: Adaptation and Learning to Achieve End-to-end Performance Objectives," *IEEE Communications Magazine*, vol. 44, pp. 51–57, December 2006. ©2006 IEEE.

H. Arslan (Ed.), Cognitive Radio, Software Defined Radio, and Adaptive Wireless Systems, 17–41.

to, but distinct from, previous work in cognitive radios and cross-layer design. A practical discussion of the implementation of a cognitive network and important areas of future work close the chapter.

2.1.1 Definition

Cognitive networks were first described by us in [1] as

> ... a network with a cognitive process that can perceive current network conditions, and then plan, decide and act on those conditions. The network can learn from these adaptations and use them to make future decisions, all while taking into account end-to-end goals.

The cognitive aspect of this definition is similar to that used to describe cognitive radio and broadly encompasses many simple models of cognition and learning. More critical to the definition are the network and end-to-end aspects. Without the network and end-to-end scope, the system is perhaps a cognitive radio or layer, but not a cognitive network. Here, end-to-end denotes all the network elements involved in the transmission of a data flow. For a unicast transmission, this might include such elements as subnets, routers, switches, virtual connections, encryption schemes, mediums, interfaces, and waveforms. The end-to-end goals are what give a cognitive network its network-wide scope, separating it from other adaptation approaches, which have only a local, single element scope.

2.1.2 Motivation and Requirements

The overall goal of any technology is that it meet some need in the best way possible for the least cost. With the first half of this goal in mind, a cognitive network should provide, over an extended period of time, better end-to-end performance than a non-cognitive network. Cognition can be used to improve such end-to-end objectives as resource management, Quality of Service (QoS), security, access control, or throughput. Cognitive networks are only limited in application by the adaptability of the underlying network elements and the flexibility of the cognitive process.

In examining the second half of the goal, the cost must justify the performance. Cognitive network costs are measured in terms of communications and processing overhead, architecture roll-out and maintenance expenses, and operational complexity. These costs must be outweighed by the performance improvement the cognitive network provides. For certain environments, such as static wired networks with predictable behavior, it may not make sense to convert to cognitive operation. Other environments, such as heterogeneous wireless networks, may be ideal candidates for cognition.

Cognitive networks should use observations (or proxy observations) of network performance as input to a decision making process and then provide output in the form of a set of actions that can be implemented in the modifiable elements of the networks. Ideally, a cognitive network should be forward-looking, rather than reactive, and attempt to adjust to problems before they occur. Additionally, the architecture of a cognitive network should be extensible and flexible, supporting future improvements, network elements, and goals.

Cognitive networks require a Software Adaptable Network (SAN) to implement the actual network functionality and allow the cognitive process to adapt the network. Similarly to cognitive radio, which depends on a Software Define Radio (SDR) to modify aspects of radio operation (e.g. time, frequency, bandwidth, code, spatiality, waveform), a SAN depends on a network that has one or more modifiable elements. Practically, this means that a network must be able to modify one or several layers of the network stack in its member nodes. A simple example of a SAN could be a wireless network with directional antennas (antennas with the ability to direct their maximum receive or transmit gain to various points of rotation). A more complex example would incorporate more modifiable aspects at various layers of the protocol stack, such as Medium Access Control (MAC) algorithms or routing control.

2.1.3 A Simple Example

As an example of the need for end-to-end rather than just link adaptations, consider an ad-hoc data session between a source node, S_1, and a destination node, D_1, as shown in Figure 2.1. The source node must route traffic through intermediate nodes R_1 and R_2 acting as regenerative relays. Node S_1 performs a link adaptation by choosing the relay node based on the set of minimum hop routes to D_1 and the probability of link outage. For this simple network, nodes R_1 and R_2 are both in the set of minimum hop relays on routes to D_1. Therefore, node S_1 selects the link on which to transmit by observing the outage probabilities on the links to R_1 and R_2 and selecting the link with the lower outage probability. From the standpoint of the link layer in node S_1, this guarantees that the transmitted packets have the highest probability of

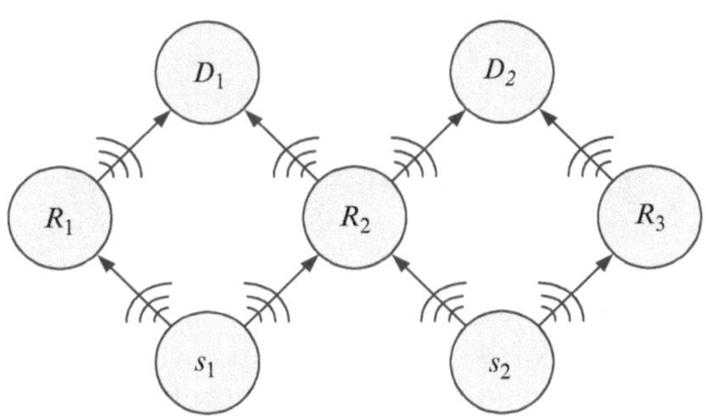

Fig. 2.1. Simple relay network for illustrating the need for cognition with an end-to-end scope.

arriving correctly at the relay node. However, it does not guarantee anything about the end-to-end performance, i.e. the total outage probability from S_1 to D_1.

In contrast to the link adaptation, the cognitive network uses observations from all nodes to compute the total path outage probabilities from S_1 to D_1 through R_1 and R_2. This shows the benefit of a more global view as well as another advantage to the cognitive network, the learning capability. Suppose that the learning mechanism measures throughput from the source to its destination in order to judge the effectiveness of previous decisions, and suppose that nodes S_1 and S_2 are both routing their traffic through R_2 because this satisfies the minimum outage probability objective.

Now suppose that R_2 becomes congested because of a large volume of traffic coming from S_2. This becomes apparent to the cognitive process in the throughput reported by S_1 and S_2, though the cognitive process is not explicitly aware of the congestion. Nevertheless, it is able to infer from the reduced throughput and its past experiences that there may be a problem. The cognitive process is then able to respond to the congestion, perhaps by routing traffic through R_1 and/or R_3. This example illustrates the potential of cognitive networks in optimizing end-to-end performance as well as reacting to unforeseen circumstances. The cognitive network goes beyond the purely algorithmic approach of the underlying routing protocol and finds efficient operating points even when unexpected events occur.

2.2 Foundations and Related Work

Having defined a cognitive network, it is helpful to review some existing research areas that are related to the cognitive network concept. We take a look at two areas in particular, cognitive radio and cross-layer design.

2.2.1 Cognitive Radio

Shared Attributes with Cognitive Networks

The 50% correlation in nomenclature would itself imply some degree of commonality, and it can certainly be argued that research in cognitive radio has sparked the formulation of the cognitive network concept. What cognitive radios and cognitive networks do share is the cognitive process that is the heart of the performance optimizations. An essential part of the cognitive process is the capability to learn from past decisions and use this learning to influence future behavior. Both are goal-driven and rely on observations paired with knowledge of node capabilities to reach decisions. Knowledge in cognitive radio is contained within a modeling language such as Radio Knowledge

Representation Language (RKRL) [2]. A network-level equivalent must exist for the cognitive network to be goal oriented and achieve context awareness, two attributes that it shares with a cognitive radio.

A cognitive radio requires tunable parameters which define the optimization space of the cognitive process. These tunable parameters are ideally provided by an SDR. The concept of the SAN is the cognitive network analog of SDR. Therefore, both technologies employ a software tunable platform that is controlled by the cognitive process.

Differences from Cognitive Networks

Cognitive networks are clearly delineated from cognitive radios by the scope of the controlling goals. Goals in a cognitive network are based on end-to-end network performance, whereas cognitive radio goals are localized only to the radio's user. These end-to-end goals are derived at run-time from operators, users, applications, and resource requirements in addition to any design-time goals. This difference in goal scope from local to end-to-end enables the cognitive network to operate more easily across all layers of the protocol stack. Current research in cognitive radio emphasizes interactions with the physical layer, which limits the direct impact of changes made by the cognitive process to the radio itself and other radios to which it is directly linked or with which it may interfere. Agreement with other radios on parameters that must match for successful link communication is reached through a process of negotiation. Since changes in protocol layers above the physical layer tend to impact more nodes in the network, the cognitive radio negotiation process would have to be expanded to include all nodes impacted by the change. However, because the negotiation process is unable to assign precedence to radios' desires without goals of a broader scope, achieving agreement among multiple nodes may be a slow process. For the same reason, the compromise can be expected to result in sub-optimal network performance. In contrast, whether the network components are acting in a cooperative or selfish manner, all cognitive network actions are referenced back to the end-to-end network goals.

Another significant difference between cognitive radios and cognitive networks is the degree of heterogeneity that is supported. Cognitive networks are applicable to both wired and wireless networks whereas cognitive radios are only used in wireless networks. Since the cognitive network may span wired and wireless mediums, it is useful for optimizing performance for these heterogeneous types of networks, which are generally difficult to integrate.

The fact that a cognitive network is composed of multiple nodes also adds a degree of freedom in how the cognitive processing is performed compared to cognitive radio. A cognitive network has the option to implement a fully distributed, partially distributed, or centralized cognitive process.

2.2.2 Cross-layer Design

Shared Attributes with Cross-layer Design

Designs that violate the traditional layered approach by direct communication between non-adjacent layers or sharing of internal information between layers are called cross-layer designs [3]. Cognitive networks indirectly share information that is not available externally in the strictly layered architecture. Therefore, cognitive networks do implement cross-layer designs.

The common theme between these two concepts is that in both, observations are made available for adaptations at layers other than the layer providing the observation. In a cognitive network, protocol layers provide observations of current conditions to the cognitive process. The cognitive process then determines what is optimal for the network and changes the configurations of network elements' protocol stacks.

Differences from Cross-layer Design

Despite similarities, cognitive networks reach far beyond the scope of cross-layer designs. The cognitive network can support trade-offs between multiple goals and in order to do so performs Multiple Objective Optimization (MOO), whereas cross-layer designs typically perform single objective optimizations. Cross-layer designs perform independent optimizations that do not account for the network-wide performance goals. Trying to achieve each goal independently is likely to be sub-optimal, and as the number of cross-layer designs within a node grows, conflicts between the independent adaptations may lead to adaptation loops [4]. This pitfall is avoided in a cognitive network by jointly considering all goals in the optimization process.

The ability to learn is another significant difference. The cognitive network learns from prior decisions and applies the learning to future decisions. Cross-layer designs are memoryless adaptations that will respond the same way when presented with the same set of inputs, regardless of how poorly the adaptation may have performed in the past. The ability to learn from past behavior is particularly important in light of the fact that our understanding of the interaction between layers is limited.

Finally, like cognitive radio, the scope of the goals and observations sets cognitive networks apart from cross-layer design. The observations used by the cognitive process span multiple nodes and the optimization is performed with the goals of all nodes in mind, whereas cross-layer design is node-centric. This global information allows the cognitive process to adapt in ways that simply are not possible when nodes have limited visibility into the state of other nodes in the network, as is the case with cross-layer design.

2.2.3 Recent Work

The concept of a cognitive networks is an emerging research field. The idea of adding cognition to a network has in the past been reserved for individual aspects of the network, such as "smart" antennas or "smart" packets. All this changed with the introduction of the cognitive radio by Mitola in [2]. His concept of putting intelligence into radio operation caught the imagination and attention of the research community. Eventually the concept worked its way from radios into the larger network.

Recent research can be divided into two categories: cognitive radio networks and cognitive networks. In the first category, we begin with work from Mitola and his original thesis on cognitive radio. Here, he mentions how cognitive radios could interact within the system-level scope of a cognitive network [2]. Neel continues this line of thinking in [5], where he investigates modeling networks of cognitive radios as a large, multiplayer game to determine convergent conditions. This kind of thinking is also observed in Haykin's paper on cognitive radio [6], where he examines multiuser networks of cognitive radios as a game.

The focus of cognitive radio networks, as with cognitive radios, is primarily on MAC and physical (PHY) layer issues, but now operating with some end-to-end objective. In a cognitive radio network, the individual radios still make most of the cognitive decisions, although they may act in a cooperative manner. Currently suggested applications for cognitive radio networks include cooperative spectrum sensing [7,8] and emergency radio networks [9]. From a more general perspective, Raychaudhuri et al. [10] present an architecture for cognitive radio networks.

Perhaps the first mention of a cognitive network rather than a cognitive radio network comes from Clark et al. [11]. Clark proposes a network that can

> assemble itself given high level instructions, reassemble itself as requirements change, automatically discover when something goes wrong, and automatically fix a detected problem or explain why it cannot do so.

According to Clark, this would be accomplished with the use of a Knowledge Plane (KP) that transcends layers and domains to make cognitive decisions about the network. The KP will add intelligence and weight to the edges of the network, and context sensitivity to its core. Saracco also observed these trends in his investigation into the future of information technology [12], postulating that the change from network intelligence controlling resources to having context sensitivity will help "flatten" the network by moving network intelligence into the core and control further out to the edges of the network.

Cognitive networks differ from cognitive radio networks in that the action space of the former extends beyond the MAC and PHY layers and the network may consist of more than just wireless devices. Furthermore, cognitive networks may be less autonomous than a cognitive radio network, with the network elements cooperating to achieve goals, using a centralized cognitive process or a parallelized process that runs across several of the network elements. However, despite these differences, the definition of cognitive networks given in Section 2.1.1 encompasses both cognitive radio networks and cognitive entworks.

More recently, Mähönen discusses cognitive networks in the context of future Internet Protocol (IP) networks and cognitive trends in a series of papers. In his earliest paper, he discusses cognitive networks with respect to future mobile IP networks, arguing that the context sensitivity of these networks could have as interesting an application as cognitive radios [13]. He then examines cognitive networks as part of a larger paper on cognitive trends [14]. He discusses how cognitive radios may be just a logical subset of cognitive networks. He also brings up the idea of a Network Knowledge Representation Language (NKRL) to express and communicate high-level goals and policies.

Several research groups have proposed cognitive network-like architectures. These architectures can be categorized into two objectives: the first centers on using cognition to aid in the operation and maintenance of the network, while the second centers on cognition to solve "hard" problems, problems that do not have a feasible solution other than the use of cognition.

Falling into the first category, the End-to-End Reconfigurability Project II (E^2R II) [15] is designing an architecture that will allow the seamless reconfiguration of a network in order to allow for universal end-to-end connectivity. Although E^2R II is an ambitious project with many facets, the overarching goal is one of maintaining connectivity to the user. This is similar to the goal of the m@ANGEL platform [16], which attempts to provide an cognitive network-like architecture for mobility management in a heterogeneous network. Both of these architectures are focused on the operation and maintenance of 4G cellular and wireless networks.

In contrast, the Center for Telecommunications Value-Chain Research (CTVR) at Trinity College [17] has presented a proposal for a cognitive network platform that consists of reconfigurable wireless nodes. Although focused on wireless operation, these nodes are able to solve a variety of problems by modifying or changing the network stack based on observed network behaviors. The possible objectives of these networks can extend beyond mobility management and connectivity. Similar to the CTVR work but less dependent on the wireless focus, Mähönen proposes a general architecture utilizing a collaborative Cognitive Resource Manager (CRM) that provides cognitive behavior from a toolbox of machine learning tools such as neural networks, clustering, coloring, genetic algorithms, and simulated annealing. The work in this chapter describes also falls under this objective, attempting to provide a general cognitive architecture capable of solving a variety of hard problems, rather than being tied to network operation issues.

2.3 Implementation

In order to synthesize the preceding concepts and components into an actual cognitive network, we investigate how a cognitive network should be implemented. We construct a framework for the cognitive process and identify the critical features of this architecture.

A common model of cognition is the three-level theory [18]. The model is often summarized as consisting of behavioral, functional, and physical layers. The behavioral level determines what observable actions the system produces, the functional layer determines how the system processes the information provided to it, and the physical layer comprises the neuro-physiology of the system.

From this concept, we draw a three-layer framework, with each layer roughly corresponding to the layers in the model described above. At the top layer are the goals of the system and elements in the network that define the behavior of the system. These goals feed into the cognitive process, which computes the actions the system takes. The SAN is the physical control of the system, providing the action space for the cognitive process. This framework is illustrated in Figure 2.2.

In our framework, we consider a cognitive process which consists of one or more cognitive elements, operating in some degree between selfish autonomy and full cooperation. If there is a single cognitive element, it may still be physically distributed over one or more nodes in the network. If there are multiple elements, they may be distributed over a subset of the nodes in the network, on every node in the network, or several cognitive elements may reside on a single node. In this respect, the cognitive elements operate in a manner similar to software agents.

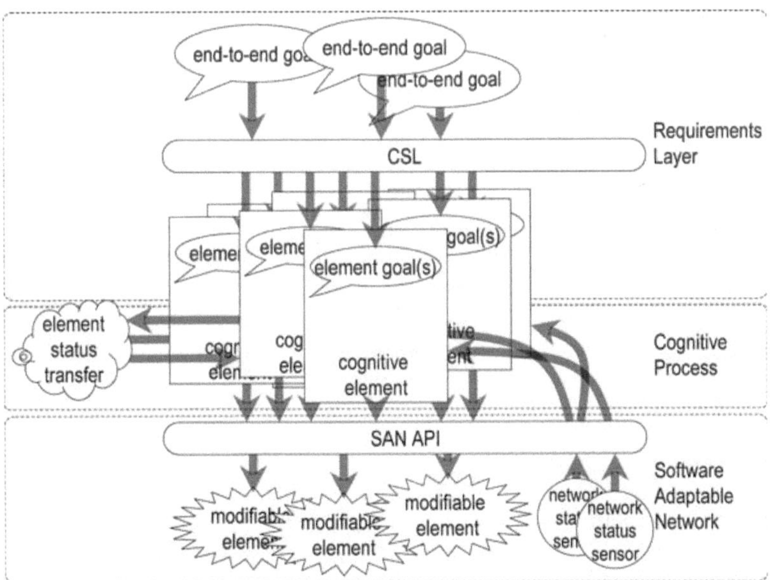

Fig. 2.2. The cognitive network framework.

2.3.1 User/Application/Resource Requirements

The top-level component of the cognitive network framework includes the end-to-end goals, Cognitive Specification Language (CSL) and cognitive element goals. Without end-to-end goals guiding network behavior, undesired consequences may arise. For instance, optimizing a network element without an end-to-end scope can cause a negative effect on the performance elsewhere in the network or node. This is a problem with many cross-layer designs and is explored in some depth in [4], which illustrates unintended end-to-end interactions in a MAC/PHY cross-layer design.

Like most engineering problems, there is likely to be a trade-off for every goal optimized on. When a cognitive network has multiple objectives it will not be able to optimize all metrics indefinitely, eventually reaching a point at which one metric cannot be improved without adversely affecting another. In order to determine this Pareto optimal front (the set of actions from which no goal can be improved without worsening another), each cognitive element must have an understanding of all end-to-end goals and their constraints.

To connect the goals of the top-level users of the network to the cognitive process, an interface layer must be developed. In a cognitive network, this role is performed by the CSL, providing behavioral guidance to the elements by translating the end-to-end goals to local element goals. Less like the RKRL proposed by Mitola for cognitive radio and more like a QoS specification language [19], the CSL maps end-to-end requirements to underlying mechanisms. Unlike a QoS specification language, the mechanisms are adaptive to the network capabilities, as opposed to fixed. Furthermore, a CSL must be able to adapt to new network elements, applications, and goals, some of which may not even be imagined yet. Other requirements may include support for distributed or centralized operation, including the sharing of data between multiple cognitive elements.

The scope of the cognitive network is broader than that of the individual network elements; it operates within the scope of a data flow, which may include many network elements. For a distributed cognitive process, the cognitive elements associated with each flow or network element may act selfishly and independently (in the context of the entire network) to achieve local goals, or act in an altruistic manner to achieve network-wide goals. The job of converting the end-to-end goals to these local element goals is often a difficult problem.

2.3.2 Cognitive Process

There does not seem to be a common, accepted definition of what cognition means when applied to communication technologies. The term cognitive, as used by this chapter, follows closely in the footsteps of the definition used by

Mitola in [2] and the even broader definition of the FCC. The former incorporates a spectrum of cognitive behaviors, from goal-based decisions to proactive adaptation. Here, we associate cognition with *machine learning*, which is broadly defined in [20] as any algorithm that "improves its performance through experience gained over a period of time without complete information about the environment in which it operates." Underneath this definition, many different kinds of artificial intelligence, decision making, and adaptive algorithms can be placed, giving cognitive networks a wide scope of possible mechanisms to use for learning.

Learning serves to complement the objective optimization part of the cognitive process by retaining the effectiveness of past decisions under a given set of conditions. Determining the effectiveness of past decisions requires a feedback loop to measure the success of the chosen solution in meeting the objectives defined. This is retained in memory so that when similar circumstances are encountered in the future, the cognitive process will have some idea of where to start or what to avoid.

The effect of a cognitive process's decisions on the network performance depends on the amount of network state information available to it. In order for a cognitive network to make a decision based on end-to-end goals, the cognitive elements must have some knowledge of the network's current state and other cognitive element states. If a cognitive network has knowledge of the entire network's state, decisions at the cognitive element level should be at least as good, if not better (in terms of the cognitive element goals) than those made in ignorance. For a large, complex system such as a computer network, it is unlikely that the cognitive network would know the total system state. There is often a high cost to communicate this information beyond those network elements requiring it, meaning a cognitive network will have to work with less than a full picture of the network status.

Filtering and abstraction may be used to further reduce the amount of information that must be exchanged and to avoid unnecessary triggering of the cognitive process. Filtering means that observations made by the node may be held back from the cognitive process if they are deemed irrelevant. Thus, the nodes themselves make some determination of what is important to the cognitive process. Filtering rules may be identified at design time with additional rules specified in real-time as the cognitive process determines its sensitivity to various types of observations and disseminates filtering rules accordingly. The goal of abstraction is to reduce the number of bits required to represent an observation. Observations or collections of observations made by a node are reported to the cognitive process at a higher level of abstraction than what is available within the node. Abstractions may also be specified at design time with real-time adaptations by the cognitive process. The reductionism resulting from filtering and abstraction carries risk because it may mask information that the cognitive process needs to operate correctly. Therefore, care should be taken in defining the abstractions or filtering.

2.3.3 Software Adaptable Network

The SAN consists of the Application Programming Interface (API), modifiable network elements, and network status sensors. The SAN is really a separate research area, just as the design of the SDR is separate from the development of the cognitive radio, but at a minimum the cognitive process needs to be aware of the API and the interface it presents to the modifiable elements. Just like the other aspects of the framework, the API should be flexible and extensible. Continuing the analogy with SDKs, an existing system that is analogous to the API is the Software Communications Architecture (SCA) used in the Joint Tactical Radio System (JTRS).

Another responsibility of the SAN is to notify the cognitive process of the status of the network (to what level and detail is a function of the filtering and abstraction being applied). The status of the network is the source of the feedback used by the cognitive process, and is composed of status sensor observations and communication with other cognitive elements. Possible observations may be local, such as bit error rate, battery life or data rate, non-local, such as end-to-end delay and clique size, or compilations of different local observations.

The modifiable elements can include any object or element used in a network, although it is unlikely that all elements in a SAN would be modifiable. Each modifiable element should have public and private interfaces to the API, allowing it to be manipulated by both the SAN and the cognitive process. Modifiable elements are assumed to have a set of states that they can operate in, and a "solution" for a cognitive process consists of a set of these states that, when taken together, meet the end-to-end requirements of the system. At any given instant the set of all possible combinations of states S can be partitioned into two subsets. The first, S', contains all possible combinations of sets that meet the end-to-end requirements and the second, $\bar{S}'$, consists of all combinations that do not meet these requirements. Of those in S', some may meet the requirements better than others, making them preferred solutions.

A cognitive network attempts to reach a set of states S'. This means that, should the network be in a state in $\bar{S}'$, or some sub-optimal state in S', the cognitive process attempts to move the system state to an optimal solution. With cognitive control over every element, the cognitive process can potentially set the system to any state; an ideal cognitive process could set the state to the optimal solution. If the system has only a few points of cognitive control, or chooses not to exercise all its control, then the cognitive process has to use the functionality and interactions of the non-cognitive aspects of the network to set the system state. Like the hole at the bottom of a funnel, certain system states will be basins of attraction, pulling the system towards them from a variety of starting states. If a system has several attractors and some are more optimal than others, then a few points of cognitive control may be enough to draw the system out of one attractor and into another. This is analogous to a watershed, in which moving the source of water a few miles may be enough to change what river the water will finally flow into.

2.4 A Cognitive Network for Multicast Lifetime

To illustrate the effect of these critical design decisions on a network, we present a cognitive network approach to maximizing a multicast flow's lifetime. By investigating even a simple cognitive network for a real-world problem, some of the subtleties of the design process can be explored. In this manner, the following cognitive network problem should be viewed as an illustrative case study.

Many factors may affect the expected lifetime of a network connection in a wireless network. For instance, traffic congestion can cause timeouts in upper layer protocols, interference can cause loss of connectivity at the PHY layer, and mobility can cause unexpected disconnections in traffic routing. However, for mobile and portable devices, one of the chief factors in determining the lifetime of a connection is the energy remaining in the batteries of the mobile nodes.

This example focuses on a cognitive network with control over the transmission power, antenna directionality, and routing tables of the network nodes. This is not the first investigation into lifetime routing in wireless networks; a large body of work on power-efficient routing exists in the literature. Gupta's survey [21] provides an excellent comparison of several power-efficient multicast algorithms for omnidirectional antennas. Weiselthier et al. [22] have examined this problem using directional antennas. A complete review of the related literature and an investigation using Mixed Integer Linear Program (MILPs) for determining the optimal lifetimes can be found in [23]. Although primarily designed to illustrate the cognitive network concept, this work is the first to provide a distributed, cognitive network approach to multicast lifetime routing that incorporates energy efficiency considerations, directional antennas, and a Signal to Interference and Noise Ratio (SINR) sufficiency requirement.

2.4.1 Problem Description

A wireless network is made up of a collection of network elements with varying energy capacity. Some elements may be battery powered, with limited capacity, while others may be less mobile, with large, high capacity batteries. The lifetime of a data path, however, is limited by the radio utilizing the largest fraction of its battery capacity. By minimizing the utilization of this bottleneck radio, the lifetime of the path can be maximized. Furthermore, we consider a network where radios are equipped with directional antennas, which are useful to reduce interference, improve spatial multiplexing, and increase range.

We model a network consisting of a set of radios $N = \{1, 2, \ldots, n\}$, in which the objective is to create a maximum lifetime multicast tree between source S and destination set D. As described earlier, the cognitive network controls three modifiable network parameters: the radio transmission power (contained in the elements of vector $\mathbf{pt}$), the antenna directionality (angles are contained in the elements of vector ϕ), and element routing tables (contained in each node of the multicast tree T). The states of the modifiable elements are part of the action set A, of which the action vector $\mathbf{a}$ contains the current state of each modifiable element.

In the model used here, the lifetime of a radio is inversely proportional to the utilization of the radio's battery,

$$\mu_i = \frac{pt_i}{ca_i} \tag{2.1}$$

where pt_i is radio i's transmission power and ca_i is the remaining energy capacity of its battery. The lifetime of a data path is limited by the radio utilizing the largest fraction of its battery capacity, so over the entire multicast tree T, the lifetime will be inversely proportional to the utilization of the max-utilization radio

$$\mu_T = \max_{j \in T} \{\mu_j\} \tag{2.2}$$

The network consists of radios with fully directional antennas in receive mode[2] (each element transmits omnidirectionally and receives directionally) with a fixed beamwidth θ that can take on a boresight angle $\phi \in [0, 2\pi)$. Figure 2.3 illustrates the operation of an ad-hoc network with directional antennas in receive mode.

When radio i transmits, the signal experiences gain factor gb within the main beam of the antenna [25]

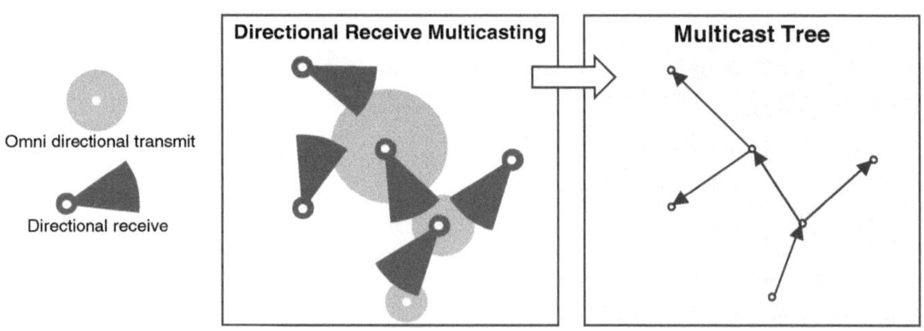

Fig. 2.3. The directional receive multicast operation. The shaded areas extending from the radios represent regions of increased gain.

[2] An argument for using directional reception rather than transmission can be found in [24].

$$gb = \frac{2\pi}{\theta} \tag{2.3}$$

Some energy leaks outside the main beam in sidelobes. The fraction that ends up in the beam is $pct \in (0, 1)$ and the fraction outside the beam is $(1 - pct)$. We also consider a path loss attenuation factor, proportional to

$$gp_{ij} = \frac{1}{d(i,j)^{\alpha}} \tag{2.4}$$

where $d(i, j)$ is the euclidean distance between source i and destination j and α is the path loss exponent. Combining these gains and attenuations, the overall gain from a transmission by radio i received at radio j is

$$g_{ij}(\phi_j) = \begin{cases} gb \cdot gp_{ij} \cdot pct & \phi_j \in a(i,j) \pm \frac{\theta}{2} \\ gp_{ij} \cdot (1 - pct) & otherwise \end{cases} \tag{2.5}$$

where $a(i, j)$ is the angular function between radios i and j.

A radio j can correctly receive information from radio i if the power received from the desired transmitter is greater than all other power and noise received by some SINR factor. We define the vector **pr** to be the power received at every radio in the tree from their parent radio,

$$pr_j(pt_i, \phi_j) = pt_i \cdot g_{ij}(\phi_j) \tag{2.6}$$

There is an entry in this vector for every radio in the tree, with the exception of the source radio ($|pr| = |T| - 1$). We then define vector **no** to be the minimum required power to overcome the interference and noise received at every element,

$$no_j(\mathbf{pt}, \phi_j, T) = \left(\sum_{k \neq i} pr_k(pt_i, \phi_j) + \sigma_j \right) \gamma_j \tag{2.7}$$

where σ_j is the thermal noise and γ_j is the SINR requirement for a particular radio. The vectors **pr** and **no** combine to give the network constraint,

$$\mathbf{pr} - \mathbf{no} \geq \mathbf{0} \tag{2.8}$$

2.4.2 Cognitive Network Design

The cognitive network framework encompasses a wide spectrum of possible implementations and solutions. This approach allows the framework to be a method for approaching problems in complex networks, rather than a specific solution. The framework sits on top of existing network layers, processes, and protocols, adjusting elements of the SAN to achieve an end-to-end goal. In this section, we show how a cognitive network that solves the multicast lifetime

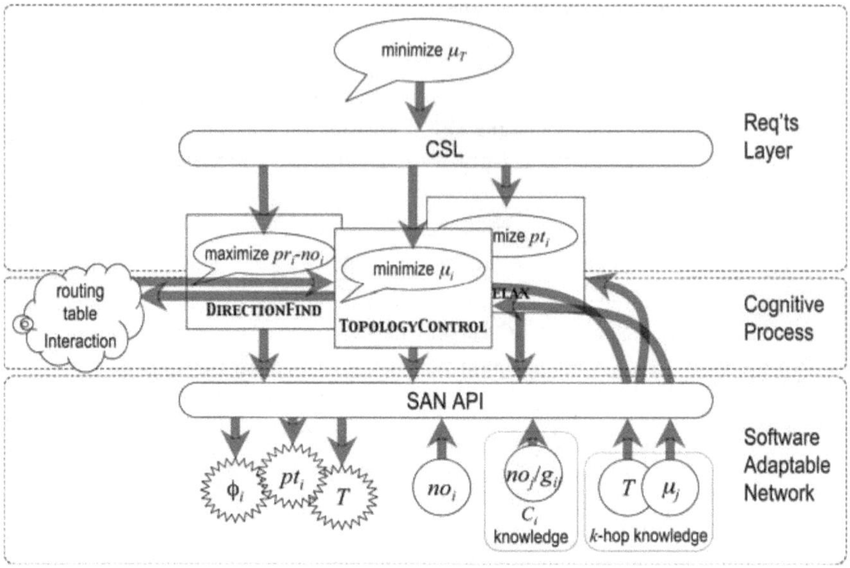

Fig. 2.4. The components of the multicast lifetime cognitive network as they fit into the framework.

problem fits into the framework. We examine each layer, showing how the requirements layer provides goals to the cognitive elements, how the cognitive process performs the feedback loop, and identify the functionality of the SAN. The ideas in this section are illustrated in Figure 2.4.

The cognitive process consists of three cognitive elements that distribute the operation of the cognitive process both functionally and spatially: PowerControl, DirectionControl, and RoutingControl. PowerControl adjusts the PHY transmission power (pt_i), DirectionControl adjusts the MAC spatial operation (ϕ_i), and RoutingControl adjusts the network layer's routing functionality (T). The SAN network status sensors provide each cognitive element with the knowledge of each radio's battery utilization in its *k-hop neighborhood.* The k-hop neighborhood of a radio is defined to be every radio reachable in the routing tree via at most k hops, following the routing tree both up and down branches.

Requirements Layer

The cognitive network investigated here is associated with a single objective optimization as its end-to-end goal. As such, the performance of an action vector is only dependent on the life-span of the multicast flow. The end-to-end objective is defined in Equation 2.9 as a cost function, where the lifetime of a flow is increased as $C(\mathbf{a})$ is minimized.

$$C(\mathbf{pt}, \phi, T) = \begin{cases} \mu_T & \mathbf{pr} - \mathbf{no} \geq \mathbf{0} \\ \infty & \text{otherwise} \end{cases} \tag{2.9}$$

Each of the modifiable elements affects the calculation of this cost: transmission power affects the lifetime directly; antenna orientation and routing table influence the lifetime indirectly by affecting the required transmission power.

The requirements layer transforms the end-to-end objective into a goal for each cognitive element through the CSL. Although these objectives are localized (each element only adapts a single modifiable element) the state of all modifiable elements affects the cognitive element's performance.

PowerControl's objective is to minimize the transmission power on every radio subject to the system constraint. This means that a radio will attempt to transmit at the minimum power that connects it to all of its children through the local control of pt_i. The objective can be represented by the utility function

$$u_i^{\mathrm{PC}}(\mathbf{pt}) = -\left(\max_{j \in \mathcal{C}_i} \left\{ \frac{no_k}{g_{ij}} \right\} - pt_i \right)^2 \tag{2.10}$$

which is maximized when the transmitting radio has exactly the power needed to reach the child radio with the greatest noise and least gain factor. $\mathcal{C}_i$ is the set of child radios that receive from radio i in the multicast tree.

The objective of DirectionControl is to maximize the receiving radio's SINR by controlling the directional angle of the antenna beam ϕ_i locally at every antenna. One form that the utility can take is

$$u_i^{\mathrm{DC}}(\mathbf{pt}, \phi_i) = pr_i - no_i \tag{2.11}$$

By rotating the directional antenna, the radio can increase the gain from the parent radio, while attenuating interfering signals.

The objective of RoutingControl is to minimize each radio's battery utilization by manipulating the routing tree (T) used by the network. The utility can be expressed as

$$u_i^{\mathrm{RC}}(pt_i) = \frac{1}{\mu_i} \tag{2.12}$$

By manipulating the children radios that it has to transmit to, a radio can reduce its transmission power and battery utilization.

Cognitive Process

The cognitive process consists of the three cognitive elements described above, each operating on every radio in the network. In this section, we discuss the strategies utilized by these elements to achieve the above objective goals and identify the critical design decisions used by each cognitive element.

Algorithm 1 RELAX($\mathbf{pt}, \phi, T$) $\rightarrow \hat{\mathbf{pt}}$

while not at $\hat{\mathbf{pt}}$ **do**
  for $i = 1 \ldots n$ **do**
    $pt_i = \max_{j \in C_i}\{\frac{no_j}{g_{ij}}\}$
  end for
end while

PowerControl The PowerControl cognitive element uses a strategy called RELAX. RELAX moves the transmission power of the elements of a tree to a minimum but sufficient power state (referred to as $\hat{\mathbf{pt}}$) for a given tree structure. Nodes do this by increasing or decreasing transmission power until it just meets the SINR sufficiency requirements of all their children. This means that parent node i will iteratively increase or decrease its transmission power according to the amount of interference and noise observed by the child j with the maximum amount of noise and interference. Algorithm 1 describes this process more formally.

RELAX is similar to the asynchronous iterative power control algorithm presented by Yates [26]. That paper proved, for a cellular network consisting of multiple handsets communicating with a base-station, if Equation 2.8 is *feasible* (meaning that there exists a solution), RELAX will find the optimal $\hat{\mathbf{pt}}$. Yates' work is for the reverse of the problem we consider – a cellular network is comprised of many nodes transmitting to a single base station. In contrast, our work considers a multicast wireless network with a set of parent nodes transmitting to many children. However, it is easy to show Yate's results still hold.

DirectionControl The second cognitive element's behavior is DirectionControl. DirectionControl moves the directional antenna to the orientation that maximizes the received SINR from a node's parent node. There are several direction-finding algorithms in the literature [27] and DirectionControl can implement one of these. If a node is a part of the multicast routing tree, it directs its antenna such that the power received from the parent is maximized with respect to the amount of interference and noise. If a node is not part of the multicast tree, it directs the antenna towards any source from which it can receive with the greatest SINR. For clarity, we will delineate these two tree structures: the first, called the *functional tree*, consists of just elements in the multicast routing tree and the second, called the *structural tree*, includes every element in the system that can receive a signal that meets the SINR requirement.

RoutingControl RoutingControl attempts to minimize the utilization of the radio batteries by approximating a Steiner tree for the utilization metric. RoutingControl uses the CHILDSWITCH strategy described in Algorithm 2. CHILDSWITCH begins by determining if it is operating on the *max-utilization* radio (the radio with maximum battery utilization) of its k-hop neighborhood,

Algorithm 2 $\textsc{ChildSwitch}(\mathbf{pt}, \phi, T) \rightarrow (\hat{\mathbf{pt}}, \hat{\phi}, T')$

if $\mu_i = \max_{n \in N_i}\{\mu_n\}$ **then** {*is on a max-util. node*}
  $\mu_{minmax} = \mu_i$ {*record the config. as the min-max*}
  $minmax = i$
  $j = \text{argmax}_{c \in C_i}\{no_c\}$ {*record the max-power child*}
  for $n \in N_i \quad n \neq j;\ n \notin B_j$ **do** {*every valid neighbor*}
    $C_n = C_n \cup \{j\}$ {*add max-power child*}
    $\hat{\phi} = \mathsf{DirectionControl}(\phi)$
    $\hat{\mathbf{pt}} = \textsc{Relax}(\mathbf{pt})$
    $\mu_{max} = \text{argmax}_{n \in N_i}\{\mu_n\}$ {*record max-util.*}
    if $\mu_{max} < \mu_{minmax}$ **then** {*max-util. is least*}
      $\mu_{minmax} = \mu_{max}$ {*record it as the min-max*}
      $minmax = max$
    end if
    $C_n = C_n \backslash \{j\}$ {*remove max-power child*}
  end for
  $C_{minmax} = C_{minmax} \cup \{j\}$ {*change to min-max config.*}
end if

by comparing its battery utilization against every k-hop neighbor's battery utilization. If it is, the radio becomes the *control-radio* and takes control over the routing tables of every element in the k-hop neighborhood. It then identifies which of the children radios in the functional tree requires the greatest amount of power to reach (the *max-power child*). The control-radio then attempts to detach the max-power child from itself and re-attach it as the child of another radio (by changing the routing table of a k-hop neighbor so that it becomes the new parent) in the k-hop neighborhood, in order to reduce the k-hop neighborhood's maximum utilization.

Valid choices for a new parent for the max-power child include all radios in the k-hop neighborhood of the structural tree, except for children of the max-power child. By using the structural tree rather than the functional tree, new radios in the network can be brought sensibly into the functional tree. After assignment, ChildSwitch waits until Relax converges and DirectionControl selects the correct beam angle. When Relax converges, ChildSwitch on the control-radio compares the utilization of all radios in the k-hop neighborhood against its initial utilization. The process is then repeated for the remaining valid radios, with the control-radio remembering the best (minimum) max-utilization configuration, and upon completion setting the routing table to this configuration.

This process repeats indefinitely until the max-utilization control-radios are no longer able to move their max-power children to configurations that lower the max-utilization radio of their k-hop neighborhood. In a synchronous network, in which only one RoutingControl control-radio performs ChildSwitch at a time, the network will (except in rare cases) converge to a single set of max-utilization radios.

Software Adaptable Network

The SAN provides an interface to the three modifiable network elements and the status of the network. The reported status is the local noise, maximum transmission power required to reach its children, k-hop battery utilization and k-hop routing tree.

The required transmission power, battery utilization of child radios, and routing tree status can be discovered and reported via a variable power handshaking scheme. In a synchronous manner, radios one by one send a HELLO message addressed to all children. The children each responded with an ACK message to the parent. The parent then decreases its transmission power and sends a new HELLO message until it fails to receive an ACK from some child. The parent radio then stops decreasing its power and returns to the previous power level, which is the maximum transmission power required to reach all its children. These HELLO and ACK messages can also transfer information about each radio's battery utilization and the routing tree within the k-hop neighborhood. In contrast to these non-local measurements, the amount of local noise can be calculated through the local SINR measurement.

2.4.3 Results

To determine the effectiveness of this cognitive network, we developed a simulation of the cognitive network. The simulation was written in Matlab, and consisted of nodes placed with a uniform random distribution in a square 2-D map with density 0.1 nodes/unit^2. There is a single source node and a variable number of receivers. The beam width θ is 30°, the path loss exponent α is 2, and 30% of the transmitted power is assumed to leak out through sidelobes $(1-pct)$. Each wireless node was given a battery with a random capacity (ca_i) uniformly distributed between 0 and 300 units of energy. The SINR sufficiency requirement is set to 1, meaning that the received power must be greater than the noise and interference to satisfy Equation 2.8.

The normalized lifetime of a path is calculated as the ratio of the lifetime obtained by the cognitive network to the optimal lifetime for the same set of source/destinations, capacities, and node positions. The optimal lifetime was determined using Wood's MILP [24]. Knowing the optimal solution is useful, since it allows a true "apples-to-apples" comparison between different scenarios, resulting in an accurate gauge of how effective the cognitive network is.

Underlying the cognitive network, one of two different generic multicast routing algorithms was used. The first, GREEDY, uses a greedy algorithm to create the multicast tree. GREEDY forms the multicast tree from the source node, adding minimum utilization nodes until a spanning tree has been formed. Utilization is estimated for every pair of nodes as the ratio of the (non-interference) transmission power required to reach each node to the

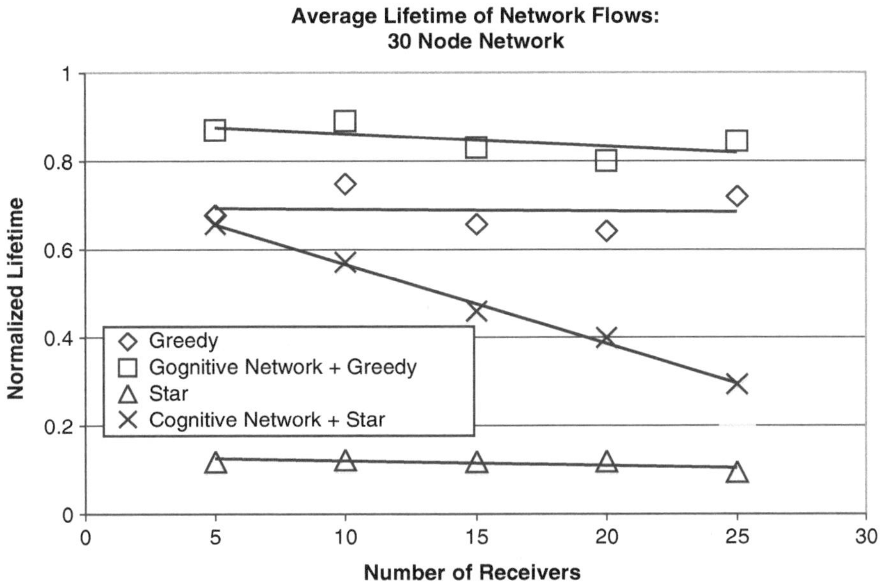

Fig. 2.5. Normalized lifetime of 30 node networks before and after cognitive network improvement. 1-Hop knowledge is used here.

node's battery capacity. GREEDY then prunes off branches until it has the minimum tree required to reach every destination. The other multicast algorithm used is STAR, which implements a 1-hop broadcast star from the source to every destination.

For a given scenario (consisting of node count, location, and battery capacity) both GREEDY or STAR were run, individually. The resultant tree topology and parent/child information from each algorithm were handed to RELAX and DirectionControl, which determine pt_i and ϕ_i, maximizing the lifetime for this route. The full cognitive process, including RoutingControl was then run on the route determined by GREEDY and STAR until it converged to a single set of max-utilization nodes. The lifetime of the resultant tree was then calculated. Finally, both the non-cognitive and cognitive lifetimes for a scenario were compared against the optimal lifetime obtained from the MILP, providing a normalized lifetime in $(0, 1)$, where 1 represents an optimal lifetime for that particular scenario.

These performance improvements are illustrated quantitatively in Figure 2.5 and qualitatively in Figure 2.6. Figure 2.5 illustrates the improvement in average lifetime produced by the cognitive process, and Figure 2.6 shows an example multicast tree and corresponding lifetime, both with and without the cognitive process.

These results show that the simplicity of STAR leads to sub-optimal performance, with at worst case less than 40% of the average lifetime of GREEDY. However, it also confirms that the cognitive network can make a significant

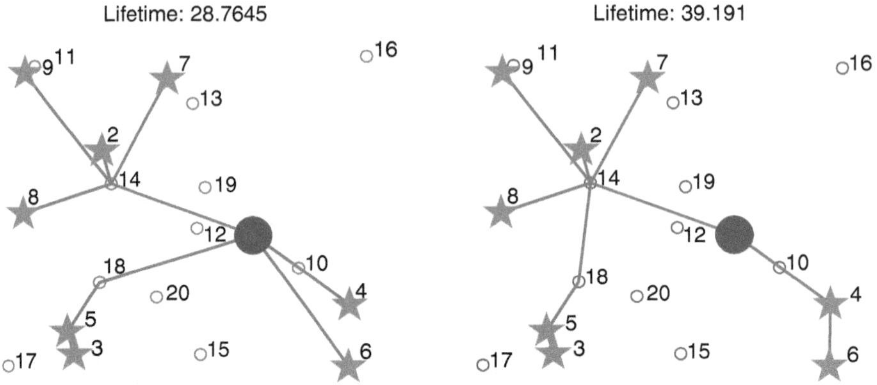

Fig. 2.6. Multicast tree topology and lifetime (source is circle, destinations are stars) for a sample scenario as first chosen by GREEDY (left) and then after improvement by cognitive element adaptations (right).

improvement on the average lifetime of the flow by using a 1-hop neighborhood – over 125% improvement in the STAR case. The GREEDY algorithm alone achieves much longer lifetimes, but the cognitive network is still able to improve it by 5–15%. In both routing algorithms, lifetimes remained steady or decreased as the number of multicast receivers increased. The cognitive network was able to improve the lifetime of the connection for all receiver counts and neighborhood sizes.

2.5 Future Questions and Research Areas

The previous sections make a case for the "what, why, and how" of cognitive networks. We now examine major issues that need to be addressed in order to move from concept to reality.

There is an implicit assumption in this chapter that the cognitive network implements configuration changes synchronously. The details of actually making this happen with high reliability are likely to be complex. The implications of nodes' switching configuration at different times may be worse than if no adaptation had been performed at all. Also, the varying topology of the network means that not all nodes will receive notification of configuration changes at the same time. A possible approach is to require nodes to be synchronized to some common time reference and to issue configuration changes with respect to the time reference. However, this adds complexity to the nodes and still does not guarantee that messages will not be lost, resulting in stranded nodes. It also forces the network to delay its adaptation to the conditions that spawned the configuration change.

Due to the autonomy of each, there is potential conflict between what a cognitive radio and a cognitive network each control if there is not an integrated architecture. One approach is to turn all control over to the cognitive network, but this is probably unwise. The reason is that the cognitive network has to limit its observations as much as possible just to make cognitive processing for a network feasible. This leaves much detailed local information out of the cognitive network picture. This detailed local information may be used by the cognitive radio to further optimize its performance outside the bounds of what is controlled by the cognitive network. To do this, the cognitive radio must know what it is allowed to change and what is in the hands of the cognitive network. A potential solution is to allow the cognitive network to establish regulatory policy for the cognitive radio in real-time, leaving the cognitive radio to perform further optimization under the constraints established by the cognitive network policy.

2.6 Conclusion

This chapter laid the groundwork for the concept of a cognitive network and proposed a definition for the term. Additionally, the cognitive network concept was compared against both cognitive radio and cross-layer design. Finally, a framework for cognitive networks was presented, and critical themes and issues were identified in the design and implementation of a cognitive network. While a significant amount of work remains to be done to make cognitive networks a reality, the rising complexity of networks and the need to manage this complexity makes the concept timely and attractive.

Although computer networks are becoming increasingly ubiquitous, the ability to manage and operate them is not becoming increasingly easier. Cognitive networks, with their promise to self-adapt to meet end-to-end objectives, are an emerging technology that will deal with this increasing complexity. This chapter presented three critical properties that designers need to trade off when architecting a cognitive network. These critical properties will provide design guidelines for future research into and implementation of cognitive networks.

References

1. R. W. Thomas, L. A. DaSilva, and A. B. MacKenzie, "Cognitive networks," in *Proc. of IEEE DySPAN 2005*, pp. 352–360, November 2005.
2. J. Mitola, *Cognitive Radio: An Integrated Agent Architecture for Software Defined Radio.* PhD thesis, Royal Institute of Technology (KTH), 2000.
3. V. Srivastava and M. Motani, "Cross-layer design: A survey and the road ahead," *IEEE Communications Magazine*, vol. 43, no. 12, pp. 112–119, 2005.
4. V. Kawadia and P. R. Kumar, "A cautionary perspective on cross-layer design," *IEEE Wireless Communications*, vol. 12, no. 1, pp. 3–11, 2005.
5. J. O. Neel, J. H. Reed, and R. P. Gilles, "Convergence of cognitive radio networks," in *Proc. of IEEE WCNC 2004*, vol. 4, pp. 2250–2255, 2004.

6. S. Haykin, "Cognitive radio: Brain-empowered wireless communication," *IEEE Journal on Selected Areas in Communication*, vol. 23, pp. 201–220, February 2005.
7. G. Ganesan and Y. Li, "Cooperative spectrum sensing in cognitive radio networks," in *Proc. of IEEE DySPAN 2005*, pp. 137–143, 2005.
8. S. M. Mishra, A. Sahai, and R. W. Brodersen, "Cooperative sensing among cognitive radios," in *Proc. of IEEE ICC 2006*, vol. 4, pp. 1658–1663, 2006.
9. P. Pawelczak, R. V. Prasad, L. Xia, and I. G. M. M. Niemegeers, "Cognitive radio emergency networks – requirements and design," in *Proc. of IEEE DySPAN 2005*, pp. 601–606, 2005.
10. D. Raychaudhuri, N. B. Mandayam, J. B. Evans, B. J. Ewy, S. Seshan, and P. Steenkiste, "CogNet: An architectural foundation for experimental cognitive radio networks within the future Internet," in *Proc. ACM/IEEE MobiArch'06*, pp. 11–16, December 2006.
11. D. D. Clark, C. Partridge, J. C. Ramming, and J. T. Wroclawski, "A knowledge plane for the Internet," in *Proc. of SIGCOMM '03*, pp. 3–10, ACM Press, 2003.
12. R. Saracco, "Forecasting the future of information technology: How to make research investment more cost-effective," *IEEE Communications Magazine*, vol. 41, pp. 38–45, December 2003.
13. P. Mähönen, J. Riihijärvi, M. Petrova, and Z. Shelby, "Hop-by-hop toward future mobile broadband IP," *IEEE Communications Magazine*, vol. 42, no. 3, pp. 138–146, 2004.
14. P. Mähönen, "Cognitive trends in making: Future of networks," in *Proc. of IEEE PIMRC 2004*, vol. 2, pp. 1449–1454, 2004.
15. D. Bourse, M. Muck, O. Simon, N. Alonistioti, K. Moessner, E. Nicollet, D. Bateman, E. Buracchini, G. Chengeleroyen, and P. Demestichas, "End-to-end reconfigurability (E^2R II): Management and control of adaptive communication systems." Presented at IST Mobile Summit 2006, June 2006.
16. P. Demestichas, V. Stavroulaki, D. Boscovic, A. Lee, and J. Strassner, "m@ANGEL: Autonomic management platform for seamless cognitive connectivity to the mobile internet," *IEEE Communications Magazine*, vol. 44, no. 6, pp. 118–127, 2006.
17. P. Sutton, L. E. Doyle, and K. E. Nolan, "A reconfigurable platform for cognitive networks," in *Proc. of CROWNCOM 2006*, 2006.
18. P. N. Johnson-Laird, *The Computer and the Mind.* Harvard University Press, 1988.
19. J. Jin and K. Nahrstedt, "QoS specification languages for distributed multimedia applications: A survey and taxonomy," *IEEE Multimedia*, vol. 11, no. 3, pp. 74–87, 2004.
20. M. A. L. Thathachar and P. S. Sastry, *Networks of Learning Automata.* Kluwer Academic Publishers, 2004.
21. S. K. S. Gupta and B. Wang, "Energy-efficient multicast protocols," in *Resource Management in Wireless Networking* (M. Cardei, I. Cardei, and D.-Z. Du, eds.), Kluwer Academic Publishers, 2004.
22. J. E. Wieselthier, G. D. Nguyen, and A. Ephremides, "Energy-aware wireless networking with directional antennas: The case of session-based broadcasting and multicasting," *IEEE Transactions on Mobile Computing*, vol. 1, pp. 176–191, July–September 2002.

23. K. N. Wood, *Directional Communications to Improve Multicast Lifetime in Ad Hoc Networks.* Doctoral dissertation, Virginia Polytechnic Institute and State University, Blacksburg, VA, September 2006.
24. K. Wood and L. A. DaSilva, "Optimization of multicast routing for network lifetime with directional listening," in *Proc. of BROADNETS 2005*, pp. 494–500, October 2005.
25. J. E. Wieselthier, G. D. Nguyen, and A. Ephremides, "Energy-limited wireless networking with directional antennas: The case of session-based multicasting," in *Proc. of IEEE INFOCOM 2002*, vol. 1, pp. 190–199, 2002.
26. R. D. Yates, "A framework for uplink power control in cellular radio systems," *IEEE Journal on Selected Areas in Communications*, vol. 13, pp. 1341–1347, September 1995.
27. H. L. V. Trees, *Optimum Array Processing.* No. 4 in Detection Estimation and Modulation Theory, Wiley-Interscience, 2002.

3

Cognitive Radio Architecture

Joseph Mitola III

The MITRE Corporation[1]

Cognitive Radio Architecture: The Engineering Foundations of Radio XML, Wiley 2006

3.1 Introduction

Cognitive radio has evolved to include a wide range of technologies for making wireless systems more flexible via more flexible transceiver platforms and enhanced computational intelligence. Dynamic spectrum access networks [1, 2] evolved rapidly from regulatory rulings of the past few years [7]. In addition, research towards context-aware services has resulted in interdisciplinary integration of complementary but somewhat isolated technologies: perception, planning and machine learning technologies from artificial intelligence on the one hand, and on the other hand software radio technologies that had come to include self-description in the extensible markup language, Radio XML [3–5]. The first significant radio-domain application for such smarter radios was the autonomous sharing of pooled spectrum [6], which the US Federal Communications Commission (FCC) endorsed relatively soon thereafter to encourage the development of secondary spectrum markets [7]. The original visionary formulation of the ideal Cognitive Radio (iCR) remains in development. iCR was formulated as an autonomous agent that perceives the user's situation (shopping or in distress) to proactively assist the user with wireless information services, particularly if the user is too busy or otherwise occupied to go through the tedium of using the cell phone, such as when in personal distress [8]. At the 2004 Dagstuhl workshop [9], cognitive radio was extended to Cognitive Wireless Networks (CWN), which has become a research area with its own conference on cognitive radio oriented wireless networks, CrownCom [10].

[1] The author's affiliation with The MITRE Corporation is provided for identification purposes only and should not be interpreted as the endorsement of the material by The MITRE Corporation or any of its sponsors.

H. Arslan (Ed.), Cognitive Radio, Software Defined Radio, and Adaptive Wireless Systems, 43–107.

This chapter summarizes the architecture of the ideal cognitive radio (iCR) that is more fully developed in the foundation text [11], particularly with respect to the critical machine learning technologies. The iCR architecture includes both isolated radio devices acting on behalf of a user and Cognitive Wireless Networks (CWN), both of which may incorporate machine perception – vision, speech, and other language skills – to ground the user continuously in a <Scene/>[2] that includes significant physical (space–time), social, and Radio Frequency (RF) aspects. Thus grounded, the iCR's embedded intelligent agent can respond more accurately to the user's current situation, interpreting location awareness signals (e.g. GPS) more astutely and focusing radio resources on the user's specific information needs.

Modern radio resources include not just dynamic radio spectrum [12] and air interface channels, but also increasingly fine-grained three-dimensional space–time resources created by directional antennas with and without Multi-Input Multi-Output (MIMO) processing [13]. The move to higher bandwidths for hot spot technologies signals an industry trend. The resulting <RF/> environments increasingly take on the characteristics of complex adaptive systems (CAS) of cooperating radio devices and networks rapidly evolving to more effectively manage the RF environment to address the user's needs for Quality of Information (QoI) given the social setting in which the user and device happen to be situated.

3.1.1 Ideal CRs Know Radio Like TellMe® Knows 800 Numbers

Long distance directory assistance in the US generally is answered by an interactive speech understanding system, an algorithm that may say "Toll Free Directory Assistance powered by TellMe®" [14]. "Please say the name of the listing you want." If you travel like I do, it may say "OK, United Airlines. If that is not what you wanted press 9, otherwise wait while I look up the number." Ninety-nine point nine percent of the time TellMe gets it right, replacing thousands of directory assistance operators of yore. TellMe, a speech-understanding system, achieves such a high degree of success by its focus on just one task: finding a toll-free telephone number. Narrow task focus is one of the keys to such a successful computationally intelligent user interface application.

The Cognitive Radio Architecture (CRA) of this chapter frames the functions, components, and design rules of Cognitive Wireless Networks (CWNs), in some sense the conceptual offspring of TellMe. CWNs are emerging in research settings as real-time, focused applications of radio and computational intelligence technologies. CWNs differ from the more general Artificial Intelligence (AI) based services like intelligent agents, computer speech, and computer vision in degree of focus. Like TellMe, CWNs focus on very narrow

[2] Such closed XML tags highlight concepts with ontological roles in organizing cognitive radio architecture.

tasks. Broader than TellMe, the task is to adapt radio-enabled information services to the specific needs of a specific user to achieve high QoI. TellMe, a network service, requires substantial network computing resources to serve thousands of users at once. CWNs, on the other hand, may start with a Cognitive Radio (CR) in your purse or on your belt, a cell phone on steroids, focused on the narrow task of creating from the myriad available wireless information networks and resources just what is needed by just one user, you. TellMe interacts with anybody, but each CR is self-aware and owner-aware via sensory perception and Autonomous Machine Learning (AML) technologies, earning the term "cognitive." Each CR fanatically serves the needs and protects the personal information of just one owner via the CRA using its audio and visual sensory perception and AML.

TellMe is here and now, while CWNs are emerging in global wireless research centers and industry forums like the Software-Defined Radio (SDR) Forum and Wireless World Research Forum (WWRF). This chapter summarizes CRA systems architecture challenges and approaches, emphasizing CR as a technology enabler for rapidly emerging commercial CWN services and generation-after-next military communications, based on the foundation technologies computer vision, computer speech, AML, and SDR.

3.1.2 CRs See What You See, Discovering RF Uses, Needs, and Preferences

In 2002, GRACE (Graduate Robot Attending Conference) [15], an autonomous mobile robot with a CRT for a face entered the International Joint Conference on Artificial Intelligence (IJCAI). It completed the mobile robot challenge by finding the registration desk; registering by talking to the receptionist; following the signs that said "ROBOTS" this way and "HUMANS" the other way; when called on giving a five-minute talk about herself; and then answering questions. She was the first to complete this challenge first articulated in the 1980s. There were no joysticks and no man behind the curtain: just a robot that can autonomously see, hear, and interact with the people and the environment to accomplish a specific task.

Compared to GRACE, the standard cell phone is not too bright. Although the common cell phone has a camera, it lacks GRACE's vision algorithms so it does not know what it is seeing. It can send a video clip, but it has no perception of the visual scene in the clip. If it had GRACE-like vision algorithms, it could perceive the visual scene. It could tell if it were at home, in the car, at work, shopping, or driving up the driveway on the way home. If GRACE-like vision algorithms show it that you are entering your driveway in your car, a Cognitive SDR could learn to open the garage door for you wirelessly. Thus, you would not need to fish for the garage door opener, yet another wireless gadget. In fact, you do not need a garage door opener anymore, once CRs enter the market. To open the car door, you will not need a key fob either.

As you approach your car, your personal CR perceives the common scene and, as trained, synthesizes the fob RF transmission and opens the car door for you.

Your CR perceives visual scenes continuously searching visual–RF correlations, cues to your needs for wireless services. A CR radio learns to open your garage door when you arrive home from your use of the garage door opener. When first you open the garage door with the wireless garage-door opener, your CR correlates the visual and RF scenes: owner's hand on device, then RF signal in the ISM band, and then the garage door opens. The next time, your CR verifies through reinforcement learning that your hand on the button, the RF signal, and the opening of the garage door form a sequential script, a use-case. The third time, your cognitive radio detects the approach to the garage door and offers to complete the RF use case for you, saying, "I see we are approaching the garage. Would you like me to open the door for us?" Thereafter, it will open the garage door when you drive up the driveway unless you tell it not to. It has transformed one of your patterns of RF usage, opening the garage door; into a cognitive (self-user perceptive) service, off-loading one of your daily tasks. Since the CR has learned to open the garage door, you may un-clutter your car by just one widget, that door opener.

Since your CR learned to open the garage door by observing your use of the radio via AML, you did not pay the cell phone company, and you did not endure pop-up advertising to get this personalized wireless service. As you enter the house with arms full of packages, your CR closes the garage door and locks it for you, having learned that from you as well. For the CR vision system to see what you see, today's Bluetooth earpieces evolve to CR Bluetooth glasses, complete with GRACE-like vision.

CRs do not attempt everything. They learn about your radio use patterns because they know a lot about radio and about generic users and legitimate uses of radio. CRs have the a priori knowledge needed to detect opportunities to assist you with your use of the radio spectrum accurately, delivering that assistance with minimum intrusion. TellMe is not a generic speech understanding system and CR is not a generic AI service in a radio.

Products realizing the visual perception of this vignette are realizable on laptop computers today. Reinforcement learning (RL) and Case-based Reasoning (CBR) are mature AML technologies with radio network applications now being demonstrated in academic and industrial research settings as technology pathfinders for CR [5] and CWN [16]. Two or three Moore's law cycles or three to five years from now, these vision and learning algorithms will fit in your cell phone. In the interim, CWNs will begin to offer such services, offering consumers new tradeoffs between privacy and ultra-personalized convenience.

3.1.3 Cognitive Radios Hear What you Hear, Augmenting your Personal Skills

Compared to GRACE, the cell phone on your waist is deaf. Although your cell phone has a miCRophone, it lacks GRACE's speech understanding technology, so it does not perceive what it hears. It can let you talk to your daughter, but it has no perception of your daughter, nor of the content of your conversation. If it had GRACE's speech understanding technology, it could perceive your speech dialog. It could detect that you and your daughter are talking about common subjects like homework, or your favorite song. With CR, GRACE-like speech algorithms would detect your daughter saying that your favorite song is now playing on WDUV. As an SDR, not just a cell phone, your CR tunes to FM 105.5 so that you can hear "The Rose." With your CR, you no longer need a transistor radio. Your CR eliminates from your pocket, purse or backpack yet another RF gadget. In fact, you may not need iPOD®, GameBoy® and similar products as high-end CR's enter the market. Your CR will learn your radio listening and information use patterns, accessing the songs, downloading games, snipping broadcast news, sports, stock quotes as you like as the CR re-programs its internal SDR to better serve your needs and preferences. Combining vision and speech perception, as you approach your car your CR perceives this common scene and, as you had the morning before, tunes your car radio to WTOP to your favorite "Traffic and weather together on the eights." With GRACE's speech understanding algorithms, your CR recognizes such regularly repeated catch phrases, turning up the volume for the traffic report and then turning it down or off after the weather report, avoiding annoying commercials and selecting relevant ones. If you actually need a tax deduction, it will record ***those*** radio commercials for your listening pleasure at tax time when you need them.

For AML, CRs need to save speech, RF, and visual cues, all of which may be recalled by the user, expanding the user's ability to remember details of conversations and snapshots of scenes, augmenting the skills of the <Owner/>.[3] Because of the brittleness of speech and vision technologies, CRs try to "remember everything" like a continuously running camcorder. Since CRs detect content such as speakers' names, and keywords like "radio" and "song," they can retrieve some content asked for by the user, expanding the user's memory in a sense. CRs thus could enhance the personal skills of their users such as memory for detail.

---

[3] Semantic Web: Researchers may formulate iCR as sufficiently speech-capable to answer questions about <Self/> and the <Self/> use of <Radio/> in support of its <Owner/>. When an ordinary concept like "owner" has been translated into a comprehensive ontological structure of computational primitives, e.g. via Semantic Web technology [3], the concept becomes a computational primitive for autonomous reasoning and information exchange. Radio XML, an emerging CR derivative of the eXtensible Markup Language, XML, offers to standardize such radio-scene perception primitives. They are highlighted in this brief treatment by <Angle-brackets/>. All iCR know of a <Self/>, a <Name/>, and an <Owner/>. The <Self/> has capabilities like <GSM/> and <SDR/>, a self-referential computing architecture, which is guaranteed to crash unless its computing ability is limited to real-time response tasks [3]; this is appropriate for CR but may be inappropriate for general purpose computing.

High performance dialog and audio–video retrieval technologies are cutting-edge but not out of reach for suitably narrow domains like TellMe and customization of wireless services. Casual dialog typically contains anaphora and ellipsis, using words like "this" and "that" to refer to anonymous events like playing a favorite song. Although innovative, speech research systems already achieve similar dialogs in limited domains [17]. When the user says, "How did you do that?" the domain of discourse is limited to the <Self/> and its contemporaneous actions. Since CR can do only one or two things at once, the question, "How did you do that?" has only one primary semantic referent, playing the song. Reasoning using analogy, also cutting edge, is no longer beyond the pale for tightly limited domains like CR and thus is envisioned in the CRA.

3.1.4 CRs Learn to Differentiate Speakers to Reduce Confusion

To further limit combinatorial explosion in speech, CR may form speaker models, statistical summaries of the speech patterns of speakers, particularly of the <Owner/>. Speaker modeling is particularly reliable when the <Owner/> uses the CR as a cell phone to place a phone call. Contemporary speaker recognition algorithms differentiate male from female speakers with high (>95%) probability. With a few different speakers to be recognized (e.g. fewer than 10 in a family) and with reliable side information like the speaker's telephone number, today's algorithms recognize individual speakers with 80 to 90% probability. Speaker models can become contaminated, such as erroneously including both <Owner/> and <Daughter/> speech in the <Owner/> model. Insightful product engineering could circumvent such problems, rendering <Owner/> interactions as reliable as TellMe® over the next few years.

Over time, each CR learns the speech patterns of its <Owner/> in order to learn from the <Owner/> and not be confused by other speakers. CR thus leverages experience incrementally to achieve increasingly sophisticated dialog. Directional miCRophones are rapidly improving to Service Video Tele-Conference (VTC) markets. Embedding these VTC miCRophones into CR "glasses" would enable CR to differentiate user speech from backgrounds like radio and TV. Today, a 3 GHz laptop supports this level of speech understanding and dialog synthesis in real time, making it likely available in a cell phone in three to five years.

Today, few consumers train the speech recognition systems embedded in most laptop computers. Its too much work and the algorithms do not take dictation well enough. Thus, although speech recognition technology exists, it is not as effective at the general task of converting speech to text as TellMe is in finding an 800 number. The CR value proposition, overcomes this limit by embedding machine learning so your CR continually learns about you by analyzing your voice, speech patterns, visual scene, and related use of the RF spectrum from garage door openers to NOAA weather, from cell phone and walkie-talkie to wireless home computer network. Do you want to know if your child's plane is in the air? Ask your CR and it could find "NiftyAir 122 Heavy cleared for takeoff by Dulles Tower." Again, in order to customize services for you, the <Owner/>, the CR must both know a lot about radio and learn a lot about you, the <Owner/>, recording and analyzing personal information and thus placing a premium on trustable privacy technologies. Increased autonomous (and thus free) customization of wireless service include secondary use of broadcast spectrum. The CRA therefore incorporates speech recognition.

3.1.5 More Flexible Secondary Use of Radio Spectrum

Consider a vignette with Lynne' the <Owner/> and Barb, the <Daughter/>. Barb drives to Lynne's house in her car. Coincidentally, Lynne' asks Genie, the CR <Self/> "Can you call Barb for me?"
Genie: "Sure. She is nearby so I can use the TV band for a free video call if you like."
Lynne': "Is that why your phone icon has a blue TV behind it?"
Genie: "Yes. I can connect you to her using unused TV channel 43 instead of spending your cell phone minutes. The TV icon shows that you are using free airtime as a secondary user of TV spectrum. I sent a probe to her cognitive radio to be sure it could do this."
Lynne': "OK, thanks for saving cell time for me. Let me talk to her." Barb's face appears on the screen.
Barb: "Wow, where did you come from?" Barb had never seen her cell phone display her Mom in a small TV picture in real time before, only in video clips.
Lynne': "Isn't this groovy. Genie, my new cognitive radio, hooked us up on a TV channel. It says you are nearby. Oh, I see you are out front and need help with the groceries. Here I come."

In 2004, the US Federal Communications Commission (FCC) issued a Report and Order that radio spectrum allocated to TV, but unused in a particular broadcast market could be used by CR as secondary users under Part 15 rules for low power devices, e.g. to create ad hoc networks. SDR Forum member companies have demonstrated CR products with these elementary spectrum-perception and use capabilities. Wireless products – military and commercial – realizing the FCC vignettes already exist. Complete visual and speech perception capabilities are not many years distant. Productization is underway. Thus, the CRA emphasizes CR spectrum agility, but in a context of enhanced perception technologies, a long-term growth path.

3.1.6 SDR Technology Underlies Cognitive Radio

To conclude the overview, take a closer look at the enabling radio technology, SDR. Samuel F B Morse's code revolutionized telegraphy in the late 1830s, becoming the standard for "telegraph" by the late 1800s. Thus when Marconi and Tesla brought forward wireless technology in 1902, Morse code was already a standard language for HF communication. Today as then, a radio includes an antenna, a RF power amplifier to transmit, and RF conversion to receive; along with a modulator/demodulator (modem) to impart the code to and from the RF channel; and a coder–decoder (codec) to translate information from human-readable form to a form coded for efficient radio transmission. Today as then, RF conversion depended on capacitors and inductors to set the radio frequency, but then some devices were the size of a refrigerator, while today they can be chip-scale devices. Then, the modulator consisted of the proverbial telegraph key, a switch to open and close the transmission circuit for on-off-keyed (OOK) data encoding. Morse code, a series of short (dits) or long marks (dahs) and spaces – sounds and silence – is still the simplest, cheapest way to communicate across a continent, and Morse code over HF radio still is used today in remote regions from the Australian outback to Africa and Siberia. Then and now, the "coder" was the person who had memorized the Morse code, manually converting dit-dit-dah-dit from and to the letters of the alphabet. Radio engineers almost never abandon an RF band (HF) or mode (Morse code). Instead, the use morphs from mainstream to a niche market like sports, amateur radio, remote regions, or developing economies. Today there are nice user interfaces and digital networking, but radio engineering has not taken anything away. At the relatively low data rates of mobile radio (<1 Mbps), networking (routing and switching) is readily accomplished in software, unlike wired networks where data rates reach gigabits per second and dedicated hardware is needed for high speed switching.

The essential functional blocks of radio have not changed for a century and are not likely to change either because the laws of physics define them: antenna, RF conversion, power amplification, modem, and codec. Today, however, miCRoelectronics technologies enable one to pack low power RF, modem, and codecs into single-chip packages while antennas fit neatly into the palm of your hand. Today, there are a myriad of modems evolved from the single RF of Morse to the sharing of RF bands in frequency, time, and code-space. The manual codec has evolved to include communications security (COMSEC) coding, authentication, and multi-layered digital protocol stacks. Cognitive radio embraces all the broad classes of modulation, each with unique modems, codecs and most importantly content, the reason people use the radio, after all.

The SDR Forum and Object Management Group (OMG) have standardized software architecture for wireless plug and play of the myriad band-mode combinations: the Software Communications Architecture (SCA) and Software Radio Architecture (SRA), respectively. But, the real enabler for SDR is the increasingly programmable analog RF of SDR: antennas, RF conversion, and amplifiers. Historically, the analog RF had fixed frequency and bandwidth, optimized for a small RF band such as 88 to 108 MHz for FM broadcast, 850–950 MHz for cell phones and 1.7 to 1.8 GHz for Personal Communications

Systems (PCS), a third generation cellular band. Today's cellular radios typically include three chip-sets, one optimized for first-generation "roaming" where infrastructure is not well built out, one for second generation digital service such as GSM, and one for PCS or NexTel®.

Each of these chip-sets accesses only the narrow band needed for the service, so today's cell phones can't open the garage door, not without another (expensive) chip set. In 1990–95, DARPA demonstrated SPEAKeasy II, the first SDR with continuous RF from 2 MHz to 2 GHz in just three analog RF bands: HF (2–30 MHz), mid-band (30–500 MHz) and high band (0.5–2 GHz).

MiCRo Electro-Mechanical Systems (MEMS) technology makes it possible to reprogram analog RF components digitally, so a cell phone could some day synthesize the garage door opener as the new digitally controlled analog RF MEMS technology emerges. RF MEMS digitally controls analog RF devices [18]. In some RF MEMS devices, a controller commands a micro-scale motor to move the interdigitated fingers of a capacitor, changing its analog value and hence changing the RF center frequency of the analog radio circuit. As the fingers move in and out by a few microns, the RF resonant frequency changes up and down by MHz. As this technology matures and enters service, RF chipsets will be reconfigurable across radio bands and modes, realizing affordable nearly ideal SDRs. FM, and TV/Broadcasts inform large markets with news, sports, weather, music, and the like. From boom box to weather radio, people around the world still depend on AM, FM and TV broadcasts for such information. In the past, you had to buy a specialized radio receiver and tune it manually to the station you like. With RF MEMS SDR, you tell the CR what you want to hear and it finds it for you. Your approval or disapproval constitutes training of the AML algorithms that tuned the MEMS SDR for your user-specific content preferences.

RF MEMS have been demonstrated to reduce the size, weight, and power of analog RF subsystems by two to three orders of magnitude, and by over 1000:1 in some cases, but they have been slow to enter markets because of lower than necessary reliability, a focus of both academic and commercial RF MEMS research and development. To facilitate the insertion of RF MEMS and other enabling technologies, the CRA embraces hardware abstraction.

3.1.7 Privacy is Paramount

A CR that remembers all your conversations for several years needs only a few hundred gigabytes of data memory, readily achieved in wearable CR-PDA even today. Many such conversations will be private, and some will include credit card numbers, social security numbers, bank account information, and the like. When my laptop was stolen with five years of tax returns, the process of dealing with identity theft was daunting and not foolproof. How can one trust a CR with all that personal information? Why would it need to remember all that stuff anyway?

One value proposition of CWNs is the reduction of tedium. Thus, asking the new owner to program the CR or to train it for an hour in the way that one is supposed to train the speech recognition system in a new laptop would be to increase tedium, not to decrease it. CR therefore aggregates experience, reprocessing the raw speech, vision, and RF data during sleep cycles so that it learns from experience with minimum tedious training interactions with the user. Although based solidly on contemporary RL and CBR technology, task-focused introspective learning for nearly unsupervised dialog acquisition, e.g. via text mining tools [19] is on the cutting edge of autonomous product development while the more general problem of minimally supervised dialog acquisition in general is at the cutting edge of language research. Thus, CR products will always "cheat" the way TellMe cheats; CR products pick a small, workable set of tasks that consumers will pay for and use, mini-killer apps. The resulting revenue streams build technology for increasingly capable tasks, evolving towards the vision-RF-dialog skills of the previous vignettes. However, to learn this way, CR really must remember all the raw data – all your keystrokes, emails, and conversations, to learn your use patterns and preferences autonomously, thus capturing private personal data.

If CR must remember your private personal data, then it must protect that information. Finger print readers are not perfect, as is any single Information Assurance (IA) measure, so CR may use a mix of IA measures. Candidate IA measures include soft biometrics like face and voice recognition along with more obtrusive measures like iris recognition. Layers of Public Key Infrastructure (PKI), GSM-like randomized challenges and signed responses with network validation of identity, and battery backup of IA protection skills, e.g. that erase all user data when the CR detects that it is being physically compromised, e.g. by the unexplained removal of screws of its case. Privacy is paramount, and practical products must protect personal information, identity, medical information, and the like with high reliability. Thus, a mix of soft biometrics likes face and voice recognition coupled with selective hard biometrics like a fingerprint reader, PKI, and other encryption methods. Given the limits of speech and visual perception technologies, CRs employ a large fraction of their sensory perception resources recognizing the face, voice, and daily habits of the <Owner/>. Some robots accumulate stimuli in a way that simulate human emotion, e.g.: happiness or distress. If the robot detects its <Owner's> voice and face, then it knows what to expect based on having learned the owners' patterns. If the voice and face are not recognized, then the CR might become defensive, protecting the owner's data and potentially erasing it rather than divulge personal data to someone the <Owner/> has not previously authorized. Embedding a backup battery deep within the motherboard and embedding sensors in screws in the motherboard might dissuade all but the most sophisticated criminals from stealing such CRs. Therefore, CRA explicitly includes hardware and software facilities to implement trustable protection of privacy.

3.1.8 Military Applications Abound

Military applications of CR in CWNs abound. It is easy to imagine realistic vignettes where radios relay the commander's change to an operations order in his own words, "Coalition partners are now located at grid square 76-11, so hold your fire. Rendezvous at Checkpoint Charlie at 1700." There might be little doubt about the authenticity of an order if it can be recalled and distributed digitally, authenticated and suitably protected to military standards, of course. Tactical military radio communications are notoriously noisy. Thus, a radio that conveys such critical information error-free and in the voice of the commander could reduce the fog of war, potentially saving lives.

With autonomous machine learning skills, military iCRs would learn coalition RF use patterns. Autonomously re-programming of their SDR transceivers, coalition CRs could learn to connect commanders directly with each other, avoiding the need for dedicated military radio operators per se and either reducing the size of a squad from ten to nine or enhancing the squad's capabilities by the 10% no longer needed to just operate the radio. Although one can never completely replace the flexibility and insight of skilled people, as iCRs offload mundane radio operation tasks from the radio operator, the team's effectiveness will increase, beneficial in the short run even if it takes decades to realize "Radar O'Reilly" in software.

Although the Phraselator [20] experiment showed the promise of real-time language translation in a handheld device for coalition operations, a Phraselator is yet another widget like the garage door opener. Envisioned iCR offer a flexible hardware platform in which to embed Phraselator algorithms invoked by language identification algorithms that detect non-native language and hence the need for real-time translation. Since iCR is about enhancing the effectiveness of communications, language translation embedded in CR to translate when and where needed certainly has the potential to enhance communications among coalition partners who speak different languages, again reducing the fog of war and improving the likelihood of success.

The CRA is not specifically designed for military applications, but its open and evolutionary nature enable a wide range of commercial and military applications.

3.1.9 Quality of Information (QoI) Metric

QoI concerns the information that meets a specific user's need at a specific time, place, physical location and social setting. If information is available, then the quality, quantity, timeliness and suitability may be measured. One expression for QoI [11] is given in Equation 1:

$$\text{QoI} = \text{Quantity} * \text{Timeliness} * \text{Validity} * \text{Relevance} * \text{Accuracy} * \text{Detail}$$

Equation 1 QoI metric

If there is no information, then **Quantity** = 0 as does QoI. If all the required information is present, then Quantity = 1.0. Since different users require different information to be fully satisfied, this user-dependent parameter is at best difficult to measure.

Timeliness must be defined in terms of the iCR user's time line along which the information would be used. If the information is needed immediately, then the quality may be characterized as inversely proportional to excessive time delay. To avoid division by zero, one may consider timeliness to be 1.0 if the information is available before a minimum delivery time:

Tmin (time, place, social-setting, topic)

For simplicity, let's adopt the convention that a situation is a specifiable subspace of time, place, and social-setting. The concept of a social setting must be defined in terms that the user accepts, such as "shopping" or "getting mugged." Suppose the shortest time delay in such a setting is ε so the maximum contribution of timeliness to QoI would be $1/\varepsilon$. If timeliness is normalized by ε, then maximum timeliness would be 1.0.

If **validity** is +1 if true and −1.0 if false, with the possibility of fuzzy set membership, then the validity value is an element of [−1, 1], and QoI may be positive or negative. Information that is known to be or that winds up being false has a qualitatively different kind of value than information the validity of which is unknown (validity = 0). Information of unknown validity may safely be ignored, so the QoI value of zero seems appropriate. Information that turns out to be false may in fact be misleading, yielding negative results because the user behaved in accordance with the falsehood. This is the sense of negative QoI.

Relevance is the degree to which the information corresponds to the need, measured in terms of precision and recall. In information retrieval, recall is the fraction of relevant documents retrieved from a corpus by a query and precision is the fraction of documents actually retrieved that turn out to be relevant. Recall of 1.0 indicates that all relevant documents are retrieved, while precision of 1.0 indicates that no irrelevant documents have been retrieved. Adapting this well known metric to QoI, one may define relevance as the product of precision and recall. This metric may not be ideal for information retrieval purposes, but it suffices in its role as a QoI metric that can be used to give an iCR feedback from its user by observing user behavior (e.g. asking for more or apparently not using items retrieved).

Accuracy refers to the quantitative aspects of the information. Quantitative errors include factual correctness (e.g. spell the President's name right) and numerical errors. Numerical accuracy reflects numerical error of the information represented with arbitrary precision, while QoI precision reflects the least error that it is possible to represent in a given numerical string. These could be differentiated, but for simplicity, one may measure whether the precision in which the number is expressed supports the required accuracy. If the accuracy required by the user is met, the value of the accuracy metric is 1.0. The rate of degradation of the accuracy metric from 1.0 may be linear, quadratic, exponential, fractal, or defined by table look-up, provided it falls in the range (0,1].

Finally if sufficient detail is provided to justify the information delivered, then Detail = 1.0, gradually dropping to zero if no elaborating detail is provided.

Consider the following example:

<Query>Name of the largest state in the USA </Query>
<Quantity>name</Quantity><Timeliness> in the next few seconds</Timeliness ><Validity>Must be true</Validity >
<Accuracy>Name must be spelled correctly</Accuracy ><Detail/>
</Query>

Equation 2 Illustrative information query

Since <Detail/> is null, the user is not asking for any special supporting information. In response to this query, the name "Texas" was valid until "Alaska" became a state. The user didn't specify a time frame so <Present/> may be assumed, but if the <User/> happens to be interested in history, such an assumption would not adequately reflect the user's QoI needs. In addition, Texas remains the largest state in the contiguous lower 48 states, so the geospatial scope might render Texas as accurate. A high QoI response from a CWN might provide name both Alaska and Texas with the associated validity. If such a complete answer were provided quickly and were spelled right, then QoI = 1.0. If the query were met an hour later because the iCR couldn't reach the cell phone network or WLAN for that length of time, then the QoI is less than 1.0. The amount of degradation from 1.0 depends on the urgency of the need. If the user were playing Trivial Pursuit with a few friends, then the penalty for time delay might not be great. If the user were playing "Who Wants to be a Millionaire?" on TV and asked the iCR for help as a phone-in, then even a few minutes of delay could yield unacceptably low QoI.

Given a working definition of QoI, the iCR could automatically manipulate the parameters of the air interface (s) as a function of the user's specific needs for QoI.

3.1.10 Architecture

Architecture is a comprehensive, consistent set of ***design rules*** by which a specified set of ***components*** achieves a specified set of ***functions*** in products and services that evolve through multiple design points over time [9]. This section introduces the fundamental design rules by which SDR, sensors, perception, and AML may be integrated to create Aware, Adaptive, and Cognitive Radios (iCR's) with better Quality of Information (QoI) through capabilities to Observe (sense, perceive), Orient, Plan, Decide, Act and Learn (the OOPDAL loop) in RF and user domains, transitioning from merely adaptive to demonstrably cognitive radio, CR.

This section develops five complementary perspectives of architecture called CRA I through CRA V. CRA I defines six functional components, black boxes to which are ascribed a first level decomposition of iCR functions and among which important interfaces are defined. One of these boxes is SDR, a proper subset of iCR. One of these boxes performs cognition via the <Self/>, a self-referential subsystem that strictly embodies finite computing (e.g. no while or until loops) avoiding the Gödel–Turing paradox.

CRA II examines the flow of inference through a cognition cycle that arranges the core capabilities of ***ideal*** CR (iCR) in temporal sequence for a logical flow and circadian rhythm for the CRA. CRA III examines the related levels of abstraction for iCR to sense elementary sensory stimuli and to perceive Quality of Service (QoS)-relevant aspects of a <Scene/> consisting of the <User/> in an <Environment/> that includes <RF/>. CRA IV of the foundation text [11] examines the mathematical structure of this architecture, identifying mappings among topological spaces represented and manipulated to preserve set-theoretic properties. Finally, CRA V of the foundation text [11] reviews SDR architecture, sketching an evolutionary path from the SCA/SRA to the CRA. The CRA <Self/> provided in CRA Self .xml of that text expresses in Radio XML (RXML) the CRA introduced in this chapter along with a priori knowledge for autonomous machine learning.

3.2 CRA I: Functions, Components and Design Rules

The ***functions*** of iCR exceed those of SDR. Reformulating the iCR <Self/> as a ***peer*** of its own <User/> establishes the need for added functions by which the <Self/> accurately perceives the local scene including the <User/> and autonomously learns to tailor the information services to the specific <User/> in the current RF and physical <Scene/>.

3.2.1 iCR Functional Component Architecture

The SDR components and the related cognitive components of iCR appear in Figure 3.1. The cognition components describe the SDR in Radio XML so that the <Self/> can know that it is a radio and that its goal is to achieve high QoI tailored to its own users. RXML intelligence includes a priori radio background and user stereotypes as well as knowledge of RF and space–time <Scenes/> perceived and experienced. This includes both structured reasoning with iCR peers and CWNs, and ad hoc reasoning with users, all the while learning from experience.

The detailed allocation of functions to components with interfaces among the components requires closer consideration of the SDR component as the foundation of CRA.

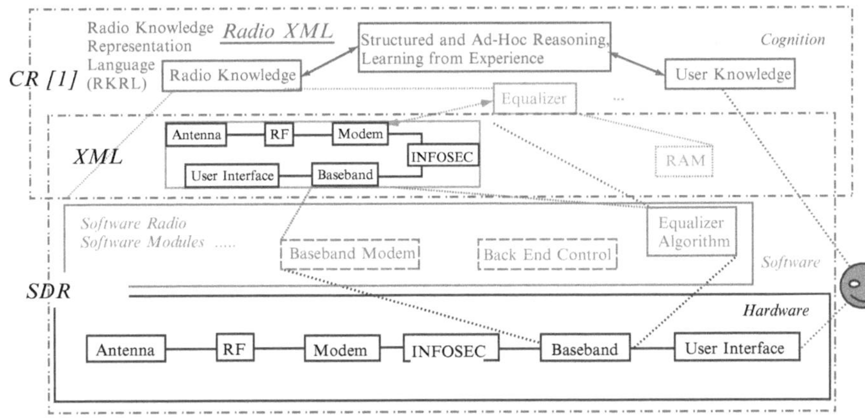

Figure © Dr. Joseph Mitola III, used with permission

Fig. 3.1. The cognitive radio architecture augments software-defined radio with computational intelligence and learning capacity.

3.2.2 SDR Components

SDRs include a hardware platform with RF access and computational resources, plus at least one software-defined personality. The SDR Forum has defined its Software Communications Architecture (SCA) and the Object Management Group (OMG) has defined its Software Radio Architecture (SRA), similar fine-grain architecture constructs enabling reduced cost wireless connectivity with next-generation plug and play. These SDR architectures are defined in Unified Modeling Language (UML) object models [21], CORBA Interface Design Language (IDL) [22], and XML descriptions of the UML models. The SDR Forum's SCA [23] and OMG SRA [24] standards describe the technical details of SDR both for radio engineering and for an initial level of wireless air interface ("waveform") plug and play. The SCA/SRA was sketched in 1996 at the first DoD-inspired MMITS Forum, developed by the US DoD in the 1990s and the architecture is now in production for the US military [25]. This architecture emphasizes plug-and-play wireless personalities on computationally capable mobile nodes where network connectivity is often intermittent at best.

The commercial wireless community [26], on the other hand, led by cell phone giants Motorola, Ericsson and Nokia envisions a much simpler architecture for mobile wireless devices, consisting of two APIs, one for the service provider and another for the network operator. They define a knowledge plane in the future intelligent wireless networks that is not dissimilar from a distributed CWN. That forum promotes the business model of the user –> service provider –> network operator –> large manufacturer –> device, where the user buys mobile devices consistent with services from a service provider, and the technical emphasis is on ***intelligence in the network***. This perspective no doubt will yield computationally intelligent networks in the near- to mid-term.

The CRA developed in this text, however, envisions the computational intelligence to create ad hoc and flexible networks with the ***intelligence in the mobile device***. This technical perspective enables the business model of user –> device –> heterogeneous networks, typical of the Internet model where the user buys a device (e.g. a wireless laptop) that can connect to the Internet via any available Internet Service Provider (ISP). The CRA builds on both the SCA/SRA and the commercial API model but integrates Semantic Web intelligence in Radio XML for mobile devices to enable more of an Internet business model to advance. This chapter describes how SDR, iCR, and iCR form a continuum facilitated by RXML.

3.2.3 iCR Node Functional Components

The simplest CRA is the minimalist set of functional components of Figure 3.2. A functional component is a black box to which functions have been allocated, but for which implementing components do not exist. Thus, while the Applications component is likely to be primarily software, the nature of those software components is yet to be determined. User Interface functions, on the other hand, may include optimized hardware, e.g. for computing video flow vectors in real time to assist scene perception. At the level of abstraction of the figure, the components are functional, not physical.

These functional components are

1. The **user sensory perception (User SP)** interface includes haptic, acoustic, and video sensing and perception functions,
2. The local **environment** sensors (location, temperature, accelerometer, compass, etc.),
3. The **system applications** (media independent services like playing a network game),
4. The **SDR** functions (which include RF sensing and SDR radio applications),

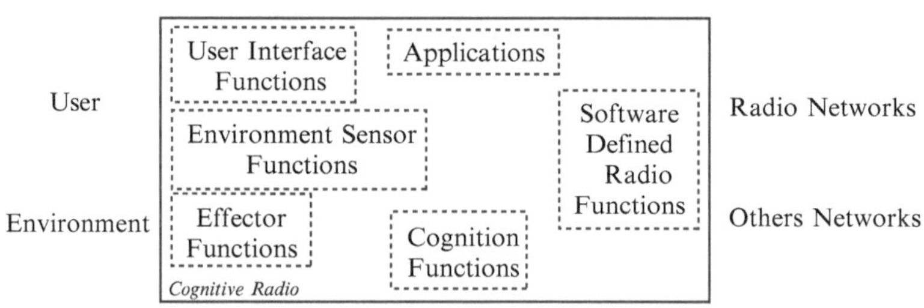

Figure © Dr. Joseph Mitola III, used with permission

Fig. 3.2. Minimal adaptive, aware, cognitive radio (iCR) node architecture.

5. The **cognition** functions (symbol grounding for system control, planning, learning) and
6. The ***local*** **effector** functions (speech synthesis, text, graphics, and multimedia displays).

These functional components are embodied on an iCR-Platform, a hardware realization of the six functions. In order to support the capabilities described in the prior chapters, these components go beyond SDR in critical ways. First, the user interface goes well beyond buttons and displays. The traditional user interface has been partitioned into a substantial user sensory subsystem and a set of local effectors. The user sensory interface includes buttons (the haptic interface) and miCRophones (the audio interface) to include acoustic sensing that is directional, capable of handling multiple speakers simultaneously and to include full motion video with visual scene perception. In addition, the audio subsystem does not just encode audio for (possible) transmission; it also parses and interprets the audio from designated speakers such as the <User/> for a high performance spoken natural language interface. Similarly, the text subsystem parses and interprets the language to track the user's information states, detecting plans and potential communications and information needs unobtrusively as the user conducts normal activities. The local effectors synthesize speech along with traditional text, graphics, and multimedia displays.

Systems applications are those *information services* that define value for the user. Historically, voice communications with a phone book, text messaging, and the exchange of images or video clips comprised the core value proposition of systems applications for SDR. These applications were generally integral to the SDR application, such as data services via GPRS, which is really a wireless SDR personality more than an information service. iCR systems applications break the service out of the SDR waveform so that the user need not deal with details of wireless connectivity unless that is of particular interest. Should the user care whether he plays the distributed video game via 802.11 or Bluetooth over the last 3 m? Probably not. The typical user might care if the iCR wants to switch to 3G at $5 per minute, but a particularly affluent user might not care and would leave all that up to the iCR.

The Cognition component provides all the cognition functions from the semantic grounding of entities from the perception system to controlling the overall system through planning and initiating actions, learning user preferences and RF situations in the process.

Each of these subsystems contains its own processing, local memory, integral power conversion, built-in-test (BIT) and related technical features.

3.2.4 The Ontological <Self/>

iCR may consist of the six functional components User SP, Environment, Effectors, SDR, Sys Apps, and Cognition. Equation 3 describes those components of the <Self/>, enables external communications and internal reasoning about the <Self/>, using the RXML syntax.

<Self/>
<iCR-Platform/>
<Functional-Components>
    <User SP/><Environment/><Effectors/><SDR/><Sys Apps/>
    <Cognition/>
< /Functional-Components>
< /Self>

Equation 3 The iCR <Self/> is defined to be an ideal cognitive radio (iCR) Platform, consisting of six functional components using the RXML syntax

Given the top-level outline of these functional components along with the requirement that they be embodied in physical hardware and software (the "Platform"), the six functional components are defined ontologically in Equation 3. In part, this equation states that the hardware–software platform and the functional components of the iCR are independent. Platform-independent computer languages like Java are well understood. This ontological perspective envisions platform-independence as an architecture design principle for iCR. In other words, the burden is on the (software) functional components to adapt to whatever RF-hardware-OS platform might be available.

3.2.5 Design Rules Include Functional Component Interfaces

These functional components of Table 3.1 imply associated functional interfaces. In architecture, design rules may include a list of the quantities and types of components as well as the interfaces among those components. This section addresses the interfaces among the six functional components.

The iCR N-Squared Diagram of Table 3.1 characterizes iCR interfaces. These constitute and initial set of iCR Applications Programmer Interfaces – iCR API's. In some ways these API's augment the established SDR APIs. For example, the Cognition API brings a planning capability to SDR. This is entirely new and much needed in order for basic ACAR's to accommodate even the basic ideas of XG.

In other ways, these API's supersede the existing SDR APIs. In particular, the SDR user interface becomes the User Sensory and Effector API. User Sensory API's include acoustics, voice, and video, while the effectors include speech synthesis to give the iCR <Self/> its own voice. In addition, wireless applications are growing rapidly. Voice and short message service become an ability to exchange images and video clips with ontological tags among wireless users. The distinctions between cell phone, PDA, and game box continue to disappear.

These interface changes enable the iCR to sense the situation represented in the environment, to interact with the user and to access radio networks on behalf of the user in a situation-aware way.

Table 3.1. iCR N-Squared Diagram characterizes internal interfaces between functional processes.

From\To	User SP	Environment	Sys Apps	SDR	Cognition	Effectors
User SP	1	7	13 PA	19	25 PA	31
Environment	2	8	14 SA	20	26 PA	32
Sys Apps	3	9	15 SCM	21 SD	27 PDC	33 PEM
SDR	4	10	16 PD	22 SD	28 PC	34 SD
Cognition	5 PEC	11 PEC	17 PC	23 PAE	29 SC	35 PE
Effectors	6 SC	12	18	24	30 PCD	36

Legend*: *P* – *Primary*; *A* – *Afferent*; *E* – *Efferent*; *C* – *Control*; *M* – *Multimedia*; *D* – *Data*; *S* – *Secondary*; *Others not designated P or S are ancillary

The Information Services API *consists of interfaces 13–18, 21, 27, and 33*

The Cognition API *consists of interfaces 25–30, 5, 11, 23, and 35*

***Interface Notes follow the numbers of the table* :**

 # Interface Note

1. **User SP–User SP** Cross-media correlation interfaces (video-acoustic, haptic-speech, etc.) to limit search and reduce uncertainty (e.g. if video indicates user is not talking, acoustics may be ignored or processed less aggressively for command inputs than if user is speaking)
2. **Environment–User SP** Environment sensors parameterize user sensor-perception. Temperature below freezing may limit video;
3. **Sys Apps–User SP** Systems Applications may focus scene perception by identifying entities, range, expected sounds for video, audio, and spatial perception processing
4. **SDR–User SP** SDR applications may provide expectations of user input to the perception system to improve probability of detection and correct classification of perceived inputs
5. **Cognition–User SP** This is the ***primary control efferent*** path from cognition to the control of the user sensory perception subsystem, controlling speech recognition, acoustic signal processing, video processing, and related sensory perception. Plans from Cognition may set expectations for user scene perception, improving perception.
6. **Effectors–User SP** Effectors may supply a replica of the effect to user perception so that self-generated effects (e.g. synthesized speech) may be accurately attributed to the <Self/>, validated as having been expressed, and/or cancelled from the scene perception to limit search.
7. **User SP–Environment** Perception of rain, buildings, indoor/outdoor can set GPS integration parameters
8. **Environment–Environment** Environment sensors would consist of location sensing such as GPS or Glonass; temperature of the ambient; light

level to detect inside versus outside locations; possibly smell sensors to detect spoiled food, fire, etc. There seems to be little benefit to enabling interfaces among these elements directly.

9. **Sys Apps–Environment** Data from the systems applications to environment sensors would also be minimal.
10. **SDR–Environment** Data from the SDR personalities to the environment sensors would be minimal.
11. **Cognition–Environment** (Primary Control Path) Data from the cognition system to the environment sensors controls those sensors, turning them on and off, setting control parameters, and establishing internal paths from the environment sensors.
12. **Effectors–Environment** Data from effectors directly to environment sensors would be minimal.
13. **User SP–Sys Apps** Data from the user sensory perception system to systems applications is a ***primary afferent path*** for multimedia streams and entity states that effect information services implemented as systems applications. Speech, images, and video to be transmitted move along this path for delivery by the relevant systems application or information service to the relevant wired or SDR communications path. Sys Apps overcomes the limitations of individual paths by maintaining continuity of conversations, data integrity, and application coherence, e.g. for multimedia games. While the cognition function sets up, tears down, and orchestrates the systems applications, the primary API between the user scene and the information service consist of this interface and its companions, the environment afferent path; the effector efferent path; and the SDR afferent and efferent paths.
14. **Environment–Sys Apps** Data on this path assists systems applications in providing location-awareness to services.
15. **Sys Apps–Sys Apps** Different information services interoperate by passing control information through the cognition interfaces and by passing domain multimedia flows through this interface. The cognition system sets up and tears down these interfaces.
16. **SDR–Sys Apps** This is the primary afferent path from external communications to the iCR. It includes control and multimedia information flows for all the information services. Following the SDR Forum's SCA, this path embraces wired as well as wireless interfaces.
17. **Cognition–Sys Apps** Through this path the iCR <Self/> exerts control over the information services provided to the <User/>.
18. **Effectors–Sys Apps** Effectors may provide incidental feedback to information services through this afferent path, but the use of this path is deprecated. Information services are supposed to control and obtain feedback through the mediation of the cognition subsystem.
19. **User SP–SDR** Although the sensory perception system may send data directly to the SDR subsystem, e.g. in order to satisfy security rules that user biometrics must be provided directly to the wireless security

subsystem, the use of this path is deprecated. Perception subsystem information is supposed to be interpreted by the cognition system so that accurate information can be conveyed to other subsystems, not raw data.

20. **Environment–SDR** Environment sensors like GPS historically have accessed SDR waveforms directly, such as providing timing data for air interface signal generation. The cognition system may establish such paths in cases where cognition provides little or no value added, such as providing a precise timing reference from GPS to an SDR waveform. The use of this path is deprecated because all of the environment sensors including GPS are unreliable. Cognition has the capability to de-glitch GPS, e.g. recognizing from video that the <Self/> is in an urban canyon and therefore not allowing GPS to report directly, but reporting on behalf of GPS to the GPS subscribers location estimates based perhaps on landmark correlation, dead reckoning, etc.
21. **Sys Apps–SDR** This is the primary efferent path from information services to SDR through the services API.
22. **SDR–SDR** The linking of different wireless services directly to each other is deprecated. If an incoming voice service needs to be connected to an outgoing voice service, then there should be a bridging service in Sys Apps through which the SDR waveforms communicate with each other. That service should be set up and taken down by the Cognition system.
23. **Cognition–SDR** This is the primary control interface, replacing the control interface of the SDR SCA and the OMG SRA.
24. **Effectors–SDR** Effectors such as speech synthesis and displays should not need to provide state information directly to SDR waveforms, but if needed, the cognition function should set up and tear down these interfaces.
25. **User SP–Cognition** This is the primary afferent flow for the results from acoustics, speech, images, video, video flow and other sensor-perception subsystems. The primary results passed across this interface should be the specific states of <Entities/> in the scene, which would include scene characteristics such as the recognition of landmarks, known vehicles, furniture and the like. In other words, this is the interface by which the presence of <Entities/> in the local scene is established and their characteristics are made known to the Cognition system.
26. **Environment–Cognition** This is the primary afferent flow for environment sensors.
27. **Sys Apps–Cognition** This is the interface through which information services request services and receive support from the iCR platform. This is also the control interface by which Cognition sets up, monitors, and tears down information services.
28. **SDR–Cognition** This is the primary afferent interface by which the state of waveforms, including a distinguished RF-sensor waveform is made known to the Cognition system. The cognition system can establish primary and backup waveforms for information services enabling the services

to select paths in real time for low latency services. Those paths are set up, monitored for quality and validity (e.g. obeying XG rules) by the cognition system, however.

29. **Cognition–Cognition** The cognition system as defined in this six component architecture entails (1) orienting to information from <RF/> sensors in the SDR subsystem and from scene sensors in the user sensory perception and environment sensors, (2) planning, (3) making decisions, and (4) initiating actions, including the control over all of the resources of the <Self/>. The <User/> may directly control any of the elements of the systems via paths through the cognition system that enable it to monitor what the user is doing in order to learn from a user's direct actions, such as manually tuning in the user's favorite radio station when the <Self/> either failed to do so properly or was not asked.
30. **Effectors–Cognition** This is the primary afferent flow for status information from the effector subsystem, including speech synthesis, displays, and the like.
31. **User SP–Effectors** In general, the user sensory-perception system should not interface directly to the effectors, but should be routed through the cognition system for observation.
32. **Environment–Effectors** The environment system should not interface directly to the effectors. This path is deprecated.
33. **Sys Apps–Effectors** Systems applications may display streams, generate speech, and otherwise directly control any effectors once the paths and constraints have been established by the cognition subsystem.
34. **SDR–Effectors** This path may be used if the cognition system establishes a path, such as from an SDR's voice track to a speaker. Generally, however, the SDR should provide streams to the information services of the Sys Apps. This path may be necessary for legacy compatibility during the migration from SDR through iCR to iCR but is deprecated.
35. **Cognition–Effectors** This is the primary efferent path for the control of effectors. Information services provide the streams to the effectors, but cognition sets them up, establishes paths, and monitors the information flows for support to the user's <Need/> or intent.
36. **Effectors–Effectors** These paths are deprecated, but may be needed for legacy compatibility.

The above information flows aggregated into an initial set of iCR APIs define an Information Services API by which an information service accesses the other five components (ISAPI consisting of interfaces 13–18, 21, 27, and 33). They would also define a Cognition API by which the cognition system obtains status and exerts control over the rest of the system (CAPI consisting of interfaces 25–30, 5, 11, 23, and 35). Although the constituent interfaces of these APIs are suggested in the table, it would be premature to define these APIs without first developing detailed information flows and interdependencies, which are defined in this chapter and analyzed in the remainder of this chapter. It would also be premature to develop such APIs without a clear idea of the kinds of RF and User domain knowledge and performance that are expected of the iCR architecture over time. These aspects are developed in the balance of the text, enabling one to draw some conclusions about these API's in the final chapters.

A fully defined set of interfaces and APIs would be circumscribed in RXML. For the moment, any of the interfaces of the N-squared diagram may be used as needed.

3.2.6 Near Term Implementations

One way to implement this set of functions is to embed into an SDR a reasoning engine such as a rule base with an associated inference engine as the Cognition Function. If the Effector Functions control parts of the radio, then we have the simplest iCR based on the simple six component architecture of Figure 3.1. Such an approach may be sufficient to expand the control paradigm from today's state machines with limited flexibility to tomorrow's iCR control based on reasoning over more complex RF states and user situations. Such simple approaches may well be the next practical steps in iCR evolution from SDR towards iCR.

This incremental step doesn't suggest how to mediate the interfaces between multi-sensory perception and situation-sensitive prior experience and a priori knowledge to achieve situation-dependent radio control that enables the more sophisticated information services of the use cases. In addition, such a simple architecture does not pro-actively allocate machine learning functions to fully understood components. For example, will autonomous machine learning require an embedded radio propagation modeling tool? If so, then what is the division of function between a rule base that knows about radio propagation and a propagation tool that can predict values like RSSI? Similarly, in the user domain, some aspects of user behavior may be modeled in detail based on physics, such as movement by foot and in vehicles. Will movement modeling be a separate subsystem based on physics and GPS? How will that work inside of buildings? How is the knowledge and skill in tracking user movements divided between physics-based computational modeling and the symbolic inference of a rule base or set of Horn clauses [34] with a Prolog engine? For that matter, how will the learning architecture accommodate a variety of learning methods like neural networks, PROLOG, forward chaining, SVM if learning occurs entirely in a cognition subsystem?

While hiding such details may be a good thing for iCR in the near term, it may severely limit the mass customization needed for iCRs to learn user patterns and thus to deliver RF services dramatically better than mere SDRs. Thus, we need to go "inside" the cognition and perception subsystems further to establish more of a fine-grained architecture. This enables one to structure the data sets and functions that mediate multi-sensory domain perception of complex scenes and related learning technologies that can autonomously adapt to user needs and preferences. The sequel thus pro-actively addresses the embedding of Machine Learning (ML) technology into the radio architecture.

Next, consider the networks. Network-independent SDRs retain multiple personalities in local storage, while network-dependent SDRs receive alternate personalities from a supporting network infrastructure – CWNs. High-end SDRs both retain alternate personalities locally and have the ability to validate and accept personalities by download from trusted sources. Whatever architecture emerges must be consistent with the distribution of RXML knowledge aggregated in a variety of networks from a tightly-coupled CWN to the Internet, with a degree of <Authority/> and trust reflecting the pragmatics of such different repositories.

The first two sections of this chapter therefore set the stage for the development of CRA. The next three sections address the cognition cycle, the inference hierarchies, and the SDR architecture embedded both into the CRA with the knowledge structures of the CRA.

3.2.7 The Cognition Components

Figure 3.1 above, shows three computational-intelligence aspects of CR:

1. Radio Knowledge – RXML:RF
2. User Knowledge – RXML:User
3. The Capacity to Learn

The minimalist architecture of Figure 3.2, and the functional interfaces of the subsequent table do not assist the radio engineer in structuring knowledge, nor does it assist much in integrating machine learning into the system. The fine-grained architecture developed in this chapter, on the other hand, is derived from the functional requirements to fully develop these three core capabilities.

3.2.8 Radio Knowledge in the Architecture

Radio knowledge has to be translated from the classroom and engineering teams into a body of computationally accessible, structured technical knowledge about radio. Radio XML is the primary enabler and product of this foray into formalization of radio knowledge. This text starts a process of RXML definition and development that can only be brought to fruition by industry over time. This process is similar to the evolution of the Software Communications Architecture (SCA) of the SDR Forum [23]. The SCA structures the technical knowledge of the radio components into UML and XML. RXML will enable the structuring of sufficient RF and user world knowledge to build advanced wireless-enabled or enhanced information services. Thus while the SRA and SCA focus on building radios, RXML focuses on using radios.

The World Wide Web is now sprouting with computational ontologies some of which are non-technical but include radio, such as the open CYC ontology [27]. They bring the radio domain into the Semantic Web, which helps people know about radio. This informal knowledge lacks the technical scope, precision and accuracy of authoritative radio references such as the ETSI documents defining GSM and ITU definitions, e.g. of 3GPP.

Not only must radio knowledge be precise, it must be stated at a useful level of abstraction, yet with the level of detail appropriate to the use-case. Thus, ETSI GSM in most cases would over-kill the level of detail without providing sufficient knowledge of the user-centric functionality of GSM. In addition, iCR is multi-band, multi-mode radio (MBMMR), so the knowledge must be comprehensive, addressing the majority of radio bands and modes available to a MBMMR. Therefore, in the development of CR technology below, this text captures radio knowledge needed for competent CR in the MBMMR bands from HF through millimeter wave. This knowledge is formalized with precision that should be acceptable to ETSI, the ITU and Regulatory Authorities (RAs) yet at a level of abstraction appropriate to internal reasoning, formal dialog with a CWN or informal dialog with users.

This kind of knowledge is to be captured in RXML:RF.

The capabilities required for an iCR node to be a cognitive entity are to sense, perceive, orient, plan, decide, act, and learn. To relate ITU standards to these required capabilities is a process of extracting content from highly formalized knowledge bases that exist in a unique place and that bear substantial authority, encapsulating that knowledge in less complete and therefore somewhat approximate form that can be reasoned with on the iCR node and in real time to support RF-related use cases. Table 3.2 illustrates this process.

The table is illustrative and not comprehensive, but it characterizes the technical issues that drive an information-oriented iCR node architecture. Where ITU, ETSI ... (meaning other regional and local standards bodies) and CWN supply source knowledge, the CWN is the repository for authoritative knowledge derived from the standards bodies and Regulatory Authorities (RAs), the <Authorities/>. A user-oriented iCR may note differences in the interpretation of source knowledge from <Authorities/> between alternate CWNs, precipitating further knowledge exchanges.

3.2.9 User Knowledge in the Architecture

Next, user knowledge is formalized at the level of abstraction and degree of detail necessary to give the CR the ability to acquire from its Owner and other designated users, the user-knowledge relevant to information services incrementally. Incremental knowledge acquisition was motivated in the introduction to AML by describing how frequent occurrences with similar activity sequences identifies learning opportunities. AML machines may recognize these opportunities for learning through joint probability statistics <Histogram/>. Effective use-cases clearly identify the classes of user and the specific knowledge learned to customize envisioned services. Use cases may

Table 3.2. Radio knowledge in the node architecture.

Need	Source Knowledge	iCR Internalization
Sense RF	RF Platform	Calibration of RF, noise floor, antennas, direction
Perceive RF	ITU, ETSI, ARIB, RA's	Location-based table of radio spectrum allocation
Observe RF (Sense and Perceive)	Unknown RF	RF sensor measurements & knowledge of basic types (AM, FM, simple digital channel symbols, typical TDMA, FDMA, CDMA signal structures)
Orient	XG-like policy	Receive, parse, and interpret policy language
	Known Waveform	Measure parameters in RF, Space and Time
Plan	Known Waveform	Enable SDR for which licensing is current
	Restrictive Policy	Optimize transmitted waveform, space–time plan
Decide	Legacy waveform, policy	Defer spectrum use to legacy users per policy
Act	Applications layer	Query for available services (White/yellow pages)
	ITU, ETSI, ..., CWN	Obtain new skills encapsulated as download
Learn	Unknown RF	Remember space–time-RF signatures; discover spectrum use norms and exceptions
	ITU, ETSI, ..., CWN	Extract relevant aspects such as new feature

also supply sufficient initial knowledge to render incremental AML not only effective, but also – if possible – enjoyable to the user.

This knowledge is defined in RXML:User. As with RF knowledge, the capabilities required for an iCR node to be a cognitive entity are to observe (sense, perceive), orient, plan, decide, act, and learn. To relate a use case to these capabilities, one extracts specific and easily recognizable <Anchors/> for stereotypical situations observable in diverse times, places, and situations. One expresses the anchor knowledge in using RXML for use on the iCR node.

3.2.10 Cross-domain Grounding for Flexible Information Services

The knowledge about radio and about user needs for wireless services must be expressed internally in a consistent form so that information services relationships may be autonomously discovered and maintained by the <Self/> on behalf of the <User/>. Relationships among user and RF domains are shown in Figure 3.3).

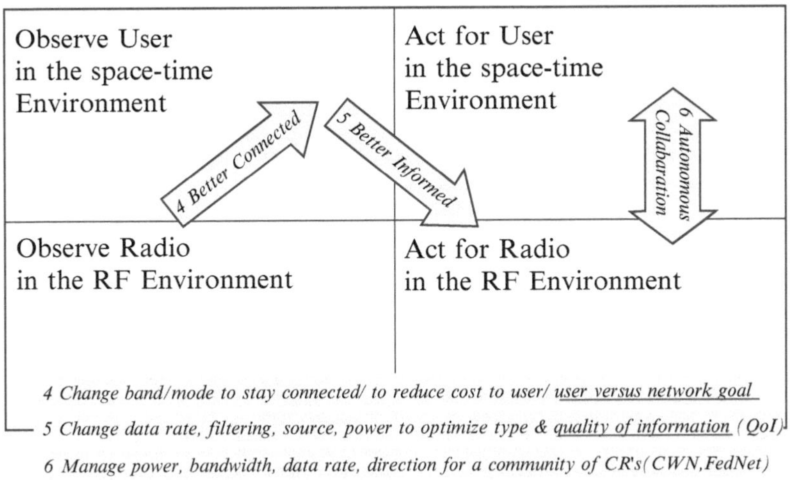

Figure © Dr. Joseph Mitola III, used with permission

Fig. 3.3. Discovering and maintaining services.

Staying better connected requires the normalization of knowledge between <User> and <RF> domains. If, for example, the <User/> says "What's on one oh seven – seven," near the Washington, DC area, then the dynamic <User/> ontology should enable the CR to infer that the user is talking about the current FM radio broadcast, the units are in MHz and the user wants to know what is on WTOP. If it can't infer this, then it should ask the user or discover by first dialing a reasonable default, such as 107.7 FM, a broadcast radio station and asking "Is this the radio station you want?" Steps 4, 5, and 6 in the figure all benefit from agreement across domains on how to refer to radio services. Optimizing behavior to best support the user requires continually adapting the <User> ontology with repeated re-grounding of terms in the <User/> domain to conceptual primitives and actions in the <RF/> domain.

The CRA facilitates this by seeding the speech recognition subsystem with the most likely expressions a particular <User/> employs when referring to information services. These would be acquired from the specific users via text and speech recognition, with dialogs oriented towards continual grounding by posing yes/no questions, either verbally or in displays or both, obtaining reinforcement verbally or via haptic interaction or both. The required degree of mutual grounding would benefit from specific grounding-oriented features in the iCR information architecture, developed below.

The process of linking user expressions of interest to the appropriate radio technical operations sometimes may be extremely difficult. Military radios, for example, have many technical parameters. For example, a "channel" in SINCGARS consists of de-hopped digital voice in one context (voice communications) or a 25 kHz band of spectrum in another context (and that may be either an FM channel into which its frequency hop waveform has hopped or an FDMA channel when in single channel mode). If the user says "Give me the Commander's Channel" the SINCGARS user is talking about a "de-hopped CVSD voice stream." If the same user a few seconds later says "This sounds awful. Who else is in this channel?" the user is referring to interference with a collection of hop sets. If the CR observes "There is strong interference in almost half of your assigned channels," then the CR is referring to a related set of 25 kHz channels. If the user then says "OK, notch the strongest 3 interference channels" he is talking about a different subset of the channels. If in the next breath the user says "Is anything on our emergency channel?" then the user has switched from SINCGARS context to <Self/> context, asking about one of the cognitive military radio's physical RF access channels. The complexity of such exchanges demands cross-domain grounding; and the necessity of communicating accurately under stress motivates a structured Natural Language (NL) and rich radio ontology aspects of the architecture developed further below.

Thus, both commercial and military information services entail cross-domain grounding with ontology oriented to NL in the <User> domain and oriented to RXML formalized a priori knowledge in the <RF> domain. Specific methods of cross-domain grounding with associated architectural features include:

1. ***<RF> to <User> Shaping dialog*** to express precise <RF> concepts to non-expert users in an intuitive way, such as
   a. Grounding: "If you move the speaker box a little bit, it can make a big difference in how well the remote speaker is connected to the wireless transmitter on the TV."
   b. iCR Information Architecture: Include facility for *rich set of synonyms* to mediate cognition-NL-synthesis interface (<Antenna>≅ <Wireless-remote-speaker>≅ "Speaker box").
2. ***<RF> to <User> Learning jargon*** to express <RF> connectivity opportunities in <User> terms.
   a. Grounding: "tee oh pee" for "WTOP," "Hot ninety two" for FM 97.7, "Guppy" for "E2C Echo Grand on 422.1 MHz."
   b. iCR Information Architecture: *NL-visual facility for single-instance update of user jargon.*
3. ***<User> to <RF> Relating values to actions:*** Relate <User> expression of values ("low cost") to features of situations ("normal") that are computable (<NOT> (<CONTAINS><Situation><Unusual/>)) and that relate directly to <RF> domain decisions
   a. Grounding: Normally wait for free WLAN for big attachment; if situation is <unusual>, ask if user wants to pay for 3G.
   b. iCR Information Architecture: *Associative inference hierarchy* that relates observable features of a <Scene> to user sensitivities, such as <Late-for-work>=><Unusual>; "The President of the company

needs this" =><Unusual> because "President" =><VIP> and <VIP> is not in most scenes.

3.2.11 Self-Referential Components

The Cognition component must assess, manage, and control all of its own resources, including validating downloads. Thus, in addition to <RF> and <User> domains, RXML must describe the <Self/>, defining the iCR architecture to the iCR itself in RXML.

3.2.12 Self-Referential Inconsistency

This class of self-referential reasoning is well known in the theory of computing to be a potential black hole for computational resources. Specifically, any Turing-Capable (TC) computational entity that reasons about itself can encounter unexpected Gödel–Turing situations from which it cannot recover. Thus TC systems are known to be "partial" – only partially defined because the result obtained when attempting to execute certain classes of procedure are not definable because the computing procedure will never terminate.

To avoid this paradox, CR architecture mandates the use of only "total" functions, typically restricted to bounded-minimalization [28]. Watchdog "step-counting" functions [29] or timers must be in place in all its self-referential reasoning and radio functions. The timer and related computationally indivisible control construct is equivalent to the computer-theoretic construct of a step-counting function over "finite minimalization." It has been proven that computations that are limited with certain classes of reliable watchdog timers on finite computing resources can avoid the Gödel–Turing paradox or at least reduce it to the reliability of the timer. This proof is the fundamental theorem for practical self-modifying systems.

Briefly: If a system can compute in advance the amount of time or the number of instructions that any given computation should take, then if that time or step-count is exceeded, the procedure returns a fixed result such as "Unreachable in Time T." As long as the algorithm does not explicitly or implicitly re-start itself on the same problem, then with the associated invocation of a tightly time- and computationally-constrained alternative tantamount to giving up, it

(a) is not Turing capable, but
(b) is sufficiently computationally capable to perform real-time communications tasks such as transmitting and receiving data as well as bounded user interface functions, and
(c) is not susceptible to the Turing–Gödel incompleteness dilemma and thus
(d) will not crash because of consuming unbounded or unpredictable resources in unpredictable self-referential loops.

This is not a general result. This is a highly radio domain-specific result that has been established only for isochronous communications domains in which

(a) Processes are defined in terms of a priori tightly bounded time epochs such as CDMA frames and SS7 time-outs and
(b) For every situation, there is a default action that has been identified in advance that consumes O(1) resources, and
(c) The watchdog timer or step-counting function is reliable.

Since radio air interfaces transmit and receive data, there are always defaults such as "repeat the last packed" or "clear the buffer" that may degrade the performance of the overall communications system. A default has O (1) complexity and the layers of the protocol stack can implement the default without using unbounded computing resources.

3.2.13 Watchdog Timer

Without the reliable watchdog timer in the architecture and without this proof to establish the rules for acceptable computing constructs on cognitive radios, engineers and computer programmers would build CRs that would crash in extremely unpredictable ways as their adaptation algorithms get trapped in unpredictable unbounded self-referential loops. Since there are planning problems that can't be solved with algorithms so constrained, either an unbounded community of CR's must cooperatively work on the more general problems or the CN must employ a Turing capable algorithm to solve the harder problems (e.g. NP-hard with large N) off line. There is also the interesting possibility of trading off space and time by remembering partial solutions and re-starting NP-hard problems with these sub-problems already solved. While it doesn't actually avoid any necessary calculations, with O(N) pattern matching for solved subproblems, it may reduce the total computational burden, somewhat like the FFT which converts $O(N^2)$ steps to O(N log N) by avoiding the re-computation of already computed partial products. This class of approach to parallel problem solving is similar to the use of pheromones by ants to solve the traveling salesman problem in less than $(2^N)/M$ time with M ants. Since this is an engineering text, not a text on the theory of computing, these aspects are not developed further here, but it suffices to show the predictable finiteness and proof that the approach is boundable and hence compatible with the real-time performance needs of cognitive radio.

This timer-based finite computing regime also works for user interfaces since users will not wait forever before changing the situation e.g. by shutting the radio off or hitting another key; and the CR can always kind of throw up its hands and ask the user to take over.

Thus, with a proof of stability based on the theory of computing, the CRA structures systems that not only can modify themselves, but can do it in such a way that they are not likely to induce non-recoverable crashes from the "partial" property of self-referential computing.

3.2.14 Flexible Functions of the Component Architecture

Although this chapter develops the six-element component architecture of a particular information architecture and one reference implementation, there are many possible architectures. The purpose is not to try to sell a particular architecture, but to illustrate the architecture principles. The CRA and research implementation, CR1 [11], therefore, offer open-source licensing for non-commercial educational purposes.

Table 3.3 further differentiates architecture features.

These functions of the architecture are not different from those of the six-component architecture, but represent varying degrees of instantiation of the six components. Consider the following degrees of architecture instantiations:

Cognition functions of radio entail the monitoring and structuring knowledge of the behavior patterns of the <Self/>, the <User>, and the environment (physical, user situation and radio) in order to provide information services, learning from experience to tailor services to user preferences and differing radio environments.

Adaptation functions of radio respond to a changing environment, but can be achieved without learning if the adaptation is pre-programmed.

Awareness functions of radio extract usable information from a sensor domain. Awareness stops short of perception. Awareness is required for adaptation, but awareness does not guarantee adaptation. For example,

Table 3.3. Features of iCR to be organized via architecture.

Feature	Function	Examples (RF; vision; speech; location; motion)
Cognition	Monitor & Learn	Get to know user's daily patterns & model the local RF scene over space, time, and situations
Adaptation	Respond to changing environment	Use unused RF, protect owner's data
Awareness	Extract information from sensor domain	Sense or perceive
Perception	Continuously identify knowns, unknowns and backgrounds in the sensor domain	TV channel; Depth of visual scene, identity of objects; location of user, movement and speed of <Self/>
Sensing	Continuously sense & pre-process single sensor-field in single sensory domain	RF FFT; Binary vision; binaural acoustics; GPS; accelerometer; etc.

embedding a GPS receiver into a cell phone makes the phone more location-aware, but unless the value of the current location is actually used by the phone to do something that is location-dependent, the phone is not location-adaptive, only location aware. These functions are a subset of the CRA that enable adaptation.

Perception functions of radio continuously identify and track knowns, unknowns and backgrounds in a given sensor domain. Backgrounds are subsets of a sensory domain that share common features that entail no particular relevance to the functions of the radio. For a CR that learns initially to be a single-Owner-radio, in a crowd, the Owner is the object that the radio continuously tracks in order to interact when needed. Worn from a belt as a Cognitive Wireless Personal Digital Assistant (CWPDA), the iCR perception functions may track the entities in the scene. The non-Owner entities comprise mostly irrelevant background because no matter what interactions may be offered by these entities, the CR will not obey them, just the perceived owner. These functions are a subset of the CRA that enable cognition.

The sensory functions of radio entail those hardware and/or software capabilities that enable a radio to measure features of a sensory domain. Sensory domains include anything that can be sensed, such as audio, video, vibration, temperature, time, power, fuel level, ambient light level, sun angle (e.g. through polarization), barometric pressure, smell, and anything else you might imagine. Sensory domains for vehicular radios may be much richer if less personal than those of wearable radios. Sensory domains for fixed infrastructure could include weather features such as ultra-violet sunlight, wind direction and speed, humidity, traffic flow rate, or rain rate. These functions are a subset of the CRA that enable perception.

The Platform Independent Model (PIM) in the Unified Modeling Language (UML) of SDR [30] provides a convenient, industry-standard computational model that an iCR can use to describe the SDR and computational resource aspects of its own internal structure, as well as describing facilities that enable radio functions. The general structure of hardware and software by which a CR reasons about the <Self/> in its world is also part of its architecture defined in the SDR SCA/SRA as resources.

3.3 CRA II: The Cognition Cycle

The Cognitive Radio Architecture (CRA) consists of a set of design rules by which the cognitive level of information services may be achieved by a specified set of components in a way that supports the cost-effective evolution of increasingly capable implementations over time [11]. The cognition subsystem of the architecture includes an inference hierarchy and the temporal organization and flow of inferences and control states, the cognition cycle.

3.3.1 The Cognition Cycle

The cognition cycle developed for CR1 [11] is illustrated in Figure 3.4. This cycle implements the capabilities required of iCR in a reactive sequence. Stimuli enter the cognitive radio as sensory interrupts, dispatched to the cognition cycle for a response. Such an iCR continually observes (senses and perceives) the environment, orients itself, creates plans, decides, and then acts. In a single-processor inference system, the CR's flow of control may also move in the cycle from observation to action. In a multi-processor system, temporal structures of sensing, preprocessing, reasoning, and acting may be parallel and complex. Special features synchronize the inferences of each phase. The tutorial code of [7] all works on a single processor in a rigid inference sequence defined in the figure. This process is called the Wake Epoch because the primary reasoning activities during this large epoch of time are reactive to the environment. We will refer to "sleep epochs" for power down condition, "dream epochs" for performing computationally intensive pattern recognition and learning, and "prayer epochs" for interacting with a higher authority such as network infrastructure.

During the wake epoch, the receipt of a new stimulus on any of a CR's sensors or the completion of a prior cognition cycle initiates a new primary cognition cycle. The cognitive radio observes its environment by parsing incoming information streams. These can include monitoring and speech-to-text

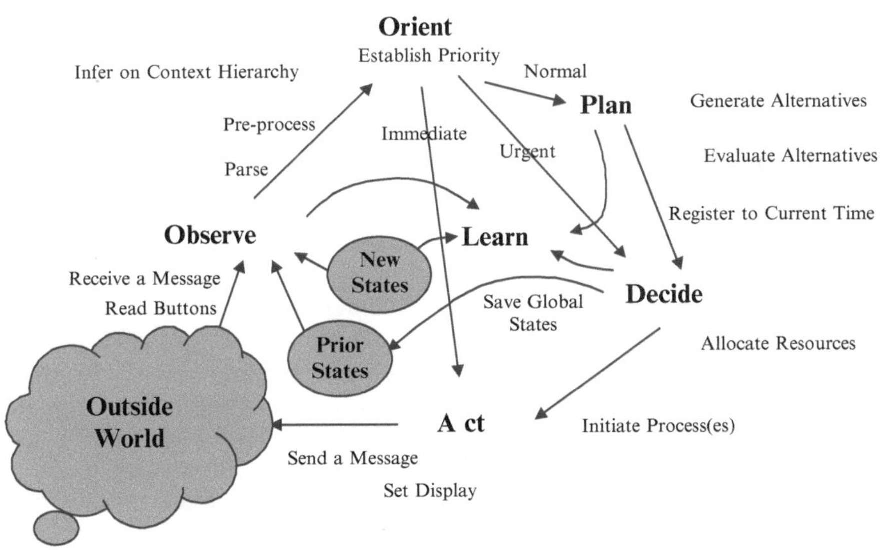

Figure © Dr. Joseph Mitola III, used with permission

Fig. 3.4. Simplified cognition cycle. The Observe – Orient – Decide – Act (OODA) loop is a primary cycle, however, learning, planning, and sensing the outside world are crucial phases necessary to be properly prepared for the OODA loop.

conversion of radio broadcasts, e.g. the weather channel, stock ticker tapes, etc. Any RF-LAN or other short-range wireless broadcasts that provide environment awareness information may be also parsed. In the observation phase, a CR also reads location, temperature, and light level sensors, etc. to infer the user's communications context.

3.3.2 Observe (Sense and Perceive)

The iCR senses and perceives the environment (via "Observation Phase" code) by accepting multiple stimuli in many dimensions simultaneously and by binding these stimuli – all together or more typically in subsets – to prior experience so that it can subsequently detect time-sensitive stimuli and ultimately generate plans for action.

Thus, iCR continuously aggregates experience and compares prior aggregates to the current situation. A CR may aggregate experience by remembering everything. This may not seem like a very smart thing to do until you calculate that all the audio, unique images, and emails the radio might experience in a year only takes up a few hundred gigabytes of memory, depending on image detail. So the computational architecture for remembering and rapidly correlating current experience against everything known previously is a core capability of the CRA. A ***novelty*** detector identifies new stimuli, using the new aspects of partially familiar stimuli to identify incremental learning primitives.

In the six-component (User SP, Environment, Effectors, SDR, Sys Apps, and Cognition) functional view of the architecture defined above, the Observe phase comprises both the User Sensory and Perception (User SP) and the Environment (RF and physical) sensor subsystems. The subsequent Orient phase is part of the Cognition component in this model of architecture.

3.3.3 Orient

The "Orient Phase" determines the significance of an observation by binding the observation to a previously known set of stimuli of a "scene."

The Orient phase contains the internal data structures that constitute the equivalent of the Short-Term Memory (STM) that people use to engage in a dialog without necessarily remembering everything with the same degree of long-term memory. Typically people need repetition to retain information over the long term. The natural environment supplies the information redundancy needed to instigate transfer from STM to Long-Term Memory (LTM). In the CRA, the transfer from STM to LTM is mediated by the sleep cycle in which the contents of STM since the last sleep cycle are analyzed both internally and with respect to existing LTM. How to do this robustly remains an important CR research topic, but the overall framework is defined in CRA.

Matching of current stimuli to stored experience may be achieved by stimulus recognition or by "binding." The orient phase is the first collection of activity in the cognition component.

Stimulus Recognition

Stimulus recognition occurs when there is an exact match between a current stimulus and a prior experience. CR1 is continually recognizing exact matches and recording the number of exact matches that occurred along with the time in number of cognition cycles between the last exact match. By default, the response to a given stimulus is to merely repeat that stimulus to the next layer up the inference hierarchy for aggregation of the raw stimuli. But if the system has been trained to respond to a location, a word, an RF condition, a signal on the power bus, etc., then it may either react immediately or plan a task in reaction to the detected stimulus. If that reaction were in error, then it may be trained to ignore the stimulus given the larger context which consists of all the stimuli and relevant internal states, including time.

Sometimes, the Orient Phase causes an action to be initiated immediately as a "reactive" stimulus-response behavior. A power failure, for example, might directly invoke an act that saves the data (the "Immediate" path to the Act Phase in the figure). A non-recoverable loss of signal on a network might invoke reallocation of resources, e.g. from parsing input to searching for alternative RF channels. This may be accomplished via the path labeled "Urgent" in the figure.

Binding

The binding occurs when there is a nearly exact match between a current stimulus and a prior experience and very general criteria for applying the prior experience to the current situation are met. One such criterion is the number of unmatched features of the current scene. If only one feature is unmatched and the scene occurs at a high level such as the phrase or dialog level of the inference hierarchy, then binding is the first step in generating a plan for behaving similarly in the given state as in the last occurrence of the stimuli. In addition to numbers of features that match exactly, which is a kind of Hamming code, Instance-Based Learning (IBL) supports inexact matching and binding. Binding also determines the priority associated with the stimuli. Better binding yields higher priority for autonomous learning, while less effective binding yields lower priority for the incipient plan.

3.3.4 Plan

Most stimuli are dealt with "deliberatively" rather than "reactively." An incoming network message would normally be dealt with by generating a plan (in the Plan Phase, the "Normal" path). Planning includes plan generation. In research-quality or industrial-strength CR's, formal models of causality must be embedded into planning tools [31]. The Plan phase should also include reasoning about time. Typically, reactive responses are pre-programmed or defined by a network (the CR is "told" what to do), while other behaviors might be planned. A stimulus may be associated with a simple plan as a function of planning parameters with a simple planning system. Open source planning tools enable the embedding of planning subsystems into the CRA, enhancing the Plan component. Such tools enable the synthesis of RF and information access behaviors in a goal-oriented way based on perceptions from the visual, audio, text, and RF domains as well as Regulatory Authority (RA) rules and previously learned user preferences.

3.3.5 Decide

The "Decide" phase selects among the candidate plans. The radio might have the choice to alert the user to an incoming message (e.g. behaving like a pager) or to defer the interruption until later (e.g. behaving like a secretary who is screening calls during an important meeting).

3.3.6 Act

"Acting" initiates the selected processes using effector modules. Effectors may access the external world or the CR's internal states.

Externally Oriented Actions

Access to the external world consists primarily of composing messages to be spoken into the local environment or expressed in text from locally or to another CR or CN using KQML, RKRL, OWL, RXML, or some other appropriate knowledge interchange standard.

Internally Oriented Actions

Actions on internal states include controlling machine-controllable resources such as radio channels. The CR can also affect the contents of existing internal models, e.g. by adding a model of stimulus-experience-response (serModel) to an existing internal model structure [13]. The new concept itself may assert related concepts into the scene. Multiple independent sources of the same concept in a scene reinforce that concept for that scene. These models may be asserted by the <Self/> to encapsulate experience. The experience may be reactively integrated into RXML knowledge structures as well, provided the reactive response encodes them properly.

3.3.7 Learning

Learning is a function of perception, observations, decisions and actions. Initial learning is mediated by the Observe-phase perception hierarchy in which all sensory perceptions are continuously matched against all prior stimuli to continually count occurrences and to remember time since last occurrence of the stimuli from primitives to aggregates.

Learning also occurs through the introduction of new internal models in response to existing models and CBR bindings. In general, there are many opportunities to integrate ML into iCR. Each of the phases of the cognition cycle offers multiple opportunities for discovery processes like <Histogram> above, as well as many other ML approaches to be developed below. Since the architecture includes internal reinforcement via counting occurrences and via serModels, ML with uncertainty is also supported in the architecture.

Finally, there is a learning mechanism that occurs when a new type of serModel is created in response to an Action to instantiate an internally generated serModel. For example, prior and current internal states may be compared with expectations to learn about the effectiveness of a communications mode, instantiating a new mode-specific serModel.

3.3.8 Self Monitoring Timing

Each of the prior phases must consist of computational structures for which the execution time may be computed in advance. In addition, each phase must restrict its computations to consume not more resources (time $\times$ allocated processing capacity) than the pre-computed upper bound. Therefore, the architecture has some prohibitions and some data set requirements needed to obtain an acceptable degree of stability of behavior for CR as self-referential self-modifying systems.

Since First Order Predicate Calculus (FOPC) used in some reasoning systems is not decidable, one cannot in general compute in advance how much time an FOPC expression will take to run to completion. There may be loops that will preclude this, and even with loop detection, the time to resolve an expression may be only loosely approximated as an exponential function of some parameters (such as the number of statements in the FOPC data base of assertions and rules). Therefore unrestricted FOPC is not allowed.

Similarly, unrestricted For, Until and While loops are prohibited. In place of such loops are bounded iterations in which the time required for the loop to execute is computed or supplied independent of the computations that determine the iteration control of the loop. This seemingly unnatural act can be facilitated by next-generation compilers and CASE tools. Since self-referential self-modifying code is prohibited by structured design and programming practices, there are no such tools on the market today. But since CR is inherently self-referential and self-modifying, such tools most likely will emerge, perhaps assisted by the needs of CR and the architecture framework of the cognition cycle.

Finally, the cognition cycle itself can't contain internal loops. Each iteration of the cycle must take a defined amount of time, just as each frame of a 3G air interface takes 10 milliseconds. As CR computational platforms continue to progress, the amount of computational work done within the cycle will increase, but under no conditions should explicit or implicit loops be introduced into the cognition cycle that would extend it beyond a given cycle time.

3.3.9 Retrospection

Since the assimilation of knowledge by machine learning can be computationally intensive, cognitive radio has "sleep" and "prayer" epochs that support machine learning. A sleep epoch is a relatively long period of time (e.g. minutes to hours) during which the radio will not be in use, but has sufficient electrical power for processing. During the sleep epoch, the radio can run machine learning algorithms without detracting from its ability to support its user's needs. Machine learning algorithms may integrate experience by aggregating statistical parameters. The sleep epoch may re-run stimulus–response sequences with new learning parameters in the way that people dream. The sleep cycle could be less anthropomorphic, employing a genetic algorithm to explore a rugged fitness landscape, potentially improving the decision parameters from recent experience.

3.3.10 Reaching Out

Learning opportunities not resolved in the sleep epoch can be brought to the attention of the user, the host network, or a designer during a prayer epoch. The sleep and prayer epochs are possibilities.

3.4 CRA III: The Inference Hierarchy

The phases of inference from observation to action show the flow of inference, a top-down view of how cognition is implemented algorithmically. The inference hierarchy is the part of the algorithm architecture that organizes the data structures. Inference hierarchies have been in use since Hearsay II in the 1970s, but the CR hierarchy is unique in its method of integrating machine learning with real-time performance during the Wake Epochs. An illustrative inference hierarchy includes layers from atomic stimuli at the bottom to information clusters that define action contexts as in Figure 3.5.

The pattern of accumulating elements into sequences begins at the bottom of the hierarchy. Atomic stimuli originate in the external environment including RF, acoustic, image, and location domains among others. The atomic symbols extracted from them are the most primitive symbolic units in the domain. In speech, the most primitive elements are the phonemes. In the exchange of textual data (e.g. in email), the symbols are the typed characters. In images, the atomic symbols may be the individual picture elements (pixels) or they may be small groups of pixels with similar hue, intensity, texture, etc.

Sequence	***Level of Abstraction***
Context Cluster	***Scenes*** in a play, Session
Sequence Clusters	***Dialogs***, Paragraphs, Protocol
Basic Sequences	***Phrases***, video clip, message
Primitive Sequences	***Words***, token, image
Atomic Symbols	***Raw Data,*** Phoneme, pixel
Atomic Stimuli	External Phenomena

Figure © Dr. Joseph Mitola III, used with permission

Fig. 3.5. Standard Inference Hierarchy.

A related set of atomic symbols forms a primitive sequence. Words in text, tokens from a speech tokenizer, and objects in images (or individual image regions in a video flow) are the primitive sequences. Primitive sequences have spatial and/or temporal coincidence, standing out against the background (or noise), but there may be no particular meaning in that pattern of coincidence. Basic sequences, on the other hand, are space–time–spectrum sequences that entail the communication of discrete messages.

These discrete messages (e.g. phrases) are typically defined with respect to an ontology of the primitive sequences (e.g. definitions of words). Sequences cluster together because of shared properties. For examples, phrases that include words like "hit," "pitch," "ball," and "out" may be associated with a discussion of a baseball game. Knowledge Discovery and Data Mining (KDD) and the Semantic Web offer approaches for defining, or inferring the presence of such clusters from primitive and basic sequences.

A scene is a context cluster, a multi-dimensional space–time–frequency association, such as a discussion of a baseball game in the living room on a Sunday afternoon. Such clusters may be inferred from unsupervised machine learning, e.g. using statistical methods or nonlinear approaches such as Support Vector Machines (SVM).

Although presented above in a bottom-up fashion, there is no reason to limit multi-dimensional inference to the top layers of the inference hierarchy. The lower levels of the inference hierarchy may include correlated multi-sensor data. For example, a word may be characterized as a primitive acoustic sequence coupled to a primitive sequence of images of a person speaking that word. In fact, since infants seem to thrive on multi-sensory stimulation, the key to reliable machine learning may be the use of multiple sensors with multi-sensor correlation at the lowest levels of abstraction.

Each of these levels of the inference hierarchy is now discussed further.

3.4.1 Atomic Stimuli

Atomic stimuli originate in the external environment and are sensed and preprocessed by the sensory subsystems which include sensors of the RF environment (e.g. radio receiver and related data and information processing) and of the local physical environment including acoustic, video, and location sensors. Atomic symbols are the elementary stimuli extracted from the atomic stimuli. Atomic symbols may result from a simple noise-riding threshold algorithm, such as the squelch circuit in RF that differentiates signal from noise. Acoustic signals may be differentiated from simple background noise this way, but generally the result is the detection of a relatively large speech epoch which contains various kinds of speech energy. Thus, further signal processing is typically required in a preprocessing subsystem to isolate atomic symbols.

The transformation from atomic stimuli to atomic symbols is the job of the sensory preprocessing system. Thus, for example, acoustic signals may be transformed into phoneme hypotheses by an acoustic signal pre-processor. However, some speech-to-text software tools may not enable this level of interface via an API. To develop industrial strength CR, contemporary speech-to-text and video processing software tools are needed. Speech to text tools yield an errorful transcript in response to a set of atomic stimuli. Thus, the speech to text tool is an example of a mapping from atomic stimuli to basic sequences. One of the important contributions of architecture is to identify such maps and to define the role of the level mapping tools.

Image processing software available for the Wintel-Java development environment JBuilder has the ability to extract objects from images and video clips. In addition, research such as that of Goodman et al. defines algorithms for what the AAAI calls cognitive vision [32].

But there is nothing about the inference hierarchy that forces data from a pre-processing system to be entered at the lowest level. In order for the more primitive symbolic abstractions such as atomic symbols to be related to more aggregate abstractions, one may either build up the aggregates from the primitive abstractions or derive the primitive abstractions from the aggregates. Since people are exposed to "the whole thing" by immersion in the full experience of life – touch, sight, sound, taste, and balance – all at once, it seems possible – even likely – that the more primitive abstractions are somehow derived through the analysis of aggregates, perhaps by cross-correlation. This can be accomplished in a CRA sleep cycle. The idea is that the wake cycle is optimized for immediate reaction to stimuli, such as our ancestors needed to avoid predation, while the sleep cycle is optimized for introspection, for analyzing the day's stimuli to derive those objects that should be recognized and acted upon in the next cycle.

Stimuli are each counted. When an iCR that conforms to this architecture encounters a stimulus, it both counts how many such stimuli have been encountered and resets a timer to zero that keeps track of the time since the last occurrence of the stimulus.

3.4.2 Primitive Sequences: Words and Dead Time

The accumulation of sequences of atomic symbols forms primitive sequences. The key question at this level of the data structure hierarchy is the sequence boundary. The simplest situation is one in which a distinguished atomic symbol separates primitive sequences, which is exactly the case with white space between words in typed text. In general, one would like a machine-learning system to determine on its own that the white space (and a few special symbols, etc.) separates the keyboard input stream into primitive sequences.

3.4.3 Basic Sequences

The pattern of aggregation is repeated vertically at the levels corresponding to words, phrases, dialogs, and scenes. The data structures generated by PDA Nodes create the concept hierarchy of Figure 3.5. These are the reinforced hierarchical sequences. They are reinforced by the inherent counting of the number of times each atomic or aggregated stimulus occurs. The phrase level typically contains or implies a verb (the verb to-be is implied if no other verb is implicit).

Unless digested (e.g. by a sleep process), the observation phase hierarchy accumulates all the sensor data, parsed and distributed among PDA Nodes for fast parallel retrieval. Since the hierarchy saves everything and compares new data to memories, it is a kind of memory-base learning technique. This is a memory-intensive approach, taking a lot of space. When the stimuli retained are limited to atomic symbols and their aggregates, the total amount of data that needs to be stored is relatively modest. In addition, recent research shows the negative effects of discarding cases in word pronunciation. In word pronunciation, no example can be discarded even if "disruptive" to a well developed model. Each exception has to be followed. Thus in CR1, when multiple memories match partially, the most nearly exact match informs the orientation, planning, and action.

Basic sequences are each counted. When an iCR that conforms to this architecture encounters a basic sequence, it both counts how many such sequences have been encountered and resets a timer to zero that keeps track of the time since the last occurrence.

3.4.4 Natural Language in the CRA Inference Hierarchy

In speech, words spoken in a phrase may be co-articulated with no distinct boundary between the primitive sequences in a basic sequence. Therefore, speech detection algorithms may reliably extract a basic sequence while the parsing of that sequence into its constituent primitive sequences may be much less reliable. Typically, the correct parse is within the top ten candidates for contemporary speech-to-text software tools. But the flow of speech signal processing may be something like:

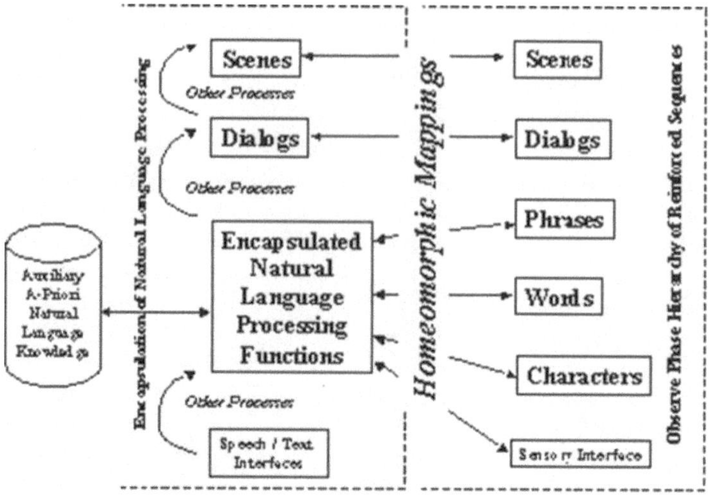

Figure © Dr. Joseph Mitola III, used with permission

Fig. 3.6. Natural language encapsulation in the observation hierarchy.

1. Isolate a basic sequence (phrase) from background noise using an acoustic squelch algorithm.
2. Analyze the basic sequence to identify candidate primitive sequence boundaries (words).
3. Analyze the primitive sequences for atomic symbols.
4. Evaluate primitive and basic sequence hypotheses based on a statistical model of language to rank-order alternative interpretations of the basic sequence.

So a practical speech processing algorithm may yield alternative strings of phonemes and candidate parses "all at once." NLP tool sets may be embedded into the CRA inference hierarchy as illustrated in Figure 3.6. Speech and/or text channels may be processed via natural language facilities with substantial a priori models of language and discourse. The use of those models should entail the use of mappings among the word, phrase, dialog, and scene levels of the observation phase hierarchy and the encapsulated component(s).

It is tempting to expect cognitive radio to integrate a commercial natural language processing system such as IBM's ViaVoice or a derivative of an NLP research system such as SNePS [33], AGFL [34], or XTAG [35] perhaps using a morphological analyzer like PCKimmo [36]. These tools both go too far and not far enough in the direction needed for CRA. One might like to employ existing tools using a workable interface between the domain of radio engineering and some of the above natural language tool sets. The definition of such cross-discipline interfaces is in its infancy. A present, one cannot just express a radio ontology in Interlingua and plug it neatly into XTAG to get a working cognitive radio. The internal data structures that are used in radio mediate the performance of radio tasks (e.g. "transmit a waveform"). The data structures of XTAG, AGFL, etc. mediate the conversion of language from one form to another. Thus, XTAG wants to know that "transmit" is a verb and "waveform" is a noun. The CR needs to know that if the user says "transmit" and a message has been defined, then the CR should call the SDR function *transmit()*. NLP systems also need scoping rules for transformations on the linguistic data structures. The way in which domain knowledge is integrated in linguistic structures of these tools tends to obscure the radio engineering aspects.

Natural language processing systems work well on well-structured speech and text, such as the prepared text of a news anchor. But they do not work well yet on noisy, non-grammatical data structures encountered when a user is trying to order a cab in a crowded bar. Thus, less-linguistic or meta-linguistic data structures may be needed to integrate core cognitive radio reasoning with speech and/or text-processing front ends. The CRA has the flexibility illustrated in the figure above for the subsequent integration of evolved NLP tools. The emphasis of this version of the CRA is a structure of sets and maps required to create a viable cognitive radio architecture. Although introducing the issues required to integrate existing natural language processing tools, the text does not pretend to present a complete solution to this problem.

3.4.5 Observe-Orient Links for Scene Interpretation

CR may use an algorithm-generating language with which one may define self-similar inference processes. In one example, the first process (Proc1) partitions characters into words, detecting novel characters and phrase boundaries as well. Proc2 detects novel words and aggregates known words into phrases. Proc3 detects novel phrases, aggregating known phrases into dialogs. Proc4 aggregates dialogs into scenes, and Proc5 detects known scenes. In each case, a novel entity at level N will be bound in the context of the surrounding known entities at that level to the closest match at the next highest level, $N + 1$. For example at the word-phrase intersection of Proc2, would map the following phrases:

Equation 4: "Let me introduce Joe"
Equation 5: "Let me introduce Chip"

Since "Chip" is unknown while "Joe" is known from a prior dialog, integrated CBR matches the phrases, binding <Chip>=<Joe>. In other words, it will try to act with respect to Chip in the way it was previously trained (at the dialog level) to interact with Joe. In response to the introduction, the system may say "Hello, Chip, How are you?" mimicking the behavior it had been trained with respect to Joe previously. Not too bright, but not all that bad either for a relatively simple machine learning algorithm.

There is a particular kind of dialog that is characterized by reactive world knowledge in which there is some standard way of reacting to given speech-act

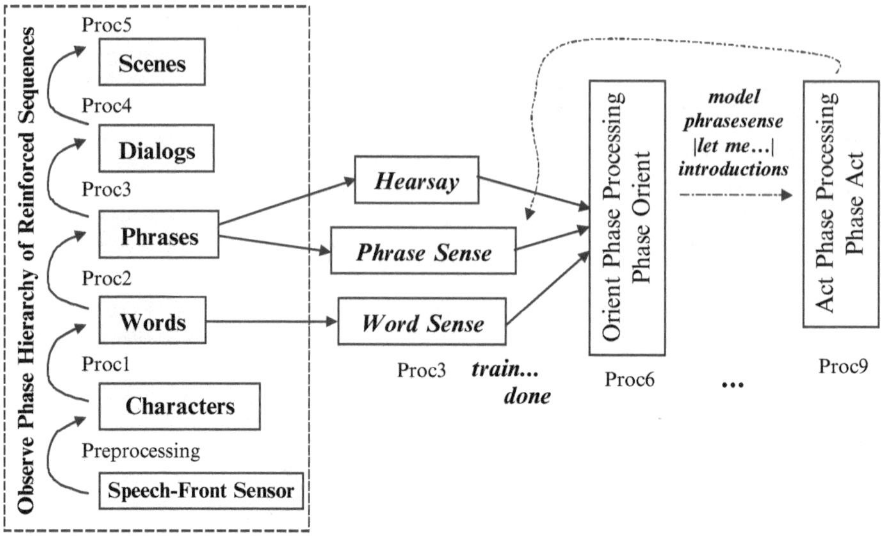

Figure © Dr. Joseph Mitola III, used with permission

Fig. 3.7. The inference hierarchy supports lateral knowledge sources.

inputs. For example, when someone says "Hello," you may typically reply with "Hello" or some other greeting. The capability to generate such rote responses is pre-programmed into a lateral component of the Hearsay knowledge source (KS). The responses are not pre-programmed, but the general tendency to imitate phrase level dialogs is a pre-programmed tendency that can be overruled by plan generation, but that is present in the orient-phase, which is Proc6.

Words may evoke a similar tendency towards immediate action. What do you do when you hear the words "Help!!" or "Fire, fire, get out, get out!!" You, the CR programmer, can capture reactive tendencies in your CR by pre-programming an ability to detect these kinds of situations in the Word-sense knowledge source, as implied by Figure 3.7). When confronted with them (which is preferred), CR should react appropriately if properly trained, which is one of the key aspects of this text. To cheat, you can pre-program a wider array of stimulus-response pairs so that your CR has more a priori knowledge, but some of it may not be appropriate. Some responses are culturally conditioned. Will your CR be too rigid? If it has too much a priori knowledge, it will be perceived by its users as too rigid. If it doesn't have enough, it will be perceived as too stupid.

3.4.6 Observe-Orient Links for Radio Skill Sets

Radio knowledge may be embodied in components called radio skills. Radio knowledge is static, requiring interpretation by an algorithm such as an

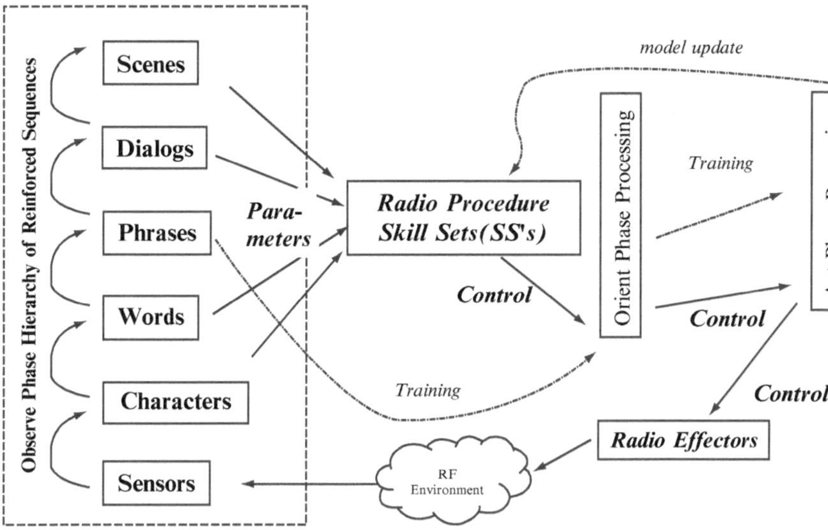

Figure ©1999 Dr. Joseph Mitola III, used with permission

Fig. 3.8. Radio skills respond to observations.

inference engine in order to accomplish anything. Radio skills, on the other hand, are knowledge embedded in serModels through the process of training or sleeping/dreaming. This knowledge is continually pattern-matched against all stimuli in parallel. That is, there are no logical dependencies among knowledge components that mediate the application of the knowledge. With FOPC, the theorem-prover must reach a defined state in the resolution of multiple axioms in order to initiate action. In contrast, serModels are continually compared to the level of the hierarchy to which they are attached, so their immediate responses are always cascading towards action. Organized as maps primarily among the wake-cycle phases "observe" and "orient," the radio procedure skill sets (SS's) control radio personalities as illustrated in Figure 3.8.

These skill sets may either be reformatted into serModels directly from the a priori knowledge of an RKRL frame, or they may be acquired from training or sleep/dreaming. Each skill set may also save the knowledge it learns into an RKRL frame.

3.4.7 General World Knowledge

An iCR needs substantial knowledge embedded in the inference hierarchies. It needs both external RF knowledge and internal radio knowledge. Internal knowledge enables it to reason about itself as a radio. External radio knowledge enables it to reason about the role of the <Self/> in the world, such as respecting rights of other cognitive and not-so-cognitive radios.

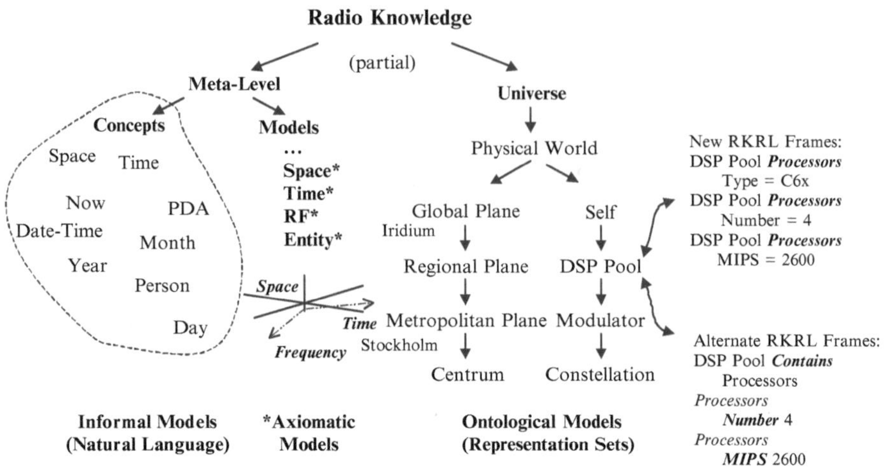

Figure ©1999, 2000 Dr. Joseph Mitola III, used with permission

Fig. 3.9. External radio knowledge includes concrete and abstract knowledge.

Figure 3.9 illustrates the classes of knowledge an iCR needs to employ in the inference hierarchies and cognition cycle. It is one thing to write down that the Universe includes a Physical World (there could also be a spiritual world, and that might be very important in some cultures). It is quite another thing to express that knowledge in a way that the iCR will be able to use that knowledge effectively. Symbols like "Universe" take on meaning by their relationships to other symbols and to external stimuli. In this ontology, meta-level knowledge consists of ***abstractions***, distinct from existential knowledge of the physical Universe. In RXML, this ontological perspective includes all in a universe of discourse, <Universe> expressed as follows:

<Universe>
<Abstractions><Time><Now/></Time><Space><Here/> </Space> ...<RF/>...
<Intelligent-Entities/> ... </Abstractions>
<Physical-universe>... <Instances/> of Abstractions ... </Physical-universe>
< /Universe>

Equation 6 the universe of discourse of iCR consists of abstractions plus the physical universe

Abstractions include informal and formal meta-level knowledge from unstructured knowledge of concepts to the more mathematically structured models of space, time, RF, and entities that exist in space–time. To differentiate "now" as a temporal concept from "Now" as the Chinese name of a plant, the CRA includes both the a priori knowledge of "now" as a space–time locus, <Now/> as well as functions that access and manipulate instances of the concept <Now/>. <Now/> is axiomatic in the CRA, so code refers to "now" (as n.o.w) in planning actions. The architecture allows an algorithm to return the date-time code from Windows to define instances of <Now/>. Definition-by-algorithm permits an inference system like the cognition subsystem to reason about whether a given event is in the past, present, or future. What is the present? The present is some region of time between "now" and the immediate past and future. If you are a paleontologist, "now" may consist of the million year epoch in which we all are thought to have evolved from apes. If you are a rock star, "now" is probably a lot shorter than that to you. How will your CR learn the user's concept of now? The CRA design offers an axiomatic treatment of time, but the axioms were not programmed into the Java explicitly. THE CRA aggregates knowledge of time by a temporal CBR that illustrates the key principles. The CRA does not fix the definition of <Now/> but enables the <Self/> to define the details in an <Instance> in the physical world about which it can learn from the user, paleontologist or rock star.

Given the complexity of a system that includes both a multi-tiered inference hierarchy and the cognition cycle's observe–orient–plan–decide–act sequence with AML throughout, it is helpful to consider the mathematical structure of these information elements, processes, and flows.

The mathematical treatment, CRA IV, is provided elsewhere [5].

3.5 CRA V: Building the CRA on SDR Architectures

A Cognitive Radio is a Software Radio (SWR) or Software-Defined Radio (SDR) with flexible formal semantics based entity to entity formal messaging via RXML and integrated machine learning of the self, the user, the environment, and the "situation." This section reviews SWR, SDR, and the Software Communications Architecture (SCA) or Software Radio Architecture (SRA) for those who may be unfamiliar with these concepts. While it is not necessary for an iCR to use the SCA/SRA as its internal model of itself, it certainly must have some model, or it will be incapable of reasoning about its own internal structure and adapting or modifying its radio functionality.

3.5.1 SWR and SDR Architecture Principles

Hardware-defined radios such as the typical AM/FM broadcast receiver convert radio to audio using radio hardware, such as antennas, filters, analog de-modulators, and the like. SWR is the ideal radio in which the Analog to Digital Converter (ADC) and Digital to Analog Converter (DAC) convert digital signals to and from radio frequencies (RF) directly, and all RF

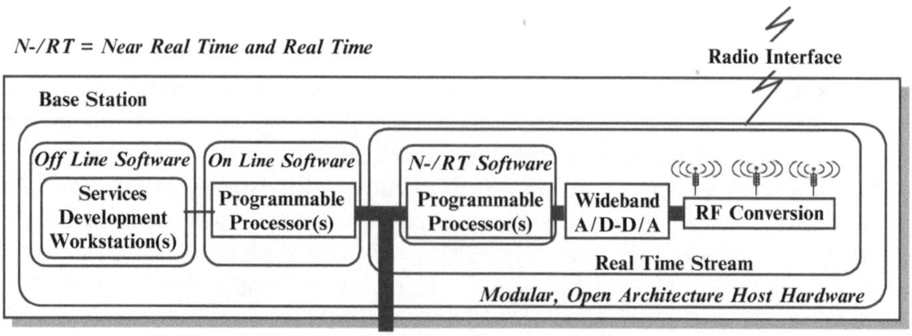

Figure © 1992 Dr. Joseph Mitola III, used with permission

Fig. 3.10. SWR principle applied to cellular base station.

channel modulation, demodulation, frequency translation and filtering are accomplished digitally. For example, modulation may be accomplished digitally by multiplying sine and cosine components of a digitally sampled audio signal (called the "baseband" signal, e.g. to be transmitted) by the sampled digital values of a higher frequency sine wave to up-convert it, ultimately to RF.

Figure 3.10 shows how SDR principles apply to a cellular radio base station. In the ideal SWR, there would be essentially no RF conversion, just ADC/DAC blocks accessing the full RF spectrum available to the (wideband) antenna elements. Today's SDR base-stations approach this ideal by digital access (DAC and ADC) to a band of spectrum allocations, such as 75 MHz allocated to uplink and downlink frequencies for third-generation services. In this architecture, RF conversion can be a substantial system component, sometimes 60% of the cost of the hardware, and not amenable to cost improvements through Moore's Law. The ideal SDR would access more like 2.5 GHz from, say 30 MHz to around 2.5 GHz, supporting all kinds of services in television (TV) bands, police bands, air traffic control bands – you name it. Although considered radical when introduced in 1991 [37] and popularized in 1995 [38], recent regulatory rulings are encouraging the deployment of such "flexible spectrum" use architectures.

This ideal SWR may not be practical or affordable, so it is important for the radio engineer to understand the tradeoffs [again, see [39] for SDR architecture tradeoffs]. In particular, the physics of RF devices (e.g. antennas, inductors, filters) makes it easier to synthesize narrowband RF and intervening analog RF conversion and Intermediate Frequency (IF) conversion. Given narrowband RF, the hardware-defined radio might employ baseband (e.g. voice frequency) ADC, DAC, and digital signal processing. The Programmable Digital Radios (PDR) of the 1980s and 90s used this approach. Historically, this approach has not been as expensive as wideband RF (antennas, conversion), ADCs and DACs. Handsets are less amenable to SWR principles than the base station (Figure 3.11). Base stations access the power grid. Thus, the fact

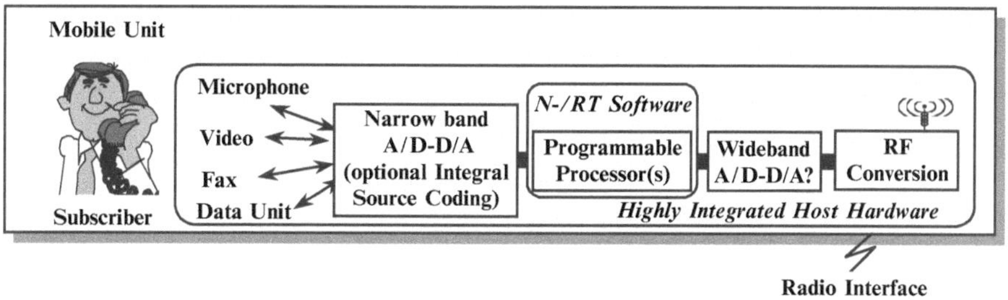

Figure © 1992 Dr. Joseph Mitola III, used with permission

Fig. 3.11. Software radio principle – "ADC and DAC at the antenna" may not apply.

that wideband ADCs, DACs, and DSP consume many watts of power is not a major design driver. Conservation of battery life, however, is a major design driver in the handset.

Thus, insertion of SWR technology into handsets has been relatively slow. Instead, the major handset manufacturers include multiple single-band RF chip sets into a given handset. This has been called the Velcro radio or slice radio.

Since the ideal SWR is not readily approached in many cases, the SDR has comprised a sequence of practical steps from the baseband DSP of the 1990s towards the ideal SWR. As the economics of Moore's Law and of increasingly wideband RF and IF devices allow, implementations move upward and to the right in the SDR design space (Figure 3.12). This space consists of the combination of digital access bandwidth and programmability. Access bandwidth consists of ADC/DAC sampling rates converted by the Nyquist criterion or practice into effective bandwidth. Programmability of the digital subsystems is defined by the ease with which logic and interconnect may be changed after deployment. Application-Specific Integrated Circuits (ASICs) cannot be changed at all, so the functions are "dedicated" in silicon. Field Programmable Gate Arrays (FPGAs) can be changed in the field, but if the new function exceeds some parameter of the chip, which is not uncommon, then one must upgrade the hardware to change the function, just like ASICs. Digital Signal Processors (DSPs) are typically easier or less expensive to program and are more efficient in power use than FPGAs. Memory limits and instruction set architecture (ISA) complexity can drive up costs of reprogramming the DSP. Finally, general purpose processors, particularly with Reduced Instruction Set Architectures (RISC) are most cost-effective to change in the field. To assess a multi-processor, such as a cell phone with a CDMA-ASIC, DSP speech codec, and RISC miCRo-controller, weight the point by equivalent processing capacity.

Where should one place an SDR design within this space? The quick answer is so that you can understand the migration path of radio technology

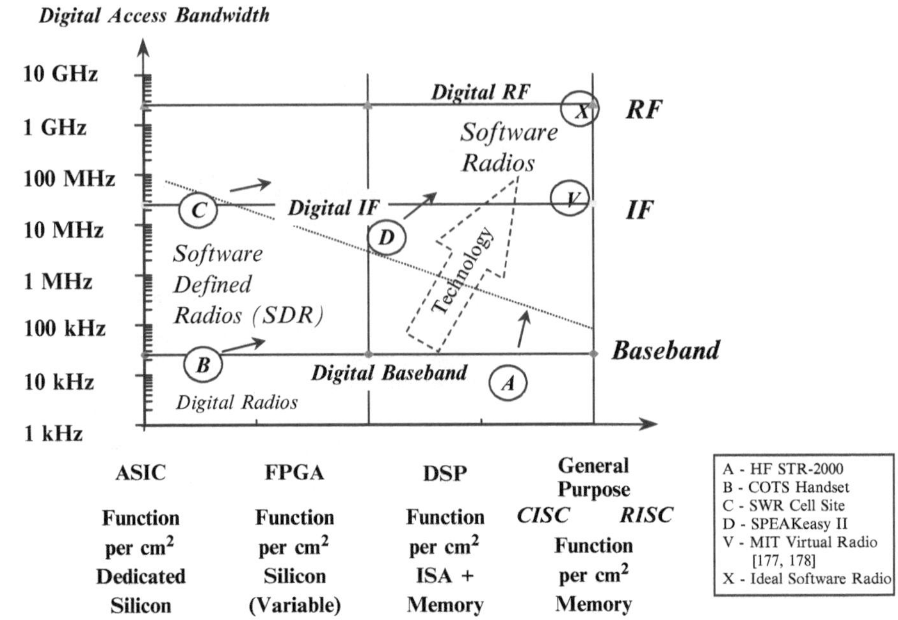

Figure © 1996–2003 Dr. Joseph Mitola III, used with permission

Fig. 3.12. SDR design space shows how designs approach the ideal SWR.

from the lower left towards the upper right, benefiting from lessons learned in the early migration projects captured in *Software Radio Architecture* [39].

This section contains a very brief synopsis of the key SDR knowledge you will need in order to follow the iCR examples of this text.

3.5.2 Radio Architecture

The discussion of the software radio design space contains the first elements of radio architecture. It tells you what mix of critical components are present in the radio. For SDR, the critical hardware components are the ADC, DAC, and processor suite. The critical software components are the user interface, the networking software, the information security (INFOSEC) capability (hardware and/or software), the RF media access software, including the physical layer modulator and demodulator (modem) and Media Access Control (MAC), and any antenna-related software such as antenna selection, beamforming, pointing and the like. INFOSEC consists of Transmission Security, such as the frequency hopping spreading code selection, plus Communications Security encryption.

The SDR Forum defined a very simple, helpful model of radio in 1997, shown in Figure 3.13. This model highlights the relationships among radio functions at a tutorial level. The CR has to "know" about these functions, so

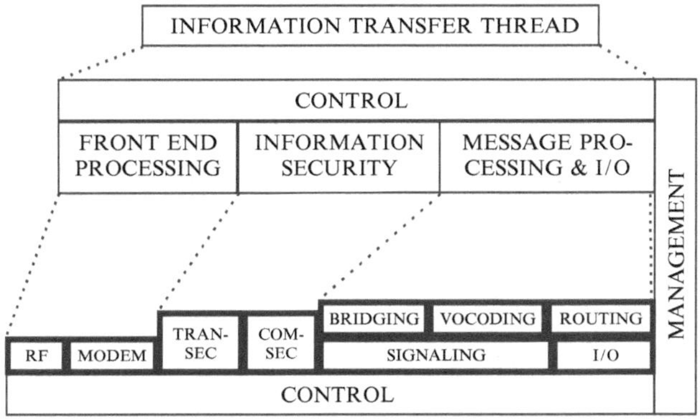

Figure ©1997 SDR Forum, used with permission

Fig. 3.13. SDR Forum (MMITS) information transfer thread architecture.

every CR must have an internal model of a radio of some type. This one is a good start because it shows both the relationships among the functions and the typical flow of signal transformations from analog RF to analog or with SDR, digital modems, and on to other digital processing including system control of which the user interface is a part.

This model and the techniques for implementing a SWR and the various degrees of SDR capability are addressed in depth in the various texts on SDR [40–43].

3.5.3 The SCA

The US DoD developed the Software Communications Architecture (SCA) for its Joint Tactical Radio System (JTRS) family of radios.

The architecture identifies the components and interfaces shown in Figure 3.14. The API's define access to the physical layer, to the Media Access Control (MAC) layer, to the logical link layer (LLC), to security features, and to the input/output of the physical radio device. The physical components consist of antennas and RF conversion hardware that are mostly analog and that therefore typically lack the ability to declare or describe themselves to the system. Most other SCA-compliant components are capable of describing themselves to the system to enable and facilitate plug and play among hardware and software components. In addition, the SCA embraces POSIX and CORBA.

The model evolved through several stages of work in the SDR Forum and Object Management Group (OMG) into a UML-based object-oriented model of SDR (Figure 3.15). Waveforms are collections of load modules that provide wireless services, so from a radio designer's perspective, the waveform is the key application in a radio. From a user's perspective of a wireless PDA, the radio waveform is just a means to an end, and the user doesn't want to

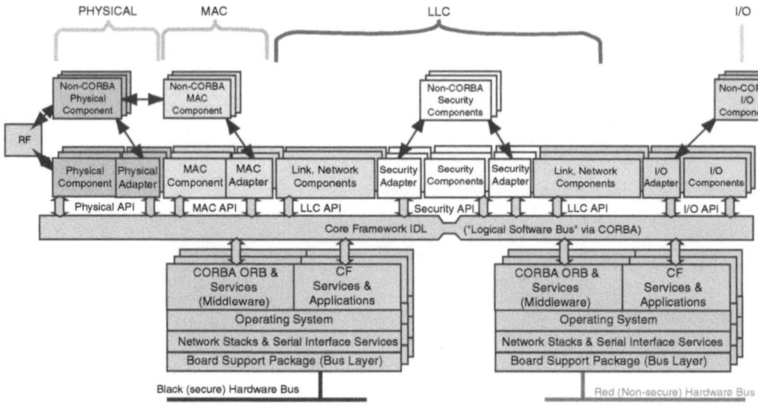

Figure ©2004 SDR Forum, used with permission

Fig. 3.14. JTRS SCA Version 1.0 [© SDR Forum, reprinted with permission].

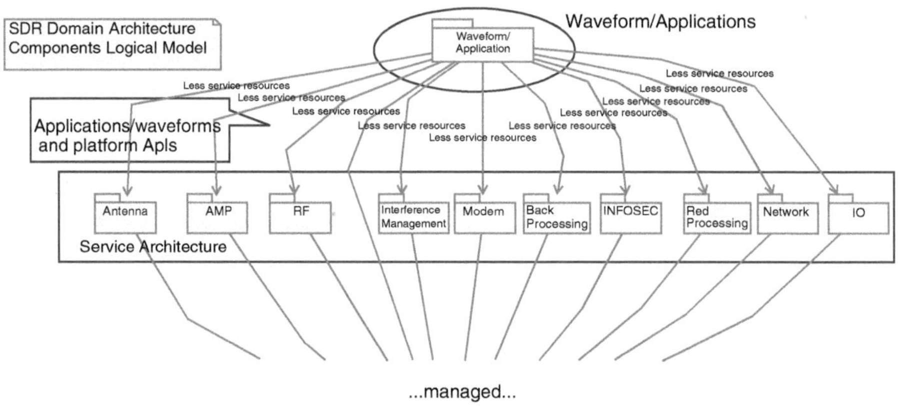

Figure ©2004 SDR Forum, used with permission

Fig. 3.15. SDR Forum UML model of radio services © SDR Forum, used with permission.

know or have to care about waveforms. Today, the cellular service providers hide this detail to some degree, but consumers sometimes know the difference between CDMA and GSM, for example, because CDMA works in the US, but not in Europe. With the deployment of the third generation of cellular technology (3G), the amount of technical jargon consumers will need to know is increasing. So the CR designer is going to write code (Java code in this book) that insulates the user from those details, unless the user really wants to know.

In the UML model, Amp refers to amplification services, RF refers to RF conversion, interference-management refers to both avoiding interference and filtering it out of one's band of operation. In addition, the jargon for US military radios is that the "red" side contains the user's secret information, but

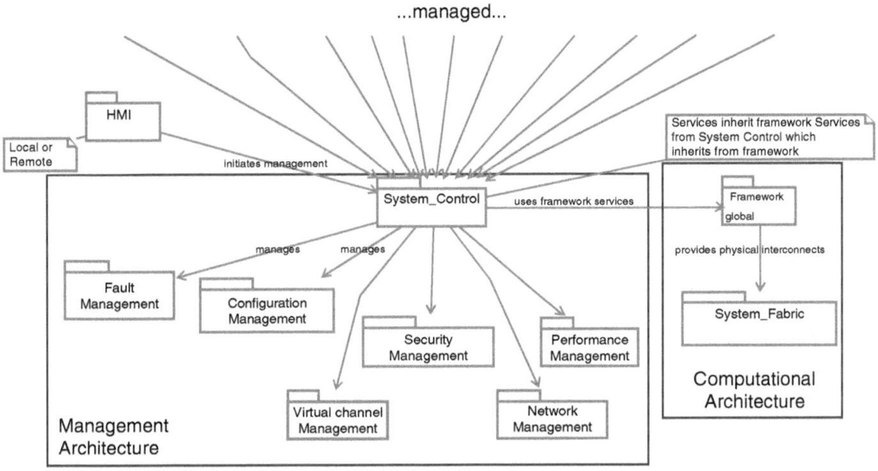

Fig. 3.16. SDR Forum UML management and computational architectures © 2004 SDR Forum, used with permission.

when it is encrypted it becomes "black" or protected, so it can be transmitted. Black processing occurs between the antenna and the decryption process. Notice also in the figure that there is no user interface. The UML model contains a sophisticated set of management facilities, illustrated further in Figure 3.16, to which Human Machine Interface (HMI) or user interface is closely related.

Systems control is based on a framework that includes very generic functions like event logging, organized into a computational architecture, heavily influenced by CORBA. The management features are needed to control radios of the complexity of 3G and of the current generation of military radios. Although civil sector radios for police, fire, and aircraft lag these two sectors in complexity and are more cost-sensitive, baseband SDRs are beginning to insert themselves even into these historically less technology-driven markets.

Fault management features are needed to deal with loss of a radios processors, memory, or antenna channels. CR therefore interacts with fault management to determine what facilities may be available to the radio given recovery from hardware and/or software faults (e.g. error in a download). Security management is increasingly important in the protection of the user's data by the CR, balancing convenience and security which can be very tedious and time consuming. The CR will direct Virtual Channel Management and (VCM) will learn from the VCM function what radio resources are available, such as what bands the radio can listen to and transmit on and how many it can do at once. Network management does for the digital paths what VCM does for the radio paths. Finally, SDR performance depends on the availability of analog and digital resources, such as linearity in the antenna, Millions of Instructions Per Second (MIPS) in a processor, and the like.

3.5.4 Functions–Transforms Model of Radio

The self-referential model of a wireless device used by the CRA and used to define the RKRL and to train the CRA is the functions–transforms model illustrated in Figure 3.17. In this model, the radio knows about sources, source coding, networks, INFOSEC, and the collection of front-end services needed to access RF channels. Its knowledge also extends to the idea of multiple channels and their characteristics (the channel set), and that the radio part may have both many alternative personalities at a given point in time, and that through evolution support, those alternatives may change over time.

Since CR reasons about all of its internal resources, it also must have some kind of computational model of analog and digital performance parameters and how they are related to features it can measure or control. MIPS, for example, may be controlled by setting the clock speed. A high clock speed generally uses more total power than a lower clock speed, and this tends to reduce battery life. Same is true for the brightness of a display. The CR only "knows" this to the degree that it has a data structure that captures this information and some kind of algorithms, pre-programmed and/or learned, that deal with these relationships to the benefit of the user. Constraint languages may be used to express interdependencies, such as how many channels of a given personality are supported by a given hardware suite, particularly in failure modes. CR algorithms may employ this kind of structured reasoning as a specialized Knowledge Source (KS) when using case-based learning to extend its ability to cope with internal changes.

The ontological structure of the above may be formalized as follows:

<SDR>
<Sources/><Channels/><Personality>
<Source-Coding-Decoding/><Networking/><INFOSEC/>

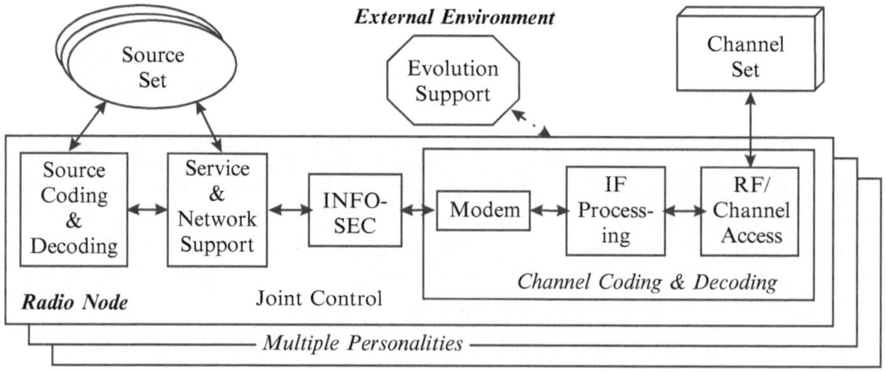

Figure © 1996 Dr. Joseph Mitola III, used with permission

Fig. 3.17. Function–transforms model of a wireless node.

```
    <Channel-Codec><Modem/><IF-Processing/><RF-Access/>
    </Channel-Codec>
    < /Personality>
  <SDR-Platform/><Evolution-Support/>
< /SDR>
```

Equation 7 Defines SDR subsystem components

While this text does not spend a lot of time on the computational ontology of SDR, semantically based dialogs among iCRs about internal issues like downloads may be mediated by developing the RXML above to more fully develop the necessary ontological structures.

3.5.5 Architecture Migration: From SDR to iCR

Given the CRA and contemporary SDR architecture, one must address the transition of SDR, possibly through a phase of iCRs toward the ideal CR. As the complexity of hand-held, wearable, and vehicular wireless systems increase, the likelihood that the user will have the skill necessary to do the optimal thing in any given circumstance goes down. Today's cellular networks manage the complexity of individual wireless protocols for the user, but the emergence of multi-band multi-mode iCR moves the burden for complexity management towards the PDA. The optimization of the choice of wireless service between the "free" home WLAN and the for-sale cellular equivalent moves the burden of radio resource management from the network to the WPDA.

3.5.6 Cognitive Electronics

The increasing complexity of the PDA-user interface also accelerates the trend towards increasing the computational intelligence of personal electronics. iCR is in some sense just an example of a computationally intelligent personal electronics system. For example, using a laptop computer in the bright display mode uses up the battery power faster than when the display is set to minimum brightness. A cognitive laptop could offer to set the brightness to low level when it was turned on in battery powered mode. It would be even nicer if it would recognize operation aboard a commercial aircraft and therefore automatically turn the brightness down. It should learn that my preference is to set the brightness low on an aircraft to conserve the battery. A cognitive laptop shouldn't make a big deal over that, and it should let me turn up the brightness without complaining. If it had an ambient light sensor or ambient light algorithm for an embedded camera, it could tell that a window shade is open, so I have to deal with the brightness. By sensing the brightness of the *on-board aircraft* scene and associating my control of the brightness of my display with the brightness of the environment, a hypothetical cognitive laptop could learn do the right thing in the right situation.

How does this relate to the CRA? For one thing, the CRA could be used as-is to increase the computational intelligence of the laptop. In this case, the self is the laptop and the PDA knows about itself as a laptop, not as a WPDA. It knows about its sensors suite, which includes at least a light level sensor if not a camera through the data structures that define the <Self/>. It knows about the user by observing keystrokes and mouse action as well as by interpreting the images on the camera, e.g. to verify that the Owner is still the user since that is important to building user-specific models. It might build a space–time behavior model of any user or it might be a one-user laptop. Its actions then must include the setting of the display intensity level. In short, the CRA accommodates the cognitive laptop with suitable knowledge in the knowledge structures and functions implemented in the map sets.

3.5.7 When Should a Radio Transition towards Cognition?

If a wireless device accesses only a single RF band and mode, then it is not a very good starting point for cognitive radio. It's just too simple. Even as complexity increases, as long as the user's needs are met by wireless devices managed by the network(s), then embedding computational intelligence in the device has limited benefits. In 1999, Mitsubishi and AT&T announced the first "four-mode handset." The T250 operated in TDMA mode on 850 or 1900 MHz, in first generation Analog Mobile Phone System (AMPS) mode on 850 MHz, and in Cellular Digital Packet Data (CDPD) mode on 1900 MHz. This illustrates early development of multi-band, multi-mode, multimedia (M3) wireless. These radios enhanced the service provider's ability to offer national roaming, but the complexity was not apparent to the user since the network managed the radio resources in the handset.

Even as device complexity increases in ways that the network does not manage, there may be no need for cognition. There are several examples of capabilities embedded in electronics that typically are not heavily used. Do you use your laptop's speech recognition system? What about its IRDA port? If you were the typical user circa 2004, you didn't use either capability of your Windows XP laptop all that much. So complexity can increase without putting a burden on the user to manage that complexity if the capability isn't central to the way in which the user employs the system.

For radio, as the number of bands and modes increases, the SDR becomes a better candidate for the insertion of cognition technology. But it is not until the radio or the wireless part of the PDA has the capacity to access multiple RF bands that cognition technology begins to pay off. With the liberalization of RF spectrum use rules, the early evolution of iCR may be driven by RF spectrum use etiquette for ad hoc bands such as the FCC use case. In the not-too-distant future, SDR PDAs could access satellite mobile services, cordless telephone, WLAN, GSM, and 3G bands. An ideal SDR device with these capabilities might affordably access three octave bands from 0.4 to 0.96 GHz (skip the air navigation and GPS band from .96 to 1.2 GHz),1.3 to 2.5 GHz,

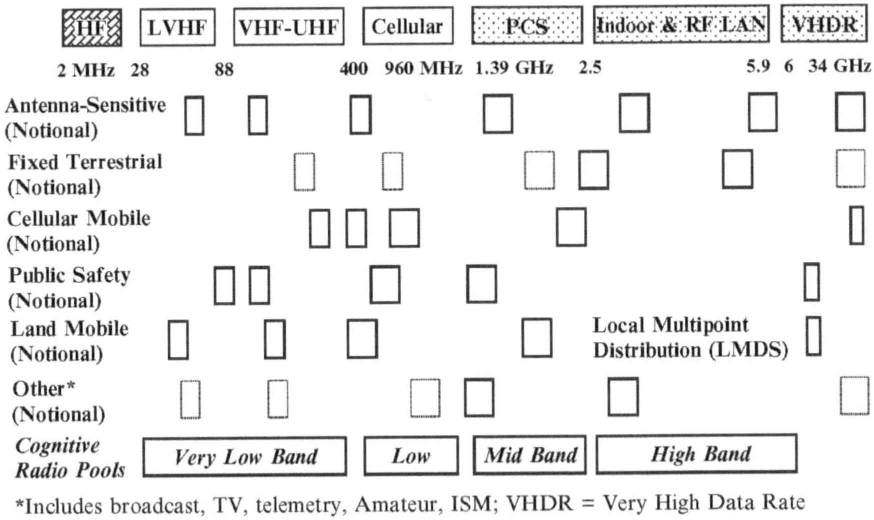

Figure © 1997 Dr. Joseph Mitola III, used with permission

Fig. 3.18. Fixed spectrum allocations versus pooling with cognitive radio.

and from 2.5 to 5.9 GHz (Figure 3.18). Not counting satellite mobile and radio navigation bands, such radios would have access to over 30 mobile sub-bands in 1463 MHz of potentially sharable outdoor mobile spectrum. The upper band provides another 1.07 GHz of sharable indoor and RF LAN spectrum. This wideband radio technology will be affordable first for military applications, next for base station infrastructure, then for mobile vehicular radios and later for handsets and PDAs. When a radio device accesses more RF bands than the host network controls, it is time for CR technology to mediate the dynamic sharing of spectrum. It is the well-heeled conformance to the radio etiquettes afforded by cognitive radio that makes such sharing practical [44].

3.5.8 Radio Evolution towards the CRA

Various protocols have been proposed by which radio devices may share the radio spectrum. The US FCC Part 15 rules permit low power devices to operate in some bands. In 2003, a Rule and Order (R&O) made unused television (TV) spectrum available for low power RF LAN applications, making the manufacturer responsible for ensuring that the radios obey this simple protocol. DARPA's NeXt Generation (XG) program developed a language for expressing spectrum use policy [45]. Other more general protocols based on peek-through to legacy users have also been proposed [33].

Does this mean that a radio must transition instantaneously from the SCA to the CRA? Probably not. The simple six-component iCR architecture may be implemented with minimal sensory perception, minimal learning, and no autonomous ability to modify itself. Regulators want to hold manufacturers responsible for the behaviors of such radios. The simpler the architecture, the simpler the problem of explaining it to regulators and of getting concurrence among manufacturers regarding open architecture interfaces that facilitate technology insertion through teaming. Manufacturers who fully understand the level to which a highly autonomous CR might unintentionally reprogram itself to violate regulatory constraints may decide they want to field aware-adaptive (AA) radios, but may not want to take the risks associated with a self-modifying CR's just yet.

Thus, one can envision a gradual evolution towards the CRA beginning initially with a minimal set of functions mutually agreeable among the growing community of iCR stakeholders. Subsequently, the introduction of new services will drive the introduction of new capabilities and additional API's, perhaps informed by the CRA developed in this text.

3.5.9 Cognitive Radio Architecture Research Topics

The cognition cycle and related inference hierarchy imply a large scope of hard research problems for cognitive radio. Parsing incoming messages requires natural language text processing. Scanning the user's voice channels for content that further defines the communications context requires speech processing. Planning technology offers a wide range of alternatives in temporal calculus [46], constraint-based scheduling [30], task planning [31], causality modeling [32], and the like. Resource allocation includes algebraic methods for wait-free scheduling protocols [33], Open Distributed Processing (ODP), and Parallel Virtual Machines (PVM). Finally, machine learning remains one of the core challenges in artificial intelligence research [34]. The focus of this cognitive radio research, then, is not on the development of any one of these technologies per se. Rather, it is on the organization of cognition tasks and on the development of cognition data structures needed to integrate contributions from these diverse disciplines for the context-sensitive delivery of wireless services by software radio.

Learning the difference between situations in which a reactive response is needed versus those in which deliberate planning is more appropriate is a key challenge in machine learning for CR. The CRA framed the issues. The CRA goes further, providing useful KS's and related ML so that the CR designer can start there in developing good engineering solutions to this problem for a given CR applications domain.

3.5.10 Industrial Strength iCR Design Rules

The CRA allocates functions to components based on design rules. Typically design rules are captured in various interface specifications including Applications Programmers Interfaces (APIs), and Object Interfaces, such as Java's JINI/JADE structure of intelligent agents [47]. While the previous section introduced the CRA, this section suggests additional design rules by which user domains, sensory domains and radio knowledge of RF Band knowledge may be integrated into industrial-strength iCR products and systems.

The following design rules circumscribe the integration of cognitive functions with the other components of a wireless PDA within the CRA

1. The cognition function should maintain an explicit [topological] model of space–time
   a) Of the user
   b) Of the physical environment
   c) Of the radio networks
   d) Of the internal states of the radio, the <self/>
2. The CRA requires each CR to predict in advance, an upper bound on the amount of computational resources (e.g. time) required for each cognition cycle. The CR is must set a trusted (hardware) watchdog (e.g. timer) before entering a cognition cycle. If the watchdog is violated, the system must detect that event, log that event, and mark the components invoked in that event as non-deterministic.
3. The CRA should internalize knowledge as procedural skills, e.g. serModels.
   a) The CRA requires each CR to maintain a trusted index to internal models and related experience.
   b) Each CR must preclude cycles from its internal models and skills graph because a CRA conformance requires reliable detection of cycles to break cycles (e.g. via timer) to avoid Gödel–Turing unbounded resource use endemic to self-referential Turing-capable computational entities like iCRs.
4. Context that references space, time, RF, the <User/> and the <Self/> for every external and internal event shall be represented formally using a topologically valid and logically sound model of space–time–context.
5. Each CR conforming to the CRA shall include an explicit grounding map, M that maps its internal data structures onto elements sensed in the external world represented in its sensory domains, including itself. If the CR cannot map a sensed entity to a space–time–context entity with specified time allocated to attempt that map, then the entity should be designated "UNGROUNDABLE."
6. The model of the world shall follow a formal treatment of time, space, radio frequency, radio propagation, and the grounding of entities in the environment.
7. Models shall be represented in an open architecture radio knowledge representation language suited to the representation of radio knowledge (e.g. a Semantic Web derivative of RKRL). That language shall support topological properties and inference (e.g. forward chaining) but must not include unconstrained axiomatic first order predicate calculus which per force violates the Gödel–Turing constraint.
8. The cognition functions shall maintain location awareness, including
   a) the sensing of location from global positioning satellites,

   b) sensing position from local wireless sensors and networks,
   c) and sensing precise position visually,
   d) location shall be an element of all contexts,
   e) the cognition functions shall estimate time to the accuracy necessary to support the user and radio functions,
   f) the cognition functions shall maintain an awareness of the identity of the PDA, of its Owner, of its primary user, and of other legitimate users designated by the owner or primary user.
9. The cognition functions shall reliably infer the user's communications context and apply that knowledge to the provisioning of wireless access by the SDR function.
10. The cognition functions shall model the propagation of its own radio signals with sufficient fidelity to estimate interference to other spectrum users.
    a) The cognition function shall also assure that interference is within limits specified by the spectrum use protocols in effect in its location (e.g. in spectrum rental protocols).
    b) It shall defer control of the <Self/> to the wireless network in contexts where a trusted network manages interference.
11. The cognition functions shall model the domain of applications running on the host platform, sufficient to infer the parameters needed to support the application. Parameters modeled include QoS, data rate, probability of link closure (Grade of Service), and space–time–context domain within which wireless support is needed.
12. The cognition functions shall configure and manage the SDR assets to include hardware resources, software personalities, and functional capabilities as a function of network constraints and use context.
13. The cognition functions shall administer the computational resources of the platform. The management of software radio resources may be delegated to an appropriate SDR function (e.g. the SDR Forum domain manager). Constraints and parameters of those SDR assets shall be modeled by the cognition functions. The cognition functions shall assure that the computational resources allocated to applications, interfaces, cognition and SDR functions are consistent with the user communications context.
14. The cognition functions shall represent the degree of certainty of understanding in external stimuli and in inferences. A certainty calculus shall be employed consistently in reasoning about uncertain information.
15. The cognition functions shall recognize preemptive actions taken by the network and/or the user. In case of conflict, the cognition functions shall defer the control of applications, interfaces, and/or SDR assets to the Owner, to the network or to the primary user, according to appropriate priority and operations assurance protocol.

3.6 Summary and Future Directions

The progeny of TellMe® seem headed to a purse or belt near you to better sense and perceive your needs for communications services so that you can take fuller advantage of the technology by, well, by doing nothing but letting the technology adapt to you. In 2005, the technology was not capable of sensing, perceiving, and adapting to the user, but enabling technologies in machine speech and vision were maturing. Because of the FCC's rulings encouraging CR, many "cognitive radio" products capable only of sniffing TV channels and employing unoccupied radio spectrum were appearing in the marketplace.

3.6.1 Architecture Frameworks

Often technical architecture frameworks of the kind presented in this chapter accelerate the state of practice by catalyzing work across industry on plug-and-play, teaming, and collaboration. The thought is that to propel of wireless technology from limited spectrum awareness towards valuable user awareness, some architecture like the CRA will be needed. In short, the CRA articulates the functions, components, and design rules of next-generation cognitive radios. Each of the different aspects of the CRA contributes to the dialog:

1. The functional architecture identifies components and interfaces for cognitive radios with sensory and perception capabilities in the user domain, not just the radio domain.
2. The cognition cycle identifies the processing structures for the integration of sensing and perception into radio: observe (sense and perceive), orient (react if necessary), plan, decide, act, and learn.
3. The inference hierarchies suggest levels of abstraction helpful in the integration of radio and user domains into the synthesis of services tailored to the specific user's current state of affairs given the corresponding state of affairs of the radio spectrum in space and time.
4. The introduction to ontology suggests an increasing role for semantic web technologies in making the radios smarter, initially about radio and over time about the user.
5. Although not strictly necessary for CR, SDR provides a very flexible platform for the regular enhancement of both computational intelligence and radio capability, particularly with each additional Moore's law cycle.
6. Finally, this chapter has introduced the CRA to the reader interested in the cutting edge, but has not defined the CRA. The previous section suggested a few of the many aspects of the embryonic CRA that must be addressed by researchers, developers, and markets in the continuing evolution of SDR towards ubiquitous and really fun iCRs.

3.6.2 Industrial Strength Architecture

Although the CRA provides a framework for API's, it doesn't specify the details of the data structures nor of the maps. Thus, the CRA research prototype emphasizes ubiquitous learning via serModels and Case Based Reasoning [see [11]], but it doesn't implement critical features that would be required in consumer-class iCR's. Other critical aspects of such industrial-strength architectures include more capable scene perception and situation interpretation specifically addressing:

1. ***Noise***, in utterances, images, objects, location estimates and the like. Noise sources include thermal noise, conversion error introduced by the process of converting analog signals (audio, video, accelerometers, temperature, etc.) to digital form, error in converting from digital to analog form, preprocessing algorithm biases and random errors, such as the accumulation of error in a digital filter, or the truncation of a low energy signal by threshold logic. Dealing effectively with noise differentiates a tutorial demonstration from an industrially useful product.
2. ***Hypothesis management***, keeping track of more than one possible binding of stimuli to response, dialog sense, scene, etc. Hypotheses may be managed by keeping the N-best hypotheses (with an associated degree of belief), by estimating the prior probability or other degree of belief in a hypothesis, and keeping a sufficient number of hypotheses to exceed a threshold (e.g. 90 or 99% of all the possibilities), or keeping multiple hypotheses until the probability for the next most likely (2nd) hypothesis is less than some threshold. The estimation of probability requires a measurable space, a sigma-algebra that defines how to accumulate probability on that space, proof that the space obeys the axioms of probability and a certainty calculus that defines how to combine degrees of belief in events as a function of the measures assigned to the probability of the event.
3. ***Training Interfaces***, the reverse flow of knowledge from the inference hierarchy back to the perception subsystems. The recognition of the user by a combination of face and voice could be more reliable than single-domain recognition either by voice or by vision. In addition, the location, temperature, and other aspects of the scene may influence object identification. Visual recognition of the Owner outdoors in a snow storm, for example, is more difficult than indoors in an office. While the CR might learn to recognize the user based on weaker cues outdoors, access to private data might be constrained until the quality of the recognition exceeds some learned threshold.

Non-Linear Flows: Although the cognition cycle emphasizes the forward flow of perception enabling action, it is crucial to realize that actions may be internal, such as advising the vision subsystem that its recognition of the user is in error because the voice does not match and the location is wrong. Because of the way the cognition cycle operates on the self, these reverse flows from perception to training are implemented as forward flows from the perception system to the self, directed towards a specific subsystem such as vision or audition. There may also be direct interfaces from the CWN to the CR to upload data structures representing a priori knowledge integrated into the UCBR learning framework.

3.6.3 Conclusion

In conclusion, iCR seems headed for the Semantic Web, but the markets for services layered on practical radio networks will shape that evolution. Although many information-processing technologies from eBusiness Solutions to the Semantic Web are relevant to iCR, the integration of audio and visual sensory perception into SDR with suitable cognition architectures remains both a research challenge and a series of increasingly interesting radio systems designs. A CRA that is broadly supported by industry could accelerate such an evolution.

References

1. Proceedings of the IEEE 2005 Conference on Dynamic Spectrum Access Networks (DySPAN) (Hoeboken, NJ: IEEE Press), 2005
2. Proceedings of the IEEE 2007 Conference on Dynamic Spectrum Access Networks (DySPAN) (Hoeboken, NJ: IEEE Press), 2007
3. J. Mitola III, email to Professor Gerald Q. Maguire, recommending the title of the Licentiate as Cognitive Radio, January 1998
4. J. Mitola III, *Cognitive Radio: Model-Based Competence for Software Radio*, Licentiate Thesis TRITA-IT AUH 99:04 (Stockholm, Sweden: KTH, The Royal Institute of Technology), August, 1999
5. Joseph Mitola III, *Cognitive Radio: An Integrated Agent Architecture for Software Defined Radio* (Stockholm: KTH, The Royal Institute of Technology), June 2000
6. Joseph Mitola III, "Cognitive Radio for Flexible Mobile Multimedia Communications", Mobile Multimedia Communications (MoMUC 99) (NY: IEEE Press), November, 1999
7. Order FCC 03-322, Notice of Proposed Rule Making (Washington, DC: United States Federal Communications Commission), December, 2003
8. www.sdrforum.org/publicsafety.htm
9. Proceedings of the Dagtsuhl Workshop on Cognitive Radios and Networks (Aachen, DE: RWTH Aachen), October 04
10. www.crowncom.org
11. Joseph Mitola III, *Cognitive Radio Architecture* (NY: Wiley), 2006
12. www.ieee-dyspan.org
13. http://en.wikipedia.org/wiki/Multiple-input_multiple-output_communications
14. TellMe Networks (www.tellme.com) Mountain View, CA, 2005

15. Anne Watzman, "Robotic Achievements: GRACE Successfully Completes Mobile Robot Challenge at Artificial Intelligence Conference" (Pittsburg, PA: Carnegie Melon Views) 09/06/02
16. Petri Mahonen, *Cognitive Wireless Networks* (Aachen, DE: RWTH Aachen), September 04
17. Victor Zue, *Speech Understanding System* (Boston: MIT), 2005
18. A. Pisano, *MEMS 2003 and Beyond* (www.darpa.mil/mto/mems: DARPA), 1999
19. R. Michalski, I. Bratko, and M. Kubat, *Machine Learning and Data Mining* (NY: John Wiley & Sons, LTD), 1998
20. Jamie Carbonell, "Phraselator", presentation at Fordham University, December 2004
21. Eriksson and Penker, UML Toolkit (NY: John Wiley & Sons, Inc.), 1998
22. T. Mowbray and R. Malveau, *CORBA Design Patterns* (NY: John Wiley & Sons), 1997
23. www.sdrforum.org
24. www.omg.org
25. www.jtrs.mil
26. Wireless World Research Forum (www.wwrf.com), 2004
27. http://www.cyc.com/cycdoc/upperont-diagram.html
28. J. Mitola III, "Software Radio Architecture: A Mathematical Perspective", *IEEE JSAC* (NY: IEEE Press), April 98
29. R. Hennie, *Introduction to Computability* (Reading, MA: Addison-Wesley), 1997
30. www.omg.org/UML
31. J. Pearl, *Causality: Models, Reasoning, and Inference* (San Francisco: Morgan-Kaufmann), Mar 2000
32. Cognitive Vision (Palo Alto, CA: AAAI), June 2004
33. SNePS (Internet: ftp.cs.buffalo.edu:/pub/sneps/) 1998
34. Koser, et al., "read.me" www.cs.kun.nl: (The Netherlands: University of Nijmegen), March 99
35. The XTAG Research Group, A Lexicalized Tree Adjoining Grammar for English Institute for Research in Cognitive Science (Philadelphia, PA: University of Pennsylvania), 1999
36. PC-KIMMO Version 1.0.8 for IBM PC, 18-Feb-92
37. J. Mitola III "Software Radio: Survey, Critical Evaluation and Future Directions". Proc. National Telesystems Conference (NY: IEEE Press), May 92
38. J. Mitola III, "Software Radio Architecture," *IEEE Communications Magazine* (NY: IEEE Press) May 95
39. J. Mitola III, *Software Radio Architecture*, (NY: Wiley), 2000
40. Walter H. Tuttlebee, *Software Defined Radio: Enabling Technologies* (NY: Wiley), 2002
41. Jeff Reed, *Software Radio: A Modern Approach to Radio Engineering* (Englewood Cliffs, NJ: Prentice Hall), 2002
42. Joseph Mitola III and Zoran Zvonar, eds, *Software Radio Technologies* (NY: IEEE Press), 1999
43. Friedrich Jondral, *Software Radio* (Karlsruhe, Germany: Universität Karlsruhe) 1999
44. L. Esmahi et al., "Mediating conflicts in a Virtual Market Place for Telecommunications Network Services" Proceedings of the 5th Baiona Workshop on Emerging Technologies in Telecommunications (Vigo, Spain: Universidade de Vigo), 1999

45. Preston Marshall, *Remarks to the SDR Forum* (Rome, NY: SDR Forum), September 2003
46. C. Phillips, "Optimal Time-Critical Scheduling", STOC 97 (www.acm.org: ACM), 1997
47. S. K. Das et al., "Decision making and plan management by intelligent agents: theory, implementation, and applications", Proceedings of Autonomous Agents 97 (www.acm.org: ACM), 1997

4

Software Defined Radio Architectures for Cognitive Radios

Hüseyin Arslan[1] and Hasari Celebi[2]

[1] University of South Florida, Tampa, FL, USA, arslan@eng.usf.edu
[2] University of South Florida, Tampa, FL, USA, hcelebi@eng.usf.edu

4.1 Introduction

Wireless communication devices are composed of three main entities; signaling, physical hardware, and its functionalities. These three main streams, which complement each other, have evolved since the invention of the radio transmission by Guglielmo Marconi. The primitive communications devices had very simple signaling, analog hardware, and limited functionality. In time, each of these entities evolved significantly. Different signaling methods have been invented and used around the world. Furthermore, numerous different wireless communications systems and standards have been developed around the world without any global plan. Recently, the diversity in wireless systems and standards bring some issues on the surface such as interoperability and global seamless connectivity. In parallel, hardware technology evolved significantly, too. Some of the key milestones in this progress are transition from analog hardware to digital hardware and then introduction of sophisticated processors. This is followed by the development of Software Defined Radio (SDR) structures and virtual hardwares that are under development currently. SDR is envisioned initially to be a promising solution for interoperability, global seamless connectivity, multi-standard, and multi-mode issues. Also in parallel, the functionality of wireless devices is increased and they become more and more sophisticated. For instance, the cellular technology was developed to provide voice communications for mobile users initially. However, current cellular phones have multi-functionalities such as internet access, digital camera, Global Positioning System (GPS), games, personal assistance, and music player. Ever increasing demands from the users and service providers result in continuously increasing Quality of Service (QoS) requirements. This trend requires to add intelligent functionalities to the wireless devices, which introduced cognitive radio technology. Nowadays, these three streams start to merge under the umbrella of cognitive radio technology. It is an emerging technology to realize wireless devices with cognition capabilities such as learning, sensing, awareness, and reasoning. Moreover, it has capability of providing

H. Arslan (Ed.), Cognitive Radio, Software Defined Radio, and Adaptive Wireless Systems, 109–144.

global seamless connectivity and solve the interoperability issue. SDR is a key enabling technology to realize cognitive radios. Until development of cognitive radio technology, SDR has been mainly proposed to realize multi-mode and multi-standard wireless devices. However, the role of SDR in cognitive radios is very essential, which is the realization of cognition features (e.g. awareness, sensing, etc.) in cognitive radios. Hence, we discuss SDR structures in the context of cognitive radio technology in this chapter.

The main objective of this chapter is to demonstrate the tight relationship between SDR and cognitive radio. This is followed by a discussion on SDR requirements of cognitive radios. Then, we will present the current SDR enabling technologies along with their strengths and weaknesses. Finally, concluding remarks will be provided.

4.2 SDR and Cognitive Radio Relationship

As discussed in the previous chapters, one of the main characteristics of cognitive radio is the adaptability where the radio parameters (including frequency, power, modulation, bandwidth) can be changed depending on the radio environment, user's situation, network condition, geolocation, and so on. SDR can provide a very flexible radio functionality by avoiding the use of application specific fixed analog circuits and components. Therefore, cognitive radio needs to be designed around SDR. In other words, SDR is the core enabling technology for cognitive radio. One of the most popular definitions of cognitive radio, in fact, supports the above argument clearly: "A cognitive radio is an SDR that is aware of its environment, internal state, and location, and autonomously adjusts its operations to achieve designated objectives."

Even though many different models are possible, one of the simplest conceptual model that describes the relation between cognitive radio and SDR can be described as shown in Figure 4.1. In this simple model, cognitive radio is wrapped around SDR. This model fits well to the aforementioned definition of cognitive radio, where the combination of cognitive engine, SDR, and the other supporting functionalities (e.g. sensing) results in cognitive radio. Cognitive engine is responsible for optimizing or controlling the SDR based on some input parameters such as that are sensed or learned from the radio environment, user's context, and network condition. Cognitive engine is aware of the radio's hardware resources and capabilities as well as the other input parameters that are mentioned above. Hence, it tries to satisfy the radio link requirements of a higher layer application or user with the available resources such as spectrum and power. Compared to hardware radio where the radio can perform only single or very limited amount of radio functionality, SDR is built around software based digital signal processing along with software tunable Radio Frequency (RF) components. Hence, SDR represents a very flexible and generic radio platform which is capable of operating with many different bandwidths over a wide range of frequencies and using many different modulation

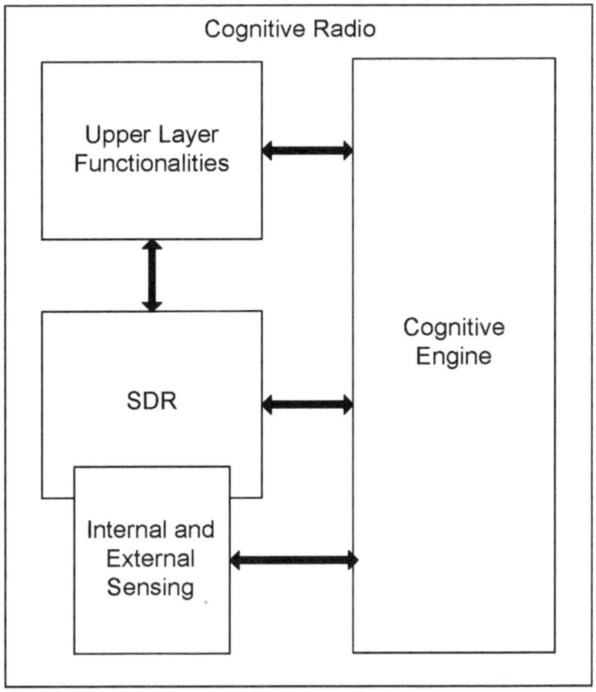

Fig. 4.1. The illustration of relationship between SDR and cognitive radio.

and waveform formats. As a result, SDR can support multiple standards (i.e. GSM, EDGE, WCDMA, CDMA2000, Wi-Fi, WiMAX) and multiple access technologies such as Time Division Multiple Access (TDMA), Code Division Multiple Access (CDMA), Orthogonal Frequency Division Multiple Access (OFDMA), and Space Division Multiple Access (SDMA).

The recent boom in the diversity of wireless standards with different options exposes interoperability and multi-mode support issues. SDR has been considered as an inherent solution to address such issues. Although SDR is naturally evolved due to the need to implement radios that can support multiple mode and standards, utilization of SDR in a cognitive radio is not limited to the aforementioned functionalities. SDR is a promising technology to introduce cognition capabilities to cognitive radios. For instance, one of the crucial cognition capabilities of cognitive radios is the dynamic spectrum management system. Spectrum sensing, optimization mechanism to utilize a specific part of the spectrum, and spectrum shaping are the main steps of dynamic spectrum management systems. In the case of spectrum, sensing devices are required to sense the spectrum, which can be either embedded into SDR internally or incorporated to SDR externally. For instance, antenna can be considered as an internal sensor whereas the video camera can be considered as an external sensor for SDR structures. In other words, SDR can have a structure like a miniature spectrum analyzer in order to provide the spectrum information to cognitive engine. Either the existing receiver front-end of SDR or a designated receiver parallel to the receiver side of SDR can be used to perform spectrum capturing. Captured spectrum is digitized by Analog-to-Digital Converter (ADC) and then the digital samples are sent to digital signal processor for the post-processing.

4.3 SDR Architectures

4.3.1 Ideal SDR Architectures

Ideal SDR architecture consists of three main units, which are reconfigurable digital radio, software tunable radio along with embedded impedance synthesizer, and software tunable antenna systems [1, 2]. This structure is illustrated in Figure 4.2. The main responsibilities of reconfigurable digital radio are performing digital radio functionalities such as different waveform generation, optimization algorithms for software tunable radio and antenna units, and controlling of these units. There are various different enabling technologies for implementation of reconfigurable digital radio, which are discussed in the later sections. Software tunable analog front-end system is limited to the components that cannot be performed digitally such as RF filters, combiners/splitters, Power Amplifier (PA), Low Noise Amplifiers (LNA), and data converters. Moreover, this unit has impedance synthesizer subsystem, which is a crucial unit to optimize the performance of software tunable analog radio systems. For instance, impedance synthesizer is used to optimize the performance of software tunable antenna systems for an arbitrary frequency plan specified by cognitive engine. The details of these SDR units are provided in the later sections.

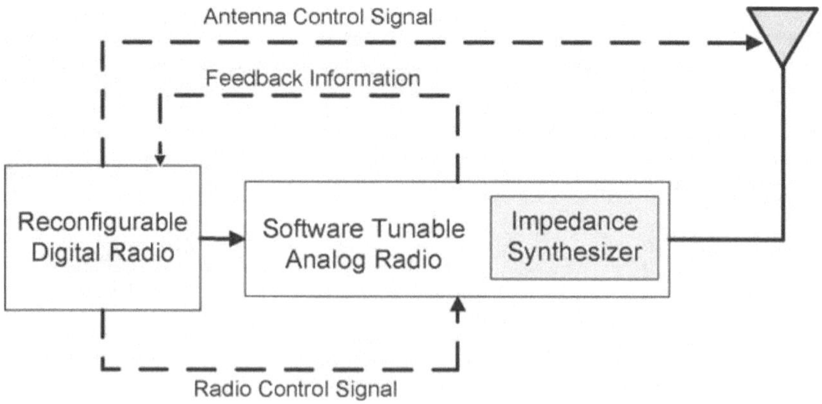

Fig. 4.2. An ideal SDR architecture.

Reconfigurable digital radio system monitors and controls the software tunable radio system continuously or periodically depending on system specifications. A basic relationship between the main units of SDR is described as follows. The cognitive engine sends radio configuration parameters to the reconfigurable digital radio so that it can reconfigure the entire radio according to the parameters. These parameters can be waveform type that needs to be generated (e.g. OFDM, CDMA, UWB), frequency plan (e.g. bandwidth, operating center frequency), and power spectrum specifications. Moreover, cognitive engine can request from reconfigurable digital radio to measure or calculate some parameters from environments such as location information of a particular user. Reconfigurable digital radio configures itself along with software tunable radio components and antenna systems. In order to optimize the performance of the these two units, reconfigurable digital radio utilizes the feedback information from software tunable radio, especially from impedance synthesizer. Based on this information, it adjusts the parameters of software tunable radio and antenna units through radio and antenna control signals, respectively. Finally, reconfigurable digital radio acknowledges cognitive engine that the specified configuration is performed.

4.3.2 Realistic SDR Architectures

Due to the current limitations (size, cost, power, performance, processing time, data converters), ideal SDR architectures are costly. There are various practical SDR platforms available in the literature. Note that it is expected that these platforms will evolve significantly in the future as the limitations towards ideal SDR platform are being removed. Hence, as reconfigurable hardware technology advances, software tunable analog radio functionalities will be implemented in reconfigurable digital radio platforms. In the following, we will provide some current and practical SDR architectures. Figure 4.3 shows an example of practical SDR architecture for the Worldwide Interoperability Microwave Access (WiMAX) networks [3]. Reconfigurable digital radio can

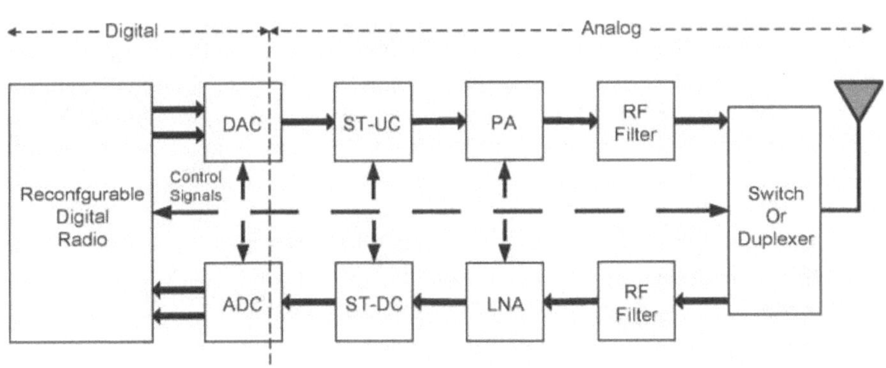

Fig. 4.3. A current practical SDR architecture.

be implemented using one of the SDR enabling technologies such as Digital Signal Processor (DSP) or Field-Programmable Gate Arrays (FPGAs). This unit mainly generates and demodulates OFDM waveform and control the radio components. Generated OFDM signal is in the form of digital Inphase (I) and Quadrature (Q) samples. Interpolation, digital filtering, Peak-to-Average-Power-Ratio (PAPR) reduction algorithms and digital Intermediate Frequency (IF) upconversion are applied to the I/Q signals prior to Digital-to-Analog Converter (DAC). Consequently, DAC converts the digital OFDM signal into the corresponding analog waveform. IF signal at the output of DAC is upconverted to a final RF stage using Software Tunable-UpConverter (ST-UC). An ST-UC generally consists of software tunable attenuators and clock synthesizers. With such capability, transmit power level and local oscillator frequency can be adjusted by reconfigurable digital radio. Furthermore, IF-to-RF upconversion can be performed in one or multiple stages depending on the performance criterion such as minimum image rejection level. For instance, the first IF signal is upconverted to the second IF and then it is upconverted to the final RF carrier in the case of two-stage upconversion. In the sequel, the RF signal is amplified using PA according to the power spectral specifications that come from digital radio. A typical PA consists of software tunable attenuators for adaptive transmit power level control. The adaptive transmit power level is an important task for adjusting link quality. The amplified RF signal is filtered out and it is radiated to the air through antenna. Depending on the duplexing method, radio can be classified into three major groups, which are Time Division Duplexing (TDD), Frequency Division Duplexing (FDD) and Half-Frequency Division Duplexing (H-FDD) radios [4]. In case of TDD radios, the transmit/receive switch is used, which is controlled by reconfigurable digital radio. On the other hand, for the FDD radios, duplexer filter is used to support simultaneous transmission and reception in different bands.

On the receiver side, the received RF signal is filtered out to suppress unwanted out-of-band signals. The filtered RF signal is amplified using LNA. This unit can have digital attenuators to protect the receiver from the signals with high power. Furthermore, it can consist of Variable Gain Amplifier (VGA), which is controlled by the reconfigurable digital radio, for RF automatic gain control. The amplified RF signal is downconverted to a low IF stage using Software Tunable-Downconverter (ST-DC), which can be performed either in one or multiple downconversion stages as well. A typical ST-DC consists of software tunable internal or external AGC with peak detector and software tunable clock synthesizer. The ST-DC as well as ST-UC can contain externally selectable IF filters to support different bandwidths. Consequently, ADC digitizes the analog IF signal and generates the corresponding digital I/Q samples. Decimation and digital filtering are applied to the samples. In the following steps, digital radio demodulates the received OFDM signal after time and frequency synchronization. Software tunable radio components in this example are implemented using programmable Application Specific Integrated Circuit (ASIC) technology. These components are configured by reconfigurable digital radio through Serial Peripheral Interface (SPI) and independent pins. Using SPI and independent configuration signals, reconfigurable digital radio writes configuration parameters into the registers that are embedded in software tunable radio components. Moreover, software tunable radio components can be powered down by the digital radio to save power. For instance, the components in transmit chain can be switched to power down mode in the TDD radios when radios receive signal and vice versa.

Another example that can be represented as a model for realistic SDR architecture is SPEAKeasy program developed by the Department of Defense (DoD) in the United States (US) [5]. The goal of SPEAKeasy program was to generate a portable, but not handheld, software programmable radio capable of emulating more than 10 existing military radios operating from 2 MHz to 2 GHz. These radio systems would be implemented in software instead of the traditional hardware and could be stored in the systems memory or downloaded from the air [6]. SPEAKeasy Phase-I was a lab based rack mounted proof-of-concept program to demonstrate multi-band and multi-mode operations between voice networks. SPEAKeasy used DSPs to perform operations and generate waveforms in software instead of the traditional method of using hardware. With the completion of Phase I, the program moved into Phase II. This phase delivered a portable radio and demonstrated that the overall concept of SDR was feasible. Several demonstrations were performed including the bridging of communications systems, in-field re-programming, and in-field repairs using Components Off The Shelf (COTS) [7]. SPEAKeasy was an important stepping stone towards the largest SDR based program to date, the Joint Tactical Radio System (JTRS). The JTRS is also a DoD initiative to develop a flexible software programmable radio technology to meet the diverse communications needs of the military. This includes transmissions of real-time voice, data, and video across the US military branches as well as with coalition forces and allies [8]. The JTRS radio technology spans from 2 MHz to 2 GHz using waveforms that are defined in software using minimal waveform specific hardware. The development of these waveforms allows for interoperability among radios, the reuse of common software across many different radios, and scalability in the number of available channels [9]. The implementation of these waveforms is based upon a Software Communications Architecture (SCA). The SCA governs the structure and architecture of the JTRS telling designers how elements of hardware and software are to operate in harmony with each other. The development of the SCA is an international effort lead by the US with help from Japan, Canada, England, and Sweden. Many different commercial and international bodies have embraced the SCA to become the open international commercial standard for SDRs [9,10]. Using this architecture with a modular design approach gives the JTRS many benefits compared with the plethora of current systems. Currently, many different radio systems must be carried, maintained, and operated. The JTRS will reduce the total communications equipments weight and footprint by eliminating multiple systems. This will also reduce maintenance costs and complexity due to the elimination of the equipments and less systems to maintain. Upgrades to the JTRS should require minimum support, including the introduction of new waveforms, due to the system being defined in software, which can be upgraded easily over the air and in the field, reducing downtime and increasing efficiency. On the commercial side, there are various efforts on building SDR platforms, mainly focussed on cellular base stations. Airnet [11] is one of these companies currently using SDR technology to implement their base stations. Airnet's AdaptaCell reduces hardware by 90 processing units handling a dozen separate channels reducing the number of DSPs and air interface boards. It also supports GSM, GPRS, and EDGE with adaptive smart antenna controls [11].

4.4 Software Tunable Analog Radio Components

Cognitive engine requires a flexible and tunable analog front-end that can respond to arbitrary tasks. To satisfy such requirements, software tunable analog radio can be considered as a platform consisting of generic software tunable components such as software tunable data converters, software tunable-filters, software tunable-amplifiers, software tunable-duplexing devices, and software tunable-multiple antenna systems. In this platform, the generic components are tuned and/or optimized to achieve arbitrary tasks given by cognitive engine through reconfigurable digital radio. This structure is illustrated in Figure 4.4. In this structure, Software Tunable-DAC (ST-DAC) and Software Tunable-ADC (ST-ADC) are assumed to be part of reconfigurable digital radio platform whereas Software Tunable-Multiple Antenna Systems (ST-MAS) are considered to be part of software tunable analog radio platform. However, they can be considered as separate blocks from both platforms depending on the implementation choice. The software tunable analog radio platform consists of Software Tunable-Filters (ST-filter), Software Tunable-PA (ST-

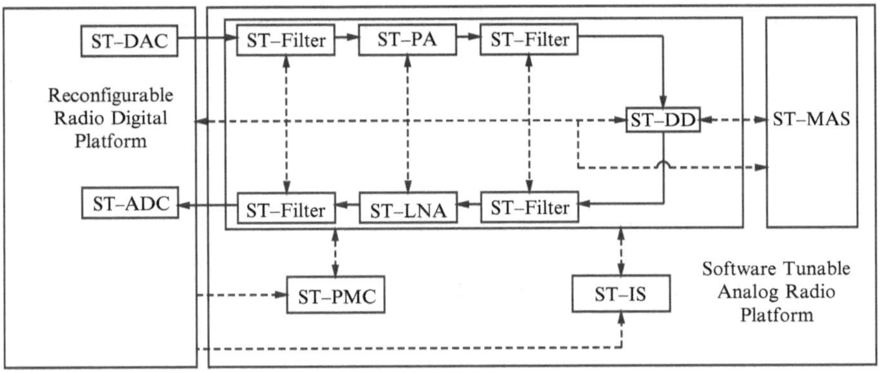

Fig. 4.4. Software tunable analog radio structure.

PA), Software Tunable-LNA (ST-LNA), Software Tunable-Duplexing Devices (ST-DD), Software Tunable-Power Management Circuitry (ST-PMC), Software Tunable-Impedance Synthesizer (ST-IS), and ST-MAS. In this section, the requirements of software tunable analog radio components along with their design challenges for realizing the above platform are discussed. Moreover, a discussion on some other state-of-art software tunable radio components such as ST-DAC, ST-ADC, ST-UC, and ST-DC is provided as well.

4.4.1 Software Tunable-Filters

Ideally, SDR requires IF and RF filters that can be optimized for given filter specifications such as arbitrary center frequency and bandwidth without compromising the performance. The desired performance parameter of such filters in SDR are low insertion loss, excellent out-of-band rejection, and high power handling. Designing such filters with the current technologies is a challenging task. However, there are some successful approaches towards this goal such as the electronically tunable filter design [12]. This filter is implemented using Low Temperature Co-fired Ceramic technology and the tuning elements are voltage-controlled dielectric capacitors so called Parascan varactors [13]. A tunable UWB Band Pass Filter (BPF) [14], which is based on ring resonator structure, is another stepping stone towards achieving the above goal. The passband of the designed UWB BPF is from 3.8 GHz to 9.2 GHz and tuning unit is a capacitor shunted to the ground at the end of the stub.

4.4.2 Software Tunable Power Amplifiers

PAs are one of the essential components of wireless communications systems. PAs are used to amplify the signal before transmission so that the signal can reach to distant receivers with a desired Signal-to-Noise Ratio (SNR). PAs can be divided into two major groups: linear and nonlinear PAs. Linear PAs have the advantage of high linearity which is very important for signals with a wide range of amplitude values. However, they suffer from poor power efficiency, limiting their applications in wireless communications systems. On the other hand, nonlinear PAs can achieve better efficiencies with poor linearity in their responses. The nonlinearity causes several problems like amplitude and phase distortion (AM-to-AM and AM-to-PM effects) and adjacent channel regrowth. Power back-off technique is widely used in current wireless communications systems (e.g. WLAN) to remedy the problems due to wide dynamic signal ranges [15, 16]. However, this technique sacrifices the efficiency and increases the power consumption. On the other hand, baseband linearization techniques are utilized to predistort the signal, and hence to compensate the nonlinear effects [17–19]. The efficiency and adaptation of linearization algorithms are, therefore, very important. Depending on the mode of operation and transmitted waveform characteristics, the linearization algorithms need to be adapted. Digital linearization techniques are often based on a feedback scheme and, therefore, able to react to drifts of the nonlinear PAs. Additionally, PAs generally have embedded digitally controlled VGA to support adaptive coverage systems in SDR. Furthermore, since PAs are the major source of power consumption and heat generation in SDR, PAs can operate in different power level modes such as high and low power modes to save power. By doing this, the temperature of SDR board can be controlled by reconfigurable digital radio. In order to achieve high power transmission, highly efficient and linear multiple PAs can be cascaded.

4.4.3 Software Tunable-Duplexing Devices

Duplexing devices are used to manage transmission and reception process in transceivers. As previously mentioned, TDD, FDD, and H-FDD are three major duplexing methods [4]. To support each of these methods in the legacy radios, different hardware components are required, which is costly. For example, transmit/receive switch is used for TDD and H-FDD radios whereas duplexer filter is employed for FDD radios. It is expected that SDR will support all these three radio structures and change its structure dynamically from one to another. Because, each duplexing method has some advantages, which cognitive engine can benefit from them. For instance, TDD radios are popularly known structure to be used for the unlicensed bands operation due to more relaxed regulations (e.g. output noise level) in these bands compared to the licensed bands [4]. By considering opportunistic spectrum usage capability of cognitive radio, it can reconfigure SDR as TDD or H-FDD radios when it operates in the unlicensed bands. When the licensed bands are utilized temporarily, cognitive engine can configure SDR as an FDD radio. Using transmit/receive switch and duplexer filter to support these three radio structures is not a practical solution. Furthermore, the current duplexer filters are designed for specific frequency plan. There is a need to develop a single hardware component or find a solution to support various duplexing methods in SDR. One way of achieving this goal is to develop a software tunable component that functions as a transmit/receive switch and duplexer filter. Alternatively, employing software tunable antenna pair eliminates the need for transmit/receive switch and duplexer filter, which the details of this approach can be found in [2].

4.4.4 Software Tunable Antenna Systems

Antennas are essential components of any radio system. An antenna radiation pattern (half power beamwidth) and gain are the two important characteristics that affect the system coverage and performance. Many of the current antennas are designed to operate for a specific frequency range and bandwidth. In SDR, it is important that the antenna would have uniform characteristics over a broad range of frequencies. In other words, ideal SDR requires software tunable antenna that its performance can be optimized for an arbitrary center frequency and bandwidth provided by cognitive engine. Also, the need for multi-band and UWB antennas is increasing. Reconfigurable antennas, smart antennas, and Multiple-Input and Multiple-Output antenna (MIMO) systems are already becoming an integral part of wireless communications systems and they will certainly take a major role in SDR.

The type of antenna that could satisfy the aforementioned SDR antenna requirements is a current research topic. However, shorted patch antenna is one of the approaches for SDR implementation due to its small size, low cost, and omni-directional patterns. In [2], shorted patch antenna based electronically tunable antenna for SDR is proposed. The antenna is controlled by antenna control unit, which is composed of Field-Effect Transistor (FET) switches and FPGAs. These switches are employed to change the electrical length of the antenna. FPGAs control these switches based on the information that is received from the DSP, which can be thought as a cognitive engine in this context. With this structure the cognitive engine can tune the antenna to a specific frequency plan. Due to the importance of antenna systems in SDR, there will be some more discussions on this subject later in this chapter.

4.4.5 Software Tunable Impedance Synthesizers

Software tunable impedance synthesizer plays a key role in the optimization of entire software tunable analog radio for achieving a task given by cognitive engine. In cognitive radios, characteristics of the radio (e.g. RF carrier frequency or bandwidth) are dynamic. It is well known that if one of the fundamental characteristics of analog radio is changed, impedance matching network of the components (at least the affected ones) needs to be optimized. For instance, input impedance of antenna in mobile systems is one of the most frequently changeable radio parameters due to the mobility. A change in the input impedance of antenna results in impedance mismatch between power module and antenna. Such mismatches can deteriorate the performance of radio significantly, even it can be damaged [20, 21]. For instance, the reflected power due to the mismatch reduces the radiated power, which deteriorate the efficiency. In such case, radio needs to increase power level resulting in the reduction of battery life. In case of not pumping additional power, QoS is degraded. Note that the reflected power can damage radio components if there is no protection circuit. In order to eliminate the problems and degradation due to the impedance mismatching, software tunable impedance synthesizers are placed between software tunable radio components to perform impedance matching whenever the software tunable analog radio is reconfigured.

The underlying topology of the most software tunable impedance synthesizers are based on generic low-pass pi-network topology as shown in Figure 4.5. This topology consists of variable inductor (L) and capacitors ($C_{1,1}, \ldots, C_{1,N_1}, C_{2,1}, \ldots, C_{2,N_2}$). The more the number of capacitors (N_1, N_2) employed, the better the tunability is achieved. By varying these values

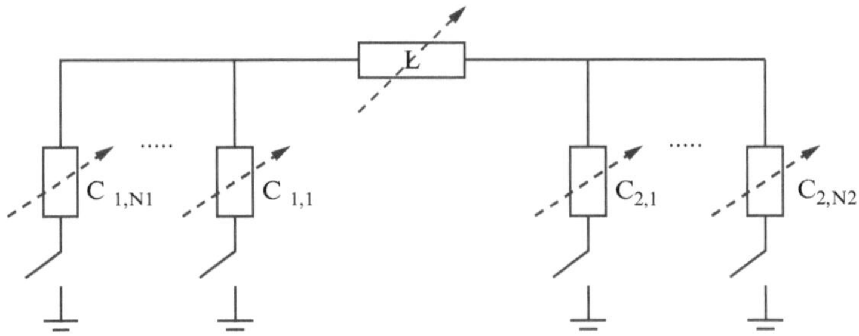

Fig. 4.5. Impedance synthesizer structure.

the impedance matching is achieved. Cognitive engine determines the values of these components. This is achieved by providing operating frequency plan and range of impedances to be matched information to reconfigurable digital radio.

4.4.6 Software Tunable-Power Management Circuitry

Power management circuitry is an integral part of complex board design that consists of discrete components. These components can require different voltage supplies in a predefined sequence. Moreover, reconfiguring radio components may require resetting the entire software tunable analog radio platform. A Software Tunable-Power Management Circuitry (ST-PMC) can be incorporated into software tunable analog radio platform to supply an arbitrary voltage to radio components.

4.4.7 Software Tunable Data Converters

The principle feature of SDR is that the capabilities and operation of radio can be reconfigured efficiently on-the-air, rather than at the time of design. Reconfigurable blocks in SDR systems offer easy changes to radio characteristics such as channel coding/decoding methods, modulation types, multiple access schemes, frequency spreading/despreading algorithms, operating carrier frequencies, and bandwidths. Traditional hardware radio requires hardware changes to modify these fundamental characteristics of radios. Data converters (ADC and DAC) constitute the interface (boundary) between the analog and digital world. Therefore, conducting more of processing digitally requires moving data converters towards antenna.

A typical ADC consists of a sampler followed by a quantizer. Sampling rate is one of the important features of ADCs. One of the limitations towards ideal SDR is the sampling rate requirements as the bandwidth of the signal before ADC must be smaller than half of the sampling rate according to the Nyquist theorem. Quantizer converts the discrete samples (with continuous amplitude levels) into bits with a word length. The word length determines the resolution of ADC and also determines the quantization error. In addition to the quantization noise, distortions due to the static and dynamic nonlinearity features of ADCs also affect the performance of ADC. Signal-to-Noise-and-Distortion (SINAD) is the ratio of the root-mean-square (rms) signal amplitude to the mean value of the root-sum-square of all other spectral components, including harmonics, but excluding DC. SINAD is a good indication of the overall dynamic performance of an ADC, because it includes all components which make up noise and distortion. Due to the distortions, the Effective Number Of Bits (ENOB) that specifies the dynamic performance of an ADC at a specific frequency, amplitude, and sampling rate will be different than what is expected from an ideal ADC that only includes the quantization noise.

Since cognitive radio requires receiving different waveforms with different operating center frequency and bandwidth, SDR needs to have a capability of reconfiguring ADC dynamically. In order to support such capability, the sampling rate, resolution, SINAD of the ADC needs to be optimized and reconfigured by reconfigurable digital radio. Reconfigurable digital radio requires to have capability of monitoring the above parameters of ADC in order to optimize its performance. This is the case as well for all the aforementioned software tunable analog radio components. For instance, sampling rate can be changed by reconfiguring clock circuit management block, which is responsible to generate and supply any type of clock signal to reconfigurable blocks. Note that the phase noise of clock signal plays an important role on the performance of SDR. Therefore, dynamically generated clock signal needs to be clean and satisfy minimum phase noise level dictated by reconfigurable digital radio.

A generalized DAC structure consists of DAC register, resistor string and followed by output buffer amplifier blocks is given in [22]. We refer to [22, 23] for the basic operation of DAC. Similar to ADC, the main characteristics of DACs are resolution, maximum sampling rate, monotonicity, dynamic range, and phase distortion. Monitoring the parameters related to these characteristics is required to support adaptive transmission mandated by cognitive engine. Based on the feedback information from DAC, its performance can be optimized by reconfigurable digital radio.

4.4.8 Software Tunable-Upconverters and Downconverters

Since the functional block diagram of ST-UC and ST-DC are identical, the discussion in this section is based on ST-UC. A basic ST-UC is composed of a mixer and software tunable frequency synthesizer, which is shown in Figure 4.6. SDR requires a mixer that support large input frequency range and it is desirable that mixer to have image rejection capability (Image Rejection Mixer (IRM)). Furthermore, the main features of software tunable frequency synthesizers are ultra-low phase noise and high precision tunability (small step size). The main source of the phase noise is the reference clock. This is generated from reconfigurable clock distributor and it is required to generate

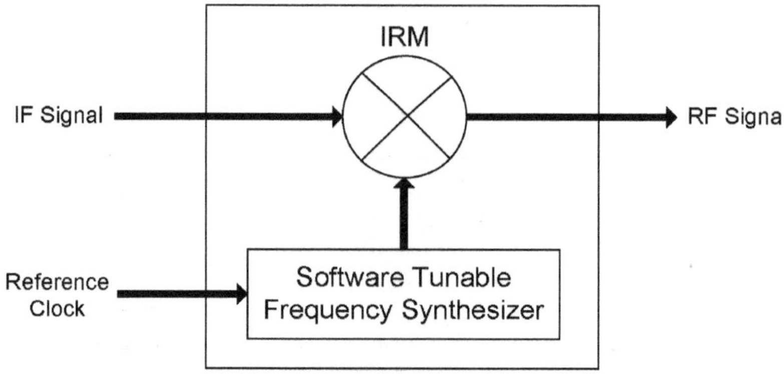

Fig. 4.6. Functional block diagram of ST-UC in SDR.

reference clock signals with ultra-low phase noise. The synthesizer lock detection signal is usually monitored by reconfigurable digital radio to make sure the synthesizer is locked to the desired frequency. In case of employing single-stage upconversion, local oscillator in the synthesizer is required to support large frequency range. Alternatively, multiple-stage upconversion can be employed in SDR. This approach provides some advantages over the single-stage case, but with the use of additional components. By employing multiple-stage upconversion in SDR, the requirements of mixer and frequency synthesizers are relaxed. For instance, local oscillators with smaller frequency range are needed to upconvert the signal to any carrier frequency. Furthermore, the performance of SDR can be improved by suppressing the harmonics at the output of ST-UC with multiple steps. Note that upconversion and downconversion have been performed digitally according to the structure in Figure 4.4. Such approach eliminates use of ST-UC and ST-DC components in transceivers.

4.5 Antenna Systems

The primitive communications systems have employed single-antenna systems. However, as the QoS requirements increase wireless communications systems require more capacity and bandwidth. In other words, wireless devices become more capacity and bandwidth intensive. Moreover, improving performance of wireless communications systems through coding and signaling has already been exploited [1]. To push the capacity, spectrum efficiency, data rate, and robustness of wireless links further, multiple antenna systems and associated technologies such as smart antennas along with beamforming, MIMO systems, and adaptive MIMO systems can be employed [1,25,26,34].

Multiple antenna systems play a significant role in cognitive radio and SDR. Multiple antennas can be used to exploit the spatial dimension of spectrum space and exploit the spatial spectrum (through beamforming) space to improve the efficiency. Spatial dimension of the spectrum space can be exploited very efficiently by multiple antenna transmitters and receivers. Each of the multi-antenna techniques in their own way directly or indirectly helps to exploit the spectrum space. In essence, multi-antenna systems can help to find spectral opportunities in the spatial domain and can help to exploit these opportunities fully.

Similarly, multiple antenna systems can be used to increase capacity, peak data rates, and range of communication. Hence, multiple antennas can allow additional capabilities to adapt the system and network parameters. For example, multiple antennas can be exploited for multiple purposes depending on the input from the sensing units. In some cases, multiple antennas can be used to increase the data rate when the channel is rich and good. In some other cases, multiple antennas can be used for diversity when the fading is significant and channel is having deep fades. Moreover, these systems can be used to compensate the shadowing effect, expand the network coverage, and improve power efficiency.

From the implementation point of view, multiple antenna systems require computationally complex digital signal processing and implementing these systems using traditional analog radio technologies is burdensome and costly. However, SDR is a promising approach to realize multiple antenna systems in wireless communications with reasonable complexity and cost [27–29]. In this section, we discuss the implementation challenges and benefits of the aforementioned systems along with SDR in mind.

4.5.1 MIMO Systems

MIMO systems have potential to improve data rate, channel capacity, diversity, and robustness of wireless communications systems [1,25,34]. Driven by multimedia based applications, anticipated future wireless applications will require high data rate capable systems. Novel techniques like MIMO offer promising choices for future high data rate systems. MIMO employs multiple antennas at the transmitter and receiver sides to open up additional subchannels in spatial domain. Since the parallel channels are established over the same time and frequency, high data rates can be achieved without the need of extra bandwidth. Due to this bandwidth efficiency, efforts are in progress to include MIMO in the standards of future Broadband Wireless Access (BWA).

MIMO channel can be described as a $K \times L$ matrix, where K is the number of transmit antennas and L is the number of receive antennas as shown in Figure 4.7. This is in contrast to the channel of a traditional Single-Input Single-Output antenna (SISO) system. Every receiver antenna captures a signal that has components from every transmitter antenna. There are two basic types of MIMO systems. The first is known as Space Time Coding (STC) and the second is known as Spatial Multiplexing (SM) [25]. The former is usually used to maximize the spatial diversity in MIMO channels whereas the latter is employed to maximize transmission rate. As a result, the way of transmitting

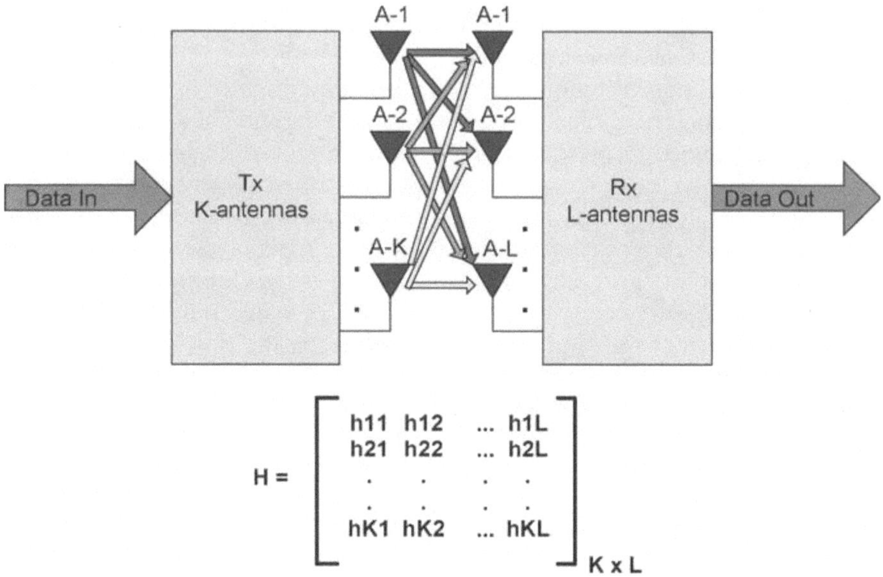

Fig. 4.7. Basic block diagram of a $K \times L$ STC-MIMO system.

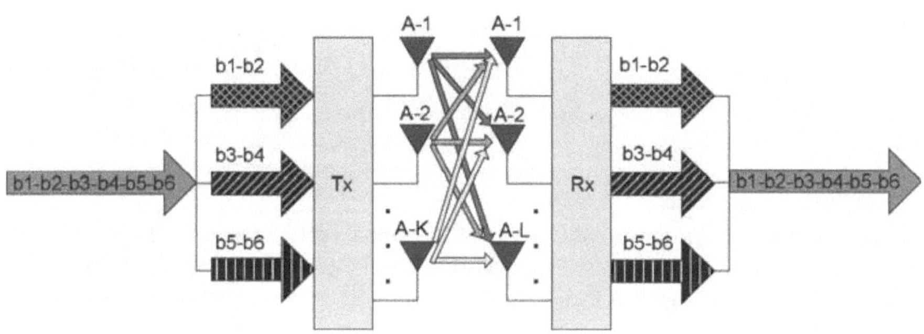

Fig. 4.8. Basic block diagram of a $K \times L$ SM-MIMO system.

information over the wireless channel is different in these two techniques. In STC-MIMO, the incoming serial data stream is replicated on every transmit antenna while in SM-MIMO the incoming serial data stream is broken up into distinct substreams that are then fed to the different transmit antennas. A basic block diagram of STC-MIMO and SM-MIMO systems are shown in Figures 4.7 and 4.8, respectively. The difference between these two types of MIMO systems can be seen by examining Figures 4.7 and 4.8. Implementing both of these computationally complex algorithms using legacy hardware analog radio is impractical and costly. Once again, SDR is a promising solution to implement both of these algorithms in reconfigurable digital radio system. Depending on the optimization criteria, cognitive engine can select the appropriate MIMO signaling.

To attain the best performance in a MIMO system, the signals received at the receive antennas need to be as uncorrelated as possible. If the signals are not completely uncorrelated, it means that the columns of the channel matrix are not linearly independent. In other words, some of the contained channel data is redundant and is not contributing to capacity. This situation causes the matrix not to be of full rank. So, performance will suffer in that case. Typically, in a rich scattering environment, the required separation of the antennas is cited to be at least half a wavelength.

MIMO and multi-antenna systems bring out a new dimension to wireless channels. The spatial dimension will be used in the future communications systems for further improvement of the bandwidth and power efficiency. However, this dimension and the related parameter estimates need to be understood well. Research on parameter estimation for fast and accurate calculation of spatial selectivity, angular spread, antenna correlation, etc. is needed. Also, further research is required on the effect of mutual coupling between antenna elements, the effect of near-field scatterers on antenna patterns and antenna correlations, exploitation of pattern selectivity when the spatial selectivity is not enough, and generation of a desired pattern selectivity between antenna elements adaptively.

As described earlier, MIMO systems can provide huge capacity and improved performance gains by exploiting spatial selectivity of the channel. However, these gains, in reality, depend heavily on the statistical properties of the channel and the correlations between antenna elements. Among the factors that affect the antenna correlation are the characteristics of the scattering environment. Therefore, an optimal way of using multiple-antenna systems depends on the situation awareness. If the transmitter knows the instantaneous channel gains (the MIMO channel matrix), it can adapt the transmission to maximize the capacity of the MIMO system. Similarly, the instantaneous antenna correlation values can be exploited to adapt the transmission. In many cases, estimation of the perfect instantaneous channel state and antenna correlation information, and feeding this information back to the transmitter might not be possible. This is the case especially when the mobility is high. Instead, other parameter measures like partial (statistical) channel information, average channel selectivity, or angular spread would be useful for adapting the transmitter and receiver. Advanced signal processing techniques to calculate this partial channel and correlation information are needed.

Note that MIMO systems employing STC methods require antenna systems to be uncorrelated. Hence, the spacing between two consecutive antenna systems need to be large enough to satisfy this requirement. Functional block diagram of generalized SDR architecture with $K \times L$ MIMO system is shown in Figure 4.5.1. In this architecture, SDR has K number of transmit and L number of antennas. Although the number of transmit and receive antenna are fixed in the conventional hardware analog radio structures, the number of transmit and receive antennas can be changed easily by powering down the required number of antenna branches. Moreover, the parameters (e.g. operating

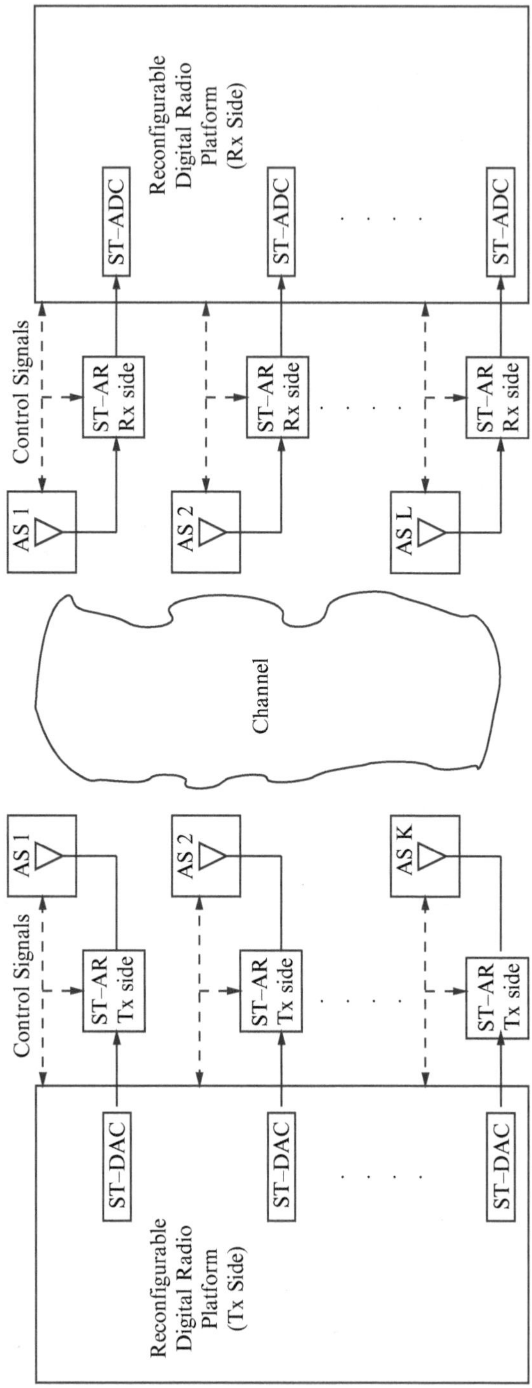

Fig. 4.9. Functional block diagram of generalized SDR architecture with $K \times L$ MIMO system (AS: antenna system, ST-AR Tx side: software tunable analog radio transmitter side, ST-AR Rx side: software tunable analog radio receiver side).

frequency) of each antenna of MIMO systems in conventional hardware analog radio structures are not reconfigurable. However, these parameters can be reconfigured in SDR structures. In SDR architecture, MIMO signaling techniques and algorithms are implemented in reconfigurable digital radio platform.

Another recent approach in multiple antenna systems is adaptive (smart) MIMO, where MIMO systems are enhanced with adaptive beamforming capability [26]. The main goal in adaptive MIMO systems is to maximize the resources offered in multiple antenna channels by using optimal algorithms. Moreover, the key design consideration for adaptive MIMO systems is the optimization of spacing between two consecutive antennas. This is because adaptive beamforming systems require antennas to be highly correlated whereas MIMO systems require them to be low correlated. The antenna spacing can be adjusted through reconfigurable antenna systems in adaptive MIMO systems. Adaptive beamforming can be easily incorporated to SDR-MIMO structure by implementing beamforming algorithms in reconfigurable digital radio [26]. Hence, the SDR-MIMO structure in Figure 4.5.1 can be used for the adaptive MIMO systems as well.

4.5.2 Smart Antennas and Beamforming

Wireless networks are interference limited systems. Smart antennas or antenna beamforming reduces the interference level significantly by exploiting the spatial dimension. Compared to using omni-directional antennas, interference is reduced partially in current wireless networks by employing sector antennas. However, this type of interference is still significant and it makes overall system interference limited.

The basic idea behind the smart antenna technology is to form the antenna beam in such a way that the main lobe is directed to the desired user while the nulls of the beam are allocated to the interfering users. A sample smart antenna pattern is shown in Figure 4.10.

A typical smart antenna system consists of array of antennas that collectively generate the desired array pattern. This transmission and reception method using antenna arrays along with advanced signal processing in reconfigurable digital hardware is referred as *beamforming*. Functional block diagram of generalized SDR architecture (transmitter side) with smart antennas capability is shown in Figure 4.11. Each antenna system has its own software tunable analog radio transmit side (ST-AR Tx side) and ST-ADC (these components are discussed in Section 4) as shown in Figure 4.11. However, the rest of circuitry, which is beamforming algorithms that are implemented in reconfigurable digital radio platform, is shared by all antenna systems. In beamforming algorithms, each user's signal is multiplied with complex weights that adjust the magnitude and phase of the signal that goes to or comes from each antenna system. As a result, these algorithms generate transmit/receive beam at the output of antenna arrays in the direction of the

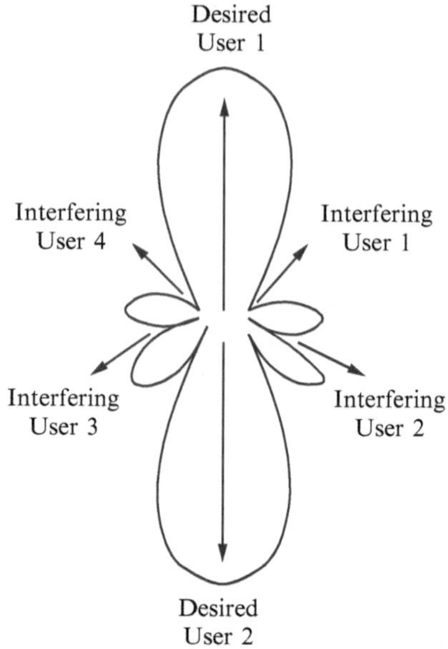

Fig. 4.10. Sample Smart Antenna Pattern.

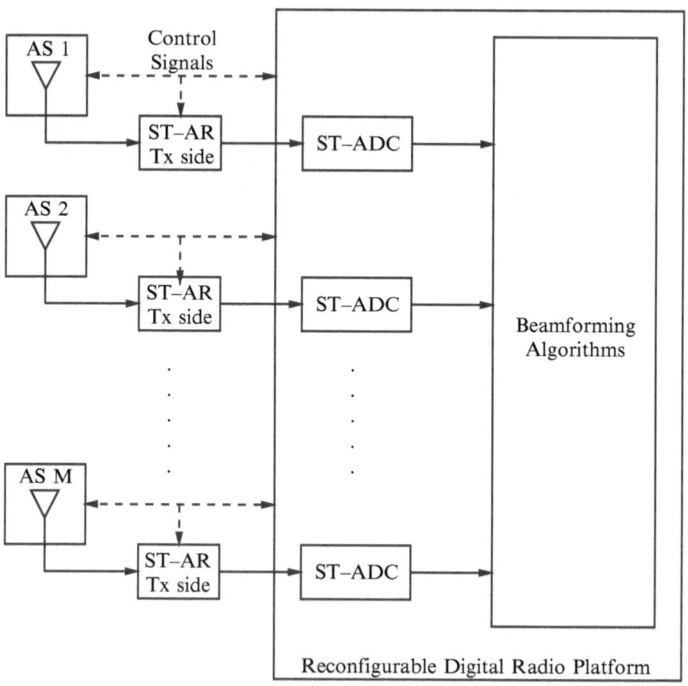

Fig. 4.11. Functional block diagram of generalized SDR architecture (transmitter side) with smart antenna capability (AS: antenna system, ST-AR Tx side: software tunable analog radio transmitter side).

desired user while minimize the interference to other users [30]. There are mainly two types of beamforming methods, which are switched and adaptive beamforming. In switched beamforming, the complex weights are selected from a library of coefficients that form beams in predefined directions. On the other hand, the complex weights are computed and adapted in real time in the case of adaptive beamforming [30].

In summary, we have discussed the advanced multiple antenna systems that require computation intensive algorithms. Such algorithms can be handled by SDR structure, which is a crucial part of cognitive radios. Cognitive radios can enhance channel capacity, diversity, robustness of wireless links and reduce the overall interference in the networks.

4.6 Reconfigurable Digital Radio Technologies

Primitive wireless systems were implemented using hardwired platform. The transition from hardwired to reconfigurable digital platform happens gradually as the digital technology evolves. This evolution starts with the introduction of Transistor–Transistor Logic (TTL) technology followed by general purpose processors (or Microprocessors) that is programmed through software. This trend continued with the development of microprocessor that is specialized in digital signal processing tasks, which is DSP. The need for high speed signal processing technology increased due to the low speed of DSP and adoption of computationally complex algorithms in the wireless systems. ASIC technology came to the scene to answer the demands on high speed signal processing. Then, inflexibility of ASIC technology motivated scientists to develop flexible and reconfigurable digital hardware technology. FPGAs technology came to the rescue, which provides both flexibility and reconfigurability. However, these features come with high power consumption and high cost. Hence, trend is moving towards development of reconfigurable, flexible, high performance, low power, small size, and low cost digital hardware platforms. In this line, there are some recent reconfigurable computing technologies such as PicoArray processors from PicoChip [31] and Adaptive Computing Machine (ACM) from QuickSilver technologies [32]. Therefore, these technologies can be categorized under two groups;

1. *Traditional Computing Technologies*: The technologies in this category is based on the idea of mapping the algorithm to a fixed set of hardware resources and/or requirements. The technologies that can be considered under this category are DSP, ASIC, and FPGAs.
2. *Adaptive Computing Technologies*: The basic idea behind this approach is to map the hardware resources and/or requirements to the algorithm requirements. Both hardware resources and algorithm requirements are dynamic. In other words, optimum amount of hardware resources are utilized to perform the given algorithm. General Purpose Processors (GPPs), PicoArray, and ACM [33] are three examples in this category.

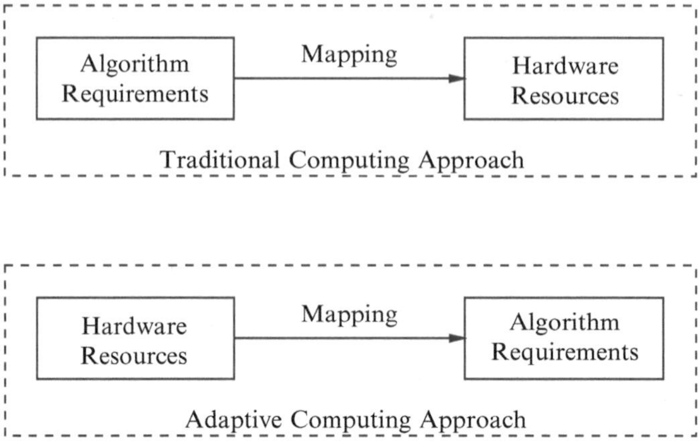

Fig. 4.12. Traditional and adaptive computing approaches.

These two approaches are illustrated in Figure 4.12. Similar to software tunable analog radio platform, ideal reconfigurable digital radio platform needs to have a capability to perform any arbitrary tasks that are given by cognitive engine. Three major areas of advance are required in order to provide dynamic processing capabilities essential to realize cognitive radio technologies. These areas are baseband component technologies, design tools and methods, and system maintenance. The challenges and concerns regarding each of these areas are summarized in [34]. In this section, we discuss the existing reconfigurable digital hardware technologies excluding the ASIC, picoArray, and ACM and provide their main architectures, pros and cons. The following are six main selection criteria that is recommended to be considered during the process of evaluating a reconfigurable digital platform for SDR applications:

- *Reconfigurability*: The ability to reconfigure a device to perform an arbitrary task that is given by cognitive engine.
- *Integration to Other Layers*: The level of complexity required to establish interface to the other layers.
- *Development Cycle*: The time that takes to develop, implement, debug, and verify a digital radio function. Ideally, cognitive engine needs to have a capability of performing the above development steps. However, it is recommended to provide bug-free verified digital radio functions to cognitive engine in reality. One way of achieving this is to store all the digital radio functions in a memory.
- *Performance*: The ability to perform a given arbitrary task within a specified time.
- *Power Consumption*: The amount of power that is dissipated for performing a given task within a specified time.
- *Cost*: The cost of reconfigurable digital radio platform is an important parameter.
- *Size*: The physical size of the processor is another important criteria.

4.6.1 Digital Signal Processors

DSP is a type of GPP that is specialized on signal processing applications. DSPs and GPPs have common functions, however, the former supports additional specialized functions such as Multiply–ACcumulate (MAC), barrel shifter, multiple memory blocks with supporting buses, and powerful Data Address Generators (DAGENs) [35]. DSPs have fixed processing architecture and they are able to execute different algorithms based on the sequential instructions that are stored in memory. However, their inherent fixed processing structure place restrictions on initial design.

Most of the current DSP processors employ modified Harvard architecture, which is shown in Figure 4.13. According to this architecture, it can be accessed to data and program memory within a single cycle using either two (program and data) or three (program and two data) buses. DSPs can process the signals either using fixed-point or floating-point arithmetic. However, most of the computations are performed using the latter format. The main advantage of floating-point over fixed-point arithmetic is the use of numbers with much larger dynamic range, which is important in many digital signal processing operations. The current DSPs have a capability of providing high speed clocks and parallel processing. However, both wireless service providers and manufacturers are beginning to acknowledge that the current DSPs cannot satisfy the requirements of SDR applications.

Although the design flow of DSP are well documented in the literature and supported by mature design tools, it cannot handle complex algorithms. For instance, multiple DSPs can be required in order to implement a complex algorithm. However, it is a tedious process to integrate multiple DSPs since the design tools for this purpose are less mature and more sophisticated. As a result, DSPs provide high reconfigurability with limited performance and

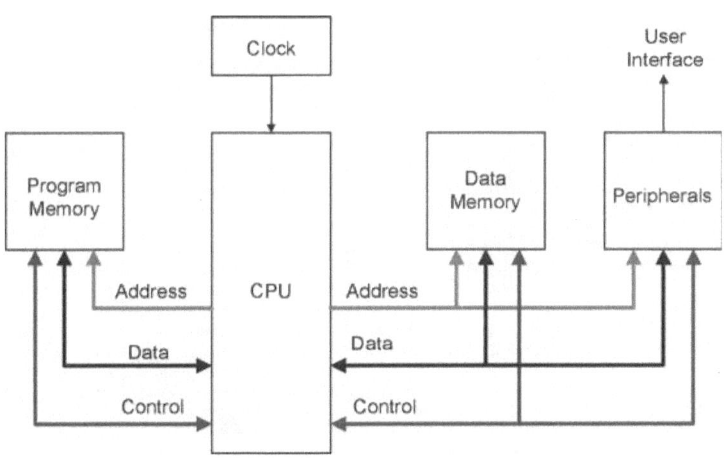

Fig. 4.13. Modified Harvard model based DSP architecture.

short development cycle. Moreover, they offer medium level of integration and consume low power. Additionally, DSPs are low cost and small size devices.

4.6.2 Field-Programmable Gate Arrays

In the beginning of 1960s, discrete logic were used to build systems consisted of many chips that are connected with wires. Modifying such systems required rebuilding the board, which took long time and it was costly. Chip manufacturers introduced Programmable Logic Device (PLD) that is a single chip and composed of an array of unconnected AND–OR gates. The PLDs contained an array of fuses that could be blown open or left closed to connect numerous inputs to each AND gate. Since PLDs could handle up to 20 logic equations designing complex systems using multiple PLDs was a challenging process. To tackle this problem, chip makers introduced Complex PLDs (CPLD) and FPGAs. A CPLD composed of bunch of PLD blocks whose inputs and outputs are governed by global interconnection matrix. CPLDs provide two levels of reconfigurability; reconfiguring the PLD blocks and interconnections between them. The structure of a CPLD is shown in Figure 4.14.

The structure of FPGAs is different than that of CPLDs. The FPGAs are composed of an array of simple and Configurable Logic Blocks (CLBs) and switches that are utilized to determine the connections between CLBs. A simplified structure of FPGAs is shown in Figure 4.15. In order to implement an algorithm in the FPGAs, each CLB is configured individually first and then switches are configured to connect or disconnect CLBs. Although there are different methods for connecting and disconnecting CLBs, the most widely used technique is based on use of RAM/flash switches. In this method, static RAM or flash bits are used to control the pass transistors for each interconnection. For instance, the switch can be closed or opened by loading bit 1 or 0, respectively. Since the current FPGAs can contain up to 10 million gates, manual control of switches is impossible. Therefore, FPGAs manufacturers provide development softwares that take logic design as input and then outputs a bitstream, which configures the switches.

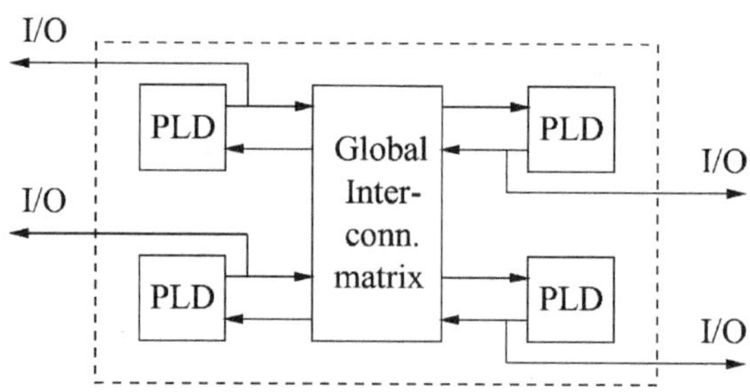

Fig. 4.14. CPLD structure.

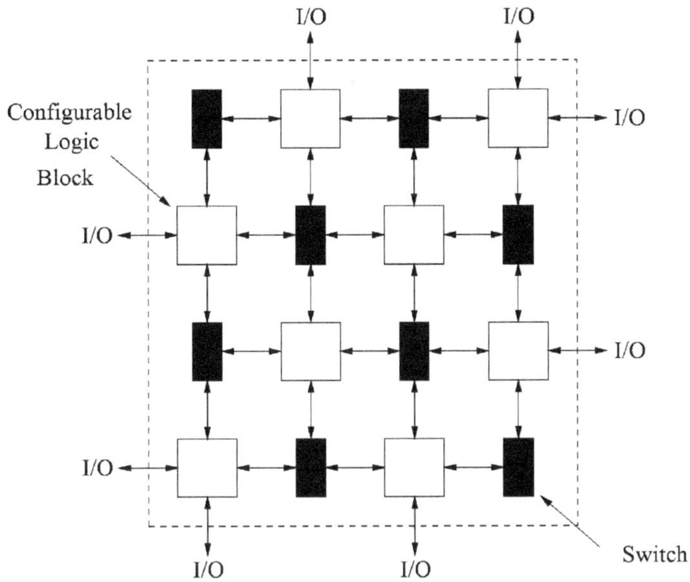

Fig. 4.15. FPGAs structure.

In the sequel, the main steps of development software for the implementation of a logic design in the FPGAs are provided. The description of logic design can be entered mainly in three ways: 1) Using high-level software such as MATLAB/Simulink. 2) Using Hardware Description Languages (HDL) such as VHDL and Verilog. 3) Using schematic editor. In the first approach, the system is designed by connecting functional blocks in high-level system design and then interface software is used to translate the design into VHDL codes. For instance, a complete OFDM transceiver can be designed in MATLAB/Simulink. In order to download this design into the FPGAs, Xilinx System Generator [36] can be used as an interface between MATLAB/Simulink and Xilinx ISE Foundation to perform automatic code generation. In the second method, the design is entered as a code using HDLs. On the other hand, the design can be drawn using a schematic editor. Regardless of the design entry method, the consequent steps are common. Once the design is entered, logic synthesizer tools transform the HDL code or schematic into a netlist. A netlist describes various logic gates and the interconnection between them. Consequently, implementation tools are used to map logic gates and interconnections into FPGAs. The mapping tool combines the netlist logic gates as a group and generates the Look-Up-Tables (LUTs). This is followed by place and route processes, which assign the collections of logic gates to the specific CLBs and determine the status of switches using routing matrices to connect the CLBs. Once the implementation process is completed, the program generates routing matrices that show the final status of switches. From routing matrices, a bitstream where ones and zeros correspond to opened or closed switch is generated. The bitstream is downloaded into physical FPGAs chip through parallel cable. The switches in FPGAs are opened or closed in response to the binary bits in the bitstream. Finally, FPGAs perform the operations specified by the design entry. The design can be verified in different ways. The classical way of verification is to insert a test signal and observe the output signal using test and measurement equipments. Another way, which is a modern way, is to generate test signals from high-level program, download the design into FPGAs, read back the results from the FPGAs, and plot the results in high-level software. For instance, MATLAB/Simulink along with Xilinx System Generator and ISE Foundation allow to verify the design using hardware co-sim, which is a second type method. Note that the aforementioned steps in the FPGAs design flow are performed by softwares automatically.

In conclusion, FPGAs have a capability of providing high degree of reconfigurability and high-level performance. Furthermore, they provide high level of integration and short development cycle. However, they are power-hungry, large size, and expensive devices.

4.6.3 General Purpose Processors

One of the reconfigurable digital hardware technologies is GPP [37–39]. Von Neumman and Harvard are two well-known GPP architectures in the literature. The main differences between these two architectures are the storage, utilization of instructions, and data. In Von Neumman architecture, instructions and data are stored into the same memory. On the other hand, instructions and data are stored into separate memory in the Harvard structures. Most of the current GPPs are based on Von Neumman architecture. However, Harvard architecture based GPPs are commonly available as well. The simplified block diagram of Von Neumman architecture is shown in Figure 4.16. The current GPPs do not have a capability to satisfy dynamic reconfiguration requirements. The main drawback of these devices is low data throughput due to all data being transferred via the processor bus. Moreover, they are power-hungry, costly and large size devices relative to the other processors.

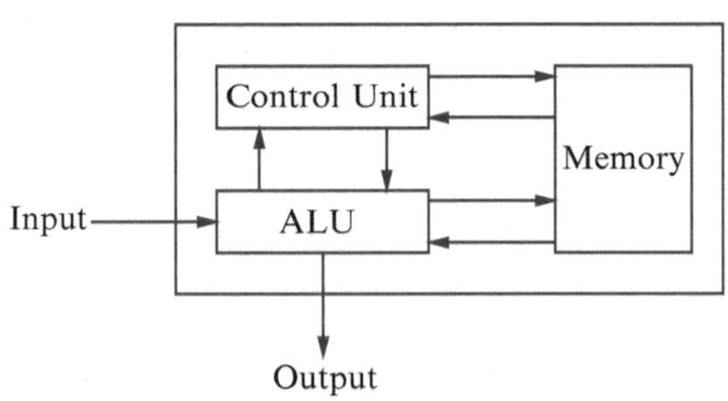

Fig. 4.16. Simplified block diagram of Von Neumman architecture.

However, they have potential advantages that cognitive engine benefits from them. Some of these advantages are [37]:

- *Experimentation*: Testing new algorithms or protocols is easy in these devices.
- *Rapid Deployment*: Cognitive radio users can easily add new devices or improvements to the existing structure through software.
- *Tight Integration with the Other Applications*: Since the upper layer applications and underlying communications systems are implemented in the same device, these upper and radio layers are tightly integrated.
- *Multi Purpose Devices*: Different devices such as fax, VoIP capabilities can be added to cognitive radios through software.
- *Improved Functionality*: These devices have the capability of allocating different channels to different communications standards.

The illustration of GPP based reconfigurable digital radio platform is provided in Figure 4.17. According to this structure, both upper layers and reconfigurable digital radio functions are implemented in GPP, which provides a tight integration between these upper layers and reconfigurable digital radio layer. In [37], such GPP based software radio architecture that is so called virtual radio is proposed. In this software radio structure, GPP is used as reconfigurable digital hardware platform and it is optimized to be suitable for software radio applications. Moreover, design tools and programming environment for this structures that is so called Spectra is developed. The Spectra is a programming tool that is used to support continuous real-time signal processing applications. This system design tool consists of three basics components [37]:

- A library for signal processing modules (e.g. convulotional encoder, Fast Fourier Transform (FFT)).
- A set of objects that is used to interconnect signal processing modules.
- A scripting language such as C++ to define the topology of system and interactions between the modules.

As a result, GPPs have the capability of providing high degree of reconfigurability with limited performance due to their inherent architectures. Their development cycle is very short and they offer high level of integration. Moreover, they are power-hungry, small size, and expensive devices.

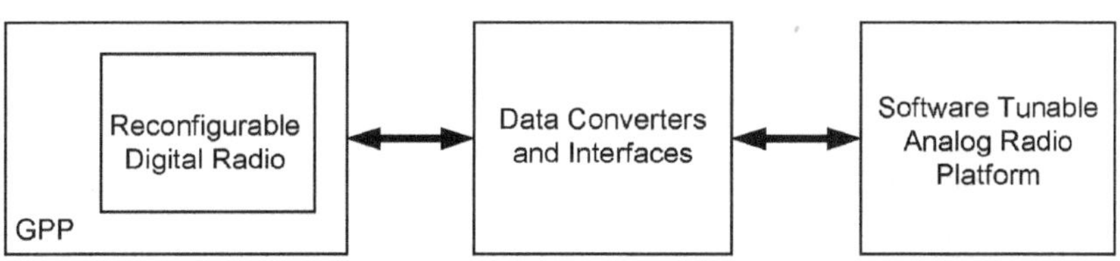

Fig. 4.17. Block diagram of a GPP based reconfigurable digital radio platform.

4.6.4 Heterogeneous Systems

It has been observed and reported that all reconfigurable technologies have some strengths and weaknesses. Another type of reconfigurable digital hardware platform is to integrate multiple different technologies to perform a certain task. It is obvious that the level of integration is low in this type of platforms since it introduces additional different type of processor. However, there are some performance gain due to partitioning the tasks between the processors. One of the well-known platforms for this approach is hybrid DSP/FPGAs platform. Many of the FPGA strengths are complementary to the weaknesses of DSPs while the strengths of the DSPs are complementary to the weaknesses of the FPGAs. A system incorporating both processors has the ability to leverage the strengths of both processors [40,41]. Such platforms are called hybrid DSP/FPGAs and the commercial hybrid DSP/FPGAs are available such as SignalWAVE from Lyrtech. This conclusion can be seen from Table 4.1, where the strengths and weaknesses of both DSPs and FPGAs are tabulated.

The system-level design flow of Lyrtech hybrid DSP/FPGAs platform is shown in Figure 4.18. As it can be seen from the figure that both processors are integrated to the MATLAB/Simulink where the design is entered. Then, each processor has its own chain of interface softwares that automatically generate the design codes. For the FPGAs, it can be either VHDL or Verilog code whereas it is C code for the DSPs. Consequently, processor-specific software converts the codes into configuration files. Then, they are downloaded into the corresponding physical chips. The following are some common design considerations for hybrid DSP/FPGAs platforms:

- Since task partitioning among the processors can affect the performance of hybrid DSP/FPGAs platform significantly, this process needs to be performed carefully.
- Interaction between two processors needs to be understood very well. Otherwise, it results in performance degradation.
- DSPs use "software design flow" whereas FPGAs use "hardware design flow." Hence, there is a little overlap between the skill set of the relevant design teams.
- Linking two processors is a complex task. These hybrid platforms are integrated at the board level (not at high level).
- High-level abstractions that are used for communication between the DSPs and FPGAs are practically nonexistent.

4.6.5 Comparison of Reconfigurable Digital Hardware Technologies

Each aforementioned technologies have their own performance criteria. For instance, Millions of Instructions Per Second (MIPS) is a common performance criteria that is used to determine the speed of DSPs. On the other

Table 4.1. Comparison of FPGAs and DSPs technologies.

	FPGAs	**DSPs**
Strengths	• Reconfigurable and flexible processing architecture • Small silicon geometry • Relatively high bandwidth • High speed clock • High performance • Low core voltage • Parallel internal architecture facilitates multiple simultaneous operations • Embedded functional blocks such as RAM and multipliers allow construction of many signal processing operations such as Finite Impulse Response (FIR) filters • User defined data word sizes allow simpler interfacing and processing of any signal data type • Many Intellectual Property (IP) cores available	• High function capability • Adaptable to a wide range of applications • Contains embedded specialized logic (e.g. mathematical units) • Efficient program branching for decision making • Low power • Relatively cheap • Relatively short development cycle • Relatively cheap debugging or maintenance
Weaknesses	• High power consumption • Lack of knowledgeable person • Less efficient branching consumes large numbers of gates for decision making • Construction of precision mathematical operations logic is difficult and consumes considerable resources • Large devices are expensive • Relatively long development cycle • Relatively costly debugging and maintenance	• Fixed processing architecture • Limited performance due to serial processing nature of design • DSPs are not useful for multitask • DSPs require high memory bandwidth • Limited parallelism that implies a low number of simultaneous operations per clock cycle

hand, the number of gates or slices that are used to perform an operation along with the clock speed is one of the common performance criteria for the FPGAs. However, the speed of processors is often specified in terms of

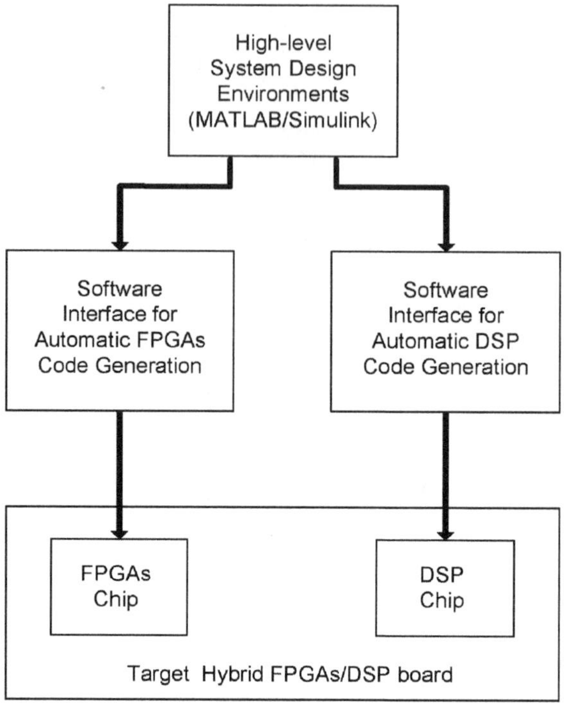

Fig. 4.18. Hybrid DSP/FPGAs architecture.

Millions of Floating-Point Operations per Second (MFLOPS). The theoretical peak MFLOPS parameter gives maximum possible speed for the processors. This parameter allows comparison of the processing speed of different processors and allows determination of the time required to perform certain algorithms.

Many different benchmarks such as SPEC, Whetstone, Dhyrstone, and Linpack are used to compare the speed of different processors. Each benchmark provides a number indicating the relative speed of processing based on testing various tasks. Since there is not any comprehensive study that compare the performance of all the technologies discussed in this section against to a benchmark, we compared these technologies based on the six criteria that is mentioned previously and the results are tabulated in Table 4.2.

4.7 Basic Digital Radio Components

In the previous section, enabling technologies for digital radio system are discussed. In this section, some of the commonly used digital radio components (functions) are presented briefly. Even though different standards and technologies require different digital radio blocks, in this section, we will only

Table 4.2. Comparison of reconfigurable digital radio technologies

Technology	Hardware	Algorithms	Comments
ASIC	Fixed	Fixed	High performance, no reconfigurability, low power, low level of integration, long development cycle, high cost, small size
DSP	Fixed	Pseudo-dynamic	Low performance, medium reconfigurability, low power, medium level of integration, short development cycle, low cost, small size
FPGAs	Pseudo-dynamic	Pseudo-dynamic	High performance, high reconfigurability, high power, medium level of integration, short development cycle, high cost large size
Hybrid DSP/FPGAs	Pseudo-dynamic	Pseudo-dynamic	High performance, high reconfigurability, medium power, low level of integration, short development cycle, medium cost, large size
GPPs	Fixed	Dynamic	High performance, low reconfigurability, high power, high level of integration, short development cycle, high cost, small size

focus on the commonly used blocks. A basic digital radio includes the following components: source encoder/decoder, channel encoder/decoder, symbol mapping/demapping, pulse shaping, digital upconversion/downconversion, synchronization, channel estimation, channel equalization, detector, soft information generator. A simplified functional block diagram of digital radio systems that contains some of these components is illustrated in Figure 4.7. There are also technology and standard specific components like FFT, Inverse Fast Fourier Transform (IFFT), randomization/derandomization, rake reception/combining, beamforming, antenna combining, MIMO, interference cancelation, various channel quality and channel parameter estimation blocks, encryption/decryption, multiplexing/demultiplexing (time, frequency, code, space division multiplexing), frequency spreading/despreading (direct-sequence, time-hopping), peak-to-average-power-ratio reduction, and so on.

Source encoding is the process of efficiently converting the information generated by source (either analog or digital signal) into a sequence of binary digits [42]. The main objective in source coding is to represent the information

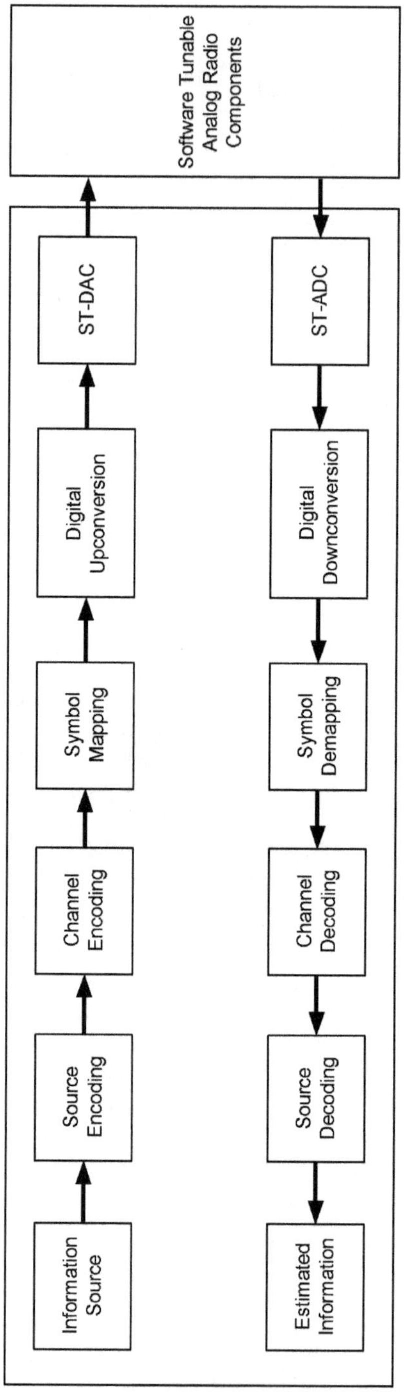

Fig. 4.19. Basic components of a digital radio system.

from source by as less binary digits as possible without compromising reconstruction fidelity. Source encoding consists of two main subprocess, which are *formatting* and *data compression* [43]. In formatting step, analog information from source is converted into the digital signal. It is obvious that if information is in digital format, there is no need for the formatting. The following step is data compression without sacrificing reconstruction fidelity. Data compression is achieved by removing the redundancy from the information. Removing redundancy in transmitted signal improves the efficiency of bandwidth utilization, which is the main advantage of using source encoding in digital communications systems. Some of the well-known source coding techniques are Pulse Code Modulation (PCM), differential PCM, delta modulation, Linear Predictive Coding (LPC), and Huffman coding. For speech (which has been the primary type of source data in cellular radio systems) vocoders are popularly used. Even though vocoders are more complex than the waveform coders, they require much lower data rates (and hence spectrally more efficient) with relatively small degradation in speech quality.

While source encoding removes the undesired redundancy, *channel encoding* introduces intelligent redundancy to the transmitted information. The introduced redundancy by the channel encoder is very important at the receiver to correct bit errors and detect the frame errors. Especially, channel encoding is very important for wireless communications systems (where channel fading and interference limit the performance) to improve the overall performance of the system and provide a desired quality of service. Channel encoding can be classified as "error detection coding" and "error correction coding." As in source encoding, there are various types of channel codes like block codes (Hamming, Cyclic Redundancy Check (CRC), Reed-solomon, etc.), convolutional codes, and turbo codes. Puncturing and depuncturing are also used along with the convolutional codes (and turbo codes) to obtain the desired code rates from the mother codes.

In wireless communications systems, Forward Error Correction (FEC) coding is often employed with interleaving to optimize the FEC performance. Interleaving is a mean of adding time diversity to a signal without adding any redundant bits. The main function of interleaving is to jumble the bits before the transmission so that the errors will be diversified at the receiver rather than having bursty errors. This way the channel decoding corrects small number of bit errors.

After the channel encoding, the binary bits are mapped to symbols using *symbol mapping* block. Although there are numerous symbol mapping schemes in the literature, the most popularly used symbol mapping techniques are Phase Shift Keying (PSK), Amplitude Shift Keying (ASK), and Frequency Shift Keying (FSK), and Quadrature Amplitude Modulation (QAM). Pulse modulation methods such as Pulse Amplitude Modulation (PAM), Pulse Position Modulation (PPM), Pulse Shape Modulation (PSM), Pulse Interval Modulation (PIM) have gained interests with the impulse radio based ultra wideband-technology. The choice of modulation is a design parameter, as there are various trades offs in choosing the appropriate modulation. One of the popular trade-offs is power versus spectral efficiency. For example low order modulations (like BPSK) are power efficient compared to higher order modulations (like 64-QAM) which are spectrally efficient. The recent wireless standards are moving in the direction of adaptive modulation (having the capability of using multiple modulations and choosing the most appropriate modulation depending on the link quality) to optimize both spectrum and power efficiency. We refer to [44] for detailed discussion on different digital modulation techniques.

As discussed earlier, the current practical SDR platforms handle the IF part of the transceiver blocks in the digital processors. The main objective of digital upconversion is to generate digital IF signal, which provides more flexibility compared to analog upconversion. Typical digital upconversion consists of digital filtering, Numerically Controlled Oscillator (NCO), and digital mixer. The main tasks of digital filtering are pulse shaping and interpolation. Cascade Integrator Comb (CIC) and programmable FIR filters are two most practical filter types that are used to perform digital filtering. NCO is used to synthesize arbitrary local oscillator frequency and digital mixer is employed to upconvert digital baseband samples to digital IF stage. Similarly, the reverse operation of digital upconversion, which is digital downconversion, is performed at receiver side.

4.8 Conclusions

In this chapter, we emphasize on SDR structures to realize cognitive radio technology. The role of SDR in cognitive radio technology is discussed along with some examples. SDR requirements of cognitive radio are identified and multiple antenna systems in the context of SDR and cognitive radio are discussed. Current SDR enabling technologies are evaluated. It can be concluded that SDR is an essential part of cognitive radio and current technologies have the capability of proofing the concept of cognitive radio technology.

References

1. B. Cetiner, H. Jafarkhani, J. Qian, H. Yoo, A. Grau, and F. De Flaviis, "Multifunctional reconfigurable mems integrated antennas for adaptive mimo systems," *IEEE Commun. Mag.*, vol. 42, no. 12, pp. 62–70, Dec. 2004.
2. S.-H. Oh, J. T. Aberle, S. Anantharaman, K. Arai, H. L. Chong, and S. C. Koay, "Electronically tunable antenna pair and novel rf front-end architecture for software-defined radios," *EURASIP J. Appl. Signal Process.*, no. 16, pp. 2701–2707, 2005.
3. F.E. Retnasothie, M. K. Ozdemir, T. Yucek, J. Zhang, H. Celebi, R. Muththaiah, "Wireless IPTV over WiMAX: Challenges and Applications," in *Proc. IEEE*

Wireless and Microwave Technology Conference, Clearwater Beach, FL, USA. Dec. 2006, to appear.

4. B. Bisla, R. Eline, and L. M. Franca-Neto, "RF System and Circuit Challenges for WiMAX," Intel, Tech. Rep., 2004.
5. R. Lackey and D. Upmal, "Speakeasy: the military software radio," *IEEE Commun. Mag.*, vol. 33, no. 5, pp. 56–61, May 1995.
6. J. Melby, "JTRS and the Evolution toward Software-defined Radio," in *Proc. IEEE International Symposium on Circuits and Systems*, vol. 2, Anaheim, CA, USA, Oct. 2002, pp. 1286–1290.
7. W. Bosner, "SPEAKeasy Military Software Defined Radio," in *Proc. International Symposium on Advanced Radio Technologies*, 1998.
8. "JTRS." [Online]. Available: http://jtrs.army.mil/
9. "Joint Tactical Radio System (JTRS) Program Review," 2003 [Online]. Available: http://spacecom.grc.nasa.gov/
10. "Going Forward with JTRS," 2005 [Online]. Available: http://www.military-information-technology.com/
11. "AdaptaCell," 2006 [Online]. Available: http://www.airnetcom.com/
12. M. Rahman and K. Shamsaifar, "Electronically Tunable LTCC based Multilayer Filter for Mobile Handset Application," in *Proc. IEEE MTT-S Microwave Symposium Digest*, vol. 3, Philadelphia, Pennsylvania, USA, June 2003, pp. 1767– 1770.
13. "Parascan Tunable Materials," 2006 [Online]. Available: http://www.paratek.com
14. H. Ishida and K. Araki, "A Design of Tunable UWB filters," in *Proc. IEEE International Workshop on Ultra Wideband Systems Joint with Conference on Ultra Wideband Systems and Technologies*, Kyoto, Japan, May 2004, pp. 424–428.
15. J. Li and M. Kavehrad, "OFDM–CDMA Systems with Nonlinear Power Amplifier," in *Proc. IEEE Wireless Communication and Networking Conference*, vol. 3, New Orleans, LA, USA, Sept. 1999, pp. 1167–1171.
16. J. Park, S. R. Park, H. J. Roh, and K. H. Koo, "Power Amplifier Back-off Analysis with AM-to-PM for Millimeter-wave OFDM Wireless LAN," in *Proc. IEEE Radio and Wireless Conference*, Waltham, MA, USA, Aug. 2001, pp. 189–192.
17. R. Koch, "Linearization: Reducing distortion in power amplifiers," *IEEE Microw.*, vol. 2, pp. 37–49, Dec. 2001.
18. S. Chang and E. J. Powers, "A Simplified Predistorter for Compensation of Nonlinear Distortion in OFDM Systems," in *Proc. IEEE Global Telecommunications Conf.*, vol. 5, San Antonio, TX, USA, Aug. 2001, pp. 3080–3084.
19. J. Y. Hassani and M. Kamarei, "A Flexible Method of LUT Indexing in Digital Predistortion Linearization of RF Power Amplifiers," in *Proc. IEEE International Symposium on Circuits and Systems*, vol. 1, Sydney, NSW, Australia, May 2001, pp. 53–56.
20. J. de Mingo, A. Valdovinos, A. Crespo, D. Navarro, and P. Garcia, "An RF electronically controlled impedance tuning network design and its application to an antenna input impedance automatic matching system," *IEEE Trans. Microwave Theory Tech.*, vol. 52, no. 2, pp. 489–497, Feb. 2004.
21. Y. Sun and J. Fidler, "Design method for impedance matching networks," *IEE Proc. Circuits, Devices Syst.*, vol. 143, no. 4, pp. 186–194, Aug. 1996.

22. "12-BIT, Quad, Ultralow Glitch, Voltage Output Digital-To-Analog Converter," 2006 [Online]. Available: http://focus.ti.com/
23. D. Comer, "A Monolithic 12-bit DAC," *IEEE Trans. Circuits Syst.*, vol. 25, no. 7, pp. 504–509, Jul. 1978.
24. S. Catreux, V. Erceg, D. Gesbert, and R. Heath Jr, "Adaptive modulation and MIMO coding for broadband wireless datanetworks," *IEEE Commun. Mag.*, vol. 40, no. 6, pp. 108–115, Jun. 2002.
25. A. Paulraj, D. Gore, R. Nabar, and H. Bolcskei, "An overview of MIMO communications: A key to gigabit wireless," *Proc. IEEE*, vol. 92, no. 2, pp. 198–218, Feb. 2004.
26. P. Xia, S. Zhou, and G. B. Giannakis, "Adaptive MIMO–OFDM based on partial channel state information," *IEEE Trans. Signal Process.*, vol. 52, no. 1, pp. 202–213, Jan. 2004.
27. J. Razavilar, F. Rashid-Farrokhi, and K. Liu, "Software radio architecture with smart antennas: A tutorial on algorithms and complexity," *IEEE J. Select. Areas Commun.*, vol. 17, no. 4, pp. 662–676, Apr. 1999.
28. A. Perez-Neira, X. Mestre, and J. Fonollosa, "Smart antennas in software radio base stations," *IEEE Commun. Mag.*, vol. 39, no. 2, pp. 166–173, Feb. 2001.
29. A. Gupta, A. Forenza, and R. Heath Jr, "Rapid MIMO–OFDM software defined radio system prototyping," in *IEEE Workshop on Signal Processing Systems*, Oct. 2004, pp. 182–187.
30. "Smart Antennas-Beamforming," 2006 [Online]. Available: http://www.altera.com
31. "PicoArray Processors," 2006 [Online]. Available: http://www.picochip.com
32. "Adaptive Computing Machine," 2006 [Online]. Available: http://www.qstech.com
33. P. Master, "The Next Big Leap in Reconfigurable Systems," in *Proc. IEEE International Conference on Field-Programmable Technology*, Hong Kong, Dec. 2002, pp. 17–22.
34. W. Tuttlebee, *Software Defined Radio: Enabling Technologies.* Wiley, 2000.
35. S. Kuo and W. Gan, *Digital signal processors: architectures, implementations, and applications.* Pearson/Prentice Hall, 2005.
36. "Xilinx System Generator for DSP," 2006 [Online]. Available: http://www.xilinx.com
37. V. Bose, M. Ismert, M. Welborn, and J. Guttag, "Virtual radios," *IEEE J. Select. Areas Commun.*, vol. 17, no. 4, pp. 591–602, Apr. 1999.
38. V. Bose, "Design and implementation of software radios using a general purpose processor," Ph.D. dissertation, MIT, Cambridge, MA,USA, June 1999.
39. P. Mackenzie, L. Doyle, D. O'Mahony, and K. Nolan, "Software Radio on general-purpose processors," in *Proc. of the First Joint IEI/IEE Symposium on Telecommunications Systems Research*, Dublin, Ireland, Nov. 2001.
40. "DSP/FPGA Development Platforms," 2006 [Online]. Available: http://www.lyrtech.com
41. "Software Defined Radio-back ground note," 2003 [Online]. Available: http://www.transtech-dsp.com
42. J. G. Proakis, *Digital Communications.* McGraw-Hill, 1995.
43. B. Sklar, "A structured overview of digital communications-review-part I," *IEEE Commun. Mag.*, vol. 21, no. 5, pp. 4–17, Aug 1983.
44. F. Xiong, *Digital Modulation Techniques.* Artech House Publishers, 2006.

5

Value Creation and Migration in Adaptive and Cognitive Radio Systems

Keith E. Nolan[1], Francis J. Mullany[2], Eamonn Ambrose[3], and Linda E. Doyle[4]

[1] Centre for Telecommunications Value-Chain Research, University of Dublin, Trinity College, Dublin 2, Rep. of Ireland, keithnolan@mee.tcd.ie, +353-1-8968443
[2] Alcatel-Lucent, Bell Labs Ireland, Blanchardstown Industrial Park, Blanchardstown, Dublin 15, Rep. of Ireland, mullany@alcatel-lucent.com, +353-1-6924245
[3] CTVR, National Institute of Technology Management, University College Dublin, Dublin, Rep. of Ireland, eamonn.ambrose@ucd.ie, +353-1-7161741
[4] Centre for Telecommunications Value-Chain Research, University of Dublin, Trinity College, Dublin 2, Rep. of Ireland, ledoyle@tcd.ie, +353-1-8962567

5.1 Introduction

In this chapter, the concept of a telecommunications value-chain is developed, leading to an exploration of the many ways in which the value-chain can be altered by reconfigurable software-defined radios, cognitive radios, and cognitive networks.

Innovative and emerging wireless communications applications involving the use of adaptive and cognitive radio technology ideally attempt to maximize and capture the value to the users while reducing the manufacturing, deployment, and upgrade costs incurred by the manufacturer and service provider(s).

The telecommunications value-chain creates consumer value through the provision of communication services over fixed and wireless networks. Porter [1] explains how value-chain analysis "divides a firm into the discrete activities it performs in designing, producing, marketing, and distributing its product". Within the telecommunications value-chain, the design and development of software-defined and cognitive radios has the potential to significantly effect consumer value. The purpose of this chapter is to firstly develop the concept of a telecommunications value-chain and secondly, explore the many ways in which the value-chain can be altered by adaptive and reconfigurable Software-Defined Radios (SDR), Cognitive Radios (CR), and cognitive networks.

H. Arslan (Ed.), Cognitive Radio, Software Defined Radio, and Adaptive Wireless Systems, 145–159.

In order to put this chapter into the correct context, Section 5.2 is a brief recap of the terms cognitive radio and cognitive networks from the authors' collective viewpoint. Section 5.3 introduces the concept of a value-chain. The topics of value-creation and value-migration are explained in Section 5.4. Section 5.5 develops an economic value model considering the net present value of a network of nodes with cognitive functionality. Section 5.6 outlines four examples and case-studies that aim to help the reader strengthen their understanding of the value-chain concepts presented in this chapter. Section 5.7 concludes. For a deeper understanding of the the concepts presented in this chapter, the reader is directed to relevant references, which are provided in the bibliography.

5.2 Cognitive Radio and Networks

For the purposes of this chapter, it is useful to define what the authors refer to as a cognitive radio and a cognitive network. Cognitive radio can be described as a node in a network with an ability to form an awareness of its environment and context, make decisions and inferences from this information combined with knowledge of the user's objectives, act in a manner that attempts to accomplish the user's objectives, and optionally learn from these experiences for possible use in the future [2, 3]. The foundation of a cognitive radio is essentially a wireless communications stack capable of being dynamically reconfigured. This may be implemented using a software-defined radio, which is a wireless device where some or all of the physical layer (PHY) and the rest of the communications stack is implemented in software, or can be configured using a software mechanism. Cognitive functionality may have an influence on all or many of the layers in a communications stack and is not just limited to the PHY only, however.

A cognitive network is a network of nodes with cognitive functionality [4,5]. These nodes may have the potential to form an awareness of each other and even combine their resources and complementary expertise in order to perform as a collaboration and even in unison as a team [6].

5.3 The Value-Chain

The term *value-chain* is used to describe the interconnected stream of organizations, activities, and capabilities that combine to generate value for customers through the production of goods and services [7]. It is a refinement of the supply chain concept, where the focus is on value generating elements as opposed to procurement and logistics. Porter identified how the activities of organizations in the chain, and in particular, interactions between organizations, could be the source of sustainable competitive advantage through the ongoing creation of customer value and through continuous reduction in the cost of creating that value. Much of the effort of supply chain or value-chain management is devoted to identifying or predicting sources of customer value, and in maximizing delivery of that value at minimal cost in the chain.

Value is typically measured in terms of revenue for the organizations within the chain, and in terms of utility for the customers. Utility is the economic term for the satisfaction derived by customers in the consumption of goods and services. In value-chains, any party that gains utility from the chain activities (whether by consumption or other means) should be considered a customer. Organizations contribute to the value of the goods and services through their activities, and generate profits by minimizing their costs and maximizing revenue, i.e. maximizing the value provided to the next organizational downstream in the chain. For the customers, the goods and services are consumed, generating utility for the customers in excess of the cost. In cellular wireless markets for example, the value of advanced high-speed video download services may not be evident to the end user until after the handset is paid for. The cost of the advanced handsets may be subsidised by the service provider or even the content provider in order to encourage the uptake of these services. With push advertising on a cellular phone being used to pay, in whole or in part, for a service consumed by the handset owner, both the advertising company and the ultimate end user are customers.

Figure 5.1 is an example of one value-chain that can be associated with a cognitive radio system. The value generating elements in this chain include cognitive functionality, dynamic spectrum access, ease of communication, and abilities to develop and deploy new applications and services with relative ease compared to a traditional fixed-architecture system model. The cognitive functionality element of this chain offers value by combining awareness, decision-making and learning capabilities with the ability to rapidly implement change in both reactive and preemptive modes of operation based on current and historical knowledge of the user's habits and anticipated communications needs. This allows the burden of modification and optimization of the system performance to be moved away from the user to the device itself.

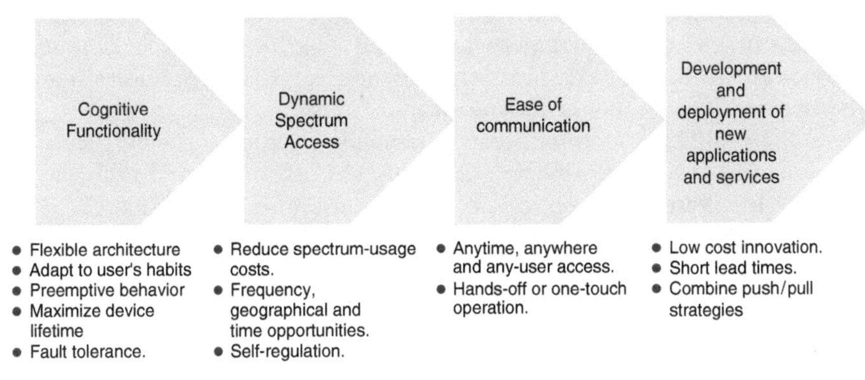

Fig. 5.1. Example of one value-chain associated with a cognitive radio system.

Cognitive functionality also provides a way to work around and fix faults that may develop in the device. In addition, this functionality can help maximize the operating lifetime of the device itself by attempting to reach a compromise between the available node resources (e.g. remaining energy, available spectrum, RF hardware) and communications demands placed on the device by the user. The value that can be derived from a more dynamic approach to spectrum access as featured in the second element of Figure 5.1, is that the probability of successful wireless communication can increase dramatically. Dynamic spectrum access techniques combined with cognitive functionality offer a means of exploiting unused or under-utilized *whitespace* spectrum segments while attempting to ensure that incumbent and non-cooperative users experience minimal interference. This ability enables the cognitive node to take advantage of the geographical and time-of-day variations in spectrum usage patterns for its own needs. From the user's perspective, the ability to communicate becomes a feature that is taken for granted as the underlying cognitive functionality and dynamic spectrum access technologies can automatically manage the spectrum access requirements. The third element of this value-chain is therefore the ease of communication that users can experience regardless of the geographical location, movement patterns, time of day, and user's technical knowledge (or lack thereof). The fourth and final element in this simplified value-chain is the value that can be derived by combining all of these features to enable both the service providers and users to quickly develop and deploy innovative revenue-generating services and applications. These new services can be both pushed onto the market by the service provider or user, and optimized quickly based on the subscriber's interest in these services in order to maximize the generated revenue.

In addition to the consumers, other non-consuming customers such as a telecommunications regulator may get value from the chain. The regulator gains utility from the creation of a telecommunications regime that meets the defined regulatory standards, which may be considered a measure of societal value. For example, frequency spectrum regulators may reserve spectrum for emergency services because of the value of such services to society. Communications regulators including the Commission for Communications Regulation (ComReg) in Ireland, also impose terms and conditions on telecommunication network operators as regards the percentage of the population covered by basic services such as mobile voice connectivity.

It can be difficult to quantify non-monetary value to society. However, when discussing value creation or migration, the involvement of multiple customers with different perceptions of value must also be considered. These different perceptions of value may also be dependent on the priority of that service according to the user. An example of this challenge is attempting to quantify the value created in public safety scenarios where through the provision of telecommunications services, the potential to more efficiently locate and co-ordinate the rescue of survivors in a disaster area can be increased. In this scenario, the potential of telecommunications is of higher societal value than the ability to establish a high quality video link. It can be difficult to quantify such societal values in economic terms, which poses a challenge for value-chain models.

The value-chain perspective can be summarized as follows:

- There are multiple customers in the chain for whom value can be created.
- Value is ultimately defined by the utility gained by consumers and other customers. Value is created when the utility of the ultimate customer(s) is increased.
- Within the chain, profits may be increased either by reducing cost or by increasing revenue gained from another organizational, even though ultimate customer value is not improved. In this way, value can migrate along the value-chain.

5.4 Value Creation and Migration

Value is not a static entity. The total value in a chain is the sum of the utility of the various customers, each having a unique perspective which may be constantly evolving. The challenge for organizations in the chain is to identify opportunities for value creation, and to ensure that they capture that value through increased profits. However, this newly created value is not easily captured – Sylwotzky [8] and Ng et al. [9] show how the dynamics of value migration within the value-chain are complex and difficult to manage. Fine [10] identifies the concept of clockspeed – the rate of change within an industry as a key driver of the dynamics of value creation and migration. Clockspeed can be measured in terms of product life cycle but it is not necessarily a fixed parameter within any value-chain. For network operators, product life is of the order of years while at the downstream end of the chain for the wireless communications application developer it might be weeks. Hence value will potentially migrate more quickly between value-chain members downstream.

The strength of a cognitive radio system is the flexibility and adaptability that it offers. This flexibility can be exploited to provide attractive features for both a service provider and user. One of these appealing features of a cognitive radio-based system, therefore, is that it has the potential for value creation and migration. A cognitive system that can adapt to the local environment, scenarios, and business models has a high value to both the service provider and user. Furthermore, this value can actually migrate either intentionally or unintentionally.

Any activities which generate increased value to any of the ultimate customers of the system are seen as value creating activities. This may be activities internal to a firm, or between existing firms. It may be the introduction of new technologies, or the deployment of existing technology in an innovative manner. Value creation, while desirable, is not always deliberate. In the European market predominately, the vastly popular mobile phone Short Message Service (SMS) in Global System for Mobile Communications (GSM) is one example of this. This facility was provided as a minor addition by phone manufacturers which has become highly valued by customers – a mechanism known as *market pull innovation*.

On the other hand, innovation can occur through *technology push*, where new technologies are offered to the market in order to stimulate demand for the associated value proposition. As an example, the availability of multimedia messaging is not generating the hoped-for fast sales growth as customers do not as of yet, assign significant value to the feature. It is the customer perception of value that is key, and no value creation takes place until the customer adopts the technology. From a service provider or innovative user viewpoint, the value of a cognitive radio and network is that new and innovative services can be offered to the public relatively quickly and on a trial basis in order to probe the market's interest in that feature. In order to initiate and sustain this activity, flexibility, an open development platform, and rapid low-cost prototyping, are necessary. The value-chain illustrated in Figure 5.1 and introduced in Section 5.3, points out that potential value from both *technology push* and *market pull* mechanisms can be leveraged in a cognitive radio system as it is designed to both handle, and facilitate change.

Moving towards a more forward-looking cognitive radio network scenario, the ability to collaborate and combine resources with other cognitive radio devices may not be of high value to a user wishing to only transmit a file to a colleague. However, the ability to exploit the potential collaboration and resource-combination features of a cognitive network in order to establish an emergency communications network in the event of a natural disaster may be of very high value to those directly affected by this disaster.

Value migration is therefore critical to firms in the value system, who wish to capture maximum value at minimum cost. The migration of value can be controlled, but not fully – it is often beyond the power of a single firm to influence. Firms must therefore understand value in the context of the complete supply chain. Two questions are key to a firm's strategic analysis of value: "When value is created who captures this and how can this be achieved?" and "when value changes (a change of target market and customer profile, for example) how to hold onto value?".

5.5 Economic Value Model

The associated microeconomics can be represented in terms of a value-chain, i.e. the sequence of suppliers and customers that are required to deliver a product or service to the end user of (in this case) communications. The value-chain can stretch from the component supplier, to the system integrator, to the service provider, and on to the end user.

Figure 5.2 is a graphical representation of a three-stage supply chain. In the context of this chapter, *A* could be network operators and *B*, Mobile Virtual Network Operators (MVNO). The end users are denoted by *C*. Equally,

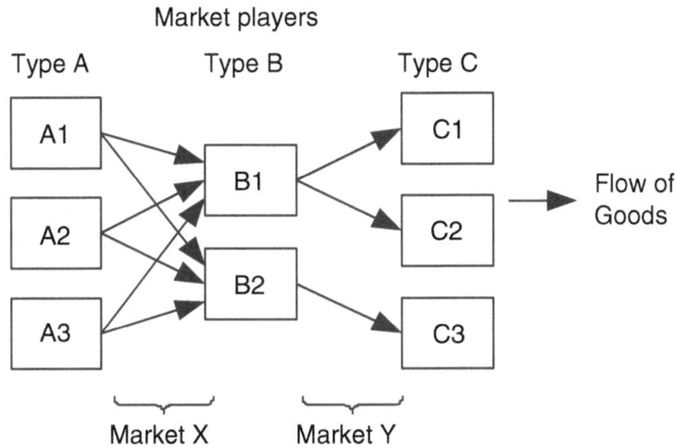

Fig. 5.2. Illustration of a three-stage supply chain.

A could be the primary spectrum users and licensees, and B wireless access operators exploiting spectrum opportunities. In this scenario, the end users, C could also act in an opportunistic manner, favoring one operator or unlicensed and free spectrum access scheme over another depending on its information conveyance needs and context. Many combinations are possible, but a viable value-chain requires that all players in the chain are achieving a surplus in their respective transactions.

Figure 5.2 is only a representation of the inter-relationships involved. A complete representation of the value-chain would include service/product capacities, market structure, measures of customer value, and market player strategies.

Regarding the latter, companies may build up business models to predict their performance. In this section, we do so for a network operator, while noting in its most general sense, a "network operator" could be as simple as a homeowner with a Wireless Local Area Network (WLAN) access point allowing access to users in the vicinity.

One typical measure of the potential success of a future project is that of net present value, i.e. the profitability of a project in today's unit of currency. Assuming a discount rate of d for future expenses and revenues and a project life of N_T years, then the Net Present Value (NPV) is given by Equation 5.1, where N_U is the number of subscribed users and $r_U(n)$ is the Average Revenue Per User (ARPU) in year n. The total CAPital EXpenditure (CAPEX) for network infrastructure is c_I, and c_S is the price paid for the spectrum used. The average handset subsidy per user is denoted by c_U and the OPerational EXpenditure (OPEX) for the network in year n is denoted by $o_I(n)$.

The NPV can therefore be described as

$$\mathrm{NPV} = N_U \sum_{n=1}^{N_T} \frac{r_U(n)}{(1+d)^n} - c_I - c_S - N_U c_U - \sum_{n=1}^{N_T} \frac{o_I(n)}{(1+d)^n} \qquad (5.1)$$

While all of the above parameters depend on the interaction in the marketplace with other network operators, suppliers and customers, the parameters N_U and $r_U(n)$ are explicitly related by the price demand curve. This is in turn, heavily dependent on the market conditions.

Other metrics exist for measuring the value of a project. One commonly used is that of Return On Investment (ROI). The ROI is the compounded annual percentage interest rate that would have yielded the same profit on the invested capital.

In order to help the reader to gain a better understanding of how the costs, number of subscribers and revenue are interrelated, consider the following simplifications and (fictitious) sample values. Assume that the number of subscribers, $r_U(n)$ and OPEX for year n, $o_I(n)$ are constant over time. Furthermore, assume that this OPEX, $o_I(n) = o_I$ can be broken down into network OPEX (usually expressed on a per annum basis as some fraction k of overall CAPEX), and per-subscriber OPEX, o_U (billing costs, customer support costs, etc.). Figure 5.1 can be simplified as follows; let

$$\hat{N}_T = \sum_{n=1}^{N_T} \frac{1}{(1+d)^n} \tag{5.2}$$

Hence the NPV, as originally stated in Equation 5.1, can now be described as

$$\text{NPV} = N_U \hat{N}_T \left(r_U - o_U - \frac{c_U}{\hat{N}_T} \right) - \left\{ c_I \left(1 + \hat{N}_T k \right) + c_S \right\} \tag{5.3}$$

Consider a scenario where there is \$200 million of CAPEX; \$100 million of spectrum license fees; network OPEX of 5 % CAPEX per annum; subscriber OPEX of \$20 per annum per user; average \$50 of handset subsidy per user; one million users; an ARPU of \$150 per annum; a discount rate of 5.5 %; and a project life of six years ($\hat{N}_T \approx 5$).

This would yield a healthy NPV of approximately \$250 million.

The above model captures the value seen by one type of player in the chain only, namely an operator. It allows one to make qualitative statements as regards the value-chain impact of cognitive radio for an operator. These are explored in further detail in the next section.

Models can be created for other players, e.g. users, content providers, etc., although some parameters can be difficult to quantify, e.g. value to end users. For yet other players such as national telecommunications regulators, it is difficult to create the model itself, given the complex mix of societal and market impacts involved. In order to put these challenges into relevant contexts, some scenarios involving different players' interpretations of value are presented in the following section.

5.6 Example Scenarios

This section describes some scenarios that aim to help the reader to gain a better understanding of the concepts described in this chapter. These scenarios involve a communications network where technological change can create opportunities for new business models. Sections 5.6.1 and 5.6.2 describe how users can gain more value through increased communication coverage and ease of use, which has potential for increased ARPU for the operator. Users may have the potential to gain financial credit from the operator by contributing the capabilities and features of their cognitive radio devices to the network infrastructure, as described in Section 5.6.3. Thus, value is increased and this migrates as the benefits are shared between user and operator. Value is not limited to the user and operator spaces but can have implications for external third parties also. As an example of this, a scenario examining value from a spectrum regulator viewpoint is outlined in Section 5.6.4.

5.6.1 Simplified Man–Machine Interface

The concept of a very simple user interface was introduced in Section 5.3. To the average non-technically minded customer, the potentially greater freedom of communication regardless of location and trajectory, using a system that is more in tune with their daily behavior patterns is what they may perceive as a high value feature. The power of cognitive radio also lies in the ability to provide a very simple interface to what can be a very complex system, where ideally the user requires no technical knowledge in order to operate the device. This is referred to in this chapter as *one-touch operation*. This customer may not place a high value on the cognitive functionality itself (or even care), but if this functionality results in universal *one-touch* or entirely hands-free operation through a simplified context-related MMI (man–machine interface), requiring little conscious effort on their part, then a potentially valuable product may be realized.

The potential benefits of cognitive radio technology to the daily lives of consumers are enormous. An ability to anticipate the wireless communications needs of the users and take preemptive action by seeking out the either the most robust or economical means of information conveyance is attractive. In addition, the ability to establish and maintain a communications link regardless of location, trajectory and wireless environment is also a desirable feature. Developing a cognitive radio device capable of one-touch or even totally hands-free and fault-tolerant operation may require increases in CAPEX and handset subsidies to help promote adoption of this system. However, the potential of this device is that it may actually help increase the ARPU as communication becomes an ability taken for granted and is used more as a result. The simple interface and increased ease of use could encourage more new subscribers to adopt the use of this technology.

We consider the economic value model described in Section 5.5 with the following sample (and fictitious) adjustments to the original set of figures for a future project involving the use of cognitive radio. In this scenario, we make conservative assumptions that the development of a fault-tolerant cognitive radio with a simplified man–machine interface requires increasing the original CAPEX (CAPEX = $300 million) and the handset subsidies required to promote initial uptake may also have to be increased to $100 per device. However, attracted by the increased ease of use and reliability, this technology may increase the customer base from 1 million to 1.2 million subscribers. In addition, the 'taken for granted' nature of the ability to communicate, combined with the increased ease of use, may be expected to result in an increase in ARPU from $150 to $175. The network OPEX is retained at their original values, although the subscriber OPEX is reduced to $15 per user as the fault-tolerance and simple MMI reduces the customer support costs. The NPV in this case is $365 million, an increase of $115 million. Although these are example figures, a key observation is that even considering the effects of a significantly increased CAPEX and handset subsidies, the NPV is still significantly greater than zero, indicating that this may be a viable project.

5.6.2 Dynamic Spectrum Access

Cognitive radio is popularly associated with enabling more efficient use of frequency spectrum. Certainly, cognitive functionality in the form of awareness, reasoning, learning, and frequency agility abilities are necessary to detect and exploit spectrum opportunities. The ability to relinquish this spectrum when the information transfer is completed or when a user with a higher priority appears in the same spectrum segment is also a necessary feature. More efficient use of spectrum also means being able to accommodate more wireless devices operating in the same spectrum segments without causing interference to each other. This interference-free coexistence ability can therefore increase the potential number of subscribers, N_U, in a network of cognitive nodes.

Figure 5.3 illustrates this concept where a network of cognitive radio devices (B) operating on a secondary opportunistic basis, co-exist with the legacy primary user network (A). The legacy users may or may not have cognitive functionality and are referred to as *non-cooperative* users. This scenario can be realized if the secondary users can avail of the *white space* spectrum (unused spectrum in the primary user spectrum allocation) [11] or transmit packets when the primary user nodes are estimated to be inactive. It is also feasible that an increased number of nodes with cognitive functionality could be supported in the same spectrum segment, and be part of the same network compared to a network of nodes without cognitive functionality [12]. In this figure, the secondary users exploit the unused spectrum segments, shown in green, on either side of the primary user spectrum illustrated in orange.

Dynamic spectrum access is of value to the average customer if this ability can enable the desirable *always-connected* feature of a cognitive radio.

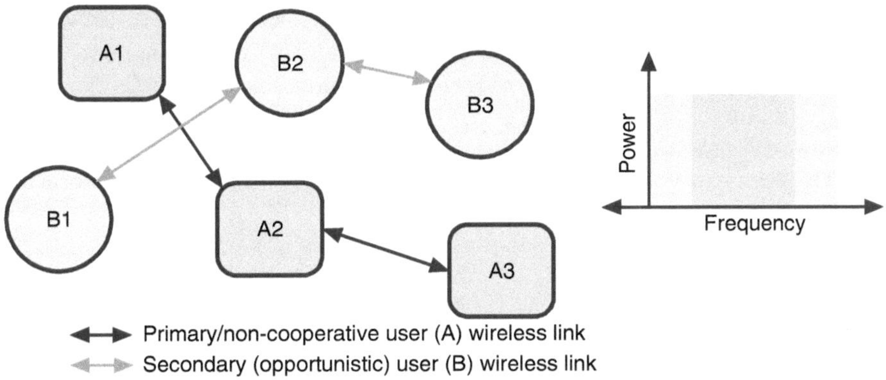

Fig. 5.3. Illustration of how the number of potential users can be increased in a spectrum segment if secondary opportunistic usage, operating on a non, or minimal interference basis, can be supported using cognitive radio technology. The primary (or *non-cooperative* users) and secondary opportunistic users, and their spectrum occupancies, are illustrated in orange and green, respectively.

The ability to utilize only as much spectrum as is deemed necessary in order to facilitate the required information transfer, or seek out spectrum with a lower cost of usage may result in extra added value to this customer in the form of lower costs in a *pay to use* spectrum market regime.

The ability to avail of unlicensed spectrum for short range, lower priority or interference-tolerant communication services helps to reduce the demand for licensed spectrum segments. This is an obvious benefit for the operator. By reducing the spectrum-usage costs while retaining the same (or greater) ARPU, a positive change in the NPV can be observed.

Considering the previous example in the context of a future cognitive radio project employing dynamic spectrum access technology, three potential scenarios exist:

- The operator may either choose to halve their current licensed spectrum allocation, thus reducing the cost of spectrum licenses, c_S.
- The operator may choose to retain their current licensed spectrum allocation but avail of unlicensed spectrum also, thus increasing the total potential available spectrum. In this case, c_S remains static.
- The operator may choose to retain exclusive usage of part of the licensed spectrum allocations, and lease the remainder to other operators for opportunistic usage in return for a flat or per-usage fee. The result of this is that the cost of spectrum licensing is reduced.

The consequences of these three approaches are that either the cost of spectrum, c_S, decreases and/or the ability to accommodate more users, N_U, increases. If the initial CAPEX estimate accounts for cognitive radio devices capable of both licensed and unlicensed spectrum usage, then it is feasible that the modification of the operator's existing business model to lower costs and increase revenue in this manner requires a relatively inexpensive software upgrade, or a change in spectrum-usage policy, governing the behavior of the cognitive radio subscriber device.

Using the example (and fictitious) values in the previous scenario, consider the scenario where the required CAPEX, c_I is \$300 million, subscriber OPEX, o_U is \$15 per user per annum, and the expected ARPU, $R_U(n)$ is \$175 (assuming that the lower cost of spectrum usage is not passed directly to the user). In addition, the number of subscribers, N_U, is 1.2 million, the cost of spectrum licensing, c_S is reduced by \$25 million to \$75 million, and the handset subsidy and discount rate remain unchanged (i.e. $c_U = \$100$, $k = 5.5\,\%$, and $\hat{N}_T \approx 5$). In this example, by reducing the cost of spectrum-usage by \$25 million, the NPV is \$390 million. This represents an increase of \$140 million over the original non-cognitive radio oriented scenario.

5.6.3 Message-Relaying and Micropayments

Customers place value on the ability to communicate. In the case of a cellular network, this ability can be lost if a customer's device is out of range of a base station. A network operator wishes to maximize revenue by accommodating users out of range of base-station range without having to significantly upgrade the current network architecture (and as a result, increase c_I and o_I). In addition, it may not always be possible to erect or move base stations to where they are deemed necessary, and human behavior patterns and concentration of users in a given geographical area may also change over time (e.g. due to new housing scheme, new roads, and rail links). In order to capture the potential revenue from these subscribers using the existing infrastructure, one option is to enhance the subscriber device abilities. In addition to the operator's main goal of increasing the revenue generated by the customers increasing demand wireless connectivity in a region, they may desire to preparing themselves for the future roll out of new wireless communications-based services. In order to do this, the operator must try and accommodate more subscribers, maximize the use of the existing network infrastructure, and ideally avoid further capital expenditure on network expansion.

One of the attractive features of cognitive radio technology also lies in the ability to deploy new services/communications abilities without incurring a significant increase in the CAPEX and OPEX. These new services may involve changing the way the existing network infrastructure is used to convey information. In this scenario, message-relaying is used to increase the effective coverage of base stations and also increase the number of subscribers that can communicate using this base station.

Cognitive radio devices can act as message relays in parallel to the regular audio/video/data communication services. Customers out of range of a base station but within range of another customer's cognitive radio device can therefore route their messages and voice calls through one or more available

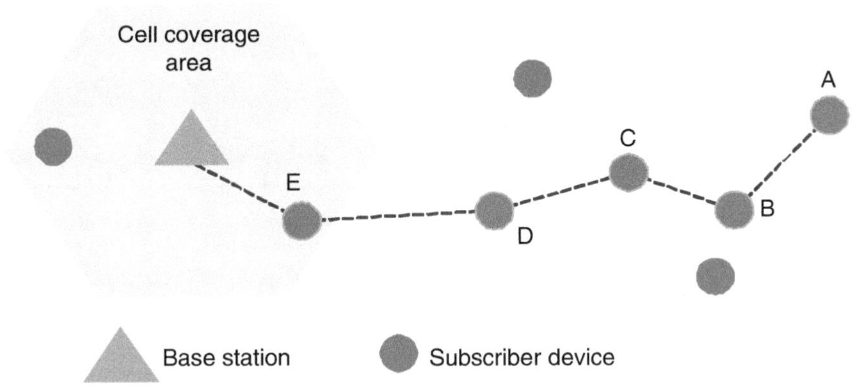

Fig. 5.4. Conceptual illustration of a message-relaying for nodes out of base station coverage.

nodes until the messages reach a base station and can continue through the rest of the network. By enhancing the cognitive functionality of each node, greater numbers of potential subscribers, N_U, may be supported per year without having to significantly upgrade the network infrastructure. This may help to keep c_I at a static or slowly rising level thus increasing the potential revenue.

This message-relay in a cellular network context is illustrated in Figure 5.4. Message traffic from **Node A** is relayed by **Node B**, **Node C** and **Node D** to **Node E**. Node E is within the base station coverage area and can complete the wireless communications link.

This method used to expand the capabilities of each consumer device can take the form of an over-the-air upgrades allowing each device to reconfigure as a message relay when required. Each device may choose to act exclusively as a message relay when not in active use by the customer, possibly encouraged by the prospect of payment for this service.

From a consumer standpoint, extra value may be extracted from the use of cognitive functionality in radio systems by being able to collect payment in return for relaying messages from other users. This small share of the total payment is referred to as a micropayment [13]. This incentive could help the 'always-connected' value of a cognitive network by encouraging devices to act as relay nodes in an ad hoc network (becoming part of the network infrastructure) thus helping to increase the possibility of communication.

However, the increase in network usage (e.g. through increased number of users N_U) will be offset by the degree to which micropayments reduce the ARPU, r_U.

5.6.4 Spectrum Regulator Value

Interference-free co-existence and more dynamic use of spectrum using cognitive functionality may be perceived as having value to an operator. As stated earlier, from a customer point of view, value is gained from these abilities if they enable the desired 'always-on, always-connected' experience, simplified MMI, etc. While it is useful for considering direct user and operator economic value, the NPV representation in this chapter does not account for this external third party value. An example of a relevant third party is a spectrum regulator, who may stand to gain from these features also but is not an operator or wireless user. From a value viewpoint, cognitive functionality could help shift some of the burden of regulation into the cognitive device and network space, as illustrated in the cognitive radio value-chain shown in Figure 5.1. Cognitive radio devices capable of operation in a non-interference manner and automatically adherence to the relevant regulatory policies helps to advance the concept of a self-regulation spectrum environment. While it is difficult to quantify the potential created value from a spectrum regulator viewpoint, it is feasible that self-regulation may reduce the need for active spectrum compliance testing. As a consequence, therefore, more time may potentially be allocated for carrying out other regulator activities. It is also possible that self-regulation can help support the argument for liberalization of the current spectrum licensing arrangements.

5.7 Conclusions

This chapter introduced the concept of a telecommunications value-chain and explored some of the many ways in which the value-chain can be altered by adaptive and reconfigurable software-defined radios, cognitive radios and cognitive networks. A description of how value can migrate and be created was also presented.

In order to help further illustrate the concepts described in this chapter, scenarios involving cognitive radio technology operating in a customer-centric network were discussed in terms of the impact on that model. The largest changes will be seen where significant improvements result in the services offered (e.g. improved coverage or simplified interfaces), since the ARPU or the number of customers may increase. Lesser value-chain impact will arise where there are changes in the internal operations of the network only, i.e. where impacts are confined to increasing handset subsidies or delayed CAPEX spend.

The business model allows one to predict the behavior of one operator. It does not directly describe the impact of competition from other operators. Thus, there are other significant impacts due to the introduction of cognitive radio that arise when the market structure itself changes. For example, if cognitive radio enables cheaper access to spectrum then new players may enter the operator market, increasing competition.

Ultimately, cognitive radio technology has significant potential to change how internal business models of the players interact with the market structures. The models and scenarios discussed here represent a first step in the analysis of the resulting value creation and migration.

Acknowledgments

This material is based upon work supported by Science Foundation Ireland under Grant No. 03/CE3/I405 as part of the Centre for Telecommunications Value-Chain Research (CTVR) at Trinity College Dublin, Ireland. The authors also wish to thank Stephen Wilkus of Alcatel-Lucent for his insights and comments.

References

1. Porter, M.E., *Competitive Advantage : Creating and Sustaining Superior Performance*, The Free Press, New York, 1985.
2. Haykin S., "Cognitive Radio: Brain-Empowered Wireless Communications", *IEEE Journal on Selected Areas in Communications,* February, 2005.
3. Fette, B. (Editor), *Cognitive Radio Technology,* Newnes, Aug. 2006,ISBN: 0-7506-7952-2.
4. Thomas, R.W., DaSilva, L.A., and Mackenzie, A.B., "Cognitive Networks", in Proceedings of the 1st IEEE International Symposium on New Frontiers in Dynamic Spectrum Access Networks, Baltimore, MD, USA, 8–11 Nov. 2005, pp. 352-60.
5. Thomas, R.W., Friend, D.H., DaSilva, L.A., and Mackenzie, A.B., "Cognitive Networks: Adaptation and Learning to Achieve End-to-End Performance Objectives", *Topics in Radio Communications, IEEE Communications Magazine,* December 2006.
6. Nolan, K.E., Doyle, L.E., "Principles of Cognitive Network Teamwork", in Proceedings of the 51st SDR Forum General Meeting and 2006 Technical Conference, Nov. 13–17, 2006, Orlando, Florida.
7. Nolan, K.E., Ambrose, E., and O'Mahony, D., "Cognitive Radio: Value Creation and Value Migration", in Proceedings of the 51st SDR Forum General Meeting and 2006 Technical Conference, November 13–17 2006, Orlando.
8. Sylwotzky, A.J., *Value-Migration,* Boston: Harvard Business School Press, 1996.
9. Ng, P.T.A., Lu, D., Li, C.K., and Harry Chan, H.Y., *Strategic Lessons of Value Migration in IT Industry.* Technovation 25, 2005, pp. 45–51.
10. Fine, Charles H., *Clockspeed: Winning Industry Control in the Age of Temporary Advantage,* New York: Perseus Books, 1998.
11. Cordeiro, C., Challapali, K., Birru, D., and Shankar, S., "IEEE 802.22: The First Worldwide Wireless Standard based on Cognitive Radios", in Proceedings of the 1st IEEE International Symposium on New Frontiers in Dynamic Spectrum Access Networks, Baltimore, MD, USA, 8–11 November 2005, pp. 328–337.
12. Nolan, K.E., Rondeau, T.W., Sutton, P., Le, B., Doyle, L.E., and Bostian, C., "Dynamic Spectrum Access and Coexistence Experiences Involving Two Independently Developed Cognitive Radio Testbeds" in Proceedings of the 2nd IEEE International Symposium on New Frontiers in Dynamic Spectrum Access Networks (DySPAN 2007), April 2007.
13. Peirce, M., and O'Mahony, D., "Flexible Real-Time Payment Methods for Mobile Communications", *IEEE Personal Communications,* Vol. 6, No. 6, December 1999, pp. 44–55.

6

Codes and Games for Dynamic Spectrum Access

Yiping Xing, Harikeshwar Kushwaha, K.P. Subbalakshmi, and R. Chandramouli

Media Security, Networking and Communications Lab,
Department of ECE,
Stevens Institute of Technology

6.1 Introduction

Cognitive radio is an emerging wireless communications paradigm in which either the network or the wireless node itself intelligently adapts particular transmission or reception parameters by sensing the environment. The goals of adaptation include maximizing spectral efficiency, minimizing interference to other cognitive radios, coexistence of licensed and unlicensed band communications, battery energy efficiency, etc. The environmental parameters that are continually sensed for adaptation include occupied radio frequency bands, user traffic, network state, etc. One promising technology that enables the implementation of a cognitive radio network is software-defined radio. The underlying theoretical principles for cognition are broadly based on signal-processing and machine-learning. The cognitive cycle [1] consists of three major components: (a) sensing of Radio Frequency (RF), (b) cognition/management, and (c) control action. In particular,

1. RF sensing:
   - estimation of the total interference in the radio environment;
   - detection of spectrum holes (or unused bands);
   - estimation of channel state information (e.g., Signal to Noise Ratio (SNR));
   - prediction of channel capacity for use by the transmitter.
2. Cognition/management:
   - spectrum management which controls the opportunistic spectrum access;
   - traffic shaping;
   - routing;
   - Quality of Service (QoS) provisioning.

H. Arslan (Ed.), Cognitive Radio, Software Defined Radio, and Adaptive Wireless Systems, 161–187.

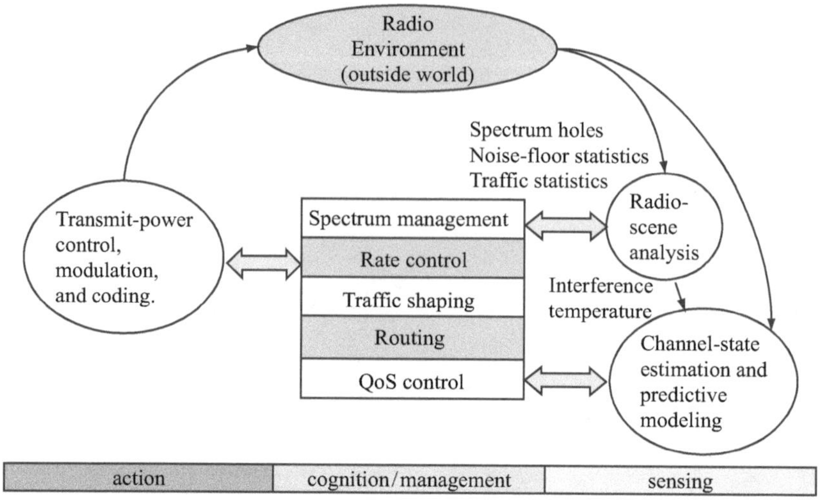

Fig. 6.1. Basic cognitive cycle.

3. Control actions:
   - transmit-power control;
   - adaptive modulation and coding;
   - transmission rate control.

Task (1) deals with spectrum sensing and prediction, hence, places emphasis on physical layer issues. In this chapter we will focus on tasks (2) and (3). We discuss these tasks for dynamic spectrum access from game and coding theoretic perspectives. Stochastic learning based solution to some of the issues in these tasks are also presented. The three tasks in a cognitive cycle are shown in Figure 6.1

Dynamic spectrum access using cognitive radios is an emerging research topic. This research motivated by the fact that enhancing spectrum efficiency is an important task of regulatory authorities worldwide. Radio spectrum is considered a scarce resource with the growing demand for spectrum-based services because a major portion of the spectrum has been allocated for licensed wireless applications. Therefore, new applications are competing for the very little spectrum that is left unlicensed. On the other hand, actual measurements taken on the 0–6 GHz band (Figure 6.2) in downtown Berkeley show that the actual utilization of the spectrum is quite low. This reiterates the argument that FCC's spectrum regulations diminish spectrum from being utilized efficiently. Other empirical studies (e.g., [8]) also seem to support the observation that spectrum is used inefficiently both in space and time.

Low utilization and increased demand for the radio resource suggest the notion of *secondary use*. This allows dynamic access to unused parts of spectrum owned by the primary license holder to become available temporarily for Secondary (non-primary) Users (SU). This dynamic access of spectrum

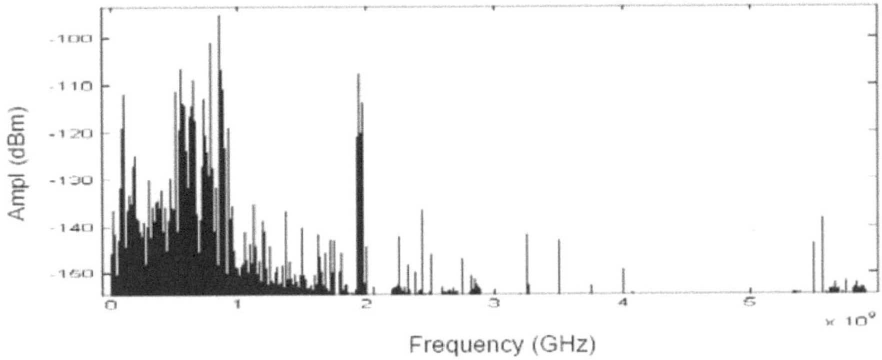

Fig. 6.2. Spectrum utilization snapshot at Berkeley.

by secondary users is one of the promising ideas that can mitigate spectrum scarcity, potentially without major changes to incumbents.

The first step in dynamic spectrum access is the detection of unused spectral bands. Therefore, a cognitive radio device measures the RF energy in a channel or monitors the received signal strength indicator to determine whether the channel is idle or not. But this approach has a problem in that wireless devices can only sense the presence of a Primary User (PU) if and only if the energy detected is above a certain threshold. It is true that one cannot arbitrarily lower the threshold as this would result in non-detection because of the presence of noise. In the feature detection approach, which has been used in the military to detect the presence of weak signals [9], the wireless device uses cyclostationary signal processing to detect the presence of primaries. If a signal exhibits strong cyclostationary properties, it can be detected at very low Signal-to-Noise Ratios (SNR) [11]. Then, the question is how to share the available spectrum efficiently and fairly.

The FCC spectrum Policy Task Force [12] has recommended a paradigm shift in interference assessment from the largely fixed operations. This facilitates real-time interactions between a transmitter and a receiver in an adaptive manner. The recommendation is based on a new metric called the *interference temperature,* which is intended to quantify and manage the sources of interference in a radio environment. The interference temperature is defined to be the RF power measured at a receiving antenna per unit bandwidth. The key idea for this new metric is that, firstly, the interference temperature at a receiving antenna provides an accurate measure for the acceptable level of RF interference in the frequency band of interest; any transmission in that band is considered to be "harmful" if it would increase the noise floor above the interference temperature threshold as shown in Figure 6.3. Secondly, given a particular frequency band in which the interference temperature is not exceeded, that band could be made available to secondary users. Hence, a secondary device might attempt to coexist with the primary, such that the presence of secondary devices goes unnoticed.

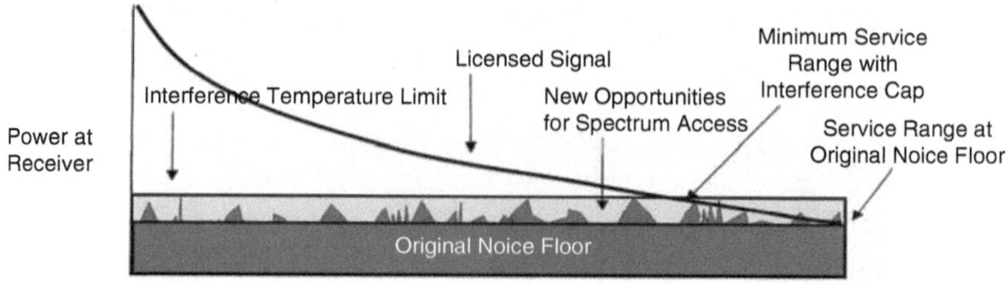

Fig. 6.3. This figure [14] shows the power that is received from the licensed transmitter as a function of distance. The spectrum agile radio has opportunities represented by the righthand side box.

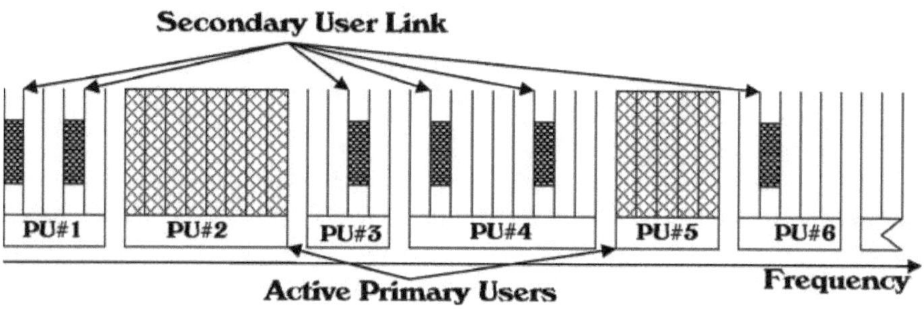

Fig. 6.4. Spectrum pooling.

Game theory plays an important role in the modelling and optimization of wireless communications among users with competing performance goals and resource constraints. For example, secondary users compete to access the available spectrum. Typically, these users do not cooperate with each other and selfishly try to maximize their own utility (e.g., throughput). This can be viewed as a non-cooperative game. Is there an equilibrium transmission strategy set for this game? If such a strategy set exists, how can it be computed by the user in a distributed fashion using only local information? These are all important questions that we address later in this chapter. We provide some game theoretic models for dynamic spectrum access and study the existence of equilibrium solution. Stochastic learning based techniques to discover the equilibrium solution is also discussed.

In the secondary usage scenario PUs are the license holders and secondary users SUs are allowed to use the band when PUs are inactive, but they are required to vacate the band as soon as PUs become active. The arrival of the PU causes the SU to loose that band and therefore link maintenance becomes a key issue. A solution to this problem has been proposed in [21] where the whole bandwidth is divided into a large number of subchannels as shown in Figure 6.4. A SU selects a set of subchannels in such a way that they lie in different primary users' bands. This minimizes the loss of subchannels upon the arrival of a particular PU as no PU can cause the complete breakdown of the SU link. But the arrival of a PU acts like an erasure for a SU link and it causes the SU to loose all the packets that are being transmitted over the subchannel which was under that particular PU's band. In counteracting the effect of PU's interference over the SU's link, a class of erasure correction codes called *LT codes* or *Digital Fountain Codes* [22, 23] play an important role in the link maintenance mechanism. We consider the use of these codes for secondary spectrum access and provide some analysis and simulations.

The chapter is organized as follows. Preliminaries of stochastic automata games are discussed in Section 6.2. Coexistence issues in dissimilar secondary radio systems is presented in Section 6.3. Section 6.4 deals with dynamic spectrum access with QoS and interference temperature constraints. Digital fountain codes for link maintenance is presented in Section 6.5.

6.2 Stochastic Learning Automata and Games

Abstractly, a learning automaton [2] can be considered to be an object that can choose from a finite number of actions. For every action that it chooses, the random environment in which it operates evaluates that action. A corresponding feedback is sent to the automaton based on which the next action is chosen. As this process progresses the automaton learns to choose the *optimal* action for that unknown environment asymptotically. The stochastic iterative algorithm used by the automaton to select its successive actions based on the environment's response defines the stochastic learning algorithm. An important property of the learning automaton is its ability to improve its performance with time while operation in an unknown environment. In this chapter, for the sake of consistency our notations follow or parallels that from standard books on game theory (e.g., [3]) and stochastic learning [4].

In multiple automata games, instead of one automaton (player) playing against the environment, N automata, say $A_1, A_2, ..., A_N$ take part in a game. Consider a typical automaton A_i described by a 4-tuple $\{S_i, r_i, T_i, \mathbf{p}_i\}$. Each player i has a finite set of actions or pure strategies, $S_i, 1 \leq i \leq N$. Let the cardinality of S_i be $m_i, 1 \leq i \leq N$. The result of each play is a random payoff to each player. Let r_i denote the random payoff to player $i, 1 \leq i \leq N$. It is assumed here that $r_i \in [0, 1]$. Define functions $d^i : \Pi_{j=1}^{N} S_j \rightarrow [0, 1], 1 \leq i \leq N$, by

$$\begin{aligned} d^i(a_1, ..., a_N) = \; & E[r_i | \text{player } j \text{ chose action } a_j, \\ & a_j \in S_j, 1 \leq j \leq N]. \end{aligned} \tag{6.1}$$

The function d^i is called the *expected payoff* function or *utility* function of player $i, 1 \leq i \leq N$. The objective of each player is to maximize its expected payoff. Players choose their strategies based on a time-varying probability distribution. Let $\mathbf{p}_i(k) = [p_{i1}(k)...p_{im_i}(k)]^t$ denote the action choice probability distribution of the ith automaton at time instance k. Then $p_{il}(k)$ denotes

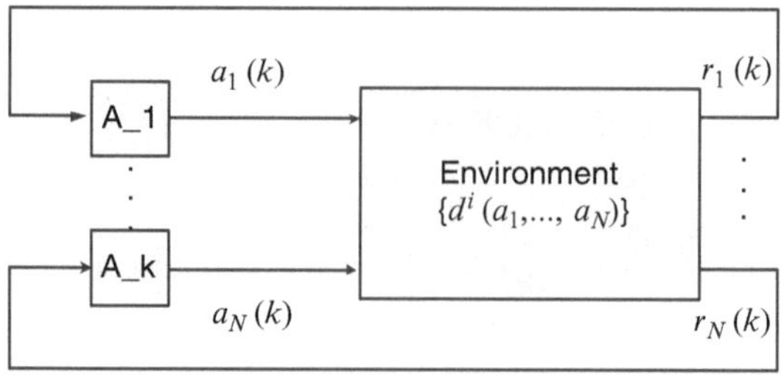

Fig. 6.5. The basic multiple automata game set-up.

the probability with which i^{th} automaton player chooses the lth pure strategy at instant k. Thus $\mathbf{p}_i(k)$ is the strategy probability vector employed by the ith player at instant k. T_i denotes the stochastic learning algorithm according to which the elements of the set $\mathbf{p}_i$ are updated at each time k, i.e., $\mathbf{p}_i(k+1) = T_i(\mathbf{p}_i(k), a_i(k), r_i(k))$, where a_i and r_i are the actual actions selected by i and the payoff received by i at $k, k = 0, 1, 2, \ldots$, respectively. The working of a learning automaton say A_i can be described as follows. Initially at $k = 1$ one of the actions is chosen by the player at random with an arbitrarily chosen initial probability. This action is then applied to the system and the response from the environment is observed. Based on the response r_i, the probabilities of actions $\mathbf{p}_i$ for the next period of time are updated according to the updating rule T_i. This process is repeated by all the N players until a stopping criterion is reached or the probability vector converges. Figure 6.5 depicts the basic automata game set-up.

We will frequently consider the situation of varying the strategy of a single player i while holding the strategies of his opponents fixed. Here, we let $s_{-i} \in S_{-i}$ denote a strategy selection for all players but i, and write (s'_i, s_{-i}) for the profile $(s_1, \ldots, s_{i-1}, s'_i, s_{i+1}, \ldots, s_N)$.

A strategy for player i is defined to be a probability vector $\mathbf{p}_i = [p_{i1}, \ldots, p_{im_i}]^t$, where player i chooses action j with probability p_{ij}. Then we can define the expected payoff for player i as g^i given by

$$\begin{aligned} g^i(\mathbf{p}_1, \ldots, \mathbf{p}_N) &= E[u_i | j^{th} \text{ player employs strategy} \\ \mathbf{p}_j, 1 \le j \le N] &= \sum_{j_1, \ldots, j_N} d^i(j_1, \ldots, j_N) \prod_{s=1}^{N} p_{sj_s}. \end{aligned} \tag{6.2}$$

Definition 1 *The N-tuple of strategies* $(\mathbf{p}_1^o, \ldots, \mathbf{p}_N^o)$ *is said to be a Nash equilibrium, if for each* i, $1 \le i \le N$, *we have*

$$\begin{aligned} & g^i(\mathbf{p}_1^o, \ldots, \mathbf{p}_{i-1}^o, \mathbf{p}_i^o, \mathbf{p}_{i+1}^o, \ldots, \mathbf{p}_N^o) \\ & \ge g^i(\mathbf{p}_1^o, \ldots, \mathbf{p}_{i-1}^o, \mathbf{p}_i, \mathbf{p}_{i+1}^o, \ldots, \mathbf{p}_N^o), \quad \forall \mathbf{p}_i \in [0,1]^{m_i}. \end{aligned} \tag{6.3}$$

A Nash equilibrium is said to be in *pure* strategies if $(\mathbf{p}_1^o, \ldots, \mathbf{p}_N^o)$ is a Nash equilibrium with each $\mathbf{p}_i^o$ being a unit probability vector. While, it is a non-degenerate mixed Nash equilibrium if the strategy assigns positive probabilities on more than one pure strategy. In general, each $\mathbf{p}_i^o$ above may be a mixed strategy and we refer to $(\mathbf{p}_1^o, ..., \mathbf{p}_N^o)$ satisfying (6.3) as a Nash equilibrium in *mixed* strategies. With this definition, when there is no pure equilibrium as may be the case in the stochastic automata game, the users can search for a mixed Nash equilibrium. It is well known that *every finite strategic-form game has a mixed strategy equilibrium* [3].

A strategy dominates another when, independent of the action taken by the other players, the first strategy leads to an outcome at least as favorable as the second. Consequently, a case can be made for never playing a strategy that is dominated. Although this is indeed true in a two-person zero-sum games and in identical payoff games, in general non-zero-sum games the use of a dominated strategy can lead to a better payoff for all players (e.g., the Prisoner's Dilemma).

In an N-person game, an outcome is said to be *Pareto-optimal* if there is no other outcome in which all players simultaneously do better. This is a modest requirement of any concept of group rationality. Unfortunately, it is not always consistent with concepts of individual rationality, such as dominance.

6.3 Coexistence of Dissimilar Secondary Radio Systems

Coexistence of dissimilar secondary radio systems in a spectral band raise two important issues: fairness and efficiency. To address these, we propose an access technique inspired by a human society model—*Homo Egualis.* Homo Egualis society consists of a group of agents who care not only about its own payoff, but also about how it compares with the payoff of others. In the human society, homo sapiens evolved in small hunter-gather groups. Such societies have no centralized structure of governance (state, judicial system, church). So the enforcement of norms depends on the voluntary participation of peers. A Homo Egualis society [5] can be modelled as follows, where the utility function of player i, u_i in an n-player game is

$$u_i = x_i - \frac{\alpha_i}{n-1} \sum_{x_j > x_i} (x_j - x_i) - \frac{\beta_i}{n-1} \sum_{x_j < x_i} (x_i - x_j), \tag{6.4}$$

where $x = (x_i, \ldots, x_j)$ are the pay-offs for each player and $0 \leq \beta_i < \alpha_i \leq 1$. $\beta_i < \alpha_i$ reflects the fact that Homo Egualis exhibits a weak urge to increase inequality when doing better than the others and a strong urge to reduce inequality when doing worse than the others. In [5] it is shown that in this model the salient behaviors in ultimatum and public goods games, where fairness does matter, can be reproduced. We proposed a distributed dynamic spectrum access scheme for dissimilar cognitive radio networks in [7] based on the Homo Egualis society model.

The 5 GHz unlicensed frequency band is a candidate for a large set of radio services, and is one of the unlicensed frequency bands that may be efficiently

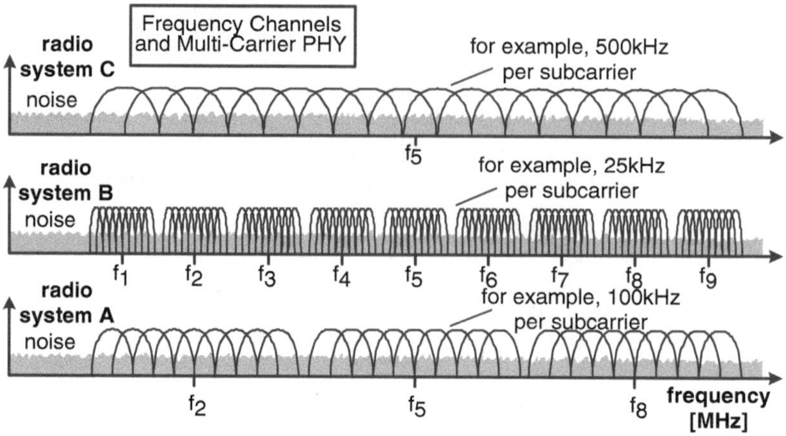

Fig. 6.6. Frequency channels and multi-carrier PHY.

used only with an established spectrum etiquette. We use the same abstract model of an unlicensed frequency band as in [38], as illustrated in Figure 6.6. Here, three different types of radio systems are assumed to operate in the band, each operating with different frequencies and channel bandwidths. The radio systems of type A operate on three frequency channels (center frequencies f_2, f_5, f_8), the radio systems of type B operate on nine frequency channels (center frequencies $f_1 \cdots f_9$), and radio system of type C operates on one frequency channel (center frequency f_5). The frequency channels overlap with each other, as indicated in the figure. The number and bandwidth of the frequency channels in Figure 6.6 do not represent any existing unlicensed band, this usage model serves without loss of generality only as an example model. Here radio system A can be compared to wireless LANs operating in the 5 GHz band (using OFDM). Radio system B represents narrow-band radio systems supporting for example a limited number of voice calls or blue tooth systems, while radio system C can be a broadband CDMA system. In our scenario instead of modelling the detailed protocols, a simplified Listen Before Talk (LBT) is used for all radio systems. A type A radio system requires the respective three frequency channels to be idle before allocating radio resources. Collisions of allocation attempts occur when more than one radio system detects the channel as idle at the same time. In simulations, when collisions happen, one of the radio systems is randomly selected to allocate the radio resource.

Two of the most representative etiquette rules [38] are as follows:

- rule ♯4: a radio system of type A, type B, or type C should apply LBT when operating;
- rule ♯6: in order to protect other radio systems most efficiently, a radio system that follows rule ♯4 should synchronize its LBT process in time across neighboring frequency channels that overlap with the same channels.

One of the most important metrics in the unlicensed band access is the average airtime per radio system. Airtime refers to the ratio of allocation time per radio system type to the reference time (say one hour) [38]:

$$\text{airtime}_{\text{type}=A,B} = \frac{1}{N_{type}} \sum_{i=1}^{N_{type}} \frac{\text{allocation time}(i)}{\text{reference time}}, \tag{6.5}$$

where N_{type} is the number of type i radio systems.

Efficiency and fairness are obviously the main goals of a spectrum etiquette. If every radio system accesses the unlicensed band in a greedy manner, then the radio system requiring broader band to operate will suffer from an unacceptable low airtime share. So one way to provision fairness to the etiquette rules would be to require each radio system to work in a cooperative manner. One option would be that each radio system i tries to contend for the spectrum with probability p_i. After the radio system has decided to contend for the spectrum it accesses the spectrum compliant to etiquette rule ♯4 described before. If perfect fairness is achieved and each radio system is maximizing its airtime share, then it can be shown that [7] there exists an access probability vector $(p_{a_{opt}}, p_{b_{opt}}, p_{c_{opt}})$ that corresponds to the strategy that no radio system can do better in terms of airtime share without harming the other coexisting radio systems. So in this sense, both the efficiency and fairness are obtained by using this optimal probability pair. Obviously, without considering any fairness issues, the most efficient access in the sense of pure spectrum utilization would be that all the users compete for the spectrum greedily. But, using this scheme some type of networks may be totally blocked out of the spectrum access. And further, if different types of radio systems have different traffic load a simple equalization of the airtime may cause low spectrum utilization. Also, exact fairness does not always apply as far as an operator is concerned, they are likely to be more concerned with revenue. So users who pay more will get more access, which means different networks may have different priorities. To properly address these problems, we need to define the fairness and efficiency carefully.

Definition 2 *Weighted fairness is achieved if the following equation holds,*

$$\frac{\text{airtime}_{\text{type}_i}}{L_i} = \frac{\text{airtime}_{\text{type}_j}}{L_j} = K \quad \forall i, j. \tag{6.6}$$

where K is a constant, and the weight $L_i = \theta_i \lambda_i$, θ_i is the priority parameter and λ_i is the traffic load for type i radio system.

Definition 3 *Given the weighted fairness of different types of radio system is satisfied, efficiency is achieved if each of the radio system's airtime is maximized.*

Note that this definition of efficiency actually corresponds to the Pareto efficiency solution in game theory. A strategy profile is Pareto optimal if some players must be hurt in order to improve the payoff of other players [6].

To obtain $(p_{a_{opt}}, p_{b_{opt}}, p_{c_{opt}})$ we need the information of all the arrival rates λ's and service rates μ's of the the radio systems which is impractical in a real access scenario. A more realistic scheme would be to allow the radio systems to learn the values p_a, p_b, and p_c themselves with only local information or measurement. We present such a Homo Egualis (HE) based technique next. The inequality aversion property of the Homo Egualis agents can be utilized to achieve fairness in the spectrum access problem. In this scheme each radio system learns the access probability p_i by itself. Here, we define $Onlinetime_i$ as the averaged cumulative "on" spectrum time per radio system of type i. Then we define $x_i = \frac{Onlinetime_i}{L_i}$, where $L_i = \theta_i \lambda_i$ is the same as used before in (6.6). The cumulative $Onlinetime_i$ is normalized by the radio system's traffic load and priority, which makes this spectrum access scheme able to adapt to different traffic loads and priority, hence achieve more efficiency and maintain our defined weighted fairness. With initial $p_i = 1$, each time the probability p_i is updated as follows:

$$p_i = \max\left(0, \min\left(1, p_i + \frac{\alpha_i}{n-1} \sum_{x_j \geq x_i} \left(\frac{x_j - x_i}{x_j}\right) - \frac{\beta_i}{n-1} \sum_{x_j < x_i} \left(\frac{x_i - x_j}{x_i}\right)\right)\right) \quad \textit{for all } j \neq i, \tag{6.7}$$

where n is the number of different radio system types, $0 < \beta_i < \alpha_i$ reflects the fact that radio system exhibits a weak urge to when doing better than others and a strong urge to reduce inequality when doing worse than the others. This forces each radio system to make an effort to efficiently use the idle spectrum while taking fairness into consideration. Here, the only local information needed is the radio system's own history of the *Onlinetime* and the *Onlinetime* of the other radio systems whose spectrum is within the same spectrum block. This can be obtained by keeping a record of the busy time of the required spectrum, which can be obtained by periodical spectrum scanning. When there are more than two different radio systems trying to coexist in the same spectrum, different radio systems can be identified by some smart technologies (e.g., we can detect the different transmitting power levels to distinguish from different radio systems). So each radio system can access the spectrum based only on its own recorded history and the local measurements performed by itself. λ_i can be estimated by historical usage record of radio system type i. The priority parameter θ_i needs to be announced by the radio system or be broadcast by some operator. If each radio system can only gather an imperfect information about the spectrum usage time of other radio systems, then there may be some degradation in the achieved weighted fairness.

We model the arrival traffic as a Poisson process. It can be seen from Figures 6.7 and 6.8 that all the different radio systems have almost the same airtime share and blocking probability by using the proposed HE access scheme, which is what is desired for fairness.

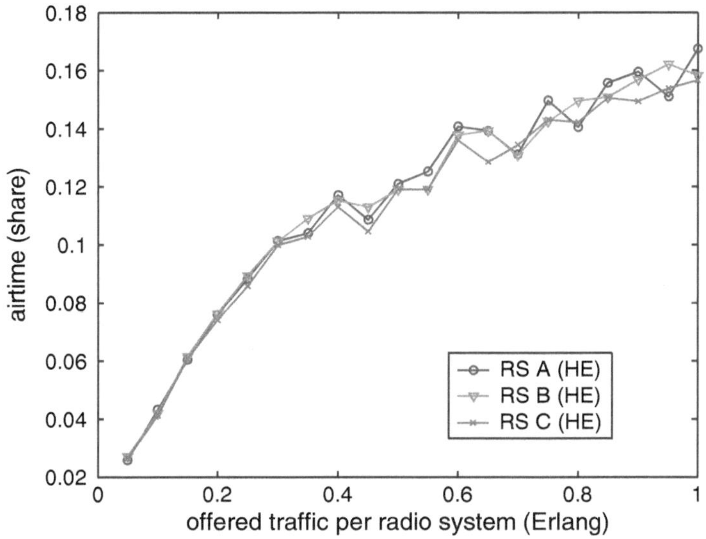

Fig. 6.7. airtime vs offered load for HE access scheme. Three different types of spectral agile radio systems with no queuing.

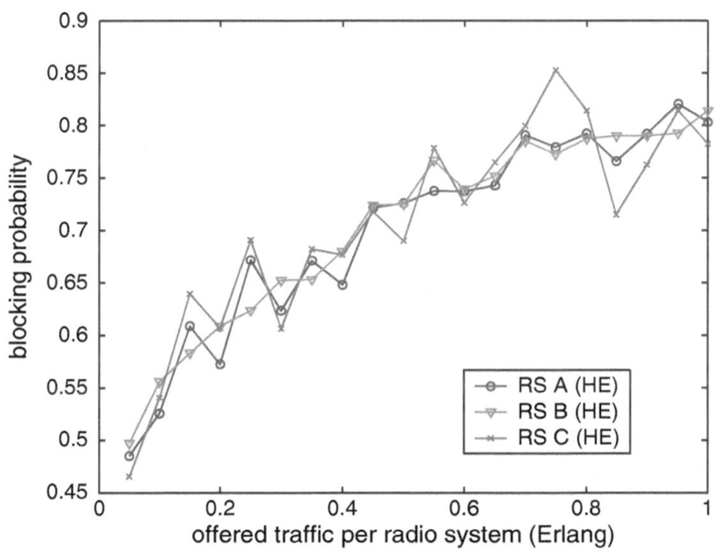

Fig. 6.8. Blocking probability vs offered load for HE access scheme. Three different types of spectral agile radio systems with no queuing.

6.4 Impact of QoS and Interference Temperature Constraints

Previous work on secondary use of radio spectrum include [7, 13, 15–18]. Let us consider a scenario similar to [18], where secondary users wish to use a local, relatively short-term data service, and all users adopt a spread spectrum signaling format, in which the transmitted power is evenly spread across the entire available band controlled by the manager. A practical realization of this model would be when secondary users (with spread spectrum signaling) and primary Direct-Sequence CDMA (DS-CDMA) system coexist in the up-link spectrum band of the primary DS-CDMA system. In the uplink, the interference temperature can be measured at the base station which is the receiver of the primary system. Hence, the number of measuring points can be significantly reduced. But of course secondary access is definitely not limited to this scenario. The spectrum can be available in a TV broadcast band or other emergency band.

6.4.1 Secondary Spectrum Sharing Model

Spectrum with bandwidth W is to be shared among M spread spectrum users, where a user refers to a transmitter and an intended receiver pair. For each i, the received signal to interference ratio (SIR) is given by

$$\gamma_i = \frac{y_i h_{ii}}{\frac{1}{L}(\sum_{j \neq i} y_j h_{ji}) + \sigma^2}, \tag{6.8}$$

where L is the normalized spreading sequence length, y_i is user i's transmission power, h_{ij} is the channel gain from user i's transmitter to user j's receiver, and σ^2 is the background noise power that is assumed to be the same for all users. In order to satisfy an interference temperature constraint, the total received power at a specified measurement point must satisfy

$$\sum_{i=1}^{M} y_i h_{i0} \leq B, \tag{6.9}$$

where h_{i0} is the channel gain from user i's transmitter to the measurement point, and $B > 0$ is a pre-defined threshold. We assume that all the secondary users adopt a spread spectrum signaling format, in which the transmitted power is evenly spread across the entire available band. This allows efficient multiplexing of data streams from different sources corresponding to different applications, and reduces the combined power-bandwidth allocation problem to a received power allocation problem. Hence, the interference temperature constraint is translated to a total received power threshold B at the measuring point.

6.4.2 Spectrum Assignment with Priority Classes

If different users have different contracts defining different priorities or throughput, then it is acceptable for a spectrum operator to provide different throughputs. Let us say that secondary links, depending on their willingness to pay belong to L priority classes. Let a_i be the priority parameter for link i. The operator problem is then to maximize the network revenue:

$$\begin{array}{lll} \max \sum_{i \in (i:x_i=1)} a_i. & & \\ \text{subject to} & & \\ SIR_i & \geq \gamma_i^t & \forall i \in (i : x_i = 1) \\ \sum_i h_{i0} y_i & \leq B & \forall i \in (i : x_i = 1) \\ y_i & > 0 & \forall i \in (i : x_i = 1) \\ y_i & \leq y_i^{max} & \forall i \in (i : x_i = 1), \end{array} \tag{6.10}$$

where x_i is a collection of binary variables, $x_i = 1$ means the ith link transmits otherwise $x_i = 0$. By maximizing this revenue, secondary users who pay more will get accessing priority over those who pay less. We model the relation between the price p_i a user paid and the priority parameter a_i as follows:

$$a_i = p_i^{\alpha}, \tag{6.11}$$

where $0 \leq \alpha \leq 1$ is an operator designable parameter. Small α corresponds to putting more emphasis on system capacity (number of active secondary links), while large α corresponds to putting more emphasis on guaranteeing service to the user paying higher price. Specifically, $\alpha = 0$ corresponds to the problem of maximizing the number of active secondary links (capacity).

6.4.3 Secondary Spectrum Sharing Potential Game

The nature of secondary spectrum sharing is temporary and distributed. Therefore a practical secondary spectrum sharing scheme must be distributed. In this section, we discuss such a distributed algorithm to solve the operator problem discussed before. This distributed process is composed of two phases. The coordination phase controls the optimal set of active secondary links which can access the spectrum, and the power control phase is to allocate transmit power to support the minimum target link SIR γ_i^t given the set of active links.

Potential Games

Suppose there are M transmitter and receiver link pairs competing for the secondary spectrum access opportunities. Let k be a time (iteration) counter and $N(\mathbf{x}(k))$ be the aggregate received power at the measuring point at time k given by

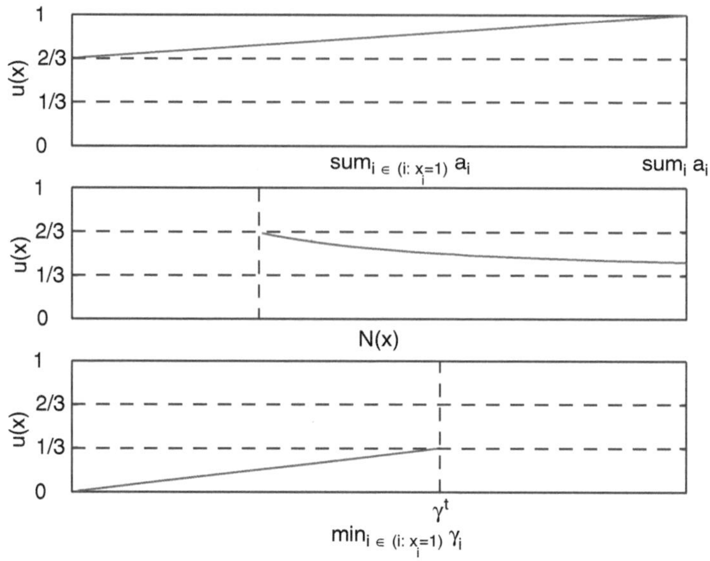

Fig. 6.9. Utility function $u_i(N(\mathbf{x}))$.

$$N(\mathbf{x}(k)) = \sum_{i=1}^{M} y_i(k) h_{i0} x_i(k). \tag{6.12}$$

In order to maximize the network revenue while keeping the aggregated received power at the measuring point under the interference threshold, we define the utility function $u_i(\mathbf{x})$ shown in Figure 6.9 for each link pair as follows:

$$u_i(x) = \begin{cases} \frac{\sum_{j \in (j:x_j=1)} a_j}{3 \sum_j a_j} + \frac{2}{3}, & N(\mathbf{x}(k)) < B, \min_{j \in (j:x_j=1)} \gamma_j > \gamma^t \\ \frac{B}{3N(\mathbf{x}(k))} + \frac{1}{3}, & N(\mathbf{x}(k)) \geq B, \min_{j \in (j:x_j=1)} \gamma_j > \gamma^t \\ \frac{\min_{j \in (j:x_j=1)} \gamma_j}{3\gamma^t}, & \min_{j \in (j:x_j=1)} \gamma_j < \gamma^t. \end{cases} \tag{6.13}$$

In the secondary spectrum sharing game each user maximizes its utility function $u_i(\mathbf{x}(k))$ by its choice of being active or not. By maximizing this utility function, the system will reach an operating point where the network revenue is maximized while satisfying QoS and interference temperature constraints. To emphasize that the ith user has control only over its own choice, we use an alternative notation $u_i(x_i, \mathbf{x}_{-i})$, where $\mathbf{x}_{-i}$ denotes that vector consisting of elements of $\mathbf{x}$ other than the ith element. And after each changing of the active link set, the distributed power control algorithm DCPC [20] will be activated to allocate the transmitting power. Results about the existence of pure strategy Nash equilibrium for the potential game, etc. can be shown to hold. A practical method to compute the Nash equilibrium would be to use the sequential play where each player maximizes its own utility function sequentially while other players' strategies are fixed. The sequential play will never converge to a solution where the total received power at the measuring point exceeds the interference temperature bound. Also, the sequential play will always converge to a solution where all the active links are supported with their target SIR. But, sequential play does not allow asynchronous updates by individual users. This may cause signaling and other overhead. To overcome this issue we consider a stochastic learning based distributed solution that is described next.

Stochastic learning techniques have been successfully used in wireless packet networks for on-line prediction, tracking and discrete power control [19,20]. It is shown to be computationally simple and efficient. Learning algorithm determines probabilistic strategies for players by considering the history of play. The probability updating algorithm used by each user is as given below:

1. Set the initial probability $p_i(0)$.
2. At every time step k, the ith user chooses $x_i(k) = 1$ or 0 (to transmit or not) according to its action probability p_i.
3. After distributed power control (DCPC [20]) phase, each player obtains a feedback $u_i(\mathbf{x}(k))$ based on the set of all actions $\mathbf{x}$.
4. Each player (i) updates its action probability according to the rule:

$$\begin{array}{ll} p_i(k+1) = p_i(k) + bu_i(k)(1 - p_i(k)) & x(k) = 1 \\ p_i(k+1) = p_i(k) - bu_i(k)p_i(k) & x(k) = 0, \end{array} \tag{6.14}$$

   where $0 < b < 1$ is the step size, and $u_i(k)$ is utility function which lies in the interval $(0, 1)$.
5. If p_i converges, stop. Otherwise, go to step 2.

The probabilistic update used in (6.14) is a stochastic learning automata updating known as linear reward-inaction (L_{R-I}) [2].

It can be proved that the proposed learning algorithm will converge to one of the Nash equilibria of the game. It will never converge to a point where the total received power at the measuring point exceeds the interference temperature bound. The algorithm will always converge to a point where all the active links are supported with their target SIR.

The learning automata algorithm needs less information and control signaling to operate than the sequential play. The sequential play requires that each player updates its strategy one by one. And at the time of decision, in order to compute the utility function, the information required includes all active users' SIRs, priorities, and the current interference temperature. Significant control signaling is also required to accomplish this process distributively. An alternative choice would be to run this sequential play at a central controller. For the learning automata, asynchronous updating is permitted. The only information needed is a feedback of the current utility function value. To compute this utility function, each active user should report its SIR and priority parameter to the measurement point which acting also as a central controller. But the trade-off is that learning automata converges much slower than the sequential play. The information needed by these algorithms can be distributed through a common control channel which has been assumed in many similar literatures in this area.

One assumption in this chapter is that the interference temperature remains constant during the secondary use of the spectrum. With mobile users, the interference temperature may vary. One solution to account this variance would be that the secondary access algorithm is triggered periodically. A short period would ensure that when the interference temperature is violated, the system would recover quickly.

Here, we first present some numerical examples for a simple secondary sharing system with only four transmitting and receiving pairs. The target SIR is selected to be $\gamma^t = 12.5$, and the noise power is $\sigma^2 = 5 \times 10^{-13}$, which approximately corresponds to the thermal noise power for a bandwidth of 1 MHz. We consider low rate data users, using a spreading gain of $L = 128$. Path gains are obtained using the simple path loss model $h_j = K/d_j^4$, where $K = 0.097$. We set the interference temperature bound and noise ratio to be $B/\sigma^2 = 60$ (we use these ratios only to illustrate clearly how the system works. Of course we can use lower ratio by reducing the target SIR). Under equal priority case $\mathbf{a} = [1, 1, 1, 1]$. When the learning automata algorithm is used, the evolution of the total received power $N(\mathbf{x})$ is shown in Figure 6.10. As discussed, we see that the total received power converges to a value below the interference temperature threshold B. The evolution of the choice probability p is shown in Figure 6.11. After convergence, only links 2 and 3 are active with equal probability initialization. It can be observed in Figure 6.10 that during the

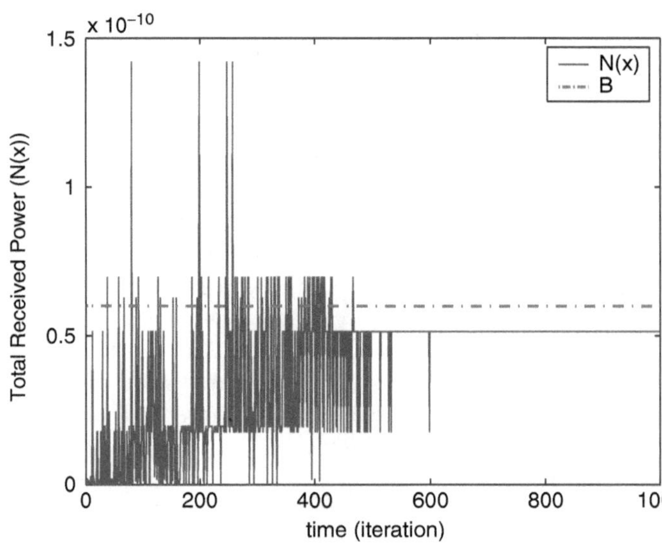

Fig. 6.10. Evolution of the total received power ($N(x)$) over time for the stochastic learning algorithm.

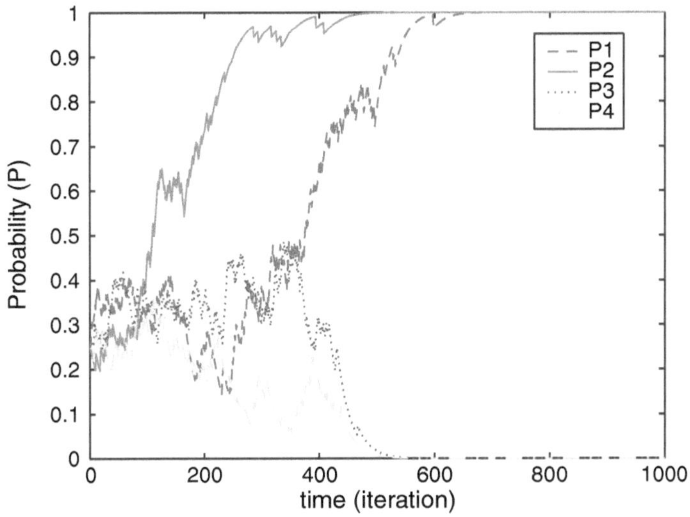

Fig. 6.11. Choice probability p of the activation strategy over time for the stochastic learning algorithm.

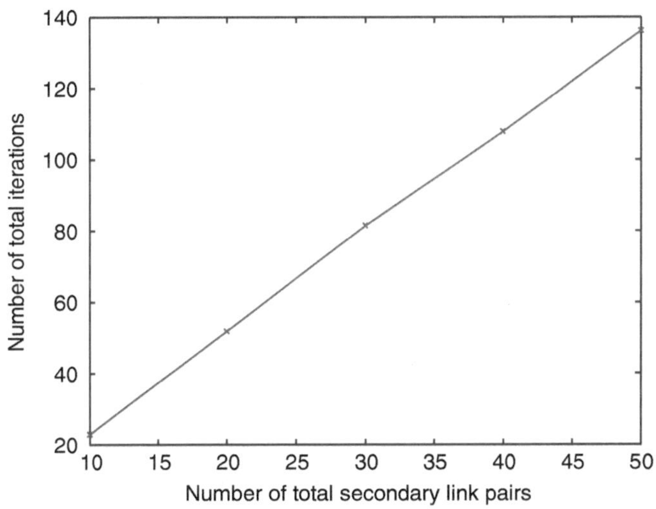

Fig. 6.12. Convergence of the sequential play

settling phase, $N(x)$ exceeds B. So, depending on the application, there should be some pre-defined extra margin to accommodate the fluctuation during the settling phase. But this may reduce the capacity of the secondary system.

Despite the suboptimal nature of the sequential play algorithm, the convergence speed is dramatically reduced as shown in Figure 6.12. It can be seen that the convergence speed is linear with respect to the number of secondary link pairs.

Note that bit errors can occur even after optimal access, power control, etc. If an error correcting code fails to correct these errors then this will lead to packet losses–erasures at higher layers. Therefore, it is important to consider erasure channels and techniques to mitigate packet losses. We discuss a strategy that addresses this in the next section.

6.5 Fountain Codes for Dynamic Spectrum Access

In this section, we discuss a new class of *erasure correction codes* for packet-based channels with erasures, called *digital fountain codes* [22–25], which can be used to provide packet-level protection at the transport layer or higher, augmenting the bit-level protection that may be provided by the medium access control (MAC) and PHY layers. These codes are capable of correcting *missing* or *lost* data. By recovering lost data packets without requiring retransmission from the sender, these codes efficiently and effectively provide reliability in data networks. And, like a water fountain producing an endless supply of water drops, any of which can be used to fill a glass, these fountain codes can generate an unlimited number of encoded output packets, any of which can be used to recover the original input packets. They have the following characteristics:

- The ability to generate a potentially limitless amount of encoded data from any original set of source data and providing reliable message delivery over extremes of low to high network losses
- The ability to recover the original data from a subset of successfully received encoded data regardless of which specific encoded data has been received
- Exceptionally fast encoding and decoding algorithms–operating at nearly symmetric speeds that grow only linearly with the amount of source data to be processed and independently of the actual amount of network loss

6.5.1 Erasure Channels

In erasure channels packet losses occur at the receiver due to a variety of reasons. For example, if the bit error correcting code fails, then the erroneous packets may not be passed on to the higher layers in the protocol stack. Therefore, the application layer sees this as an erasure. In erasure channels we can assume that a packet is either received intact or is lost.

In a wireless network, packet loss generally occurs due to following reasons:

- Packets get discarded en route to their destination for various reasons such as buffer overflows and congestion control at intermediate routers.
- Packets get corrupted due to noise and interference.
- In a secondary user environment, when a primary user occupies a channel, secondary users loose their packets on that channel.

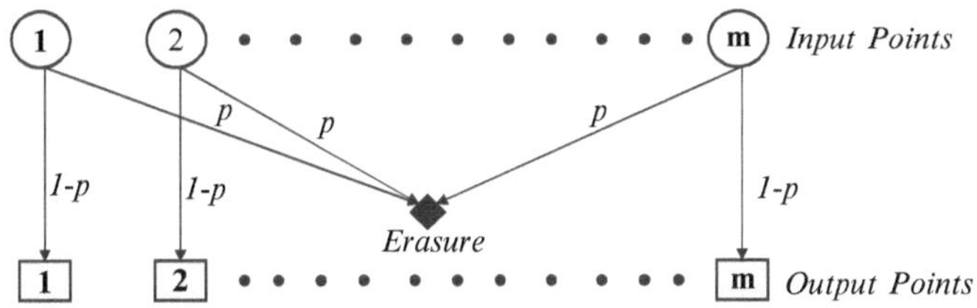

Fig. 6.13. M-ary erasure channel model.

A M-ary erasure channel is shown in Figure 6.13, where all input points $\{0, 1, 2, \ldots, m-1\}$ have a probability $1-p$ of being received correctly and a probability p of being lost or discarded (i.e., erased). If the perfect channel (i.e., $p = 0$) capacity is C then the capacity of the erasure channel is $(1-p)C$.

The traditional method for having reliable communication over such type of channels uses feedback to request for retransmission in order to make up for erased packets, but they perform poorly in the following circumstances:

- One-to-many transmission: In broadcast, if all receivers request for retransmission then the number of retransmissions at the sender is going to be very high which decreases the overall efficiency.
- Transmission over bad channels (i.e., if p is high), here again the number of retransmissions will be high causing the overall performance to suffer.
- If the sender–receiver distance is large then the effect of propagation delay due to retransmissions may be significant.

To deal with the above problems of feedback-based protocols new transmission solutions using erasure correcting codes have been proposed. So if some bits in a packet are lost or errors could not be corrected using a forward error correcting (FEC) code then it may be recovered using erasure correcting codes. It is required that these codes be capable of correcting as many erasures as possible and at the same time have fast encoding and decoding algorithms.

Since the channel capacity is $(1-p)C$ independent of the feedback channel, erasure correcting codes try to achieve this capacity without using feedback. Traditional erasure codes are $[N, K]$ block codes such as Reed-Solomon (RS) and Tornado codes [22, 23]. In these codes, any K or slightly more than K out of N outputs are sufficient to recover the original K inputs. But they are practical only for small values of K and N. Furthermore, both the codes are fixed rate codes, i.e., if the channel conditions deteriorate, additional redundant packets cannot be generated on the fly. Therefore to construct a reliable and robust transmission over erasure channels, a new class of rateless codes called digital fountain codes have been developed. In the following sections we will discuss Luby Transform (*LT Codes*) and *Raptor Codes*.

6.5.2 LT Codes

LT codes [24] that we describe here, are a new class of codes introduced by Luby in 1998 for the purpose of scalable and fault tolerant distribution of data over computer networks. The basic idea is as follows. The encoder can be thought of as a digital fountain that produces a continuous supply of water drops (i.e., *output packets*) called *output points*. Let's say the original file or message data consists of K packets or *input points*. The length of each input point can be arbitrary, from one-bit to $l > 1$-bits. Now, in order to completely decode the received stream, one needs to hold a bucket under the digital fountain and collects drops until the number of drops in the bucket is a little larger than K. This means that decoding does not depend on which packets were received but only on the number of received packets.

LT codes are rateless in the sense that the number of output points that can be generated from the original data is potentially limitless and their number can be determined on the fly. Depending on the erasure probability we can send as many output points as are needed in order for the decoder to recover the original message. The original message can be recovered from any set of N output points, for N slightly larger than K, for a $[N, K]$ code.

LT codes also have very small encoding and decoding complexities. If the original data consists of K input points then each output point can be generated, independently of all other output points, on average with $O(ln(K/\epsilon))$ point operations, and the K original input points can be recovered from any N output points with probability $1 - \epsilon$, where N is $O(\sqrt{K}\ ln^2(K/\epsilon))$ more than K, for $\epsilon > 0$. Luby describes these codes as universal due to their being near optimal for every erasure channel and being very efficient as the data length grows.

Encoding Process

For a given set $(x_1, x_2, ..., x_K)$ of input points, the output points $(y_1, y_2, ...)$ can be generated in the following steps:

1. For each output point, randomly select a degree k from a degree distribution $\Psi(k)$ on $(1, 2, ..., K)$.
2. Select uniformly at random k distinct input points from $(x_1, x_2, ..., x_K)$.
3. Set the output point equal to the bitwise XOR-sum of these k input points.

This process can be represented by a bipartite graph, connecting the input points to the output points. Here, in order for the decoder to recover the input points from the output points, it needs to know the exact graph, i.e., which k input points have been used to generate an output point. This information has to be communicated to the receiver in some manner. It could be accomplished in any of the following ways:

- Sender and receiver can use identical pseudo-random number generators, seeded by the clock if it is synchronized.

- By a pseudo-random process which can determine the degree and the connections given a key, say ν. Then the sender generates ν to compute the output point and transmits the key in the header of the packet. If packet size is much larger than the key size (32 bits or so) then it adds only a small overhead.

Decoding Process

The decoder, having received N output points and the bipartite graph, can recover the input points in the following steps. We will call output points that have been reduced to degree one as *reduced points.*

1. Find an output point y_j from the set of reduced points that is connected to an input point, say, x_i. If Reduced Points Set (RPS) is empty, this process cannot proceed further, and decoding fails.
2. Set $x_i = y_j$.
3. Add (bitwise XOR) x_i to all the output points that are connected to x_i.
4. Remove all the edges connected to the input point x_i.
5. Go to step(1) untill all $x_i^{\prime}$s are determined.

The above decoding process looks simple but it turns out that the degree distribution $\Psi(k)$ plays a critical part for the success of the decoding process. The probability that RPS is not empty depends on $\Psi(k)$. Therefore, for good decoding performance we need to choose a good degree distribution.

The Degree Distribution

The degree distribution should be designed so as to meet the following design criteria:

- Most of the packets should have low degree so as to initiate the decoding process and keep the RPS from being empty.
- Some of the packets should also have high degree so as to make sure that all input points are connected to at least one of the output points.

The Ideal Soliton Distribution

In order to satisfy the above criteria an ideal soliton distribution has been proposed

$$\Psi(k) = \begin{cases} 1/K & \text{if } k = 1 \\ 1/k(k-1) & \text{for all } k = 2, 3, \ldots, K. \end{cases}$$

The motivation for the choice of the ideal soliton distribution is that at each iteration, the rate at which the input points are covered (decoded) is the same as the rate of output points joining the RPS. This prevents the decoding process from failure by making sure that at each step there are reduced points which are ready to be decoded. And in order to avoid redundancy, this distribution causes the graph to reduce the degree of output points in a way that exactly one reduced point appears at the end of each iteration. For the ideal soliton distribution the expected degree is $ln(K)$.

In practice this distribution performs poorly because a slight deviation from its expected behavior may create a situation where there is no output point with degree one, and that will cause the decoding process to stop. In order to avoid this problem the *robust soliton distribution* has been investigated.

The Robust Soliton Distribution

The robust soliton distribution denoted by $\Omega(k)$ is defined in the following manner. It is defined by two parameters, ϵ and σ. The parameter ϵ sets a bound on the probability of decoding failure and the parameter σ adjusts the RPS. The expected size of the RPS is given by

$$R = \sigma\ ln(K/\epsilon)\sqrt{K},$$

instead of one as in the case of ideal soliton distribution. Now we define another function

$$\Phi(k) = \begin{cases} (R/K)\frac{1}{k} & \text{for } k = 1, 2, \ldots, (K/R) - 1 \\ (R/K)ln(\frac{R}{\epsilon}) & \text{for } k = K/R \\ 0 & \text{for } k > K/R \end{cases}$$

which is added to the ideal soliton distribution $\Psi(k)$ and normalized to obtain the robust soliton distribution $\Omega(k)$ as follows:

$$\Omega(k) = \frac{\Psi(k) + \Phi(k)}{Z},$$

where $Z = \sum_{k=1}^{K}(\Psi(k) + \Phi(k))$, is the normalization factor and it determines the number of output points, N, required to recover the K input points with probability $1 - \epsilon$, where $N = ZK$. Figure 6.14 shows the robust soliton distribution.

In [24], Luby's analysis explains how $\Phi(k)$ lets the decoding process to begin with a moderate size of RPS, but a spike at $k = K/R$ ensures that every input point has at least one connection with the set of output points. Luby's key result is that $N = K(1 + 2\Phi(K/R))$ output points are sufficient to recover original K input points with probability at least $1 - \epsilon$. The only disadvantage is that the decoding complexity grows as $O(K\ ln(K))$. In order to reduce the decoding complexity to $O(K)$, Raptor codes have been designed by Shokrollahi [25].

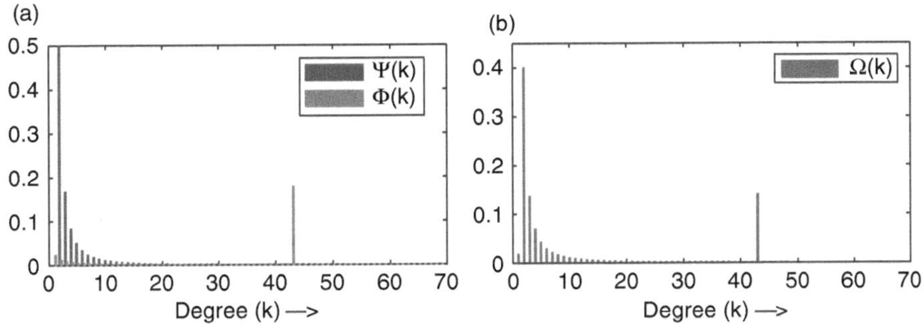

Fig. 6.14. For $K = 10000$, $\sigma = 0.2$, and $\epsilon = 0.1$, LHS figure shows the ideal soliton distribution $\Psi(k)$, and RHS shows the robust soliton distribution $\Omega(k)$.

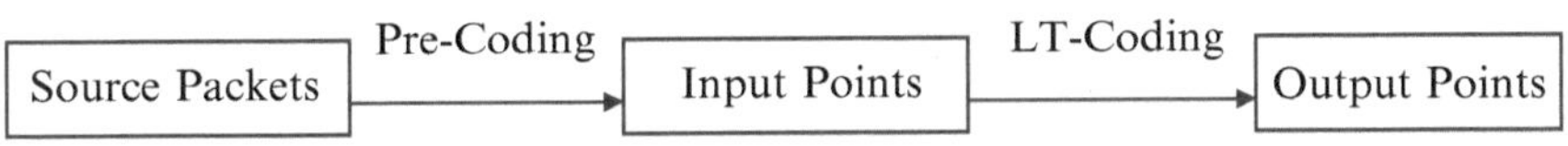

Fig. 6.15. Raptor codes.

6.5.3 Raptor Codes

Raptor codes [25], first developed by Shokrollahi, are designed to achieve linear time encoding and decoding complexity. This is accomplished by first pre-coding the source data by an appropriate outer code to generate the input points for the LT code as shown in Figure 6.15.

The main idea of pre-coding can be explained as follows. LT codes have a complexity of the order of $ln(K)$ per packet, because the average degree per output point is $ln(K)$. Raptor codes use LT codes with average degree $\bar{k} \approx 3$. The advantage of using a lower degree is that the decoder will work almost surely and will not get stuck but there is a possibility that a fraction of the input points may not be connected to any of the output points and hence they cannot be recovered by the decoding process. Let the fraction that remains unrecovered be δ, and let the original message contain M packets. Then if we use a pre-code that generates $K = M/(1 - \delta)$ packets as input points for a LT code then once slightly more than K of the output points have been received, we can recover $(1 - \delta)K$ of the input points, i.e., pre-coded packets, then we can use the pre-code decoder to recover the original message as shown in Figure 6.16. Here, $M = 9$ source packets are pre-coded to generate $K = 13$ input points for the LT Code to generate $N = 15$ output points. Here, output points have low degree and three of the input points have no connection with any of the output points, hence they cannot be recovered by LT decoding, so LT decoding will recover only ten of the input points, which are used by the pre-code decoder to deduce original nine source packets.

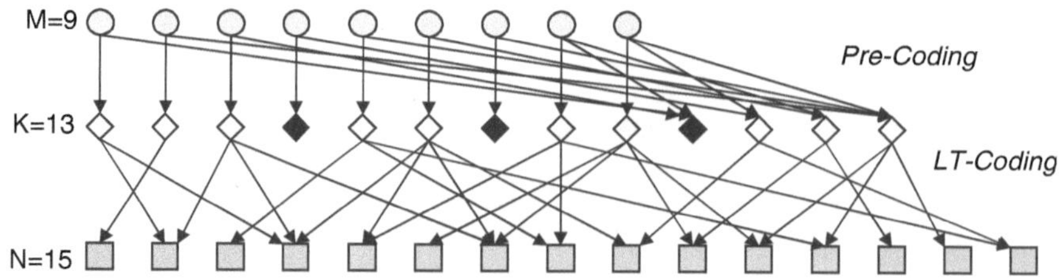

Fig. 6.16. Raptor code schematic diagram.

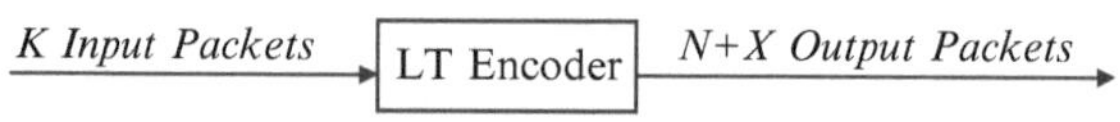

Fig. 6.17. LT encoder.

6.5.4 Secondary Usage of Spectrum using LT Codes

In the secondary user scenario, a SU selects a set of subchannels from PU's bands (Figure 6.4) to establish a Secondary User Link (SUL). A SUL is a set of subchannels that adapts itself in accordance with the PU's spectral activity on that band. The SU is required to vacate the subchannel as soon as the corresponding PU becomes active on that subchannel. This forces SU to loose packets on that subchannel. To compensate for this loss caused by PU Interference, LT codes can be used before transmitting packets on these subchannels (Figure 6.17). Let the SU have a message of K packets to transmit. It uses LT codes to encode it to $N + X$ packets (as shown in Figure 6.17), where N is the number of packets required to recover the original K packets with probability $1 - \epsilon$ and X is the redundancy required to compensate for the loss due to the interference by the PU. This in turn depends on the PU arrival probability p. So the SUL behaves like an erasure channel with erasure probability equal to p.

Assuming that one packet is transmitted per subchannel, the total number of subchannels used by SU is $S = N + X$, and this transmission will be successful only if at most X of the subchannels are captured by the PUs. Therefore, the probability of successful transmission for the SU is

$$P_{\text{success}} = \sum_{i=0}^{X} \binom{N+X}{i} p^i (1-p)^{N+X-i}. \tag{6.15}$$

Let T_p be the time required for transmission of each packet and W be the bandwidth per subchannel. Then the spectral efficiency (η) can be computed by

$$\eta = \frac{N P_{\text{success}}}{S \times W \times T_p}. \tag{6.16}$$

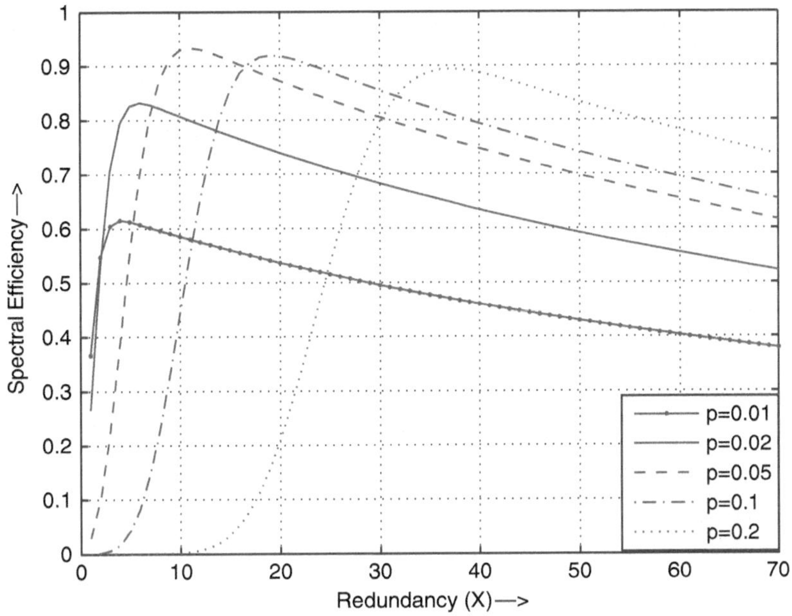

Fig. 6.18. Spectral efficiency vs redundancy with different PU arrival probability p.

For $W = 1/T_p$, the spectral efficiencies for $N = 100$ packets, with respect to subchannel redundancy X and PU arrival probability p are plotted in Figure 6.18. It can be observed that for each p there is an optimal value of redundancy X which delivers the maximum spectral efficiency.

6.6 Summary

This chapter gives an overview of some game and coding theoretic approaches to dynamic spectrum access. Several game theoretic formulations of distributed dynamic spectrum access are presented. Existence of Nash and Pareto optimal solutions to these games are discussed. Stochastic learning automaton based techniques to solve these games using local information are also presented.

Secondary spectrum access is modelled as a communication over an erasure channel. Application of LT and Raptor codes for secondary user link maintenance over these channels are discussed. Some simulation results are presented to outline the trade-off between redundancy and spectral efficiency in the presence of primary users.

References

1. J. Mitola, "Cognitive radio: Making software radios more personal," *IEEE Pers. Commun.*, Vol. 6, No. 4, pp. 13–18, Aug. 1999.
2. K.S. Narendra, and M.A.L. Thathachar, "*Learning Automata: An Introduction*," Englewood Cliffs: Prentice Hall, 1989.
3. D. Fudenberg, and J. Tirole, *Game Theory,* Cambridge: The MIT Press, 1992.
4. P.S. Sastry, V.V. Phansalkar, and M.A.L. Thathachar, "Decentralized Learning of Nash Equilibria in Multi-Person Stochastic Games With Incomplete Information," *IEEE Trans Systems, Man, and Cybernetics.*, Vol. 24. pp. 769–777, May 1994.
5. H. Gintis, *Game Theory Evolving: A Problem-Centered Introduction to Modeling Strategic Behavior*, Princeton University Press, 2000.
6. D. Fudenberg, and J. Tirole, *Game Theory,* Cambridge, MA: MIT Press, 1991.
7. Y. Xing, R. Chandramouli, S. Mangold, and S. Sankar, "Dynamic spectrum access in open spectrum wireless networks," *IEEE J. Selected Areas Communs.* Special Issue on 4G Wireless Systems, Vol. 24, No. 3, pp. 626–637, 2006.
8. http://www.sharedspectrum.com/.
9. R. Kolodziejski, and J. Betz, "Detection of Weak Random Signals in IDD Non-Gaussian Noise," *IEEE Trans. Comm.*, Vol. 48, No. 2, February 2000.
10. S. Mangold, and K. Challapali, "Coexistence of Wireless Networks in Unlicensed Frequency Bands," *Wireless World Research Forum* ♯9 Zurich Switzerland, July 2003.
11. N. Sai Shankar, C. Cordeiro, and K. Challapali, "Spectrum Agile Radios: Utilization and Sensing Architectures," *IEEE DySPAN* 2005.
12. Federal Communication Commission, "Spectrum Policy Task Force," Rep. ET Docket No. 02–135, November 2002.
13. J.C. Haartsen, L. Wieweg, and J. Huschke, "Spectrum Management and Radio Resource Management Considering Cognitive Radio Systems," *XXVIIIth URSI General Assembly*, October 2005.
14. P. Kolodzy, "Spectrum Policy Task Force: Findings and Recommendations," in *International Symposium on Advanced Radio Technologies (ISART)*, March 2003.
15. S. Haykin, "Cognitive Radio: Brain-Emprowered Wireless Communications," *IEEE JSAC*, Vol. 23, No. 2, February 2005.
16. J.M. Peha, and S. Panichpapiboon, "Real-Time Secondary Markets for Spectrum," *Telecommunications Policy*, Vol. 28, No. 7–8, August 2004, pp. 603–618.
17. A. Tonmukayakul, and M.B.H. Weiss, "Secondary Use of Radio Spectrum: A Feasibility Analysis," *TPRC*, October 2004.
18. J. Huang, R. Berry, and M.L. Honig, "Auction-based Spectrum Sharing," to appear in *ACM/Springer Mobile Networks and Applications Journal* (MONET)
19. S. Kiran, and R. Chandramouli, "An adaptive energy efficient link layer protocol using stochastic learning control." *Proc. of the IEEE Intl. Conf. Communications (ICC).*, Alaska, 2003.
20. Y. Xing, and R. Chandramouli, "Distributed discrete power control for bursty transmissions over wireless data networks," *Proc. IEEE ICC*, pp. 139–143, Vol. 1, 2004.
21. D. Willkomm, J. Gross, and A. Wolisz, "Reliable Link Maintenance in Cognitive Radio Systems," *Proc. of the IEEE Symposium on New Frontiers in Dynamic Spectrum Access Networks (DySPAN 2005)*, Baltimore (MD), November 2005.

22. D.J.C. Mackay, *Information Theory, Inference, and Learning Algorithms*, Cambridge, UK: Cambridge University Press.
23. D.J.C. MacKay, "Fountain codes," *Communications, IEE Proceedings*, Vol. 152, No. 6, December 2005, pp. 1062–1068.
24. M. Luby, "LT codes," *Proc. of the 43rd Annual IEEE Symposium on Foundations of Computer Science*, November 16–19, 2002, pp. 271–282.
25. A. Shokrollahi, "Raptor codes," *IEEE Transactions on Information Theory*, Vol. 52, No. 6, June 2006.

7

Efficiency and Coexistence Strategies for Cognitive Radio

Sai Shankar, N

Corporate R&D Systems Engineering,
Qualcomm Incorporated,
San Diego, CA 92121, USA

Abstract. *Cognitive radio devices have been considered as a key technology for next-generation of wireless communication. These devices can opportunistically utilize the wireless spectrum to achieve better individual device/user performance and improve the overall spectrum-utilization efficiency. However, allowing opportunistic use of the wireless spectrum creates new problems such as peaceful coexistence with other wireless technologies as well as understanding the influence of interference that each of these networks can create. In this chapter we model the efficiency of the cognitive radios from the Medium Access Control (MAC) analytically and study the improvement that cognitive radios can get compared to conventional radios. Then we consider the effect of peaceful coexistence with different types of cognitive radios and also consider a simple spectrum sensing protocol which is a function of the primary utilization and required opportunity.*

7.1 Introduction

Wireless networks are regulated today by government agencies wherein spectrum is allocated to a particular application. Such a static allocation has resulted in immense spectrum crunch in some parts of the spectrum and non-usage in some other parts of the spectrum. As a result the average spectral utilization is very low over most of the bands. Since spectrum is a natural resource, any wastage in spectrum impedes the growth of wireless technologies and augments the need for opportunistic usage of spectrum. Cognitive radios are seen as a way to mitigate this low spectral usage. These radios sense the medium and dynamically adapt their waveforms to comply with the compliance policies fixed by the regulatory authorities and opportunistically access portions of the spectrum that are not used by the primary systems.

As an example consider the wireless evolution in unlicensed bands. In these bands any technology that complies with the Federal Communications Commission (FCC) [10], rules for the band is allowed to operate. A good example of such a band is the Industrial Scientific and Medical (ISM) band in 2.4 GHz. There are multiple wireless technologies that are operating in these bands

H. Arslan (Ed.), Cognitive Radio, Software Defined Radio, and Adaptive Wireless Systems, 189–234.

such as IEEE 802.11 Wireless Local Area Networks (WLAN), cordless phones and Bluetooth Wireless Personal Area Networks (WPAN). Other examples of unlicensed frequency bands include the UNI II and UNI III bands where systems such as IEEE 802.11a WLAN and soon IEEE 802.11n WLAN. While unlicensed bands have opened up avenues for the advent of new technologies, their full potential is not realizable because of the presence of interference from other technologies operating in those bands. These unlicensed bands are over crowded, while, some licensed bands, such as the TV bands, are not fully utilized. This results in poor spectrum utilization.

To overcome this poor spectrum utilization, Federal Communication issued an NPRM, [6], in year 2004 which recommends opportunistic usage of wireless technologies in TV bands. The Defense Advanced Research Projects Agency (DARPA) has also started the neXt Generation (XG) Communications program to develop new technologies which allow multiple users to share the spectrum through adaptive mechanisms [11]. The US Army has also been researching the so-called "Adaptive Spectrum Exploitation" (ASE) for real-time spectrum management in the battlefield [12, 13]. Although the focus of these programs is somehow different, the basic principles are the same: if radio devices can explore the wireless spectrum and locate sparsely-used spectral bands, they can exploit these bands in an opportunistic way to improve not only the devices' performance but also the overall spectrum utilization. In the long run, such cognitive radios can also facilitate secondary markets in spectrum use (e.g, a licensee may allow secondary spectrum uses by a third party) and automated frequency coordination among different radio systems [10]. The IEEE 802.22 working group is developing a new standard to provide last mile access to rural areas by opportunistically accessing the spectrum that is not currently used by primary users in the TV bands. Details of this working group can be found at the IEEE 802.22 web site [40].

Of course, such cognitive radios cannot be realized without developing new hardware/software and changing the current spectrum allocation policies. Fortunately, the advances in Software Defined Radio (SDR) [14, 15] have enabled the development of flexible and powerful radio interfaces for supporting spectral agility. Also, the FCC ongoing reviews of the current spectrum regulations are also expediting the adoption of more flexible spectrum allocation policies for spectral agility. However, there remain many open questions that we need to answer before realizing spectral agility.[1] The first and the foremost question is to what extent the improvement can be, in terms of spectrum utilization and individual devices' performance. Without a clear understanding of the achievable improvement, one cannot justify the use of spectral agility since controlling a spectral-agile network may incur considerable amount of overhead. This leads to several implementation questions, including how individual devices discover and identify sparsely-used spectrum bands, how to

[1] Spectral agility is one of the important functions of cognitive radio. We will mainly concentrate on this functionality in this chapter.

characterize or prioritize these spectrum bands, and how and when to utilize them. Obviously, different implementations incur different amounts of control overhead. Thus, the final question is how the thus-incurred control overhead degrades the achievable improvement of using spectral agility. Without answering these questions, it is meaningless and difficult to develop spectral-agile communication protocols and networks in an effective way. IEEE formulated a new working group, in Nov. 2004, called IEEE 802.22 and named it as Wireless Regional Area Network (WRAN).

In this chapter, we address some of these questions as follows. First, we establish an analytical model and provide an upper-bound performance analysis for cognitive radio networks whose main functionality is spectral agility. The analysis sets the benchmark of an ideal spectral-agile network's performance, and thus, enables the evaluation of different implementations. Then we consider the case where different cognitive radio networks have to share the same spectral resources. After that we concentrate on a simple way that the cognitive radio can exploit the spectral opportunity. This chapter is a recap of some of the fundamental results obtained in [1–4].

7.2 System Model

We consider two types of networks, namely *primary* and *secondary*[2] networks. A primary network has exclusive access to designated spectral bands while a secondary network only accesses a spectral band when the primary network does not use that band. For example, a primary network can be any licensed-band network, and a secondary network is an unlicensed-band network such as an IEEE 802.11 wireless LAN. To realize such an opportunistic use of primary networks' spectral resources, we assume that a secondary network has spectral agility, which is enabled by the SDR. It is then a secondary network's responsibility to locate available resources, in both spectral and temporal domains, as shown in Figure 7.1.

Even though it is desirable to have the entire spectrum accessible by a secondary network, hardware limitations (such as antenna designs) usually determine the accessible range. Therefore, the "wireless spectrum" in this chapter is referred to as the portion of the wireless spectrum which can be accessed by a secondary network. The spectrum is divided into "channels", which are the smallest units of a spectral band. We assume that each secondary network only uses a single channel for basic communication, but it can also use multiple channels for better performance. For example, the SDR makes it possible to adopt a modulation scheme requiring higher bandwidth when several adjacent channels are available simultaneously. Moreover, it is also possible to use discrete channels as sub-carriers of a multi-carrier modulation scheme (such as the Orthogonal Frequency Division Multiple Access (OFDMA)) or

[2] Cognitive radios are also called as secondary radios in this chapter.

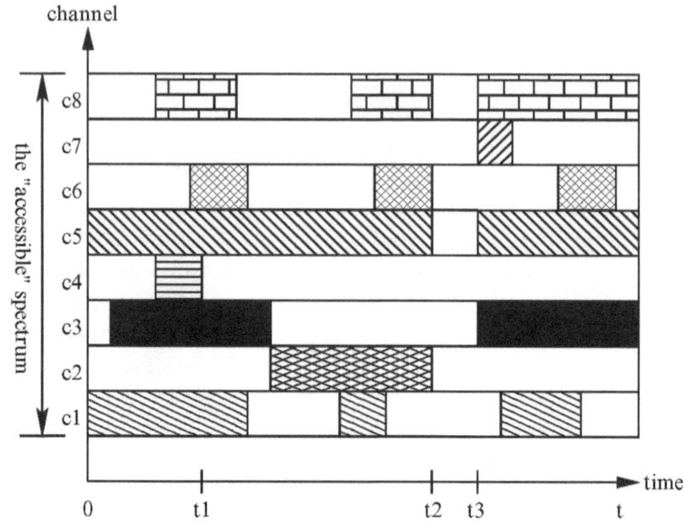

Fig. 7.1. Spectrum opportunities for spectral-agile devices.

use these channels in adaptive frequency hopping [12] for transmission in a multi-path fading environment.

We assume that the temporal usage of each channel (by the primary network/user of that channel) can be characterized by a random process. When a primary network does not always use its designated channel, it leaves some "holes" in the channel's usage schedule which may be exploited by secondary networks. As shown in Figure 7.1, the blank spots represent such holes and each hole is referred to as a *spectral opportunity* in the rest of the chapter. For example, there exists a spectral opportunity in channel 4 after $t = t_1$. Moreover, the entire spectrum is regarded as providing a spectral opportunity between $t = t_2$ and $t = t_3$. Depending on a primary network's spectrum usage pattern, the duration of a spectral opportunity can be up to several hours or even days in spectral bands reserved for emergency, or can be only few milliseconds in heavily-used spectral bands. It is relatively easy for a secondary network to use long-lasting opportunities. However, for short-lasting or "ephemeral" opportunities, a secondary network may not be able to detect their existence and then utilize them before they "expire". Therefore, we only focus on the case when spectral opportunities last in the order of seconds.

In order to exploit spectral opportunities, a secondary network has to first scan the spectrum, either periodically or randomly, to discover the opportunities. It should be noted that the problem here differs significantly from the problems of using dynamic frequency selection mechanisms in the existing systems, such as Dynamic Channel Selection (DCS) [17] in cellular networks, dynamic frequency selection(DFS) [18] in the IEEE 802.11h standard or Auto Frequency Allocation (AFA) [19] in the HiperLAN. These schemes address the problem of choosing a good channel (either a frequency in the Frequency Division Multiple Access (FDMA) system, or time slots in the Time Division Multiple Access (TDMA) system) so that transmission in that channel experiences less interference or causes less interference to other transmissions in the same channel. In our problem, a spectral-agile network seeks both spectral and temporal opportunities in the wireless spectrum, and utilizes these opportunities opportunistically. Among the found opportunities, a spectral-agile secondary network decides on which opportunities to use and when to utilize them. If and when activities of a primary network are detected, the secondary network must vacate the channel in order not to interfere with the primary network. Obvious, all wireless nodes (i.e., radio devices) in a secondary network must use the same spectral opportunity at any time instant to maintain their inter-connectivity. Therefore, the wireless nodes in a spectral-agile network also need to disseminate the information of spectral opportunities and the decision of switching among opportunities. So the MAC of the cognitive radio has to perform the above functions to access the spectrum opportunistically. Having explained the system model of cognitive radio, we will now derive the improvements that can be obtained by cognitive radios compared to conventional radios.

7.3 Spectral Utilization of Cognitive Radio

We establish a mathematical model to analyze the potential improvements gained by using spectral agility, in terms of a secondary network's spectral utilization and packet blocking/waiting time. The spectral utilization of a secondary network is measured by the total amount of time a secondary network can access a channel for transmission. One can convert the channel access time to the network's actual throughput once the underlying Medium Access Control (MAC) and modulation mechanisms are specified. Therefore, we use channel access time so as not to be confined to any specific MAC and modulation schemes. The packet blocking time is defined as the time interval during which a secondary network has no spectral opportunity to utilize (thus, it has to suspend all transmissions).

Consider that there are N primary networks with N designated channels (one designated channel per primary network), and there are M secondary networks seeking spectral opportunities. The usage pattern of the primary network in each channel is assumed to be an *i.i.d.* ON/OFF random process with independent ON- and OFF-periods. An ON-period represents that a channel is busy while an OFF-period is regarded as a potential spectral opportunity for secondary networks. To simplify our analysis, we assume that the distributions of both ON- and OFF-periods in each channel are exponentially distributed with means equal to T_{on} and T_{off}, respectively. We will explore different distributions using simple simulations at the end of this section.

In order to provide a performance upper-bound, we assume that each secondary network has an infinite amount of traffic to transmit. Moreover, each secondary network can scan a channel, switch to a channel and vacate a channel instantly without incurring any control overhead or delay. The control overhead and delays are implementation-dependent, and will certainly impact on the improvement which is not in the scope of this chapter. In order to provide a comparative feel for the performance improvement of using spectral agility, we introduce and use a "naive" secondary network which listens to a fixed channel (i.e., without spectral agility), and transmits only when that channel is not used by the primary network. The spectral utilization of such a naive secondary network can easily be computed as $\frac{T_{off}}{T_{on}+T_{off}}$, and the average blocking time is T_{on}.

7.3.1 A Special Case: $M = 1$

We first consider a special case when there is only one secondary network. As shown in Figure 7.2, the only time interval during which a secondary network has no channel for traffic transmission is when all channels are occupied by the primary networks. Such blocking intervals, denoted as t_{block}, always begin when a channel switches from an OFF-period to an ON-period and ends when one channel switches from an ON-period to an OFF-period. Therefore, t_{block} is computed as

$$t_{block} = \min_{i=1,2,\ldots,N} (T^{(i)}_{remain}), \tag{7.1}$$

where $T^{(i)}_{remain}$ is the remaining ON-period in channel i. Given that the distributions of ON-periods are independent and exponentially distributed, one can compute the distribution of t_{block} as

$$P(t_{block} = t) = \frac{N \cdot e^{-\frac{T_{on}}{N} t}}{T_{on}}. \tag{7.2}$$

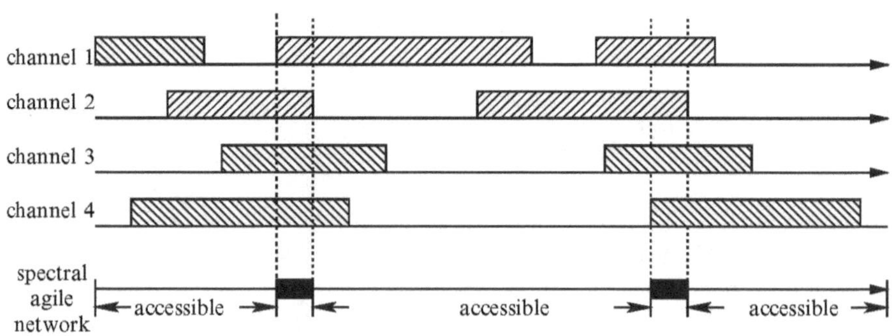

Fig. 7.2. A special case: $N = 4$.

The above equation shows that with spectral agility, a secondary network can reduce the average blocking time to $\frac{T_{on}}{N}$, as compared to T_{on} in the naive secondary network without agility. The spectral utilization of such a spectral-agile secondary network is obtained by

$$U = 1 - \frac{N(p^{N-1} \cdot \frac{T_{on}}{N})}{T_{on} + T_{off}}, \tag{7.3}$$

where $p = \frac{T_{on}}{T_{on}+T_{off}}$ is the probability that a channel is occupied by the primary network. The above equation is derived based on the fact that a blocking interval starts only if a channel switches from an OFF-period to an ON-period while all other channels have already been in the ON-periods. Equation 7.3 can be simplified further to

$$U = 1 - \left(\frac{T_{on}}{T_{on} + T_{off}}\right)^N, \tag{7.4}$$

showing that the spectral utilization of a secondary network is a simple function of the primary network's channel utilization. Finally, the improvement of the spectral utilization achieved by a spectral-agile secondary network is computed as

$$I = \frac{U}{1 - \frac{T_{on}}{T_{on}+T_{off}}} - 1, \tag{7.5}$$

as compared to the naive secondary network.

7.3.2 The General Case

Equation 7.4 shows that the spectral utilization of a spectral-agile secondary network is simply a function of the primary network's channel utilization, $\tau = \frac{T_{on}}{T_{on}+T_{off}}$. We can generalize this simple equation for the case when different channels have different utilizations, say, channel i with utilization $\tau_i = \frac{T_{on}^{(i)}}{T_{on}^{(i)}+T_{off}^{(i)}}$. Based on Equation 7.4, the fraction of time in which there are k channels available simultaneously is computed as

$$r_k = \sum_{c=1}^{\frac{N!}{k!(N-k)!}} \left[\prod_{i \in S_c^k} (1 - \tau_i) \prod_{j \in \{1,2,\cdots,N\} - S_c^k} \tau_j \right], \tag{7.6}$$

where S_c^k is a set of k channels, chosen from N channels, which are available for spectral-agile secondary networks. For example, we can set $S_1^k = \{1, 2, \ldots, k\}$, $S_2^k = \{2, 3, \ldots, k+1\}$, and so on.

To further generalize our analysis, we assume that there are $M > 1$ spectral-agile secondary networks trying to exploit available spectral opportunities. Obviously, each secondary network obtains exactly one channel if there are more than M channels available simultaneously. Otherwise, these M secondary networks have to share the idle channels. The spectral utilization of each spectral-agile network is then computed by

$$U_{agile} = \sum_{k=0}^{N} \frac{\min(M,k) r_k}{M}. \tag{7.7}$$

As we mentioned in Section 7.2, the SDR allows a radio device to dynamically use a variety of MAC and modulation schemes, depending on the underlying wireless environment. Therefore, a spectral-agile network can use multiple channels simultaneously and thus, acquire more channel access time for better performance. We explain how to analyze this type of spectral-agile networks at the end of this section.

Since there are now $M > 1$ secondary networks, each aforementioned naive secondary network (i.e., without agility) can use two approaches to choose a channel: (1) each network selects its own channel and (2) all secondary networks cooperate in a way that no more than one secondary network uses the same channel, if this is possible. The advantage of the first approach is the simplicity while the advantage of the second approach is that each secondary network obtains more channel access time.

1) Approach I: Random channel selection: Given that a secondary network chooses channel i, the probability that the other k secondary networks also choose the same channel is

$$p_k = \frac{(M-1)!}{k!(M-1-k)!} \left(\frac{1}{N}\right)^k \left(\frac{N-1}{N}\right)^{M-1-k}. \tag{7.8}$$

Therefore, the average channel access time that a spectral-agile network can acquire, given that it chose channel i, is

$$T_i = \sum_{k=0}^{M-1} p_k \frac{T_{off}^{(i)}}{(k+1)(T_{on}^{(i)} + T_{off}^{(i)})}. \tag{7.9}$$

The fraction of time in which each (no-agility) secondary network has a channel for its traffic transmission is then computed as

$$U_{random} = \frac{1}{N} \sum_{i=1}^{N} T_i. \tag{7.10}$$

2) Approach II: Coordinated channel selection: If each secondary network chooses a channel in a coordinated way to avoid more than one network using the same channel, the fraction of channel access time is computed as

$$U_{coordinated} = \frac{\sum_{c=1}^{\frac{N!}{M!(N-M)!}} \frac{1}{M} \sum_{i \in S_c^M} \frac{T_{off}^{(i)}}{T_{on}^{(i)} + T_{off}^{(i)}}}{\frac{N!}{M!(N-M)!}}. \tag{7.11}$$

Here, we simply average all the possibilities of choosing M channels from N channels for the (no-agility) secondary networks. We set $\frac{N!}{M!(N-M)!} = 1$ in case of $M > N$.

Now, we can compare the spectral utilization of a secondary network using (1) spectral agility, (2) no agility with random channel selection (Approach I) and (3) no agility with coordinated channel selection (Approach II) based on Equations 7.7, 7.10, and 7.11. We investigate two scenarios with $N = 12$ and $N = 3$, respectively. The main reason for choosing these numbers is that there are 12 (non-overlapping) channels in the 5-GHz band for the IEEE 802.11a wireless LAN and 3 (non-overlapping) channels in the 2.4-GHz band for the IEEE 802.11b wireless LAN.[3] Therefore, even though spectral agility cannot be applied immediately to the licensed bands due to the current regulations, the 802.11 wireless LAN may use spectral agility to improve performance in the crowded, unlicensed bands.

Figure 7.3 shows the case of $N = 12$ and $M = 9$ with different average channel loads generated by the primary networks. For each given channel load, we choose the loads of these 12 channels in two different ways: homogeneous

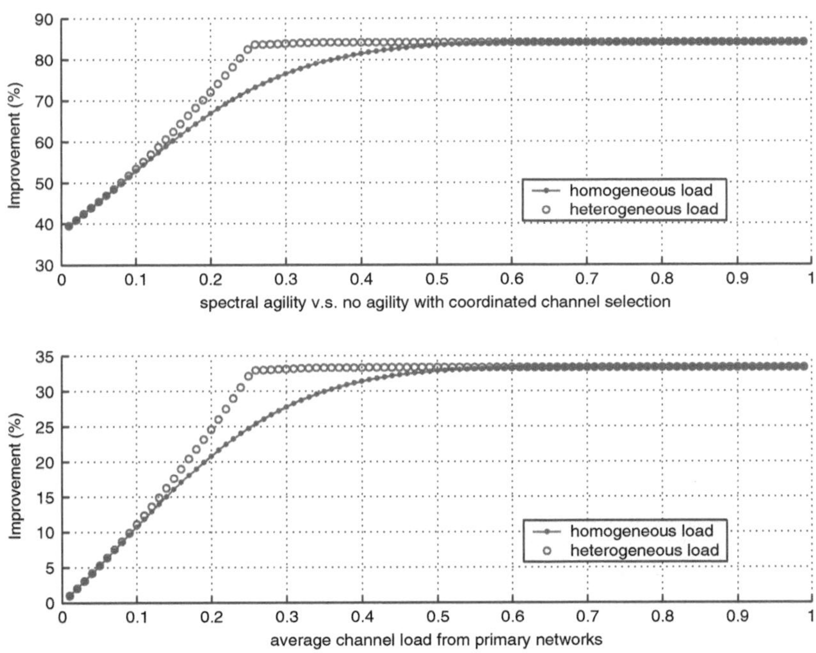

Fig. 7.3. Improvement of spectral utilization for spectral-agile networks: $N = 12$ and $M = 9$.

[3] According to the US regulation, there will be more released channels in the 5-GHz band.

and heterogeneous loads. By choosing homogeneous loads, each channel has a load equal to the average channel load. By choosing heterogeneous loads, we let each channel have a different load and the variance of these loads be maximized (i.e., the utilization of each channel differs from each other significantly). The results show that using spectral agility always achieves a higher spectral utilization of the secondary network, as compared to the case of either no agility with random channel selection or no agility with coordinated channel selection. Of course, the improvement by using spectral agility is much less (still more than 25% in most cases) as compared to the case of no agility with coordinated channel selection. However, using coordinated channel selection needs off-line channel information for selecting channels. If the channel loads are very diverse, it is possible that a secondary network may choose the busy channels (unless it scans all channels for a long enough period of time). In contrast, using spectral agility allows a secondary network to dynamically choose the channel with the least activities. Such advantages are also illustrated in Figure 7.3 as we achieve an extra 8–10% improvement in the case of heterogeneous loads when the channel load is around 0.2–0.3.

Another observation is that the improvement saturates when the average channel load of the primary network is greater than 0.5. This can be explained by Figure 7.4, which shows the fraction of time in which a secondary

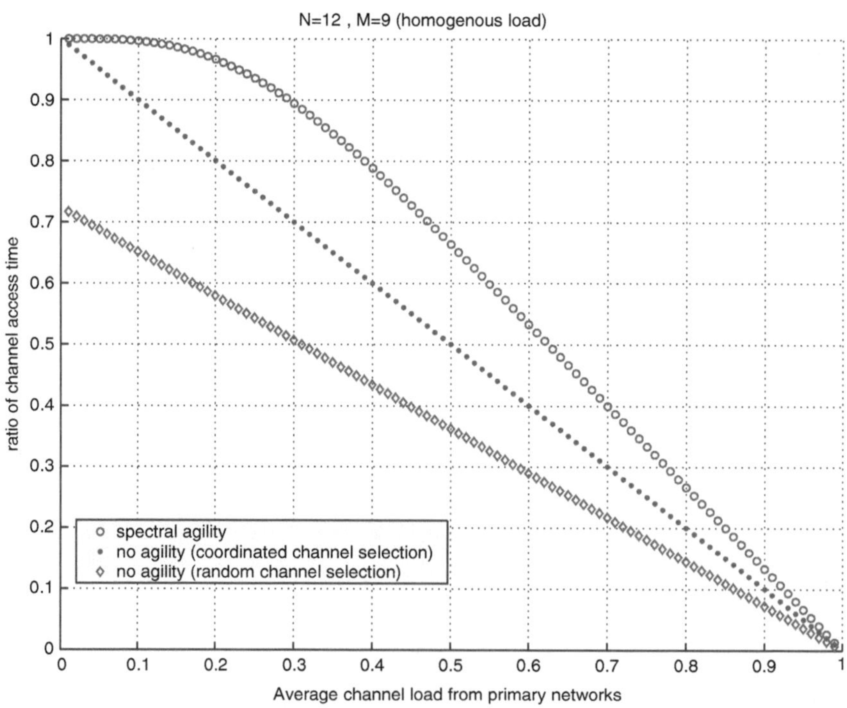

Fig. 7.4. Fraction of channel access time: $N = 12$ and $M = 9$.

network can access a channel. The fraction of time in which a secondary network can access a channel linearly decreases with the increase of the average channel load when the channel load beyond 0.3. Because of such linearity, the improvement of using spectral agility remains unchanged when the channel load increases. Figure 7.4 also suggests that when the average channel load of the primary network is very large, it does not make much sense to use spectral agility as suggested by Figure 7.3 (where it shows an 80% improvement when the load is 0.9). This is because when the channel is extremely busy, the amount of access time that each spectral-agile network can obtain is very small (less than 10% of the total time when the channel load is 0.9). Therefore, the control overhead (incurred by using spectral agility) may exhaust most of the channel access time a secondary network obtains, and thus, easily offsets the improvement gained by using spectral agility.

Next, we consider the case of $M > N$ and choose $N = 3$ and $M = 5$ as an example. Figure 7.5 shows that using spectral agility and using no agility with coordinated channel selection achieve exactly the same performance. The results make sense because when $M > N$, there are simply not enough

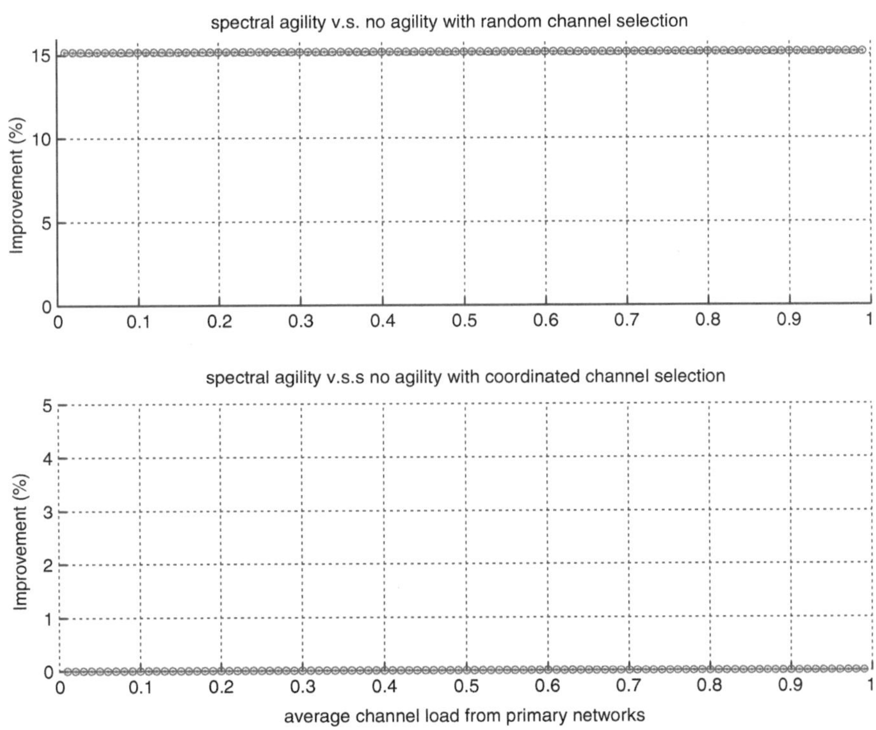

Fig. 7.5. Improvement of spectral utilization for spectral-agile networks: $N = 3$ and $M = 5$.

channels for all secondary networks (so they have to share the idle channels with each other). There are some marginal improvements by using spectral agility as compared to using no agility with random channel selection. This is simply because some idle channels may be left unused in the case of random channel selection. In fact, one can simplify both Equations 7.7 and 7.11 as

$$U_{agile} = U_{coordinated} = \frac{1}{M} \sum_{i=1}^{N} \frac{T_{off}^{(i)}}{T_{on}^{(i)} + T_{off}^{(i)}}, \tag{7.12}$$

when $M > N$.

Fortunately, field studies have shown that there are many under-utilized spectral resources in some wireless spectral band [20, 21]. Moreover, there are two additional advantages of using spectral agility that we have not yet discussed when $M > N$. First, Equation 7.2 shows that when the spectral agility is used, the average blocking time is reduced by a factor of N in the special case or reduced from $\frac{\sum T_{on}^{(i)}}{N}$ to $\frac{1}{\sum \frac{1}{T_{on}^{(i)}}}$ in the general case. Thus, even though the spectral utilization is not improved by using spectral agility when $M > N$, the packet delays are reduced significantly by using spectral agility. Another advantage is the spectral-agile network's capability of using multiple channels. In the above analysis, we assumed that a spectral-agile network (or more precisely, the wireless nodes in the network) always uses a single channel, even when more than one channel are available. We can expect that if a spectral-agile network can use all available channels, the performance must be improved. Figure 7.6 illustrates this scenario in which each spectral-agile network aggregates all available channels into a single, higher-capacity spectral opportunity. Then all spectral-agile networks use this aggregated opportunity, in contrast with using separate channels for transmission as discussed earlier.

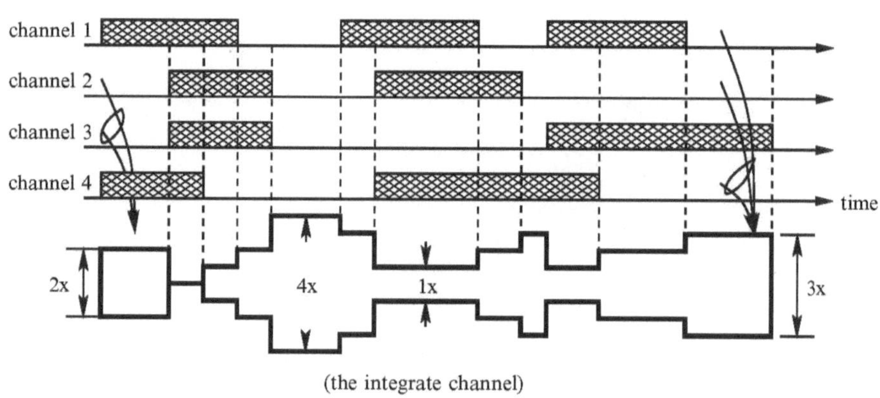

Fig. 7.6. Use of multiple channels: $N = 4$.

This will provide a multiplexing gain just as we can obtain by multiplexing several traffic flows on a high-capacity transmission line in conventional wired networks.

Before we analyze the multiplexing gain of using multiple channels, let us investigate the effects of different ON/OFF distributions on the improvement of spectral utilization, gained by using spectral agility. The main purpose of this study is to verify the applicability of our model, which is established based on the assumption of exponentially-distributed ON-/OFF periods. Here, we use Matlab to simulate the random ON/OFF periods and calculate the total time intervals of overlapping ON-periods (i.e., the blocking intervals for a spectral-agile network) for the case of $N = 3$ and $M = 1$. We use exponential (as in our earlier derivation), uniform and Rayleigh distributions. Figure 7.7 shows a very good match between our analytical results and the simple simulation results, verifying the applicability of our analytical model. The reason why the improvements are much higher is because there is only one spectral network seeking spectral opportunities, and thus, it need not share spectral opportunities with other spectral-agile networks. Also as we discussed earlier, such a large improvement in fact represents only a very small increase

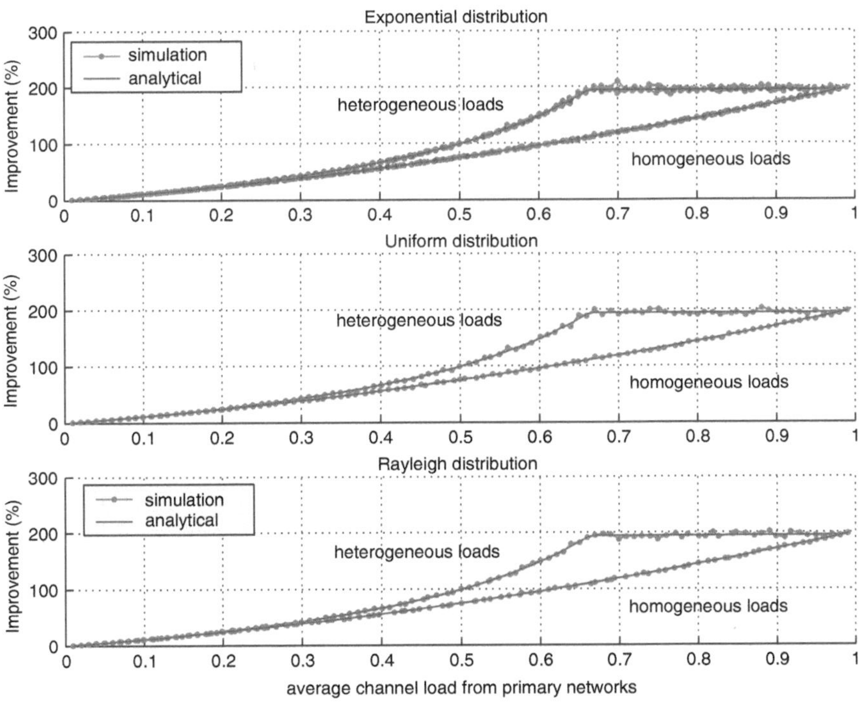

Fig. 7.7. Improvement of spectral utilization for spectral-agile networks: different ON/OFF distributions.

of channel access time (for a secondary network) if the average channel load of the primary network is extremely high.

7.3.3 The Multiplexing Gain of Using Multiple Channels

If all spectral-agile networks use the aggregated spectral opportunities, packets from all spectral-agile networks share the same "channel" with a varying transmission capacity as shown in Figure 7.6. The transmission capacity depends on how many primary networks are using the channels, and the distribution of the transmission capacity is determined by Equation 7.6. If the arrival process in each spectral-agile network is assumed to be Poisson with rate λ, the aggregated arrival process is also Poisson with rate $M\lambda$. Therefore, the system can be modelled as an M/G/1 queueing system. However, it is possible that all channels are occupied by primary networks, with probability r_0 in Equation 7.6 and for an average duration of $\frac{1}{\sum_i^N \frac{1}{T_{on}^{(i)}}}$, so that the transmission capacity is zero from the spectral-agile networks' perspectives. This blocking process is modelled as another arrival process with rate $\frac{1}{r_0}$, and the "packet" with an average service time of $\frac{1}{\sum_i^N \frac{1}{T_{on}^{(i)}}}$. The resulting M/G/1 queue with preemptive priority is illustrated in Figure 7.8(b). The average packet waiting time of a spectral-agile network (Figure 7.9) is then computed by using the results in [22] as

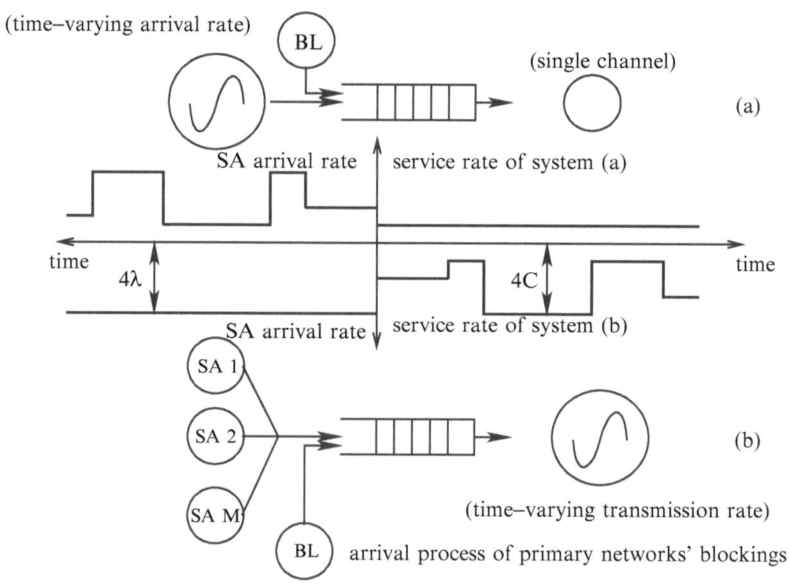

Fig. 7.8. Queueing models for statistical multiplexing gain: $N = 4$.

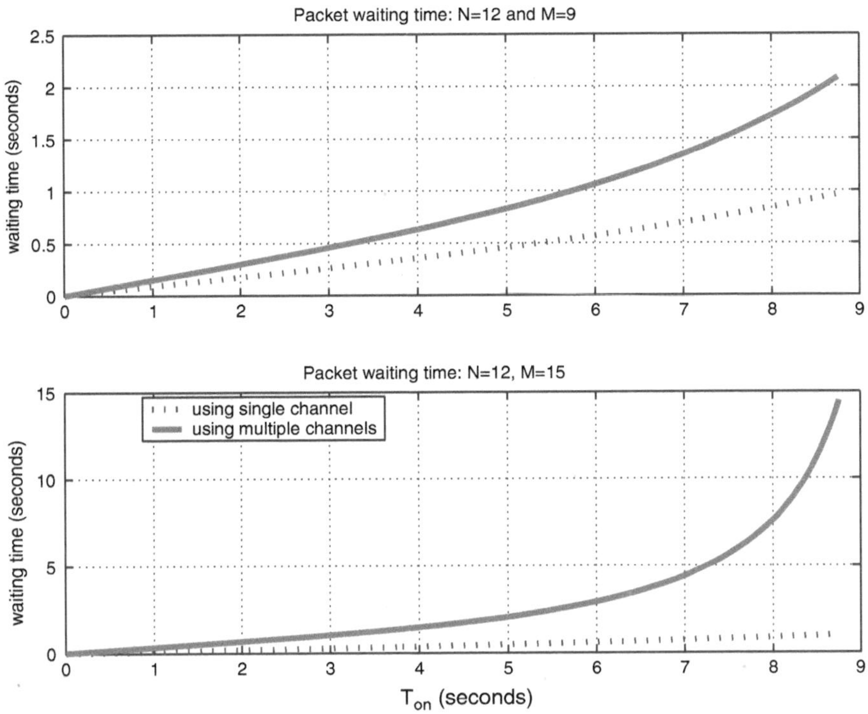

Fig. 7.9. Average waiting time of packets from a spectral-agile network: $T_{off} = 1$ and $\frac{L}{C} = 0.1$.

$$T_{SA} = \frac{\frac{1}{\mu_p}(1 - \rho_p - \rho_{SA}) + R_{SA}}{(1 - \rho_p)(1 - \rho_p - \rho_{SA})}, \tag{7.13}$$

where $\mu_p = \sum_i^N \frac{1}{T_{on}^{(i)}}$, $\rho_p = \frac{1}{r_0 \mu_p}$, ρ_{SA} represents the server utilization of the spectral-agile network, and R_{SA} represents the average residual service time seen by the packets of spectral-agile networks.

If we assume the average packet size is L and the transmission capacity of a single channel is C, ρ_{SA} is computed as

$$\rho_{SA} = \frac{M\lambda}{\mu_{SA}}, \tag{7.14}$$

where $\frac{1}{\mu_{SA}} = \sum_{i=1}^N \frac{L}{iC} r_i$ is the average service time of a packet from spectral-agile networks. Finally, the residual time R_{SA} is computed as

$$R_{SA} = \frac{1}{2}\left[M\lambda \sum_{i=1}^{N} \left(\frac{L}{iC}\right)^2 r_i + \frac{2}{r_0 \mu_p^2}\right], \tag{7.15}$$

as derived in [22].

We can use the M/G/1 queueing model with preemptive priority for the case when each spectral-agile network uses at most one channel. In this case, the "service rate" is constant (from the perspective of packets of a secondary network), and is equal to the transmission capacity of a single channel unless all the channels are occupied by the primary networks. However, packet arrivals in a channel changes with the number of active primary networks. That is, the packet arrivals in a channel are dependent on the state of the primary network's occupation of the spectrum. The less the number of idle channels, the greater the arrival rate in each idle channel. We can model this arrival process as a Markov-Modulated Poisson Process (MMPP) using Equation 7.6, but for the sake of simplicity we approximate the arrival process as a Poisson process, which gives us an M/D/1 queue with preemptive priority. In order to model it as a Poisson process, we need to calculate the average arrival rate.

1) Case I: $M < N$:

If there are at least M channels available, then the arrival rate at the $M/G/1$ queue is just λ. If the number of available channels is $M-1$, then one of the spectral-agile networks joins the channel which has already been "occupied" by another spectral-agile network. That is, multiple spectral-agile networks share one channel. The average arrival rate is then $\frac{M}{M-1}\lambda r_k$. Proceeding similarly, we have the average arrival rate computed as

$$\lambda_{new} = \sum_{i=M}^{N} r_i\lambda + \sum_{i=1}^{M-1} r_{M-i}\Big(\frac{M}{M-i}\Big)\lambda. \tag{7.16}$$

2) Case II: $M > N$:

In this case we have more spectral-agile networks than the total number of channels. If none of the channels are occupied by the primary network, then the best-case arrangement occurs when each channel has $\lceil \frac{M}{N} \rceil$ spectral-agile networks. Proceeding similarly to the previous subsection, we have the arrival rate λ computed as

$$\lambda_{new} = \sum_{i=0}^{N-1} r_{N-i}\Big(\frac{M}{N-i}\Big). \tag{7.17}$$

Finally, we can use Equations 7.13 and 7.15 with the new average arrival rate and constant packet service time $\frac{L}{C}$. The calculation is available in Appendix A.

Figure 7.13 plots the average packet waiting time of a spectral-agile when using a single channel and multiple channels. We fix the value of $T_{off} = 1\,\text{s}$ and change the value of T_{on}, so that we can vary the average channel load from the primary network in each channel. Obviously, the average packet waiting time in the case of $M < N$ is less than that in the case of $M > N$ since there are less spectral-agile networks seeking spectral opportunities in the case of $M < N$. However, the packet waiting time of using multiple channels is always less than that of using a single channel in both cases. The improvement is even more significant in the case of $M > N$ as we expect. These numerical results demonstrate the potential advantage of using multiple channels in a spectral-agile especially when $M > N$.

7.4 Coexistence and the Access Problem in Cognitive Radios

Having studied the gains that can be obtained using cognitive radio, we will now study the coexistence methodology based on the experience in unlicensed bands. Some concerns on the unlicensed spectrum usage are discussed in [32–34]. Within unlicensed frequency bands, the radio systems have to coordinate the usage of radio resources among themselves. With this open spectrum approach, there is then of course the problem of how to achieve fair and efficient sharing of radio resources between dissimilar radio systems that cannot communicate with each other. Unlicensed frequency bands are typically used by many dissimilar radio systems to provide a large set of different radio services. However, unlicensed frequency bands may be more efficiently used when the usage of the radio resources is coordinated by means of spectrum etiquette rules. A spectrum etiquette is a set of rules for radio resource management to be followed by radio systems that operate in an unlicensed band. It may help to establish a fair access to the available radio resources, in addition to a more efficient usage of the radio spectrum.

This and subsequent sections will discuss the following:

- Continuous-time Markov chain modelling of spectrum etiquette for dynamic spectrum access is presented. Systems with and without queueing are investigated.
- A fair, random channel access protocol is derived based on the Markov model.
- The channel access protocol is extended to cognitive radios.

Recently, as more and more communication protocols and commercial wireless devices are being developed to operate in crowded unlicensed spectrum bands, spectrum inefficiency is becoming a serious problem. Spectral agility is being paid considerable attention for its potential to alleviate the spectrum access inefficiency problem [35–37]. If the radio device has the flexibility of switching operating spectral bands, promising improvement of spectral efficiency is expected. Of course, such spectral agility cannot be achieved without developing new hardware/software and changing the current spectrum allocation policies. Fortunately, the advances in software defined radio (SDR) [40, 41] have enabled the development of flexible and powerful radio interfaces for supporting spectral agility. Also, the FCC's ongoing review of the current spectrum regulations is also expediting the adoption of more flexible spectrum allocation policies for spectral agility. We will show how future open spectrum scenarios can be engineered with this approach to improve spectrum efficiency and fairness in spectrum access.

7.4.1 Channel and Traffic Model

Assume a perfect knowledge of the channel. That is, a channel is either busy or idle. It is also assumed that radio systems always detect radio resource allocations of other radio systems. The offered traffic is modelled with two random processes per radio system. The arrival traffic is modelled as a Poisson random process with rate λ_i for radio system i, so the inter-arrival time is negative-exponentially distributed with mean time $\frac{1}{\lambda_i}$ ms (1 ms = 1 millisecond). The radio system access duration is also negative-exponentially distributed with mean time $\frac{1}{\mu_i}$ ms, so the departure of the radio system i is another Poisson random process with rate μ_i. And we assume that spectral scanning is performed instantaneously, so there is no scanning delay.

7.4.2 Usage Model and Etiquette Definition

The 5 GHz unlicensed frequency band is a candidate for a large set of radio services, and is one of the unlicensed frequency bands that may be efficiently used only with an established spectrum etiquette. We use the same abstract model of an unlicensed frequency band as in [38], as illustrated in Figure 7.10.

Here, two different types of radio systems are assumed to operate in the band, each operating with different frequency channel bandwidths. The radio systems of type A operate on three frequency channels (center frequencies f_2, f_5, f_8), the radio systems of type B operate on nine frequency channels (center frequencies $f_1 \ldots f_9$). The frequency channels overlap with each other, as indicated in the figure. The number and bandwidth of the frequency channels in Figure 7.10 do not represent any existing unlicensed band, this usage model serves without loss of generality only as an example model. Here, radio system A can be compared to wireless LANs operating in the 5 GHz band

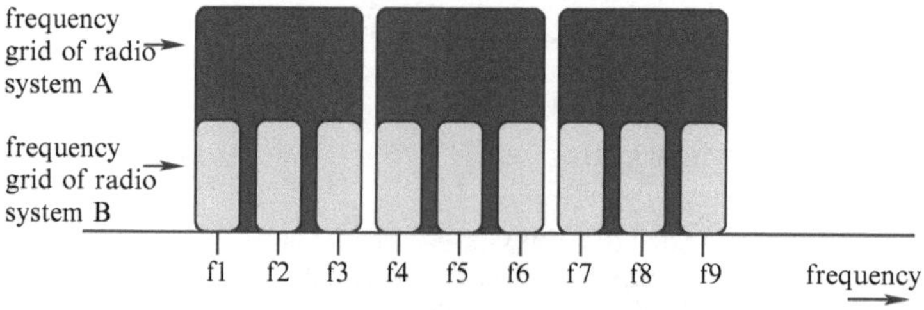

Fig. 7.10. Frequency channels used by two different types of radio systems (A, B). Each radio system represents a group of communication radio devices.

(using OFDM). Radio system B represents narrow-band radio systems supporting for example a limited number of voice calls or blue tooth systems. In our scenario instead of modelling the detailed protocols, a simplified Listen Before Talk (LBT) is used for all radio systems. A type A radio system requires the respective three frequency channels to be idle before allocating radio resources. Only if the respective channels are idle, a radio system allocates radio resources, otherwise it will be dropped, i.e., there is no queuing. The radio systems only scan their own frequencies, for example a radio system B with center frequency f_2 looks only in its frequency and not in any other frequency. Collisions of allocation attempts occur when more than one radio system detects the channel as idle at the same time. In simulations, when collisions happen, one of the radio systems is randomly selected to allocate the radio resource, the other radio systems are dropped.

Two of the most representative etiquette rules defined in [38] are as follows:

- Rule ♯4: A radio system of type A or type B should apply LBT when operating.
- Rule ♯6: In order to protect other radio systems most efficiently, a radio system B that follows rule ♯4 should synchronize its LBT process in time across neighboring frequency channels that overlap with the same A channels.

7.4.3 Markov Modelling

1) Equal traffic load without queuing: The unlicensed spectrum access problem can be modelled as a continuous-time Markov chain (Figure 7.11). Without loss of generality, we can model the two radio system access model illustrated in Figure 7.10 as a five state Markov chain as shown in Figure 7.12. The five states of the Markov chain are described in Table 7.1. The assumption here is that for each type of the radio systems, we have the same traffic load and occupation time. Later, we will relax this assumption and propose a more general model. Since the contentions of the spectrum only take place between radio systems A and B, we focus on one of the type A radio systems and the three type B radio systems whose required spectrum is within the type A

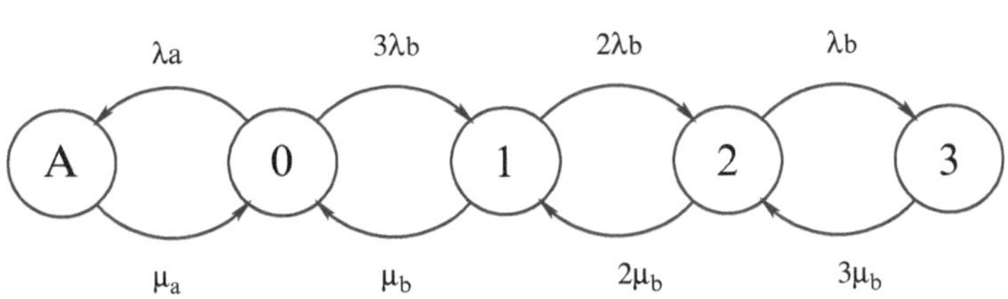

Fig. 7.11. Continuous-time Markov chain with five states to model the unlicensed spectrum access process.

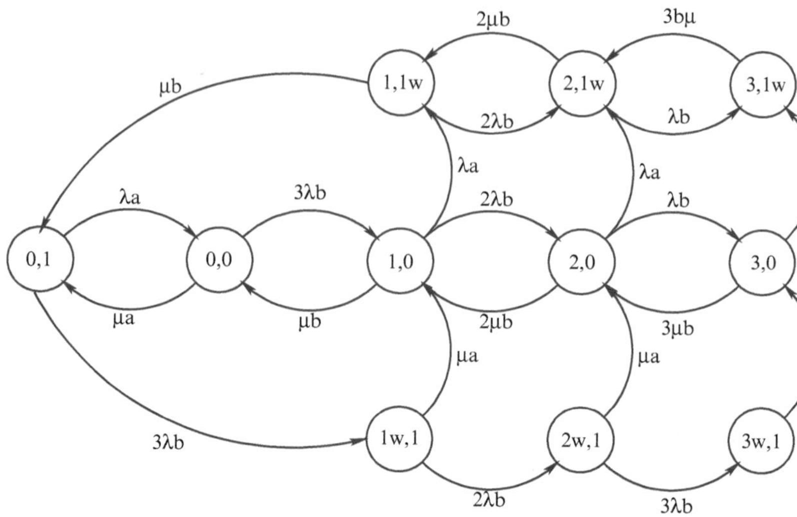

Fig. 7.12. Markov chains to model the unlicensed spectrum access process with waiting.

Table 7.1. The five states of the Markov chain.

State	Description
A	Radio system A occupies the reference spectrum range
0	All the three frequency grids are idle
1	There is only one type B radio system in the reference range
2	There are two type B radio systems in the reference range
3	There are three type B radio systems in the reference range

radio system's spectrum range. Here, we call this spectrum range as reference range. As collisions rarely happen especially with low traffic load, in this Markov model we omit the collision state.

We define an infinitesimal generator matrix **A** to characterize the transition of the states of the Markov chain. The infinitesimal generator matrix with the sum of each row equalling zero is given as follows:

$$\mathbf{A} = \begin{bmatrix} -\mu_a & \mu_a & 0 & 0 & 0 \\ \lambda_a & -\lambda_a - 3\lambda_b & 3\lambda_b & 0 & 0 \\ 0 & \mu_b & -\mu_b - 2\lambda_b & 2\lambda_b & 0 \\ 0 & 0 & 2\mu_b & -2\mu_b - \lambda_b & \lambda_b \\ 0 & 0 & 0 & 3\mu_b & -3\mu_b \end{bmatrix}. \tag{7.18}$$

Then we have,

$$\Pi\mathbf{A} = 0, \tag{7.19}$$

where $\Pi = [\Pi_A, \Pi_0, \Pi_1, \Pi_2, \Pi_3]$ is the steady state probability vector and Π_i represents the probability of being in state i. Solving recursively, we can get:

$$\Pi = [1, P_0, P_1, P_2, P_3]\, P_A, \tag{7.20}$$

where,

$$P_A = \left[\frac{\mu_a}{\lambda_a} + 1 + \frac{3\lambda_b\mu_a}{\lambda_a\mu_b} + \frac{3\lambda_b^2\mu_a}{\lambda_a\mu_b^2} + \frac{\lambda_b^3\mu_a}{\lambda_a\mu_b^3}\right]^{-1},$$
$$P_0 = \frac{\mu_a}{\lambda_a}, \quad P_1 = \frac{3\lambda_b\mu_a}{\lambda_a\mu_b}, \quad P_2 = \frac{3\lambda_b^2\mu_a}{\lambda_a\mu_b^2}, \quad P_3 = \frac{\lambda_b^3\mu_a}{\lambda_a\mu_b^3}. \tag{7.21}$$

One of the most important metrics in the unlicensed-band access is the average airtime per radio system. Airtime refers to the ratio of allocation time per radio system type to the reference time (say 1h) [38]:

$$airtime_{type=A,B} = \frac{1}{N_{type}} \sum_{i=1}^{N_{type}} \frac{allocation\ time(i)}{reference\ time}, \tag{7.22}$$

where N_{type} is the number of type i radio systems. Based on the previous Markov model, the airtime can be approximated by

$$\begin{aligned} airtime_{type_A} &= \Pi_A, \\ airtime_{type_B} &= \frac{1}{3}\Pi_1 + \frac{2}{3}\Pi_2 + \Pi_3. \end{aligned} \tag{7.23}$$

From the above equation we can see that when radio systems A and B are given the same high traffic load, $airtime_{type_B} \gg airtime_{type_A}$ which is not fair for radio system A. As the traffic load of radio system B increases, the $airtime_{type_A} \rightarrow 0$ which is unacceptable for the broader band radio system A. Some etiquette rules were proposed in [38] to mitigate this unfairness, but the improvements are limited.

2) Equal Traffic Load with Queuing: As queuing can increase the server utilization and hence the throughput in classical queueing systems, we also expect increase in terms of *airtime* share of these coexisting radio systems by introducing waiting (queuing). With queuing, instead of dropping the radio system when the channels are busy, they continue scanning until the channel(s) become idle. Of course, the tradeoff here is between the waiting time and the spectrum utilization. When there are more radio systems of the same type than the available spectrum channels, collisions will be pronounced. In order to decrease the probability of collisions, collision avoidance techniques should be used, such as random back-off, or CSMA/CA. In this chapter, since our focus is on the effect of coexistence of dissimilar radio systems on unlicensed frequency bands, a perfect collision avoidance among resource allocations is assumed.

We again consider the model shown in Figure 7.10 with equal traffic load within the same radio system. At this time, only if the respective channels are idle, a radio system allocates radio resources, otherwise it continues to scan until the channels become idle. The Markov model that characterizes this scheme is illustrated in Figure 7.13. Here, each state is (k_1, k_2), where k_i equals the number of radio system i on the spectrum block. While $k_i w$ means

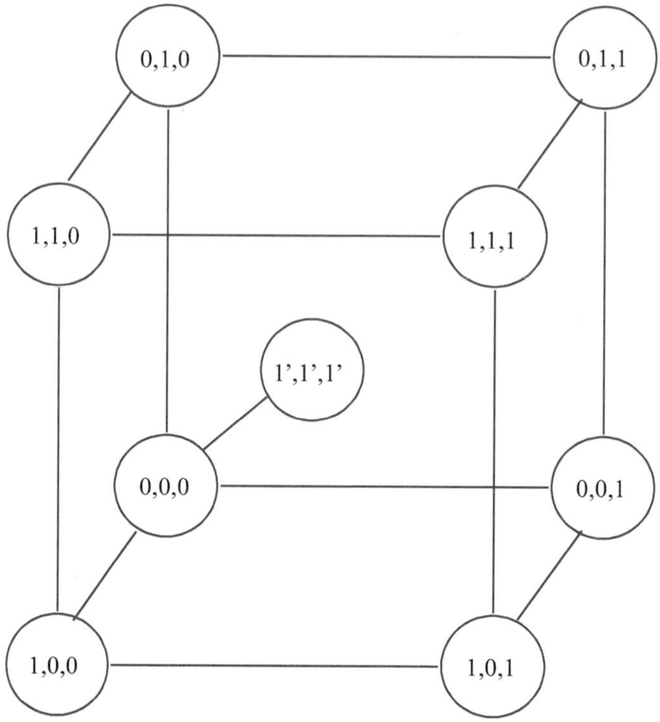

Fig. 7.13. Markov chain with nine states to model the unlicensed spectrum access process.

the number of waiting radio systems of type i. We can solve for the *airtime* share by the general method described in Section 7.4.5.

3) Markov Model for General Traffic Load: The Markov chain shown in Figure 7.13 models a more general case where within each radio system type, there may exist different traffic loads and occupation times. In Figure 7.13, a state is represented by the triplet (n_1, n_2, n_3). In this, n_1, n_2 and n_3 represent the status of the radio system of type B occupying carrier frequencies f_1, f_2 and f_3, respectively. Here, $n_i = 0$, $i = 1, 2, 3$ indicates that the radio system of type B requiring carrier frequency f_i is idling, while $n_i = 1$, $i = 1, 2, 3$ indicates that it occupies the carrier frequency f_i. State $(1', 1', 1')$ represents radio system A occupies the three frequency channels.

Here, all the states are connected by straight lines that are bi-directional. Let λ_1, λ_2, λ_3 and μ_1, μ_2, μ_3 represent the arrival rates and service rates of the three type B systems whose center frequencies are f_1, f_2 and, f_3 respectively. Let λ_a and μ_a represent the arrival and service rates of type A system respectively. The basic equation governing the above system is given by

$$\begin{aligned}
\lambda_a P_{0,0,0} &= \mu_a P_{1',1',1'}, \\
(\lambda_1+\lambda_2+\lambda_3)P_{0,0,0} &= \mu_1 P_{1,0,0} + \mu_2 P_{0,1,0} + \mu_3 P_{0,0,1}, \\
(\mu_1+\lambda_2+\lambda_3)P_{1,0,0} &= \lambda_1 P_{0,0,0} + \mu_2 P_{1,1,0} + \mu_3 P_{1,0,1}, \\
(\lambda_1+\mu_2+\lambda_3)P_{0,1,0} &= \mu_1 P_{1,1,0} + \lambda_2 P_{0,0,0} + \mu_3 P_{0,1,1}, \\
(\lambda_1+\lambda_2+\mu_3)P_{0,0,1} &= \mu_1 P_{1,0,1} + \mu_2 P_{0,1,1} + \lambda_3 P_{0,0,0}, \\
(\mu_1+\mu_2+\lambda_3)P_{1,1,0} &= \lambda_2 P_{1,0,0} + \lambda_1 P_{0,1,0} + \mu_3 P_{1,1,1}, \\
(\mu_1+\lambda_2+\mu_3)P_{1,0,1} &= \lambda_3 P_{1,0,0} + \mu_2 P_{1,1,1} + \lambda_1 P_{0,0,1}, \\
(\lambda_1+\mu_2+\mu_3)P_{0,1,1} &= \mu_1 P_{1,1,1} + \lambda_2 P_{0,0,1} + \lambda_3 P_{0,1,0}, \\
(\mu_1+\mu_2+\mu_3)P_{1,1,1} &= \lambda_1 P_{0,1,1} + \lambda_2 P_{1,0,1} + \lambda_3 P_{1,1,0}, \\
P_{0,0,0} + P_{0,0,1} + \cdots + P_{1,1,1} &+ P_{1',1',1'} = 1, \qquad (7.24)
\end{aligned}$$

where P_{n_1,n_2,n_3} represents the probability of being in state (n_1, n_2, n_3), $n_1, n_2, n_3 = 0, 1$. The above equation, with the exception of the state $(1', 1', 1')$, represents three independent M/M/1/1 queues. The solution to the above equation is given by

$$P_{n_1,n_2,n_3} = C \left[\frac{\lambda_1}{\mu_1}\right]^{n_1} \left[\frac{\lambda_2}{\mu_2}\right]^{n_2} \left[\frac{\lambda_3}{\mu_3}\right]^{n_3}, \quad n_1, n_2, n_3 = 0, 1, \qquad (7.25)$$

where C is the normalizing constant. The inclusion of the state $(1', 1', 1')$ also represents an M/M/1/1 queue that is independent of P_{n_1,n_2,n_3}. For simplicity and in order to get more insight into the analysis, we use the five state Markov model (Figure 7.12) unless stated otherwise.

7.4.4 Random Access Model

Efficiency and fairness are obviously the main goals of a spectrum etiquette. As discussed before, if every radio system accesses the unlicensed band in a greedy manner, then the radio system requiring broader band to operate will suffer from an unacceptable low airtime share. So one way to provision more fairness to the etiquette rules would be to require each radio system to work in a cooperative manner. One option would be that each radio system i tries to contend for the spectrum with probability p_i. After the radio system has decided to contend for the spectrum, it accesses the spectrum compliant to etiquette rule ♯4 described in Section 7.4.2.

This random access scheme can be approximated by slightly modifying the previous five state Markov model. Here, each radio system will only contend for the spectrum with probability p_i, so the actual traffic load to the system can be approximated by $p_i\lambda_i$. If perfect fairness is achieved then we have

$$airtime_{type_A}(p_a, p_b) = airtime_{type_B}(p_a, p_b). \qquad (7.26)$$

Then from Equation 7.23, we have

$$p_a(p_b) = \frac{1}{3}p_b P_1 + \frac{2}{3}p_b^2 P_2 + p_b^3 P_3. \tag{7.27}$$

So

$$airtime_{type_A}(p_b) = \frac{\frac{1}{3}p_b P_1 + \frac{2}{3}p_b^2 P_2 + p_b^3 P_3}{P_0 + \frac{1}{3}p_b P_1 + \frac{2}{3}p_b^2 P_2 + p_b^3 P_3 + p_b P_1 + p_b^2 P_2 + p_b^3 P_3}. \tag{7.28}$$

When the *airtime* for both radio systems A and B are the same, we can find the optimal p_b by maximizing the *airtime*. Since

$$\frac{\partial airtime_{type_A}}{\partial p_b} > 0, \tag{7.29}$$

$airtime_{type_A}$ is an increasing function of p_b. So the optimal p_b is the largest possible p_b. Depending on different λ and μ values, we have the following two cases:

- $p_a(p_b = 1) > 1$, we cannot use the maximum value of $p_b = 1$ to maximize the *airtime* function because of $0 < p_a \leq 1$. But as $p_a(p_b)$ is an increasing function of p_b, the maximum possible $p_{b_{opt}}$ can be calculated when $p_a = 1$ (which is the optimal value for p_a) from Equation 7.27.
- $p_a(p_b = 1) < 1$, we can get the maximum value of $p_{b_{opt}} = 1$ and hence by Equation 7.27 we can get $p_{a_{opt}} = p_a(p_b = 1)$.

Since $\frac{\partial airtime_{type_A}(p_a,p_b)}{\partial p_a} > 0$, $\frac{\partial airtime_{type_A}(p_a,p_b)}{\partial p_b} < 0$, $\frac{\partial airtime_{type_B}(p_a,p_b)}{\partial p_a} < 0$ and $\frac{\partial airtime_{type_B}(p_a,p_b)}{\partial p_b} > 0$, it can be shown that this $(p_{a_{opt}}, p_{b_{opt}})$ pair actually corresponds to the strategy that no radio systems can do better in terms of airtime share without harming the other coexisting radio systems. So in this sense, both the efficiency and fairness are obtained by using this optimal probability pair.

7.4.5 General Solution to *Airtime* Share and Blocking Probability

In order to analyze more complex access process, we need to deal with Markov chains with more states and more complex transition structures. Explicitly solving the balance equations to get the *airtime* share or blocking probability then becomes untractable. Therefore, we introduce a simple numerical technique described below to calculate the *airtime* shares and blocking probabilities.

1) Airtime Share: Given the traffic rates and occupying time, we can define an infinitesimal generator matrix $\mathbf{A}$ as before. Then by solving the following equation we can get the state probabilities.

$$\Pi\mathbf{A} = 0, \tag{7.30}$$

where $\Pi = [\Pi_1, \Pi_2, \ldots, \Pi_K]$ and K is the number of states in the Markov chain. The generator matrix $\mathbf{A}$ is singular so we cannot solve the state probability vector directly. But with the condition that the sum of all the steady state probabilities should be one, we can put these two conditions into the following compact equation:

$$\begin{bmatrix} \mathbf{A}^T \\ 1_{1\times K} \end{bmatrix} \Pi^T = \begin{bmatrix} 0_{K\times 1} \\ 1 \end{bmatrix}. \tag{7.31}$$

Then by defining $\mathbf{A}' = \begin{bmatrix} \mathbf{A}^T \\ 1_{1\times K} \end{bmatrix}$, $\mathbf{b} = \begin{bmatrix} 0_{K\times 1} \\ 1 \end{bmatrix}$, we have,

$$\mathbf{A}'\Pi = \mathbf{b}. \tag{7.32}$$

By using minimum mean squared error (MMSE) criterion [43], we obtain the following unique solution,

$$\Pi^T = (\mathbf{A'}^{\mathrm{T}}\mathbf{A}')^{-1}\mathbf{A'}^{\mathrm{T}}\mathbf{b}. \tag{7.33}$$

After having the state probabilities, the $airtime_i$ share for radio system i is just the weighed summation of the respective state probabilities.

2) Blocking Probability: Despite the fairness concerns, another important metric wireless users care about is the instant access probability, or the blocking probability. Our model can be considered as a finite population queuing model, where the time blocking which is the proportion of time the system spends in the blocking states is not equal to the call blocking which is the probability that an arriving call is blocked. So PASTA property does not apply here. Instead, we can compute the state probability seen by an arriving call (radio system) as follows,

$$\pi_j^* = \frac{\lambda_j \pi_j}{\sum_{k=0}^{s} \lambda_k \pi_k}, \tag{7.34}$$

where $s+1$ is the total number of states. Considering a long period of time T, on the average the system spends in state j the time $\pi_j T$. During this time there are on the average $\lambda_j \pi_j T$ call arrivals which find the system in the state j. The total number of calls arriving in time T is on the average $T\sum_{k=0}^{s} \lambda_k \pi_k$. Then the proportion of calls which find the system in the state j is as given by the above expression. When the random access probability is p_i the blocking probability experienced by the radio system i is

$$P_{BLK_i} = 1 - p_i \left[1 - \sum_{j:\ blocked\ states} \pi_j^* \right]. \tag{7.35}$$

7.4.6 Cognitive Radio Access

The proposed random access scheme makes it possible to achieve the desired fairness in open spectrum access with different types of radio systems. But with the increasing number of such devices, reducing the blocking probability and increasing the *airtime* share become a critical issue. Spectrum agility based channel access helps this cause.

1) Physical Layer Aspects of Agile Radio: Traditionally, frequency division has been adopted to divide the electromagnetic spectrum between different wireless technologies. In frequency division, portions of the spectrum are statically assigned to particular wireless technology to support their transmissions. This assignment has some inherent disadvantages with respect to spectral efficiency. For example, consider two operators operating on their licensed bands (i.e., exclusive use model). It is possible that the one wireless network operating in a particular frequency band is fully loaded, while the second may have unused resources (or vice versa). It would be profitable for both wireless network technologies, if these unused resources were shared to allow more capacity for the fully loaded operator. This form of spectral sharing is much easier to accomplish if the underlying access technique in the two operators is multi-carrier-based. In case of a single operator, the operator can access the spectrum opportunistically based on the load of the system.

In order to allow different wireless technologies to borrow/lend spectral bands from/to different wireless networks residing in the same or different spectral region, we could conceive of a simple adhoc strategy for selecting Δf (the frequency separation among carriers in a OFDM based system design). For example, two different strategies to enable spectral sharing are:

1. Different Δf for different regions
2. Common Δf for all regions

In the first strategy, different values of Δf are chosen for different spectral regions, based on the following practical considerations: 1. Δf is selected much smaller than the coherence bandwidth of the channel (to ensure that each carrier undergoes a flat fade). 2. Δf is selected big enough to prevent the use of large number of carriers (to allow practical implementation of the system via FFT/IFFT). Therefore, we could choose values of Δf that are proportional to the carrier frequency. This choice allows the system to maintain the transceiver complexity within reasonable levels. Specifically, we could have the following sub-carrier spacing:

1. $\Delta f = 25$ kHz for the low band region
2. $\Delta f = 100$ kHz for the medium band region
3. $\Delta f = 500$ kHz for the high band region

These choices were made based on coherence bandwidth measurements for a typical indoor small office channels.

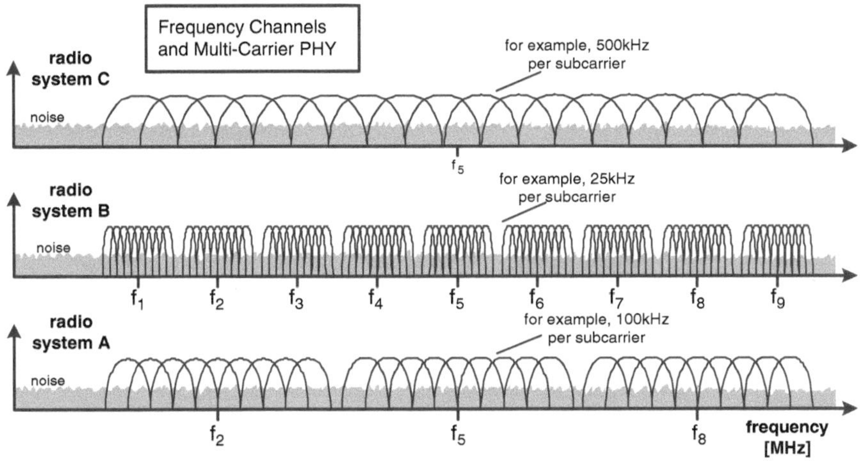

Fig. 7.14. Frequency channels and multi-carrier PHY.

In the second strategy, we suggest the use of the same carrier separation (e.g., $\Delta f = 25$ kHz) across all regions. Even though this strategy facilitates orthogonal sharing across the entire bandwidth, it results in prohibitive complexity in IFFT/FFT implementation. For example, choosing $\Delta f = 25$ kHz for a 1 GHz service in region 3 requires a 40,000-point IFFT. This can be chosen based on the current UWB standards. Figure 7.14 shows how the multi-carrier OFDM is used for agile radios.

2) MAC Layer Aspect of Agile Radio: With the advances in software defined radio, spectral-agile networks become more and more tractable. Such radio devices can dynamically utilize idle spectrum bands. One of the interesting concerns here is, given the additional freedom of carrier frequency switching, what is the gain in efficiency for radio systems with different bandwidth requirements. There are lots of ways to take advantage of this switching. To achieve the upper-bound for the agile spectrum efficiency, one way is to "pack" all the radio systems tightly together in the spectral domain. Such a packing would ensure that there is no spectral hole/white space.

We describe the "packing" behavior as follows. If one radio system releases the spectrum, the other radio systems will switch their operating frequency band so that the vacant band is occupied (if it meets the demands of at least one of the active radio systems). Then a new accessing radio system scans to find spectrum opportunities from the beginning of the spectral band, and occupies the first idle spectrum opportunity it finds. In this way, all the spectrum fragments can be saved for future accessing radio systems. A simple example given in Figure 7.15 explains this process clearly, where when radio system B_2 releases the spectrum, radio system B_4 (also could be B_3) then switches to frequency f_5. When a new accessing radio system A_2 scans the spectrum, it will occupy frequency f_7, f_8 and f_9. Without this spectral agility, it will have to queue or be dropped.

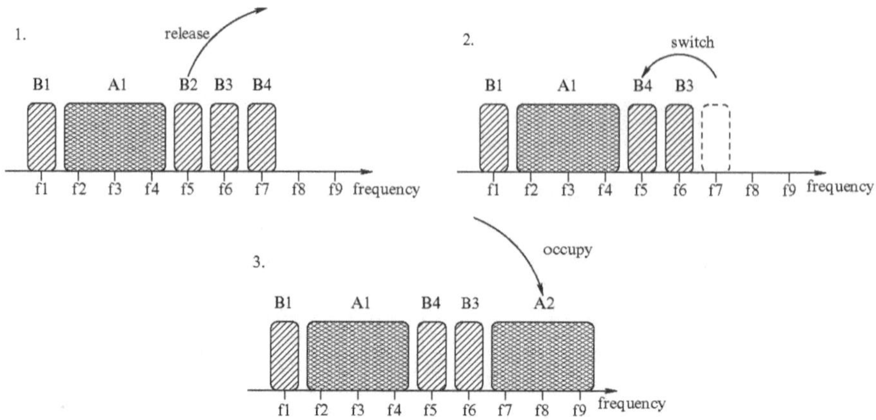

Fig. 7.15. An example for "packing" behavior.

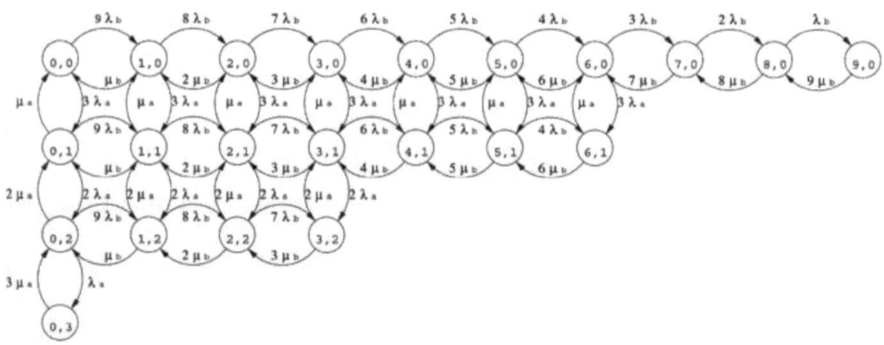

Fig. 7.16. Markov chain to model the unlicensed spectrum access process with spectral agility.

All the switching and scanning may introduce considerable signaling overhead and delay. To alleviate this, in practical schemes, we can divide the spectrum into blocks, and implement this packing in each blocks. This can considerably save the switching and scanning. Besides, software radios work very well with short scanning range.

Without loss of generality, we still use the basic scenario discussed before in Figure 7.10 to illustrate this scheme, and compare it with the other schemes. The most significant difference for this scheme is that each radio system here is an agile radio system. And it behaves in the "packing" way. The corresponding Markov model is shown in Figure 7.16. Here, each state $\Pi_j = (k_1, k_2)$, where k_i equals the number of radio system i on the spectrum block. We can solve for the *airtime* share and the blocking probability by the general method described previously.

7.4.7 The Homo Egualis (HE) Society Model Based Access Scheme

To obtain $(p_{a_{opt}}, p_{b_{opt}})$ we need the information of all the λ's and μ's which is impractical in a real access scenario. A more realistic scheme would be to allow the radio systems to learn these p_a and p_b themselves with only local information or measurement. We present such a technique next.

7.4.8 The Agent Egualis Society

In many decision-making and strategy-setting people do not behave like the self-interested "rational" actor depicted in neoclassical economics and classical game theory [39]. In a Homo egualis society, individuals have an inequality aversion. As a result altruists appear in ultimatum and public good games. As Gintis states in [39] support for Homo egualis comes from the anthropological literature, describing how Homo sapiens evolved in small hunter-gather groups. Such societies had no centralized structure of governance, so the enforcement of norms depends on the voluntary participation of peers. A Homo egualis society [39] can be modelled as follows, where the utility function of player i, u_i in an n-player game is:

$$u_i = x_i - \frac{\alpha_i}{n-1} \sum_{x_j > x_i} (x_j - x_i) - \frac{\beta_i}{n-1} \sum_{x_j < x_i} (x_i - x_j), \tag{7.36}$$

where $x = (x_i, \ldots, x_j)$ are the pay-offs for each payer and $0 \leq \beta_i < \alpha_i \leq 1$. $\beta_i < \alpha_i$ reflects the fact that Homo egualis exhibits a weak urge to inequality when doing better than the others and a strong urge to reduce inequality when doing worse than the others. In [39] it is shown that in this model the salient behaviors in ultimatum and public goods games, where fairness does matter, can be reproduced.

1) The Proposed Homo Egualis Based Access Scheme: The inequality aversion property of the Homo Egualis agents can be utilized to achieve fairness in the spectrum access problem. In this scheme each radio system learns the access probability p_i by itself. Here, we define $Onlinetime_i$ as the averaged cumulative "on" spectrum time per radio system of type i. Then we define $x_i = \frac{Onlinetime_i}{L_i}$, where L_i is a parameter proportional to the traffic load of radio system type i. The cumulative $Onlinetime_i$ is normalized by the radio system's traffic load, which makes this spectrum access scheme able to adapt to different traffic loads and hence achieve more efficiency and maintain fairness. With initial $p_i = 1$, each time the probability p_i is updated as follows,

$$p_i = \max\left(0, \min\left(1, p_i + \frac{\alpha_i}{n-1}\sum_{x_j \geq x_i}\left(\frac{x_j - x_i}{x_j}\right)\right.\right.$$
$$\left.\left. - \frac{\beta_i}{n-1}\sum_{x_j < x_i}\left(\frac{x_i - x_j}{x_i}\right)\right)\right) \text{ for all } j \neq i, \qquad (7.37)$$

where n is the number of different radio system types, $0 < \beta_i < \alpha_i$ reflects the fact that radio system exhibits a weak urge to inequality when doing better than others and a strong urge to reduce inequality when doing worse than the others. This forces each radio system to make an effort to efficiently use the idle spectrum while taking fairness into consideration. Here, the only local information needed is the radio system's own history of the *Onlinetime* and the *Onlinetime* of the other radio systems whose spectrum is within the same spectrum block. This can be obtained by keeping a record of the busy time of the required spectrum, which can be obtained by periodical spectrum scanning. When there are more than two different radio systems trying to coexist in the same spectrum, different radio systems can be identified by some smart technologies (e.g., we can detect the different transmitting power levels to distinguish from different radio systems). So each radio system can access the spectrum based only on its own recorded history and the local measurements performed by itself. While L_i can be estimated by historical usage records of radio system type i.

7.5 Numerical Results

We describe the simulation results in this section and compare it with the theoretical analysis. Equal loads on the radio systems are assumed. Figure 7.17 shows the simulated average *airtime* per radio system and the theoretical results obtained from the Markov model in Figure 7.12. We see that the proposed Markov model fits the simulation results very well. Here, radio systems are blocked when there are no idle frequency bands, and each radio system's carrier frequency is fixed. Radio systems only scan their own fixed frequency for spectrum opportunities. As one can imagine, it is seen here that with only *Rule*♯4 (LBT), the narrow bandwidth radio system B will dominate the airtime share over the broad-band radio system A. This dominance emphasizes the issue of unfairness, which is a prominent problem in coexistence of different types of radio systems.

The proposed HE based random access scheme can mitigate this dramatically, as illustrated in Figure 7.18. The theoretical *airtime* share for radio systems A and B are the same when optimal access probability pair $(p_{a_{opt}}, p_{b_{opt}})$ is used, and both of them increase with the increase of the traffic load. The HE access scheme using only local information is observed to produce a close to optimal solution. The HE access scheme is observed to produce a performance

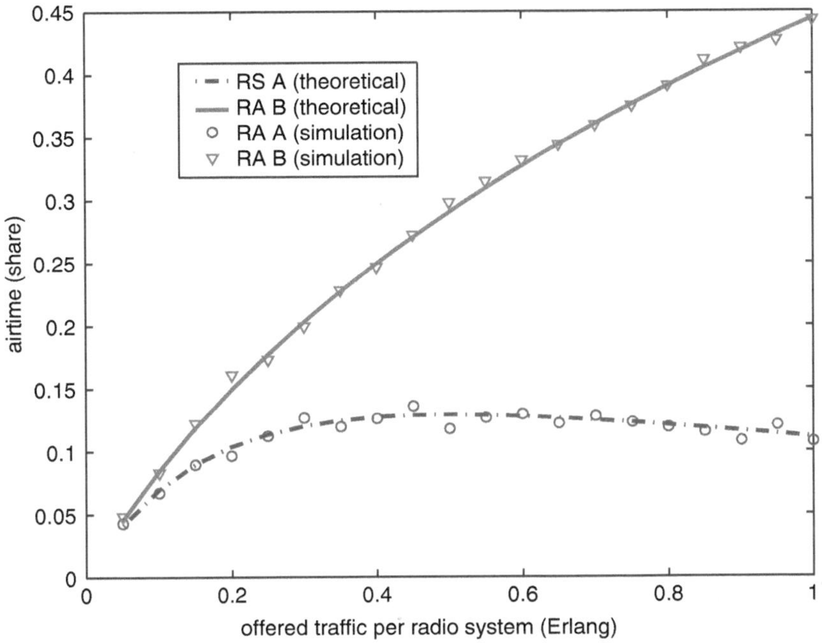

Fig. 7.17. Simulated and the Markov modelled spectrum access *airtime* under Rule ♯4. No queuing case.

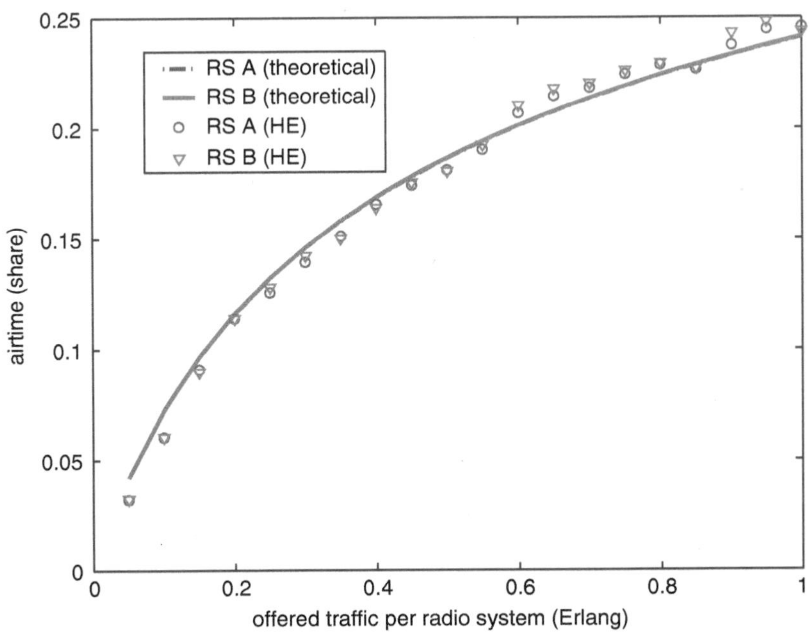

Fig. 7.18. Proposed HE access scheme compared with theoretical optimal solution in terms of *airtime*. No queuing case.

gain even higher than the optimal achievable solution using $(p_{a_{opt}}, p_{b_{opt}})$ as seen in Figure 7.18 for some traffic load. This is not surprising, because in the HE access, the p_a and p_b values change during the access, while the optimal probability solution pair $(p_{a_{opt}}, p_{b_{opt}})$ is obtained with the assumption that both of them will remain unchanged throughout the access. So when using the HE access in a real system, it may sometimes perform better than the predicted optimal results from the proposed Markov model solution of the random access scheme. Of course, for some instances, it may produce lower performance than the optimal solution as seen later.

The etiquette *Rule*♯6 can protect the broad-band radio system A by requiring a radio system of type B that follows *Rule*♯4 to synchronize its LBT process in time across neighboring frequency channels that overlap with the same reference channel [38]. But as can be seen in Figure 7.19, although the *airtime* share for radio system A increases, the cost is a significant decrease of *airtime* share for radio system B. The HE based access is seen to be much better in terms of fairness compared to the etiquette *Rule*♯6.

When each radio system is allowed to wait(queue) instead of being dropped if the desired frequency band is busy, the efficiency of the spectrum usage is expected to increase as illustrated in Figure 7.20, where only LBT etiquette is used. The theoretical result is derived from the Markov model in Figure 7.21.

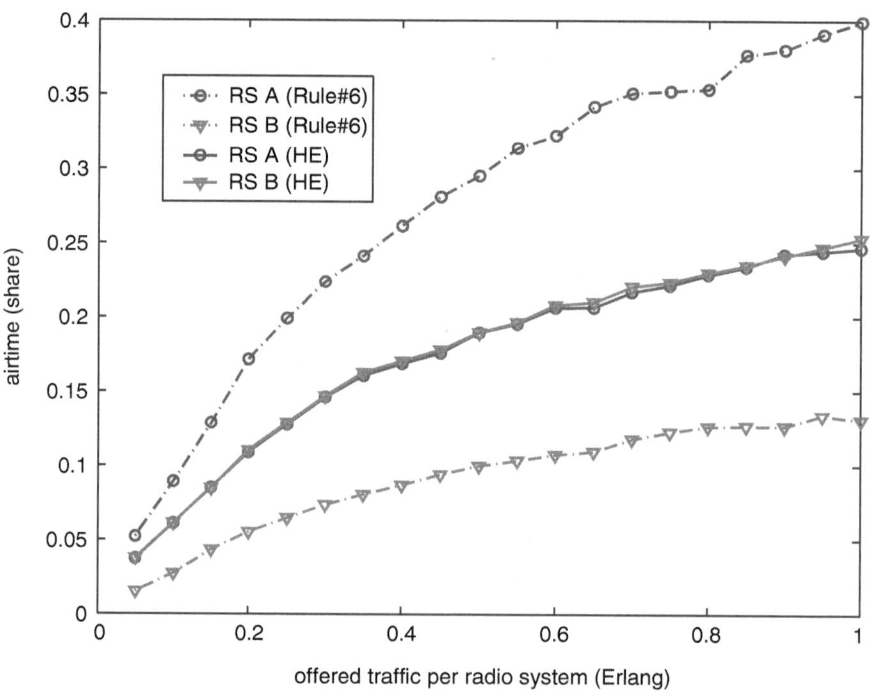

Fig. 7.19. Proposed HE access scheme compared with Rule ♯6 in terms of *airtime*. No queuing case.

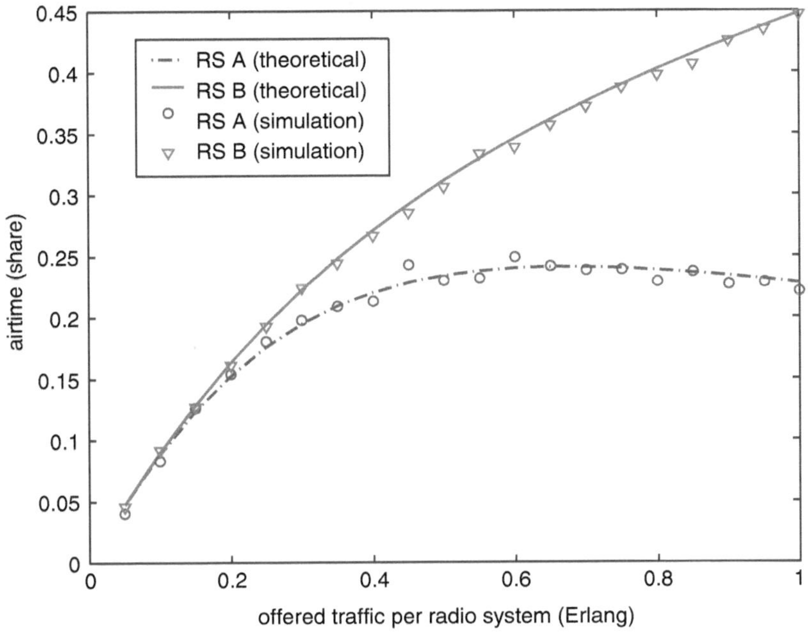

Fig. 7.20. Simulated and the Markov modelled spectrum access *airtime* under Rule ♯4. Queuing of radio systems case.

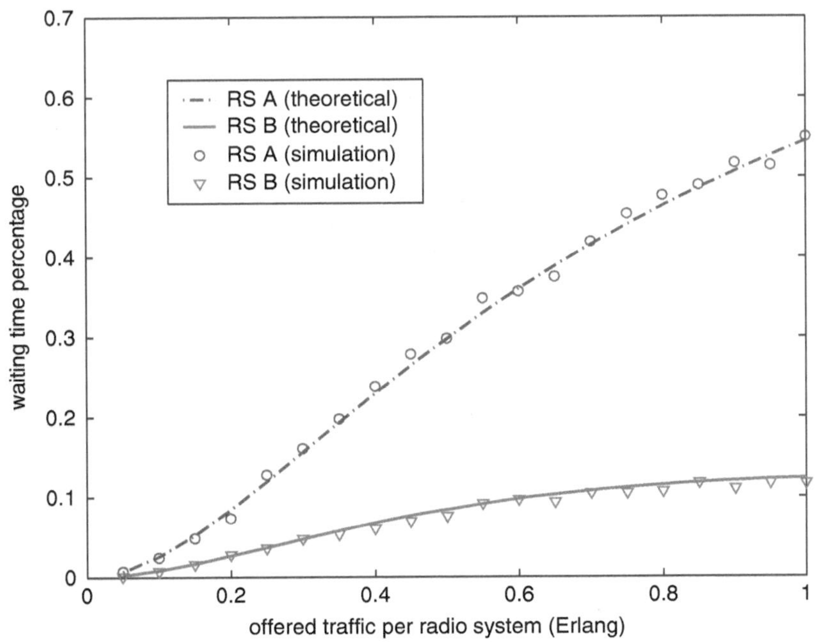

Fig. 7.21. Simulated and the Markov modelled spectrum access waiting time. Queuing of radio systems case.

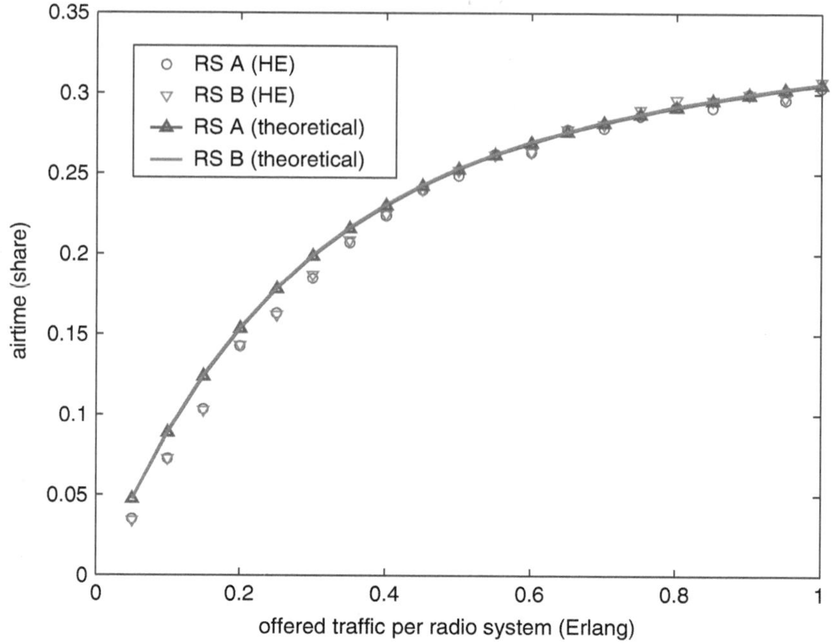

Fig. 7.22. Proposed HE access scheme compared with theoretical optimal solution in terms of *airtime*. Queuing of radio systems case.

With waiting, as can be seen, the *airtime* share for the broad-band radio system A almost doubled while it increased a little bit for radio system B. Therefore, the total utilization of the spectrum increases. The tradeoff here is the delay due to waiting, which is shown in Figure 7.21. The total waiting time is normalized to the *reference time*. And further, when there are more radio systems, collisions will be prominent and a sophisticated collision avoidance scheme is required. Figure 7.22 shows that the HE access scheme works well for this case as well. Comparing Figures 7.22 and 7.18, we can see that when using HE with waiting, the radio systems' *airtime* share increases significantly.

Agile radio systems can potentially increase the spectrum efficiency significantly as illustrated in Figure 7.23. When radio systems have the ability to switch their carrier frequency, they can pack themselves together in the spectral domain and hence will increase the overall *airtime* share. This agility can as well decrease the blocking probability experienced by the radio systems as seen in Figure 7.24. In Figure 7.24, the HE access scheme will experience higher blocking probability than the expected theoretical prediction especially at low traffic load. This phenomena can be explained by the same reason as mentioned before, which is because the p_a and p_b change during the access when using HE. So, when the traffic load is low, the optimal access probabilities are 1, but this may not be so when using HE. They may be lower than 1 for some time, which will result in a higher blocking probability as can be

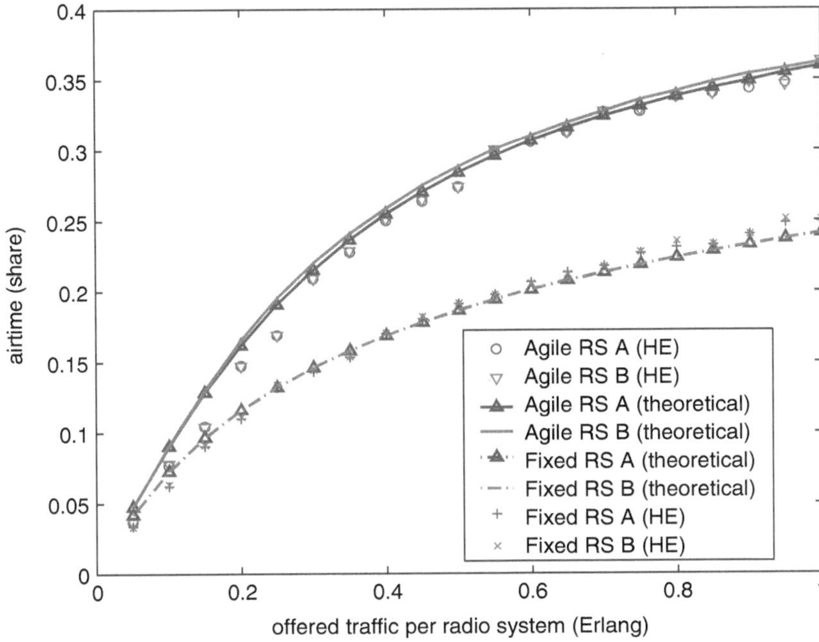

Fig. 7.23. Spectral-agile access scheme compared with original fixed access scheme in terms of *airtime*. No queuing case.

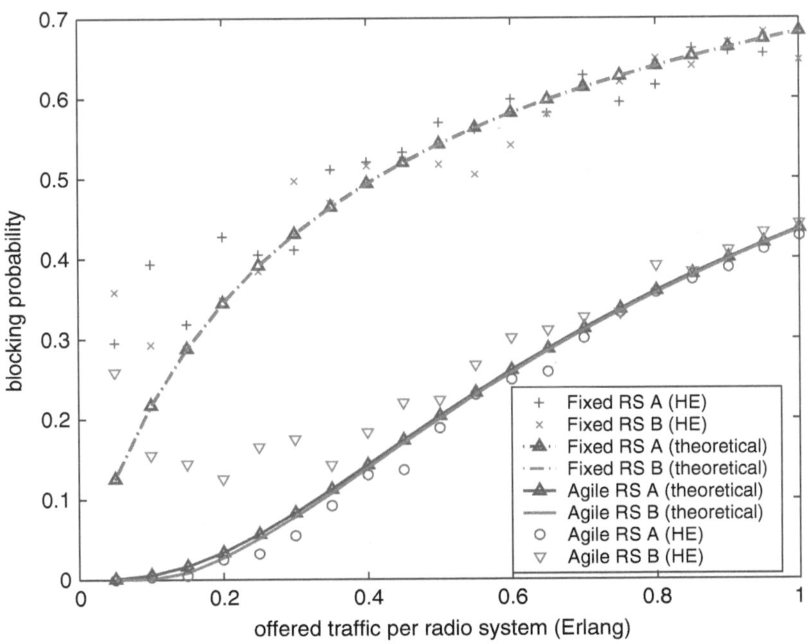

Fig. 7.24. Spectral-agile access scheme compared with original fixed access scheme in terms of blocking probability. No queuing case.

calculated through (7.35). Here, the theoretical solution is derived using the Markov model given in Figure 7.16.

Figures 7.25 and 7.26 show that the HE scheme can also work well when there are multiple different types of radio systems. Here, we considered three different types of agile radio systems. Radio systems A and B are the same as before, while radio system C represents radio system that uses broad-band transmission schemes such as UWB or spread spectrum. It requires even the whole spectrum (all nine spectrum bands) to be idle before allocating radio resources in the scenario discussed in Figure 7.10. It can be seen from Figure 7.25 that all these different radio systems almost have the same airtime share, which is what is desired for fairness. The theoretical solution is obtained from the Markov model described in Figure 7.16 by adding one more state for radio system C. Due to the same reason of the changing of access probabilities p_a, p_b and p_c in HE, the HE access scheme is observed to produce a performance gain higher than the optimal theoretical solution both in airtime share and the blocking probability in this case.

α and β are two important control parameters in the HE access scheme. Therefore, it is important to study the influence of α and β to the HE scheme. We find that when one of the radio systems uses $\alpha < \beta$, both of them will experience much lower airtime share, while as long as all of them use $\alpha > \beta$, they will almost get the same airtime performance no matter what specific α and β values they are using. Figure 7.27 illustrates one case when both

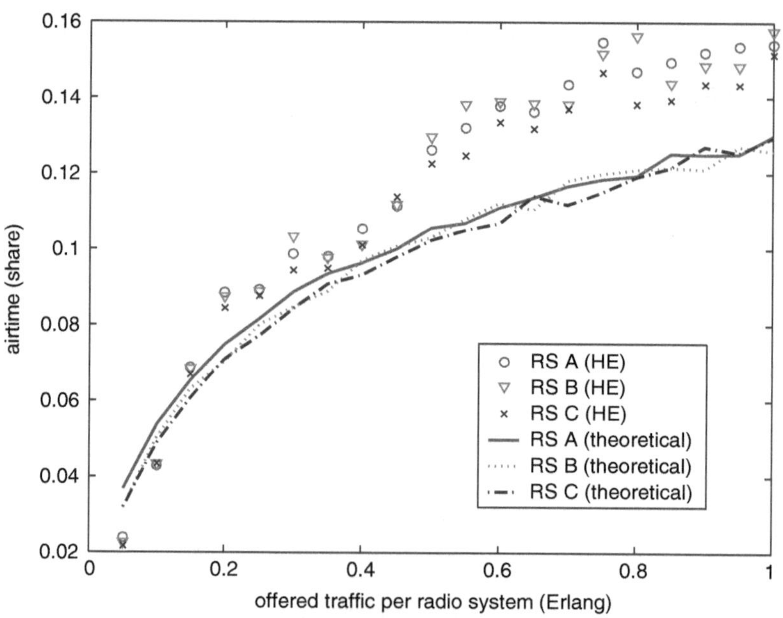

Fig. 7.25. Proposed HE access scheme compared with theoretical optimal solution in terms of *airtime*. Three different types of spectral-agile radio systems. No queuing case.

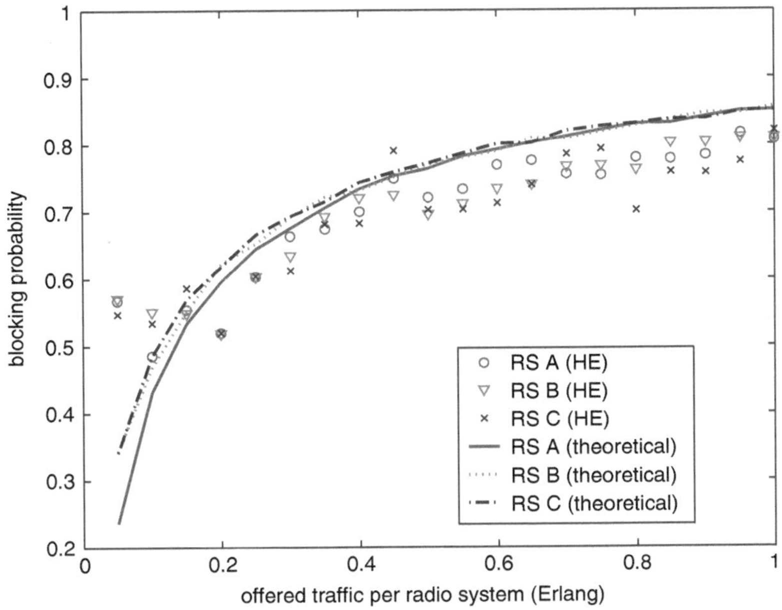

Fig. 7.26. Proposed HE access scheme compared with theoretical optimal solution in terms of blocking probability. Three different types of spectral agile radio systems. No queuing case.

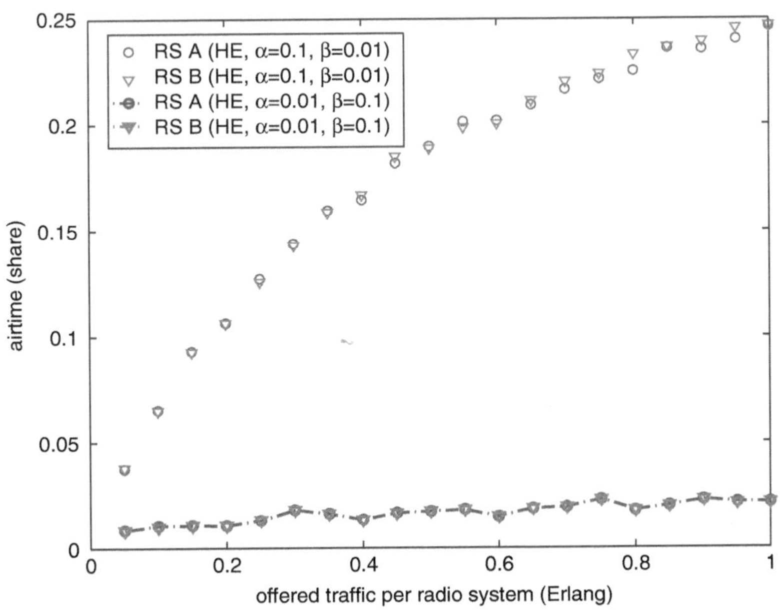

Fig. 7.27. Two original fixed radio systems use proposed HE access scheme. No queuing case. The comparison is between different parameter pairs ($\alpha = 0.1, \beta = 0.01$) and ($\alpha = 0.01, \beta = 0.1$).

of the radio systems use $\alpha = 0.01 < \beta = 0.1$ compared with both of them use $\alpha = 0.1 > \beta = 0.01$. It can be seen that when α is smaller than β, the performance for both of the two radio systems degrades significantly.

7.6 Channel Opportunity Study and the Optimal Sensing Protocol

So far, we analyze the performance of cognitive radios with respect to their efficiency as well as coexistence. Assuming that the sensing problem is solved we want the secondary user to decide whether an accessible channel is a good opportunity based on channel sensing statistics. In particular, a secondary user prefers a channel where it can finish the transmission before the primary users return. Determining whether the channel is idle or not poses a serious challenge, particularly in the context of available estimation and detection algorithms, in the design of agile radio, but there have been studies indicating the use of sensors whose prime function is to update a central server. A agile radio uses its position and accesses this central server to determine if the particular channel is idle or not. Whether sensors are used or not, the following simple algorithm is useful in estimating the white spaces of the spectrum [1].

The secondary wireless network would determine that a particular channel is an opportunity if it can find an idle time in that channel that is greater than T_{opp}. T_{opp} is the requirement in time and stems from the applications running in the secondary wireless network such as video, audio, data, etc. In order to use a particular primary channel whose idle times are greater than T_{opp}, we need a sensing protocol that would determine if the particular channel is an opportunity or not. The occupancy in a particular channel is defined as the probability that the physical layer signatures of the current primaries are present. Using simple correlation or feature detection techniques one can easily determine the presence of the primary. Let us now consider important aspects in designing a sensing protocol. A secondary wireless network has a requirement in terms of time or bandwidth (bits/s) when it is looking for opportunities or white spaces in the spectrum to transmit its data. If the bandwidth requirement is in bits/s, it is translated to a time requirement based on the physical transmission rate that the wireless system is currently using.

As indicated above, let a secondary wireless system looks for a spectral opportunity equal to T_{opp}. This is different from the channel occupancy, T_{on}, of the primary. The channel occupancy represents the actual occupancy of the channel by the primary. A secondary wireless network occupies that particular channel if and only if it determines that channel is a spectral opportunity which is given by the following equation:

$$T_{off} \geq \alpha T_{opp}, \quad \alpha > 1. \tag{7.38}$$

T_{off} in the above equation represents the time that a particular channel is not occupied by the primary. The intuition behind Equation (7.38) is to reduce the probability of collision with the primary. α is a design parameter and smaller values of α make the probability of collision with the primary higher. Reducing the collision probability has a negative effect of lowering the spectrum utilization. This would mean that the secondary wireless network is very conservative in determining that the channel is an opportunity. On the other hand, a non-conservative approach would increase the chance of colliding with primary. So one needs to choose α in optimal way that maximizes the spectrum utilization for a given probability of collision. Currently all primary channels have their value of $\alpha = \infty$ implying that the secondary wireless network cannot access this channel. The value of α is chosen to be 2 in this chapter for simplicity. It should be noted that designing the right value of α is out of scope of this chapter.

Once the sampling rate is determined, the secondary wireless network senses that channel and collects information about that channel. The results of the sampling are updated in the database maintained by the individual station or the central controller, such as access point or base station, for possible switching in the future. The initial requirement of sampling requirement comes from the secondary wireless network applications that require an opportunity of T_{opp}. Initially it is fixed at:

$$T_{sample} = \frac{T_{opp}}{2}. \tag{7.39}$$

T_{sample} represents the sampling interval and this will determine if the channel is a spectral opportunity for this secondary wireless network. This sampling interval is also called as the Nyquist opportunity determination rate as this represents the maximum rate that will be used by the secondary wireless network to determine the availability of the channel. This rate can construct the original occupancy of the channel if that channel has $E[T_{off}] = T_{opp}$. This sampling rate may not be optimal, as it may spend more time in sampling the spectral opportunity for this particular channel thus increasing the overheads arising from sampling this channel and thereby increasing the power consumption.

We outline a technique to find the optimal sampling frequency. Let X_t be a stochastic process denoting whether the channel is occupied or not at time t. X_t represents the indicator random variable. Then the fraction of time the channel is busy in an interval $[0, \tau]$ is given by

$$O_\tau = \frac{1}{\tau}\int_0^\tau X_t \, dt. \tag{7.40}$$

O_τ is actual occupancy of the channel that was currently sensed and is a random process whose realizations are different at different instants of time.

The above equation represents the continuous process. As mentioned we will be sampling the channel for regular interval given by T_{sample} whose optimal value has to be determined. Based on the sampled process, the best prediction of the channel occupancy is given by

$$\hat{O}_{\tau} = \frac{1}{n} \sum_{i=1}^{n} X_{iT_{sample}}. \tag{7.41}$$

Here, $n = \frac{1}{T_{sample}}$. Under continuity conditions of the process X_t, the process $\hat{O}_{\tau} \rightarrow O_{\tau}$ as $T_{sample} \rightarrow 0$. This would mean that the secondary wireless network is continuously sensing a particular channel and so the measurement results would yield the occupancy and availability times exactly. From $X_{iT_{sample}}$ collected over the entire measure interval, one can easily determine the $E[T_{on}]$ and $E[T_{off}]$ of the channel by noting the number of consecutive ones and zeros and looking into the transitions from 0s to 1s and vice versa. One of the most important goals is to verify reliability of the sampling process by determining the variance of the estimator. The variance of the sampled process, $\hat{O}_{\tau}$, is given by

$$\begin{aligned} Var\{\hat{O}_{\tau}\} &= Var\left\{\frac{1}{\frac{1}{T_{sample}}\sum_{i=1}^{n} X_i T_{sample}}\right\} \\ &= \left(\frac{1}{T_{sample}}\right)^2 \sum_{1 \leq i,j \leq n} Cov(X_{iT_{sample}}, X_{jT_{sample}}). \end{aligned} \tag{7.42}$$

If T_{sample} is small compared to the expected ON-period, $E[T_{on}]$, of the channel, the measurements will be dependent. Hence one needs to determine the off diagonal elements of the covariance matrix. If the process were stationary, one can replace the $Cov(X_{iT_{sample}}, X_{jT_{sample}})$ by a function f with one argument that is given by the difference in time between $(X_{jT_{sample}} - X_{iT_{sample}})$. Thus, we can rewrite Equation (7.42) as

$$Var\left\{\hat{O}_{\tau}\right\} = n^2 \sum_{1 \leq i,j \leq n} f((i-j)T_{sample}). \tag{7.43}$$

If the duration T_{sample} is not zero and has very large value, the variance of the measurement will be maximum and is equal to $\rho(1-\rho)$, where ρ represents the occupancy utilization of the channel by the primary.

Now to determine the optimal sampling interval once the mean's of ON, OFF and the variance of the OFF-periods are determined is to double the sampling interval until the measured variance of the newly measured variance is greater than the original variance using the Nyquist rate by certain bound. The bound is also dependent on how far the mean ON- and OFF-periods vary from the true ON- and OFF-period that was determined by the initial sampling interval. The above new sampling periods obtained from Equation 7.39

may not be optimal in terms of resource power utilizations. Hence, determination of the optimal number of sampling points is mandatory that characterizes the ON- and OFF-periods efficiently while conserving the wireless resources. Having obtained the mean and variance of the ON- and OFF-periods using T_{sample}, we will determine the optimal number of samples that are required to capture the characteristics of the channel. From the central limit theorem of random samples [3], we know that as the sample size is large with the number of samples, $n \rightarrow \infty$, the average of the sampled data approaches the original mean regardless of the distribution. This is expressed by the following equation:

$$P\left\{\left|\frac{O-\hat{O}}{O}\right| > \epsilon\right\} \approx 2\left(1-\phi\left(\frac{\epsilon\mu\sqrt{n}}{\sigma}\right)\right) \leq \eta. \tag{7.44}$$

In the above equation, ϵ and η are the design parameters and μ and σ denote the sample mean and sample standard deviation using T_{sample} as the sample duration. From Equation 7.44, we can easily calculate the optimal number of samples and this is given by

$$n \geq n_{optimal} = \left[\frac{\phi^{-1}\left(1-\frac{\eta}{2}\right)\sigma}{\epsilon\mu}\right]^2. \tag{7.45}$$

Figure 7.28 shows the optimal number of samples required using the approximation of the above equation. It is clear that the number of samples required to estimate the channel occupancy increases if the channel availability is very small and decreases as this time increases. The plot is for the exponential channel availability time and can be done for different distributions that have the second moment. In this numerical analysis, the value of η was set to 0.01.

7.6.1 Implementation of the Sampling Function in Practical Networks

Consider the case of infrastructure networks, wherein the Access Point (AP) or Base Station (BS) is the central controller that is responsible for the wireless resources. The AP/BS initiates a measurement request to the clients who in turn measure a particular channel and report the activity to the access point. The secondary wireless network first tunes to a particular channel and will listen to the channel for 1 s to calculate the mean ON- and OFF-periods of the channel. In a distributed Adhoc network, this sensing is periodic for a channel and may happen once in every few minutes. All devices agree upon the periodicity and sense the channel. It may turn out that they may dedicate users who take turns to sense a particular channel for the 1 s period. In case of infrastructure networks, the central controller may dedicate some wireless devices to periodically visit the channel to collect the information that is then

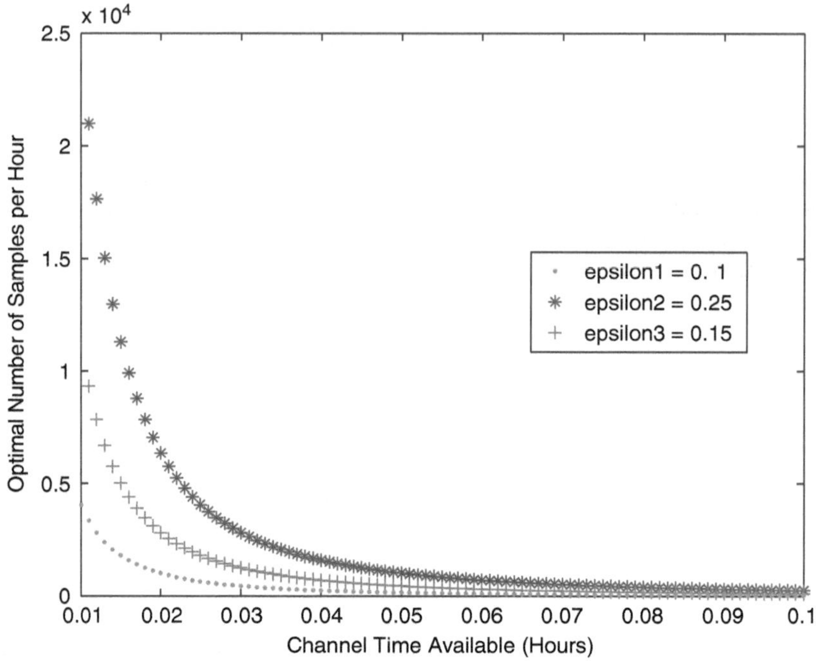

Fig. 7.28. Optimal number of samples as a function of exponential channel occupancy. The value of η is fixed at 0.01 and the three values of ϵ are 0.1, 0.15 and 0.25.

updated in its database. After determining the mean ON- and OFF-periods of the channel it will then use Equation 7.45 to sample that channel for a duration of $T_{duration}$. In the existing protocols like IEEE 802.11, IEEE 802.16 and IEEE 802.15, the AP/BS collectively chooses a channel and scans the channel for information for 1 s and then estimate the mean ON- and OFF-periods first. Then it will scan the channel at the optimal sampling rate using the estimated mean and variance by disseminating the optimal n to all the individual wireless stations. The other way is to disseminate the $^{n}/_{x}$ instead of n. Here, x represents the number of clients associated with that AP/BS. Since the value of n is obtained based on the characteristics of the particular channel it represents the sampling rate of a particular channel in order to reconstruct the occupancy properties of that channel. Hence distributing the new sampling frequency improves the radio resource usage resulting in more data traffic transmissions. In the case of Adhoc networks, there is complexity on the individual clients to calculate the value n and a dissemination protocol has to be designed that will propagate this information so that wireless clients can use radio resources efficiently.

7.7 Conclusion

In this chapter we model the efficiency of the cognitive radios from the Medium Access Control (MAC) analytically and study the improvement that cognitive radios can get compared to conventional radios. Then we consider the effect of peaceful coexistence with different types of cognitive radios and also consider a simple spectrum sensing protocol which is a function of the primary utilization and required opportunity. It is clear that from our initial analysis we have proved that there is immense benefit of cognitive radio compared to conventional radios and lots of newer research are required before this becomes a reality in shaping the lives of mankind.

References

1. N. Sai Shankar, Cordeiro, C., and Challapali, K., *Spectrum Agile Radios: Utilization and Sensing Architectures*, *IEEE DYSPAN 2005*, Washington DC, USA, Nov. 2005.
2. C.T. Chou, Sai Shankar, N., Kim, H., and Shin, K., *What and How Much to Gain From Spectral Agility*, *IEEE JSAC*, April 2007.
3. X. Liu and Sai Shankar, N., *Sensing Based Opportunistic Channel Access*, *ACM MONETS*, Aug., 2006.
4. Y. Xing, Chandramouli, R., Sai Shankar, N., and Mangold, S., "Dynamic spectrum access in open spectrum wireless networks," *IEEE JSAC*, March 2006.
5. "Facilitating opportunities for flexible, efficient, and reliable spectrum use employing cognitive radio technologies," *The FCC Notice of Proposed Rulemaking and Order – ET Docket No. 03-108.*
6. FCC (2003) Notice for Proposed Rulemaking (NPRM 03 322): Facilitating Opportunities for Flexible, Efficient, and Reliable Spectrum Use Employing Cognitive Radio Technologies. *ET Docket No. 03-108, Dec. 2003.*
7. IEEE 80.22 web site: http://grouper.ieee.org/groups/802/22/
8. Functional Requirements for IEEE 802.22 WRAN Standard, 802.22/05-0007r46, Sep. 2005.
9. Method for point-to-area prediction for terrestrial services in the frequency range 30 MHz to 3000 MHz, ITU-R P.1546-1, Oct. 11, 2005.
10. "Facilitating opportunities for flexible, efficient, and reliable spectrum use employing cognitive radio technologies," *The FCC Notice of Proposed Rulemaking and Order — ET Docket No. 03-108.*
11. http://www.darpa.mil/ato/programs/XG/rfc_vision.pdf – *Vision RFC.*
12. C. Bergstrom, Chuprun, S., and Torrieri, D., "Adaptive spectrum exploitation using emerging software defined radios," *IEEE Radio and Wireless Conference 1999*, pp. 113–116.
13. P.K. Lee, "Joint frequency hopping and adaptive spectrum exploitation," *IEEE MILCOM 2001*, Vol. 1, pp. 566–570.
14. J. Mitola, "The software radio architecture," *IEEE Communications*, Vol. 33, No. 5, 1995, pp. 26–38.
15. "White paper on regulatory aspects of software defined radio," *SDR forum document number SDRF-00-R-0050-v0.0.*

16. http://www.darpa.mil/ato/programs/XG/rfc_af.pdf – *Architecture RFC.*
17. J.B. Punt, Sparreboom, D., Brouwer, F., and Prasad, R., "Mathematical analysis of dynamic channel selection in indoor mobile wireless communication systemse," *IEEE Transactions onVehicular Technology,* Vol. 47, No. 4, Nov. 1998, pp. 1302–1313.
18. "IEEE 802.11h standard (Amendment to IEEE 802.11 Standard, 1999 Edition)," 2003.
19. J. Khun-Jush, Malmgren, G., Schramm, P., and Torsner, J., "Overview and performance of HIPERLAN type 2-a standard for broadband wireless communications," *IEEE VTC 2000-Spring, Tokyo,* Vol. 1, pp. 112–117.
20. "Additional spectrum for unlicensed devices below 900 MHz and in the 3 GHz band," *FCC ET Docket No. 02-380.*
21. "Future trends in defence antenna technology," http://www.bcba15324.pwp.blueyonder.co.uk/consulting/Bibliography/Paper-16.pdf.
22. D. Bertsekas and Gallager, R., *Data Networks,* Prentice-Hall, New York, 1992.
23. A. Papoulis, *Probability, Random Variables, and Stochastic Processes,* Third Edition, McGraw Hill, 1991.
24. H.L. Van Trees, *Detection, Estimation, and Modulation Theory: Part 1,* Wiley, 1968.
25. S. Kay, *Fundamentals of Statistical Signal Processing: Detection Theory,* Prentice Hall, 1998.
26. G. Chouinard, WRAN Reference Model Spreadsheet, IEEE 802.22-04-0002r12
27. ATSC A/74 Recommended Practice Guideline Document entitled: "ATSC Recommended Practice: Receiver Performance guidelines," Sections 4.5.2 & 4.5.3.
28. S. Shellhammer, Tawil, V., Chouinard, G., Muterspaugh, M., and Ghosh, M., Spectrum Sensing Simulation Model, IEEE 802.22-06/0028r6, June 2006.
29. Advanced Television Standards Committee, ATSC Recommended Practice: Receiver Performance Guidelines, ATSC A74, June 2004 [11] Steve Shellhammer, Performance of the Power Detector, IEEE 802.22-06/0075r0, May 2006.
30. R. Tandra, Fundamental Limits on Detection in low SNR, Master Thesis, University of California, Berkeley, Spring 2005.
31. XG WORKING GROUP, "The XG Vision." Request For Comments, version 1.0. Prepared by: BBN Technologies, Cambridge, Massachusetts, USA. July 2003. Available from: http://www.darpa.mil/ato/programs/XG/rfcs.htm.
32. D.P. Satapathy, Peha, J.M., "Spectrum sharing without licenses: opportunities and dangers," *Telecommunications Policy Research Conference (TPRC)* Section 3, pp. 15–29, 1996.
33. D.P. Satapathy, Peha, J.M., "Performance of unlicensed devices with spectrum etiquette." *Proceedings of IEEE Globecom,* Nov. 1997, pp. 414–418.
34. D.P. Satapathy, Peha, J.M., "Etiquette modifications for unlicensed spectrum: approach and impact," *Proceedings of the 48th Annual International IEEE Vehicular Technology Conference,* May 1998, Vol. 1, pp. 272–276.
35. C.T. Chou, Hyoil, K., Sai Shankar, N., and Shin, K.G., "What and how much to gain from Spectrum Agility," Submitted to *IEEE/ACM Trans. on Networking,* June 2004.
36. S. Mangold, Zhong, Z., Challapali, K., and Chou, C.T., "Spectrum agile radio: radio resource measurements for opportunistic spectrum usage". *IEEE Globecom 2004,* Dallas TX, USA, Nov. 2004.
37. K. Challapali, Birru, D., and Mangold, S., "Spectrum agile radio for broadband applications" *EETimes In Focus Article* Aug. 23.

38. S. Mangold, Challapali, K., "Coexistence of wireless networks in unlicensed frequency bands," *Wireless World Research Forum* ♯9 Zurich Switzerland, July 2003.
39. H. Gintis, *Game Theory Evolving: A Problem-Centered Introduction to Modeling Strategic Behavior.* Princeton University Press, 2000.
40. J. Mitola, "The software radio architecture," *IEEE Communications*, Vol. 33, No. 5, 1995, pp. 26–38.
41. "White paper on regulatory aspects of software defined radio," *SDR forum document number SDRF-00-R-0050-v0.0.*
42. P. Leaves, Ghaheri-Niri, S., Tafazolli, R., Christodoulides, L., Sammut, T., Staht, W., Huschke, J., "Dynamic spectrum allocation in a multi-radio environment: concept and algorithm," *Second International Conference on 3G Mobile Communication Technologies*, 2001.(Conf. Publ. No. 477), 26-28 March 2001 pp. 53–57.
43. S. Gilbert, *Introduction to Linear Algebra.* Wellesley-Cambridge Press, 1998.

7.8 Appendix

7.8.1 Calculation of Multiplexing Gain

But looking at how the system evolves, we get a good approximate solution. Assume that there is one $M/G/1$ queue with preemptive priority as before. The server capacity is constant with mean $EX = \mu$ and second moment at EX^2. We will now show how this $M/G/1$ queue with preemption will model 7.8a. This arrivals to this queue capture the SARA network arrivals into the system with N channels. Since the SARA network will get a maximum capacity determined by one channel's capacity, the service rate of the above queue is fixed to μ which is generally distributed. The arrival rate into the queue is modelled by bulk arrivals. The arrival rate is poisson with rate λ but with probability g_k we have k simultaneous arrivals into the $M/G/1$ queue. What does g_k model here. To understand the role of $g_k = P(G = k)$, we need to focus on the arrivals of SARA networks. SARA networks arrive to the queue with Poisson rate λ. The number of networks arriving into this queue depends on the channels being occupied by the primary. We consider M SARA networks trying to use N channels. Consider the first case of $M < N$. Choose an arbitrary instant when all channels are not occupied by primaries. So all M networks get their own channels. Now, consider the case where only $M-1$ channels are available. Then we will have the two arrivals to our M/G/1 system. If $\lceil \frac{M}{2} \rceil - 1$ channels are only available, then we have three arrivals in our $M/G/1$ queue. Proceeding similarly we can evaluate the probability g_M where all the SARA networks occupy the last available channel. This $M/G/1$ queue with preemption can be preempted if all the channels are occupied by primaries and so rate of preemption and the service time of the preemptive primary is the blocking time for the SARA network and is the same as in the previous section. The arrival rate into the queue is given by

$$\Lambda_{SARA} = \lambda E(G). \tag{7.46}$$

Here, $E(G) = \sum_{k=1}^{M} k g_k$. Two cases are considered with the first being $M < N$ and the second being $M > N$. The following equation represents the rate of arrival of SARA networks into the queue when $M < N$.

$$\begin{aligned}
g_1 &= \sum_{i=M}^{N} r_i \\
g_2 &= r_{M-1} \\
g_3 &= r_{\lceil M/2 \rceil - 1} \\
&\cdot \\
&\cdot \\
&\cdot \\
g_i &= r_{\lceil M/(i-1) \rceil - 1}, \quad \lceil M/(i-1) \rceil - 1 > 2.
\end{aligned} \tag{7.47}$$

The above equation becomes g_M if $\lceil M/(i-1) \rceil = 2$.

If $M > N$ then all poisson arrival instants have atleast two SARA network arrivals into our $M/G/1$ queue. The size of M is going to play a role in determining the values of g_k. The minimum value of the subscript k in g_k is given by $k = \lceil \frac{M}{N} \rceil$. The relationship of g_k to r_k is given by

$$g_{\lceil \frac{M}{N-i} \rceil} = r_{N-i}. \tag{7.48}$$

We use the above g_k to get the arrival rate as before by computing $E(G)$.

8

Enabling Cognitive Radio via Sensing, Awareness, and Measurements

Hüseyin Arslan[1] and Serhan Yarkan[2]

[1] University of South Florida, Tampa, FL, USA, arslan@eng.usf.edu
[2] University of South Florida, Tampa, FL, USA, syarkan@eng.usf.edu

8.1 Introduction

Wireless communications is established through a common medium which is highly dynamic. The elements of wireless communications systems such as nodes in a network, users, and some properties of the wireless devices themselves (e.g. battery) are dynamic as well. In order for wireless communications systems to better perform, adaptation to these dynamic conditions and elements is essential. How well a wireless system adapts to these dynamic conditions depends on the amount of the knowledge of varying parameters. It is clear that the more the knowledge, the better the adaptation.

Recently, wireless communication community meets a new concept called "cognitive radio," which is a radio that can sense, be aware of, learn, and adapt to its surrounding environment [3]. Built on the top of Software Defined Radio (SDR), cognitive radio can adapt the radio parameters with the aid of a special structure called cognitive engine. Cognitive engine can be regarded as the counterpart of human brain in human body, since the brain is the center for intelligence, as described in Chapter 14 with the same analogy.

Cognitive radio is expected to push the concept of adaptation further with the aid of its advanced attributes. The main reason behind this expectation is the fact that cognitive radio is equipped with extended sensing capabilities in addition to Artificial Intelligence (AI) sort of tools that are kept in cognitive engine.

In this chapter, we will discuss how cognitive radio can sense and be aware of major factors that affect its communications. First, we will address sensing and being aware of the wireless channel. Next, we will take network related awareness issues into consideration. In the following, we will discourse user relevant topics along with other possible measurements. Finally, we will address some major challenges and explain future research directions pertinent to the realization of cognitive radio.

H. Arslan (Ed.), Cognitive Radio, Software Defined Radio, and Adaptive Wireless Systems, 235–261.

8.2 Wireless Channel Awareness

This section discusses prominent wireless channel characteristics, relevant observable quantities, and some methods to measure them.

8.2.1 Channel Selectivity

Before introducing the channel selectivity in detail, it is appropriate to explain what selectivity means. Selectivity is a measure of how differently a channel behaves over the dimension in which the selectivity is defined. Measuring the selectivity is established by *channel coherence*, which is statistically defined, again, over the same dimension in which the selectivity is measured [8]. More formally, channel coherence over a dimension is the width of the window over which the signal is assumed as invariant. The less the width of the window, the more the selectivity on the dimension of interest. For instance, time selective channel basically means that at time instants that are close to each other, the correlation between the components of the channel is weak. Therefore, the width of the time window (in this case, the duration of the window) formally determines time selectivity. Similar concepts can be deduced by just replacing the dimension with the desired one, such as frequency selectivity and coherence bandwidth; space selectivity and coherence distance. An illustration of the concept of selectivity and coherence is shown in Figure 8.1.

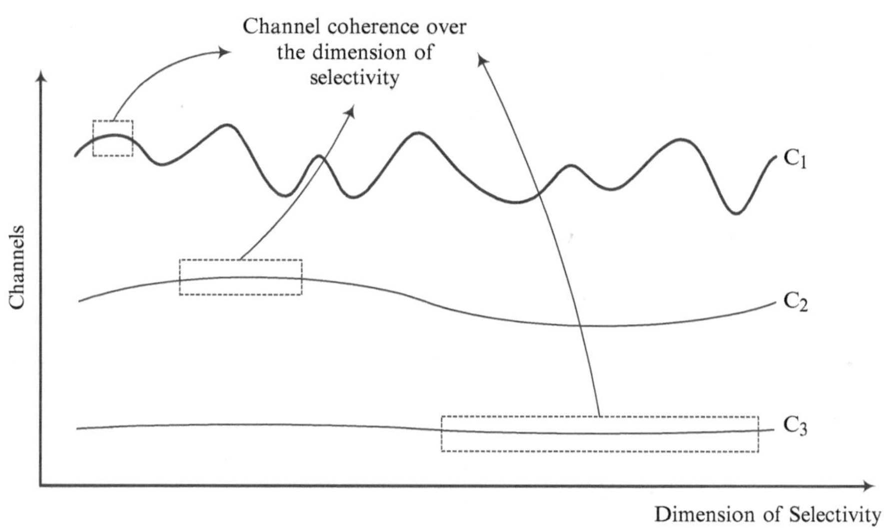

Fig. 8.1. The concept of selectivity and its measurement through channel coherence. The width of the window for $\mathbf{C}_1$ is the narrowest one, whereas that of $\mathbf{C}_3$ is the widest. Thus, for the dimension of interest, the selectivity of $\mathbf{C}_1$ is more than that of $\mathbf{C}_2$ and the selectivity of $\mathbf{C}_2$ is more than that of $\mathbf{C}_3$.

Duality

In this context, it is appropriate to mention a very important notion called *duality*. Fourier[1] transform ($\mathcal{F}\{\cdot\}$) and its inverse ($\mathcal{F}^{-1}\{\cdot\}$) allow one to see time and frequency domain interpretations of the functions as dual of each other. For functions of deterministic type, with the aid of $\mathcal{F}\{\cdot\}$ (and $\mathcal{F}^{-1}\{\cdot\}$), it can be shown that if the width (in time domain, "width" refers to "duration," whereas in frequency domain it refers to "bandwidth") of the function in one domain expands, it shrinks in the dual domain. However, the signals are of stochastic type in wireless communications. Therefore, it is better to investigate the duality in terms of stochastic processes. It is known that, a stochastic process that is defined in one domain automatically has a dual stochastic process in the dual domain with the aid of, again, $\mathcal{F}\{\cdot\}$ or $\mathcal{F}^{-1}\{\cdot\}$ [4]. In order to ease the mathematical tractability, in wireless communications, often, signals are assumed as Wide-Sense Stationary (WSS).

In order to see the connection between selectivity, coherence, and duality, stochastic linear time-varying wireless channels with WSS properties can be considered. If the autocorrelation function of a channel is calculated over one of the three domains (time, frequency, or space), the transform domain can easily be obtained with the aid of Wiener–Khintchine Theorem. A channel that spreads over the transform domain corresponds to a shrinkage in the autocorrelation function because of the duality. Statistically, the shrinkage of the autocorrelation function of a channel implies the decrease of coherence and the increase of selectivity. As can be seen, when selectivity occurs in a domain, the spreading is observed in its dual domain and vice versa. In wireless communications community, the dual domain, namely "spreading" domain, is labeled with its cause. Consider the selectivity in time. Since the dual of time is frequency because of $\mathcal{F}\{\cdot\}$, spreading occurs in the dual domain, namely in frequency. Spreading in frequency is caused by Doppler effect, therefore, selectivity in time corresponds to Doppler spread in frequency. Conversely, selectivity in frequency has its dual in time as a spreading signal. Spreading in time is caused by the delays between multipaths. Therefore, selectivity in frequency corresponds to "delay spread." Duality can be defined over space dimension as well. However, for this case, the space is transformed into a domain called "wavevector" and vice versa.

---

[1] Jean Baptiste Joseph Fourier, the French mathematician and physicist who was born on March 21, 1768, in Auxerre, France and died on May 16, 1830, in Paris, France. The representation of functions through sum of trigonometric series has been named after him [3], although [3] caused plenty of controversies. Later on, Johann Peter Gustav Lejeune Dirichlet, who was born on February 13, 1805, in Düren, and died on May 5, 1859, in Göttingen, contributed to the Fourier's theory by appending the convergence conditions, which are known as "Dirichlet [Fourier Series] conditions."

Frequency Selectivity

When electromagnetic waves are released into a physical medium, multiple replicas of the original waves arrive at the destination (or receiver) because of the objects within the environment. The replicas arrive at the receiver with different delays, amplitudes, and phases, which is known as "multipath effect." Depending on the relative distances of the objects that reflect/scatter/refract the electromagnetic waves to the receiver, the replicas spread in time and lead to some important consequences:

1. If the relative delays between multipaths are shorter than (or on the order of) the symbol duration of the transmitted signal, the receiver cannot resolve each separate multipath, therefore, it sees the superposition of the multipaths, which causes a randomly fading channel.
2. If the relative delays of multipaths exceed the symbol duration, the symbols previously transmitted will impinge on other symbols causing Inter-Symbol Interference (ISI). ISI is sometimes described by the analogy of "channel memory," since the channel remembers the previous symbols even in the presence of the new symbols.

Having the knowledge of frequency selectivity provides extremely important performance improvement to the adaptive wireless communication systems including cognitive radio. As a sample application, adaptive channel equalization in single carrier systems can be considered. For example, in Global System for Mobile communications (GSM), channel equalizers are employed to compensate for ISI. However, the number of channel taps needed for equalization might vary depending on the dispersion of the channel. Instead of fixing the number of channel taps for the worst-case channel condition, it can be changed adaptively, allowing simpler receivers with reduced battery consumption and improved performance [8].

Frequency selectivity carries slightly more importance for Orthogonal Frequency Division Multiplexing (OFDM) systems compared to single carrier systems. Even though the symbol duration is prolonged because of the use of multiple orthogonal carriers, there is still partial ISI due to the multipath effect. Therefore, a certain amount of data, which is called Cyclic Prefix (CP), is replicated and added in front of the OFDM symbols to be able to alleviate ISI degradation. Considering the fact that the multipath effect is highly environment dependent, the width of CP is chosen in such a way that it is larger than the maximum excess delay of the channel for the environment in which the wireless system operates.[2] However, alleviating ISI comes at the expense of reducing the spectral efficiency, since a certain amount of the data is repeated.

[2] Since the width of CP is determined by the maximum excess delay of the channel, the maximum excess delay of the channel must be estimated. As a rule of thumb, the maximum excess delay of the channel is computed by multiplying Root-Mean-Squared (RMS) delay spread of that environment by four [5].

Instead of choosing a CP for the worst-case multipath delay spread condition, the size of CP can be adjusted adaptively. Using adaptive CP size increases the spectral efficiency.

Channel selectivity information can be estimated directly from the received signal and/or from channel estimates that are obtained after processing the received signal. However, some other opportunities in estimating the time dispersion of the wireless channel emerge through the use of additional sensing capabilities of cognitive radio. Since time dispersion is highly environment dependent, any tool that can provide cognitive radio with information about the environment plays a crucial role in estimating the selectivity of the channel. For instance, in an indoor environment, it is very likely that the RMS delay spread of the channel is considerably lower than that in a typical outdoor environment [6]. Beside the components that can provide absolute location information such as Global Positioning System (GPS), the peripherals such as light and temperature sensors might be used to characterize whether cognitive radio is in an indoor or outdoor environment [3]. When cognitive radio is in an outdoor environment, a more descriptive sub-class of the environment[3] can be identified via different enabling technologies. For outdoor cases, there are enabling technologies for obtaining the topographical (geomorphological) information about the environment such as Digital Elevation Models (DEMs) and recently becoming popular one, Geographical Information System (GIS). These digital tools allow one to analyze the spatial information. Therefore, these sorts of tools might be very helpful for cognitive radio to comprehend the surrounding environment in terms of its topographical characteristics. Table 8.1 presents some of the techniques that have been proposed for estimating the frequency selectivity of the channel along with new ones that can be used by cognitive radio for the same purpose.

Time Selectivity

When there is a relative motion between transmitter and receiver, a physical phenomenon called Doppler effect[4] occurs. The observed frequency of the

[3] In European Co-operation in the field of Scientific and Technical research (COST) 231, there are four forms for four different environmental classes: Typical urban, bad urban, rural, and hilly terrain [7]. These four types have already been defined in its predecessor, COST 207 for GSM. The Power Delay Profiles (PDPs) of typical urban and rural environments are described via single exponential cluster with different parameters, whereas those of in hilly terrain and bad urban environments are described by two exponential clusters with different parameters. Main parameter differences between environments that have common cluster structure are the arrival times of the clusters.

[4] This phenomenon has been named in the honor of Austrian mathematician and physicist Johann Christian Andreas Doppler who was born on November 29, 1803, in Salzburg, Austria and died on March 17, 1853, in Venice, upon his discovery [8] in 1842.

Table 8.1. Measuring frequency selectivity and some adaptation options.

How To Measure	What To Adapt
Frequency domain Level Crossing Rate (LCR)	Number of equalizer taps for single carrier systems
Delay spread and Channel Impulse Response (CIR) estimation	Number of pilots and spacing for multi-carrier systems
Channel frequency correlations	Fast Fourier Transform (FFT) size for multi-carrier systems
Via digital elevation model (DEM) and GIS	Carrier spacing for multi-carrier systems
	Adaptive filtering for channel estimation
	CP length for multi-carrier systems

transmitted signal[5] at the receiver is shifted because of Doppler effect. In wireless communications, Doppler effect describes the time-varying nature of a wireless channel. Generally, impulse response of a wireless channel varies in time rapidly because of the relative motion in the channel. These rapid variations in time cause spectral broadening, which is called "Doppler spread." However, impact of the broadening depends on the bandwidth of the transmitted signal. Under the same conditions, the signals that have wider transmission bandwidths are affected less by this broadening compared to those which have narrower bandwidths. In accordance with earlier discussion about duality, it is concluded that the increase in bandwidth of the transmitted signal corresponds to a shorter symbol duration in which the variation of the channel in time becomes negligible.

Doppler spread information can be very useful in various wireless system improvements. The applications can be investigated from two perspectives: (a) transmitter/receiver improvements and (b) network improvements. For (a), practical channel estimation methods can be examined. Whether channel interpolators or channel trackers are used, contemporary channel estimation algorithms are designed to operate on the worst case Doppler spread value. It is clear that in case of having the Doppler spread information in hand, the parameters of the channel estimation algorithms can be adjusted adaptively rather than adopting a fixed scheme [8, 10]. Variable coding and interleaving schemes can also be employed depending on this information, which is directly related to the speed of the mobile [11]. For (b), the use of Doppler spread information in controlling some network algorithms can be considered.

[5] The application of Doppler effect to the light, which is a sort of electromagnetic wave, was established by the French physicist Armand Hippolyte Louis Fizeau (1819–1896) in 1848 independent of Ernst Mach (1838–1916), who also discovered the same shift in 1860. The effect was first used in calculating the relative speed of the stars by William Huggins [9].

For instance, in cellular systems, hand-off (or handover), cell assignment, and channel allocation can be established efficiently having the Doppler spread estimate [12]. Assignment of fast-moving mobiles to umbrella cells in hierarchical cell structures can be considered as a specific application. The assignment of fast-moving mobiles to umbrella cells reduces the number of hand-offs, whereas the assignment of slow-moving mobiles to microcells increases the capacity [8].

There are several approaches to estimate Doppler spread. Examining the variation and autocorrelation of channel estimates are two fundamental methods in measuring the Doppler spread. Instead of channel estimates, the envelope of the signal can also be used [13]. When the channel estimates are of interest, the variation is calculated by *differentials*. However, the results obtained after differentials are generally noisy and low-pass filtering is required for smoothing. The bandwidth of the low-pass filter depends on Doppler spread too. Therefore, in essence, this method relies on changing the bandwidth of the filter adaptively. Apart from variation of the channel estimates, the autocorrelation of the channel estimates can be used in Doppler spread estimation as well. The autocorrelation of the channel is computed over the known part of the received data. Some examples of Doppler spread estimation that use the autocorrelation of the channel estimations can be found in [13, 14]. A brief list of quantifying time selectivity and some relevant adaptation options is given in Table 8.2.

As in time dispersion, Doppler spread estimation can also be improved by additional capabilities of cognitive radio. Since Doppler shift is a function of the speed of the mobile and Angle-of-Arrival (AoA), a sensor that provides the absolute location information improves the Doppler spread estimation. GPS is one of the prominent candidates for this sort of sensing applications for cognitive radio. After several consecutive measurements,[6] the speed and angle-of-arrival (AoA) can be obtained in case the position of the base station is known.

Table 8.2. Measuring time selectivity and some adaptation options.

How To Measure	What To Adapt
Correlation of channel estimates	Channel tracker step size
Correlation of signal envelope	Coding and interleaving schemes
Variation of channel estimates	Hand-off management
Variation of signal envelope	Frequency allocation
Multiple antennas	
Positioning methods such as GPS	

[6] This can be seen with the fact that $v = d\mathbf{r}(t)/dt$, where $\mathbf{r}$ is the position vector (such as $\mathbf{r}(t) = [x\ y\ z]^T$, $(\cdot)^T$ denotes the transpose operation) and v is the speed of the mobile. Note that transmission frequency and speed of light are assumed as known quantities.

Space Selectivity

Space selectivity is caused by the arrivals (or departures) of different multipaths at the receiver (from the transmitter) at different angles. When the power of arriving (or departing) multipaths is considered over angle domain, a spread is observed. The amount of spread is directly related to the richness of the scatterers within the environment. The more the scatterers, the larger the spread. Such as in time and frequency selectivity, coherence distance is a measure of the selectivity over space. Coherence distance is inversely proportional to angular spread. Hence, the shorter the coherence distance in the space, the larger the spread over angle domain.

Although space selectivity is not studied as much as time and frequency selectivity, there is a significant interest on space selectivity in multi-antenna systems. Adaptive wireless systems and cognitive radio can make use of the information about space selectivity in several ways. For instance, the information about space selectivity can be used in adaptive multi-antenna system design. Adaptive power allocation is another method in which space selectivity is used to improve the performance [15]. Adaptive modulation and coding across multi-antenna elements are also possible depending on the channel correlations. Some other options are presented in Table 8.3.

Interference Selectivity

Apart from the three basic wireless channel dimensions (time, frequency, and space) and related selectivity issues, several other dimensions can also be discussed [see Chapter 9 in this book]. Dimension of interference and selectivity over which it is defined can be considered as one of them. However, unlike the three basic dimensions, the selectivity cannot be defined solely over interference dimension. Interference selectivity becomes clearer when it is considered along with the basic dimensions. For instance, interference can be selective over frequency. It can be either Narrow Band Interference (NBI) or Wide Band Interference (WBI) depending on the bandwidth of the interferer as illustrated in Figure 8.2. Similarly, interference conditions may change in time resulting "time selective interference," whereas change of interference conditions in space causes "space selective interference."

Table 8.3. Measuring space selectivity and some adaptation options.

How To Measure	What To Adapt
Antenna arrays	Beamforming
Environmental characterization	Smart antenna
(such as indoor/outdoor)	Adaptive Multiple-Input Multiple-Output
	(MIMO) systems
	Interference management
	Frequency allocation

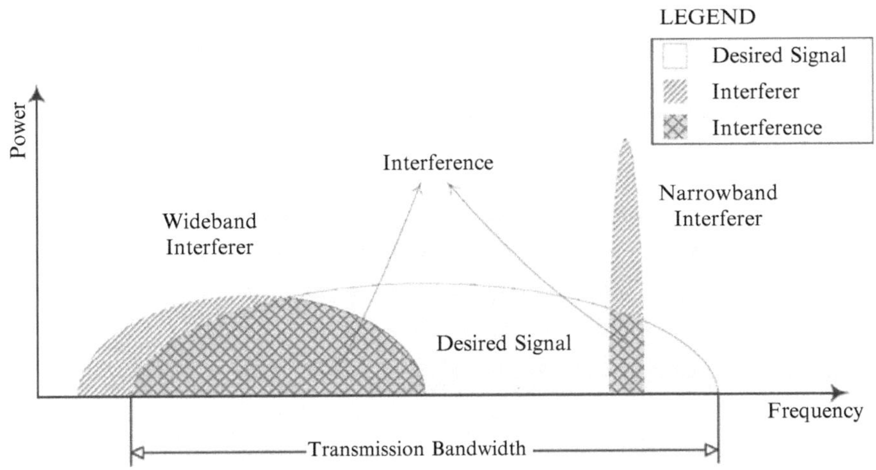

Fig. 8.2. Interference selectivity can be described via three basic dimensions: time, frequency, and space. Here, an example of interference selectivity is shown over frequency dimension. Note that the interference characteristics vary in frequency.

A cognitive radio, with enough knowledge of interference selectivity, can adapt its radio parameters to have a better transmission. As an example, a cognitive radio that uses Ultra Wide Band (UWB) scheme with OFDM can be considered. The information about the frequency selectivity of interference can be used to avoid NBI by de-activating the carriers corresponding to the interfering bands [16]. Similarly, being aware of the time selectivity of interference will give cognitive radio a chance to schedule its transmission accordingly in time as well.

Code Selectivity

Toward the latest steps of the evolution of the wireless communication systems, we witness the emergence of *code* as an extension to the basic dimension set. Therefore, selectivity over code dimension can be considered as well. Code selectivity could be a very important measure for cognitive radio systems that use codes in accessing the channel such as Direct-Sequence Spread-Spectrum (DSSS) systems using Pseudo Noise (PN) codes, Frequency Hopping (FH) systems using FH codes, and UWB systems using Time Hopping (TH) codes. Since most of the wireless systems are interference limited, the interference that is caused by statistical properties of the codes (e.g. ISI, which is caused by non-zero autocorrelation sidelobes of the codes or Multi Access Interference (MAI), which is caused by non-zero cross-correlation sidelobes of the codes) can be controlled by designing the codes appropriately. Beside interference, spectral efficiency can also be achieved by adjusting the statistical properties of the codes such as suppressing the sidelobes of auto- and cross-correlation. However, it is impossible to have codes that have both perfect auto- and cross-correlation properties. Furthermore, there is a trade-off between having sidelobes suppressed and the amount of interference created. Suppressing the sidelobes of the autocorrelation of the codes reduces ISI and increases MAI, and vice versa. Therefore, taking all these concerns into account, cognitive radio systems can increase the overall system capacity by adjusting the statistical properties of the codes adaptively depending on other system, channel, and transceiver parameters.

Other Selectivities

In this sequel, we must state that, wireless communications is not limited to the aforementioned dimensions and selectivities. There are some other aspects of wireless communications that need to be examined such as polarization, signal, and power. Although these dimensions might not be directly related to the actual wireless medium and can be regarded as elements of signal space, they have strong connections with channel space.

8.2.2 Link Quality

First and foremost condition of having a communication is to make sure that the information reaches at the receiver in a form that the receiver can reconstruct what is transmitted. As discussed in Section 8.1, the communication link exhibits different behaviors over several dimensions because of the dynamic nature of wireless communications. Thus, sensing the communication link is regarded as one of the most important tasks of cognitive radio in order to be aware of the wireless channel. In this regard, this subsection reviews some popular link quality measurement methods from the perspective of cognitive radio.

Path-Loss

Path-loss is the measure of the difference between transmitted and received power [6]. It is known that this loss increases with the transmitter–receiver separation in distance and depends on the environment, which is represented by the path-loss exponent.[7,8]

---

[7] Here, it must be stated that path-loss includes the antenna gains of the transmitter and receiver as well as the wavelength (or frequency) of the transmitted electromagnetic wave. However, the average large-scale path-loss can be approximated with the use of a function of both transmitter–receiver separation and a path-loss exponent that takes different values for different environments [6].

[8] There is also another type of path-loss, which is called frequency dependent path-loss, that manifests itself in UWB communications systems.

Path-loss information can be used in a very well-known application: adaptive power control [17]. In Code Division Multiple Accessing (CDMA) systems, when power control is not employed, all the users transmit with the same power level. Hence, the users closer to the base station cause a very high level of interference to the users which are far away from the base station. Therefore, power control algorithms are applied to adjust power levels of the users [8, and references therein]. Similar to adaptive power control, interference management based on sensing the path-loss is also possible. Apart from transmitter–receiver centric applications, path-loss information is useful for the network control. In cellular systems, hand-off (or handover) can be managed with the aid of path-loss information. As a direct consequence of the use of this information in hand-offs, adaptive channel allocation schemes can be employed, which increase the channel utilization and decreases the probability of call blocking [6]. As stated, path-loss highly depends on the environment. Therefore, the performance of the aforementioned applications can be improved in case of having information about the path-loss exponent.

In order to be able to make use of path-loss information as explained above, it must be measured. Received Signal Strength (RSS) is one of the simplest tools serving this purpose. In order to get RSS, the receiver samples the channel and averages them out.[9]

At this point, it must be stated that being aware of the location, cognitive radio can better estimate the path-loss by taking advantage of statistical propagation models. Before getting into the details of this discussion, it is appropriate to review the statistical propagation models briefly.

Statistical Propagation Models for Path-Loss – It is known that the profile of the terrain in which the communication is established has a significant impact on path-loss [6]. Various statistical propagation models for different terrain profiles are available in the literature. These models, which are based on extensive field measurements, can provide quite simple formulae for path-loss estimation in connection with the terrain of interest. For instance, the path-loss model for "urban" areas can be considered. According to Hata's model [18], the median path-loss in "urban" areas is given by the following formula:

$$L_{\text{urban}}(\text{dB}) = 69.55 + 26.16\log\left(\frac{f_c}{\text{MHz}}\right) - 13.82\log\left(\frac{h_{\text{BS}}}{\text{m}}\right) - a(h_{\text{MS}}) + \left(44.9 - 6.55\log\left(\frac{h_{\text{BS}}}{\text{m}}\right)\right)\log\left(\frac{d}{\text{km}}\right), \tag{8.1}$$

where

$$a(h_{\text{MS}}) = \left(1.1\log\left(\frac{f_c}{\text{MHz}}\right) - 0.7\right)\frac{h_{\text{MS}}}{\text{m}} - \left(1.56\log\left(\frac{f_c}{\text{MHz}}\right) - 0.8\right), \tag{8.2}$$

[9] RSS can be obtained by processing pilot signals (as in Wide Band CDMA (WCDMA)) or link layer beacon (as in IEEE 802.11). However, the duration of averages depends on many things such as system itself (having single or multiple antennas), variation of the channel, application, and so on.

and MS stands for "mobile station;" BS stands for "base station;" f_c denotes the transmission frequency; h_i denotes the effective antenna height ($i \in \{\text{MS, BS}\}$); d is the distance between MS and BS; and $a(\cdot)$ represents the correction factor. In (8.2), the correction factor is defined for small to medium scaled city.[10]

In Hata's model, we can model path-loss for other propagation environment classes as well. If one wants to estimate the path–loss for "suburban" area, (8.1) becomes

$$L_{\text{suburban}}(\text{dB}) = L_{\text{urban}}(\text{dB}) - 2\left(\log\left(\frac{f_c}{28\text{MHz}}\right)\right)^2 - 5.4, \tag{8.3}$$

whereas for "open rural" area, (8.1) becomes

$$\begin{aligned} L_{\text{open rural}}(\text{dB}) = L_{\text{urban}}(\text{dB}) - 4.78\left(\log\left(\frac{f_c}{\text{MHz}}\right)\right)^2 \\ + 18.33\left(\log\left(\frac{f_c}{\text{MHz}}\right)\right) - 40.94. \end{aligned} \tag{8.4}$$

As can be seen from (8.1), (8.3), and (8.4), obtaining the path–loss depends only on choosing the correct environmental index such as "urban," "suburban," or "open rural."[11] In conventional systems, there is no method, infrastructure, or device that can distinguish the propagation environments from each other. However, with the emergence of cognitive radio, the use of auxiliary sensing methods are brought forward to fill this gap. Hence, cognitive radio can take advantage of these extra sensing capabilities to distinguish the propagation environments from each other. In this sequel, one might wonder how cognitive radio can establish the distinguishing process. The answer of this question requires the formal definition of each propagation environment. Unfortunately, there is no formal definition for propagation environments. Nonetheless, a coarse classification of the propagation environments can still be established. In Figure 8.3, very frequently used propagation environments in the literature and their classification are shown.

Having a classification such as in Figure 8.3 will definitely be useful for cognitive radio to employ the corresponding path-loss formula. However, there is still a missing link in the chain: "How can cognitive radio understand that which of the propagation environments presented in Figure 8.3 corresponds to its surrounding environment?" Now, we are going to search for an answer to this question.

---

[10] There are several correction factors further to represent other sort of environments such as large city for different transmission frequencies, f_c.

[11] In COST 231, the extension of Hata's model is provided as well [7]. Here, the details of the specifications of Hata's model such as the limits for h, d, and f_c are not discussed. However, interested readers may refer to [6, 7, and references therein] for further details.

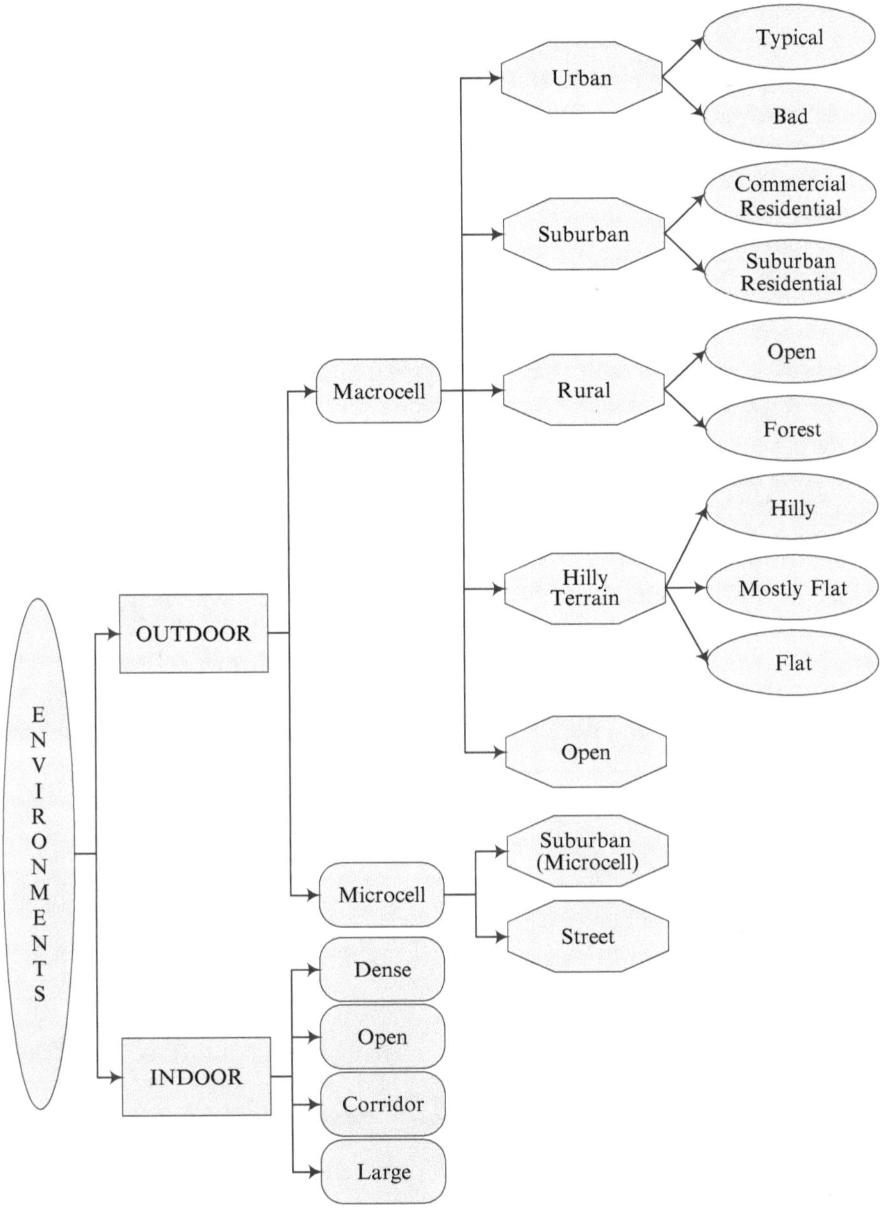

Fig. 8.3. Some popularly used propagation environments in the literature and their classification.

Environmental Characterization – In Longley–Rice model, which is one of the earliest propagation models, there are two operation modes defined depending on the availability of the terrain profile: "point-to-point mode" and "area mode" [19, 20]. Although this model is for point-to-point communication, its significance lies beneath the use of terrain profiles in estimating the path-loss. Similar to [19, 20], [21, 22] make use of a topographical database to estimate the field strength. In light of this approach, it can be said that cognitive radio can take advantage of terrain profile data to characterize the topography of its surrounding environment. Some of the basic topographical databases such as DEMs – i.e. digital representations of a topographic surface, which is used for determining properties of terrain in terms of elevation at any point, slope, aspect and extracting features of it, such as peaks, pits, and other landforms – are already available. GIS is also another very popular application which can be used for the same purpose.[12] In fact, GIS is more promising than DEMs, because it can be queried by several methods one of which is the position vector $\mathbf{r}(t) = [x, y, z]^T$. Recall that global positioning system (GPS) can provide the position vector. Thus, cognitive radio that can be provided with position information by a network or a positioning capable sensor (such as GPS) can easily extract the topographical information. Then, the digital data is processed with the aid of Spatial Interpolation Method (SIM) to obtain characteristics of the local physical environment [21, 22, 24]. The remaining part is just to find the best match for the extracted topographical data among environmental classes. As stated above, matching the data can be handled via pattern recognition and/or parallel processing capable tools that cognitive radio possesses. Upon finding the best match, a statistical model related to the matched environmental class is chosen and adaptation stage is initiated. The algorithm for this process is presented in Figure 8.4. Besides, Table 8.4 presents the possible ways of quantifying the path-loss along with adaptation options.

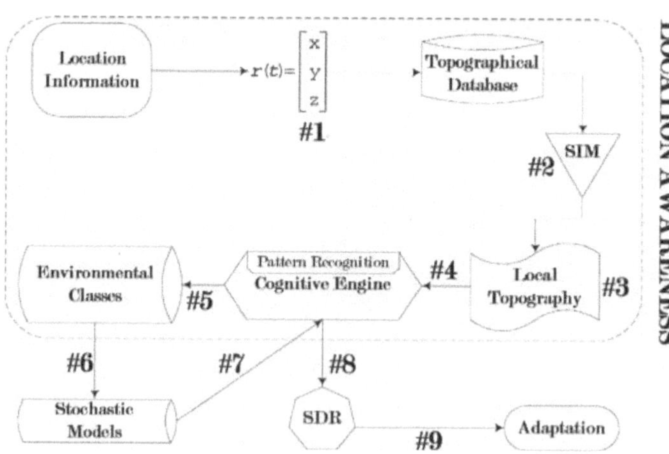

Fig. 8.4. The algorithm of location awareness and environmental characterization for cognitive radio. #*is* represent the steps of the algorithm.

[12] Note that, there are already some products in the market for mobile version of GIS applications [23].

Table 8.4. Measurement of path-loss and some adaptation options.

How To Measure	What To Adapt
RSS	Link adaptation via adaptive coding/modulation
	Hand-off (handover)
Geolocationing Methods	Channel allocation
	Interference management
	Simple distant–based power control algorithm

The impact of noise upon communications systems is known very well [34]. Noise is also taken into account in statistical wireless channel models. In order to improve the performance of wireless systems, characterizing the behavior of noise is very important. As will be shown subsequently, environments can provide some hints about the statistical behavior of noise. Hence, cognitive radio can make use of the relationship between environment and corresponding statistical behavior of the noise for its adaptation.

Noise

In most of the wireless channel models, noise is assumed as white and Gaussian distributed because of its mathematical tractability. However, in practice, several other types of noise behaviors exist too. In the literature, several studies show that offices, factories, and hospitals have impulsive noise [26]. Similarly, in outdoor environments such noise sources are observed. Thus, the performance of a system designed under the assumption of white and Gaussian distributed noise will be affected in the presence of an impulsive noise. Moreover, it is also reported that some diversity schemes such as maximal ratio combining, equal gain combining, and selection diversity are not effective in impulsive noise environments [27].

If cognitive radio knows the characteristics of the ambient noise, some of the system parameters such as coding requirements can be adjusted accordingly [26]. Besides, information about noise can be very useful in designing transceivers via adaptive modulation, optimal soft information calculation, and improved channel estimation [28,29]. From the network perspective, noise information can be used in improving several applications such as hand-off, power control, and channel allocation techniques [8].

In extracting the characteristics of the ambient noise, Signal-to-Interference Ratio (SIR) (or Signal-to-Noise Ratio (SNR) or Signal-to-Interference-plus-Noise Ratio (SINR)) are popularly used during or right after the demodulation process at the receiver. Unlike Received Signal Strength Indicator (RSSI), these quantifiers need to wait for the completion of the demodulation process.[13] However, the estimates are more reliable at this

[13] It must be stated that because these quantifiers can be read during or right after the demodulation procedure, they introduce additional complexity to the system compared to RSSI.

stage. As their names imply, these quantifiers are based on the ratio between the power level of the desired signal (S), and the power level of the unwanted signal(s) (I and/or N), where $\frac{S}{I}$, $\frac{S}{N}$, and $\frac{S}{I+N}$ denote SIR, SNR, and SINR, respectively. They can be obtained in several ways. In many new–generation wireless systems, coherent detection is employed. Since coherent detection requires estimation of channel parameters, estimated parameters can be used in calculating the signal power as well. The use of training sequences and data symbols are other two options of signal-to-interference ratio (SIR) estimation.[14] SNR (or SIR, or signal-to-interference-plus-noise ratio (SINR)) is formed by estimating the desired received signal and the impairment separately. Hence, estimation of SNR provides information about the noise. In OFDM based systems, noise power estimation is often based on the difference between the noiseless and noisy samples in frequency domain, which assumes the noise is white and Gaussian distributed. However, as stated above, noise in frequency domain can sometimes have a different power spectrum from a flat power spectrum. In such cases, looking at the estimate of the noise variance in frequency domain provides beneficial information about the noise color [28, 29]. Some of the parameters in measuring the noise and relevant adaptation options are given in Table 8.5.

There are further link quality measurements on the physical and Medium Access Control (MAC) layers too. These measurements such as Bit-Error-Rate (BER), Frame-Error-Rate (FER), and Cyclic Redundancy Check (CRC) become available after the decoding process and their reliabilities are improved significantly compared to the aforementioned ones. However, one should keep in mind that, having information at this stage comes at the expense of larger processing delays and additional computational complexity. Furthermore, some of them such as BER and FER require excessive amount of time to attain a reliable quantification.

Table 8.5. Measurement of noise and some adaptation options.

How To Measure	What To Adapt
SNR	Link adaptation via adaptive coding/modulation
SIR	Transmission frequency
SINR	Transmit power
	Hand-off
	Receiver algorithm parameters
	Bandwidth
	OFDM carriers
	Carrier assignment in OFDM Access (OFDMA)

[14] For further information, the interested readers may refer to [8, and references therein].

Network and Transport Layer Measurements

When network and transport layer are considered, observing and quantifying the wireless communication link brings about different perspectives. In these layers, transceivers perceive the communication link as a whole that includes other transceivers. Therefore, in these layers, transceivers take into account the status of other nodes in the network as well.

Packet loss is one of the basic and simple quantifier that can be used in these layers. Transceivers can get information about the quality of the link by simply counting the packets which could not be acknowledged. Similarly, Round-Trip Time (RTT) can be used in estimating the quality of the link to some extent. The buffer status of the nodes can be used in determining the congestion level of the link, which can also be regarded as a measure of link quality. Particularly in ad hoc networks, being aware of other nodes become prominent. For instance, a quantifier that indicates the power level of the other nodes introduces another metric for routing, which is known as "power aware routing" [30]. In connection with power aware routing, being aware of the locations (or relative positions) of the nodes will definitely provide these layers with an extra and very important quantifier.

With the aid of these quantifiers, cognitive radio can adjust its transmission rate to avoid congestion, increase efficiency, save energy, and so on. In addition, cognitive radio can contribute to the network optimization by combining many of them. For instance, combination of power aware routing and being aware of the locations of other nodes will provide a superior routing scheme compared to those which make use of only one routing metric. Some of the quantification parameters and related adaptation options are presented in Table 8.6.

Upper Layers

After passing through session and application layers, cognitive radio reaches at the user even though the user by itself is not a layer in the protocol stack. From the perspective of cognitive radio, the user carries a significant importance

Table 8.6. Measurement of the quality of the communication link in network and transport layer and some adaptation options.

How To Measure	What To Adapt
Packet Loss	Routing Algorithm
Routing Table Change Rate	Routing Metric
Congestion Level	Clustering Parameters
Positions of Nodes	Network Scheduling Algorithm
Power Level of Nodes	Congestion Control Parameters
RTT	Rate Control Parameters

because of its place in the cognition cycle. Therefore, cognitive radio can perceive the user as another layer that is placed on top of the traditional protocol stack and sense it. These issues will be discussed in Section 8.4 in detail.

8.2.3 Other Wireless Channel Characteristics

In addition to the topics discussed in Sections 8.2.1 and 8.2.2, there are some further parameters that have significant impact on the transmission such as being in Line-of-Sight (LOS)/Non-Line-of-Sight (NLOS).

LOS/NLOS

When the field measurements are established in order to have a statistical channel model, it is extremely important to distinguish the measurements as for LOS and NLOS. This stems from the fact that LOS channels will behave very differently compared to NLOS channels because of the presence of the direct component. LOS/NLOS distinction is very important in terms of the operation band as well, since the behavior of propagation differs in LOS and NLOS. For instance, in order to be able to establish a communication with electromagnetic waves that have wavelengths on the order of millimeters (the bands above 10 GHz), LOS is required as in 10–66 GHz portion of the physical layer part of IEEE 802.16 [31]. However, the necessity of having LOS is not required for sub-10 GHz bands. In addition, being in LOS/NLOS is very important for positioning algorithms. The error characteristics change drastically depending on being in LOS or NLOS [32].

The knowledge of being in LOS or NLOS allows cognitive radio to have some adaptation options. Cognitive radio can easily switch to an appropriate upper frequency band to achieve higher data rates in case of being in LOS or switches back to the band in which it was previously operating when there is no LOS. Ranging and positioning algorithms can be selected by cognitive radio adaptively, depending on the status of the transmission in terms of being in LOS or NLOS as well.

In order to determine whether the status of the communication is LOS or NLOS, hypothesis test is applied. Hypothesis test makes use of the mutually exclusive relationship between LOS and NLOS [32–35]. Considering the fact that the channel amplitudes of the first tap in narrowband systems follow Ricean distribution in LOS and Rayleigh distribution in NLOS, a comparison between the reference (theoretical) distributions and values observed can be established before the hypothesis test [36].[15] However, a reliable decision for the comparison approach, *a priori* knowledge about the noise level of the system is required [32]. Apart from these methods, autocorrelation characteristics of multiple channel taps have also been proposed [31].

[15] In comparing the statistics obtained to a reference one, some statistical tests such as Pearson's test statistics [36] or Kolmogorov–Smirnov test [37] can be employed.

Table 8.7. Measuring LOS/NLOS and some adaptation options.

How To Measure	What To Adapt
Channel Estimates	Transmission frequency
Geolocationing Methods	Power adjustment
	Locationing algorithms for improved accuracy
	Receiver algorithm parameters

Quantifying the transmission status in terms of being LOS or NLOS becomes possible through the use of additional sensing capabilities of cognitive radio. A list of quantification options and relevant adaptation parameters for LOS/NLOS is given in Table 8.7. As discussed in detail in Section 8.2.2, the use of DEMs is a very promising candidate for this purpose. In fact, previously, DEMs have already been used in digital domain to determine the status of being LOS or NLOS [21, 22].

8.3 Network Awareness

This section outlines what and how cognitive can sense, be aware of, and consequently adapt the network related issues. The following two perspectives are considered: being aware of the same network and other network structures.

8.3.1 Being Aware of the Same Network

This sort of awareness is necessary while cognitive radio is already communicating with some other nodes. From this point of view, being aware of its own network and its other members (nodes) will definitely improve the quality of overall network communications. However, being a member of the same network does not necessarily require that all the nodes have cognitive capabilities. Therefore, it can be said that being "entirely" aware of the same network is only possible when all the nodes in that network are cognitive radio.

As discussed in Section 8.2.2, being aware of the same network can be extended by using conventional and advanced methods such as the use of packet loss quantifiers, routing algorithms, and some other additional capabilities which are introduced by cognitive radio such as location sensing.

8.3.2 Being Aware of Other Networks

Before cognitive radio begins to communicate, it can sense not only the unoccupied bands in the spectrum, but also the signaling schemes over the air. For instance, cognitive radio can sense unoccupied slots for Time Division Multiple Accessing (TDMA)–based signaling, and furthermore, it can be aware of the network type by using several methods such as cyclostationary-based detection. This is very important for cognitive radio, since it can easily change its transmission parameters such as employing valid waveforms and relevant policies for the network sensed with the aid of SDR.

This type of sensing and awareness carries a great importance especially for emergency, disaster relief, and rescue operations. The transmission of other devices can be observed by sensing the spectrum, extracting the data from other users' transmission, processing it, comparing it with some *a priori* information (such as standard information), and making a decision about the existence of a possible network. Thus, in such an environment, cognitive radios can establish a network and connectivity for the devices which cannot easily be detected by first responders [39].

8.4 User Awareness

When cognitive radio has first been brought forward [3, 19], beside its advanced properties such as the capability of sensing the spectrum, adjusting the transmission parameters via software (or software defined radio (SDR)), the concept of "*user dimension*" has been pulled inside the radio communication domain entirely. Here, the word "entirely" is preferred, because, there are already some earlier attempts. However, these attempts are limited compared to that of cognitive radio, because, they mostly require user's intervention. Some advanced cell phones, Personal Digital Assistants (PDAs), and laptops can define several user profiles as options, but, these options are not automated. In addition, once the user chooses any of the options provided, the device cannot make modifications on the options depending on the changing conditions.

However, cognitive radio, beside the aforementioned capabilities, can learn and even make predictions about the user. One of the interesting awareness topic about the user is being aware of user's perception. When a simple voice conversation over phone is considered, some sort of adaptations become clearer depending on the environment and/or who the user is. For instance, if the user is in a crowded and noisy environment such as a stadium, the intelligibility of the voice decreases drastically. In the conventional way, the user intervention is required by just choosing the loudest voice level of the phone or warning the other party to raise his/her voice. However, cognitive radio can reduce or totally remove the user intervention by sensing the environment with the aid of its additional sensing and advanced recognition capabilities. Detecting the crowd via visual sensing and picking the phrases such as "I did not hear you," "Can you repeat it?" "Can you raise your voice?" cognitive radio is aware of the user's perception (as well as the environment) and adapts to satisfy its user's needs. Another interesting user awareness and adaptation scenario can occur while driving a car. Driving car inherently limits some of the abilities of the user such as reading and/or using the keypad of the communication device. When a text message (such as Short Message Service (SMS) or e-mail) arrives, cognitive radio can convert the text to speech without requiring user intervention. When the user wants to reply, it can do the process in reverse direction, in other words it can convert speech to text, and send the text message. In case the user wants to call the sender, it can initiate an ordinary phone conversation by recognizing the user's speech and command (such as "Call him!"). Another very important application area of user oriented sensing occurs in emergency related events. Cognitive radio that is aware of physiological state of its user can immediately establish a 911 call in case of an emergency.

Because humans have many aspects in terms of their perceptions, psychological status, and some other further characteristics, the items to be included into the list in user dimension are numerous. However, possessing very powerful abilities such as learning and sensing beside awareness and adaptation, cognitive radio can have various sensing options for other aspects of human, which have very stong relations with AI.

8.5 Other Possible Awareness Scenarios

It is certain that wireless communications cannot be limited to the topics mentioned throughout this chapter. Although a wide perspective is tried to be established, it is very difficult to put all possible aspects of the communications into text, since most of them evolve in time. However, some of the topics that are and will be of interest can still be discussed.

The security is one of the most important aspects in all types of communications. In this manner, cognitive radio can decide the appropriate security level without requiring any external intervention. Here, the discussion excludes the use of the safest level, because it comes at the expense of delay, overhead, and so on. These concerns force cognitive radio to optimize (deciding the appropriate level) the security level. In connection with optimization, cognitive radio itself can be considered as another dimension (or entity) to be sensed and aware of in the communications domain. Cognitive radio cannot attain a global optimization unless it is aware of itself. In this regard, being aware of its resources such as battery level and hardware limitations (e.g. the limits of Analog-to-Digital Converter (ADC), processor speed, memory size, and so on), the operational status such as being in hibernation, in-charge mode, communication mode, learning mode can be considered just to name few.

8.6 Challenges and Future Directions

So far, it is discussed that the capabilities of cognitive radio enable many new sensing options providing alternatives to already existing ones. A hierarchical list of some major items to be sensed by cognitive radio is presented in

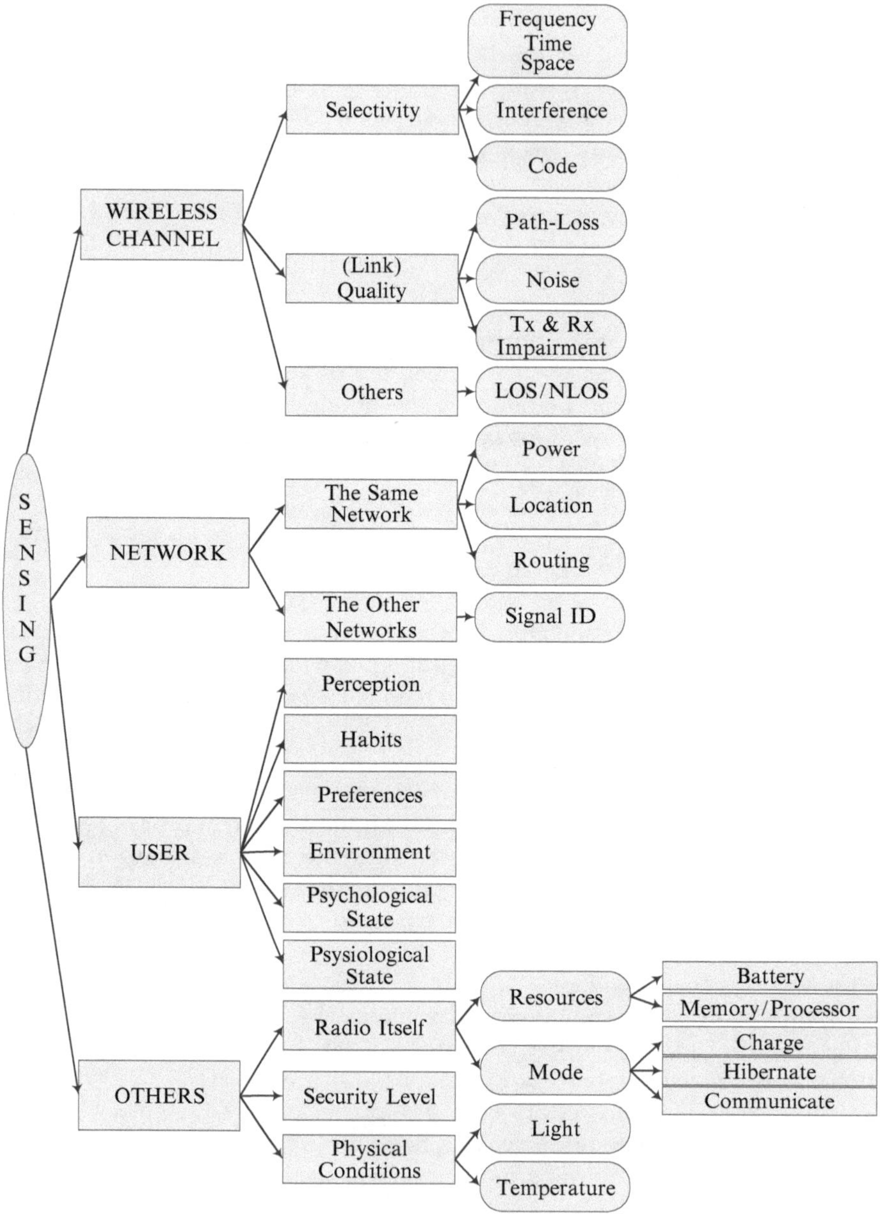

Fig. 8.5. Some major items to be sensed by cognitive radio. Please note that this represents a limited list that only provides some of the key measurements. As the cognitive radio evolves, this list will certainly evolve as well.

Figure 8.5. However, there are many challenges related to these enabling approaches. In this section, the prominent challenges and hurdles regarding to these approaches will be outlined.

First, the challenge with the general architecture of cognitive radio needs to be addressed. In order for cognitive radio to attain the point that has been discussed so far, a design, which integrates SDR and tools that handle artificial intelligence (AI) sort of operations, is required. This can be considered as the most challenging issue for cognitive radio.

Assuming that such an architecture or design is established, the complexity of the procedures in sensing mentioned throughout this chapter is quite overwhelming. Therefore, cognitive radio needs to find ways of handling this complexity issue. Cognitive radio makes use of not only existing methods and approaches, but also the new opportunities of sensing with the aid of additional sensors and devices to improve the adaptation process. Since the information traffic will increase considerably, controlling, processing, and therefore managing the resources automatically become a major concern.

Apart from general challenges, there are some other challenges peculiar to sensing, being aware, and adaptation for each topic discussed up until now. In Section 8.2.2, determining the path-loss through the use of external sensors has been discussed. The main challenge here is to find how cognitive radio can identify the environment in order to use the appropriate path-loss exponent. As explained in [3], distinguishing indoor from outdoor is a simple job by just using light and thermal sensor. However, distinguishing several indoor environments from each other, such as LOS and NLOS indoor communications requires additional effort.

In Section 8.2.3, extracting time dispersion parameters of the wireless channel with the aid of DEMs and GIS has been discussed as well. However, digital information provided by these sorts of tools must be interpreted by cognitive radio in such a way that it can understand the geomorphological characteristics of the environment and classify them as hilly terrain, urban, and so on. As can be seen, the interpretation process requires the use of advanced pattern recognition algorithms. Another aspect of geomorphological characterization is that it may require the classification of the propagation environments for cognitive radio to choose an appropriate model as discussed earlier. There are extensive statistical channel models in the literature to include various possible propagation environments. If cognitive radio can match the surrounding environment with an existing statistical model in its memory, it can easily adjust a few relevant parameters to adapt the environment. However, in order for cognitive radio to choose one of the statistical models among many of them, it should store the statistical models in a hierarchical way. Unfortunately, there is no clear-cut definition for types of propagation environments in the literature. Cognitive radio may suffer from lack of these definitions.

In quantifying time selectivity, apart from the conventional way of extracting the parameters, cognitive radio can make use of its location sensors such as GPS. However, it is not known how frequently cognitive radio must refer to the sensors to obtain the information. In addition, the capabilities of the sensors (such as acquisition time, precision, and so on) must also be taken into account.

In time selective interference, characterizing the pattern of the interference that changes in time for adapting purposes may require long observation durations. This hurdle becomes clearer in channels that are randomly accessed. Similar to interference, the relationships between code dimension and three basic dimensions need to be studied to have a comprehensive understanding of the selectivity.

The inclusion of aforementioned and possible future dimensions into the universe of cognitive radio will force researchers to examine the relationships between these dimensions. The entangled structures of these dimensions must be investigated throughly to realize the ultimate cognitive radio, which is a very challenging task.

Considering the complete adaptation and global optimization jointly unveils one of the biggest challenges in realizing cognitive radio. In order for cognitive radio to achieve both complete adaptation and global optimization, the complex nature of complete adaptation needs to be scrutinized. The analysis must encompass both already existing and recently emerging sensing options and their relevance. Upon this analysis, which is very likely to provide a very comprehensive list, the ways of attaining global optimization must be researched. Global optimization through additional sensing capabilities along with the already existing ones can be achieved by AI structure, which is, generally, referred as cognitive engine. Thus, cognitive engine must be capable of "understanding," "interpreting," and even "reasoning" via the input provided by layers, sensors, and even by its own hardware. In order for cognitive engine to "think" taking into account all input, a descriptive language, Radio Knowledge Representation Language (RKRL), is proposed in [3]. Hence, already existing and recently emerging options must be included into the knowledge space of cognitive radio in connection with the analysis.

The interactions, applications, and algorithms in a network that include both cognitive and non-cognitive radios need to be studied as well. In this regard, networks in which all the nodes are cognitive radios form another field of study.

Especially in sensing other networks, cognitive radio is challenged by technological limitations. Currently, sensing other types of networks and signaling schemes is carried out by huge devices. Beyond that, the techniques that are employed for that purpose require computationally very complex signal processing operations which are power hungry. Thus, in order for cognitive radio to have these abilities, practically simple and less complex algorithms that can perform the same operations must be developed.

Considering the fact that sensing, learning, being aware, and adaptation capabilities make cognitive radio more personal, the characterization of the user can be examined in detail. Especially sensing the user's psychological and physiological status and act on accordingly will be an interesting research for the sake of becoming more personal.

8.7 Conclusion

Cognitive radio provides new horizons to the radio communications with its advanced capabilities such as sensing, learning, being aware of, and adaptation to its surrounding environment. These advanced capabilities improve almost every aspect of wireless communications. In this chapter, some major items to be sensed by cognitive radio are discussed. It is certain that the things that cognitive radio can sense, be aware of, and measure are not limited to the items discussed here. As cognitive radio evolves, the list of the things that cognitive radio can sense, be aware of, and measure will evolve as well. However, in order for cognitive radio to come true, the hurdles in front of the new methods of sensing and measuring processes must be overcome.

References

1. J. Mitola and G. Q. Maguire Jr, "Cognitive radio: making software radios more personal," *IEEE Personal Communications*, vol. 6, no. 4, pp. 13–18, Aug. 1999.
2. H. Arslan, *Signal Processing for Mobile Communications Handbook, M. Ibnkahla, Ed. Boca Raton, FL: CRC.* CRC Press, 2004, ch. 28, Adaptation Techniques and the Enabling Parameter Estimation Algorithms for Wireless Communication Systems, pp. **28**-1–**28**-26.
3. J. B. J. Fourier, *Théorie analytique de la chaleur.* Paris: Chez Firmin Didot, Père et Fils, 1822.
4. P. Bello, "Time–Frequency Duality," *IEEE Transactions on Information Theory*, vol. 10, no. 1, pp. 18–33, Jan. 1964.
5. M. Engels, *Wireless OFDM Systems: How to Make Them Work?*, ser. The International Series in Engineering and Computer Science. Norwell, Massachusetts, U.S.A.: Kluwer Academic Publishers, 2002, vol. 692.
6. T. S. Rappaport, *Wireless Communications: Principles and Practice*, 2nd ed., ser. Prentice Hall Communications Engineering and emerging Technologies Series. New Jersey, U.S.A.: Prentice–Hall, Inc., 2002.
7. Action Committee, "231: Digital mobile radio towards future generation systems – final report," European Communities, Tech. Rep., 1999.
8. J. C. A. Doppler, *Beitrage zur fixsternenkunde.* Prag, Druck von G. Haase sohne, 1846.
9. W. Huggins, "Further Observations on the Spectra of Some of the Stars and Nebulae, with an Attempt to Determine Therefrom Whether These Bodies are Moving towards or from the Earth, Also Observations on the Spectra of the Sun and of Comet II., [1868]," *Philosophical Transactions Series I*, vol. 158, pp. 529–564, 1868.

10. C. Tepedelenlioglu and G. B. Giannakis, "On velocity estimation and correlation properties of narrow–band mobile communication channels," *IEEE Transactions on Vehicular Technology*, vol. 50, no. 4, pp. 1039–1052, July 2001.
11. D. Mottier and D. Castelain, "A doppler estimation for UMTS–FDD based on channel power statistics," in *Proc. IEEE 50th Vehicular Technology Conference, VTC–Fall*, vol. 5, Amsterdam, Netherlands, Sept. 1999, pp. 3052–3056.
12. G. P. Pollini, "Trends in handover design," *IEEE Communications Magazine*, vol. 34, no. 3, pp. 82–90, Mar. 1996.
13. L. Krasny, H. Arslan, D. Koilpillai, and S. Chennakeshu, "Doppler spread estimation in mobile radio systems," *IEEE Communications Letters*, vol. 5, no. 5, pp. 197–199, May 2001.
14. M. Morelli, U. Mengali, and G. Vitetta, "Further results in carrier frequency estimation for transmissions over flat fading channels," *IEEE communications Letters*, vol. 2, no. 12, pp. 327–330, Dec. 1998.
15. D.-S. Shiu, G. J. Foschini, M. J. Gans, and J. M. Kahn, "Fading correlation and its effect on the capacity of multielement antenna systems," *IEEE Transactions on Communications*, vol. 48, no. 3, pp. 502–513, Mar. 2000.
16. H. Arslan and M. E. Şahin, "Cognitive UWB–OFDM: Pushing ultrawideband beyond its limit via opportunistic spectrum usage," *Journal Of Communications and Networks (Special Issue on Spectrum Resource Optimization)*, vol. 10, no. 10, pp. 151–157, June 2006.
17. Şennur Ulukuş and R. D. Yates, "Adaptive power control and MMSE interference suppression," *Wireless Networks*, vol. 4, no. 6, pp. 489–496, Oct. 1998.
18. M. Hata, "Empirical formula for propagation loss in land mobile radio services," *IEEE Transactions on Vehicular Technology*, vol. VT-29, no. 3, pp. 317–325, Aug. 1980.
19. P. Rice, A. G. Longley, K. A. Norton, and A. P. Barsis, "Transmission Loss Predictions for Tropospheric Communication Circuits, Volume I," National Bureau of Standards, Technical Note 101, Jan. 1967 [Issued May 7, 1965; Revised May 1, 1966; Revised (II) January 1, 1967].
20. P. Rice, A. G. Longley, K. A. Norton, and A. P. Barsis, "Transmission Loss Predictions for Tropospheric Communication Circuits, Volume II," National Bureau of Standards, Technical Note 101, Jan. 1967 [Issued May 7, 1965; Revised May 1, 1966; Revised (II) January 1, 1967].
21. R. Edwards and J. Durkin, "Computer Prediction of Service Area for VHF Mobile Radio Networks," in *Proceedings of the IEE*, vol. 116, no. 9, Sept. 1969, pp. 1493–1500.
22. J. Durkin, "Computer prediction of service areas for VHF and UHF land mobile radio services," *IEEE Transactions on Vehicular Technology*, vol. VT-26, no. 4, pp. 323–327, Nov. 1977.
23. ESRI, "ArcPad: Mobile GIS, ESRI White Paper," ESRI, Tech. Rep., September 2004.
24. L. Mitas and H. Mitasova. (1998) Multidimensional spatial interpolation. GMS Laboratory, University of Illinois at Urbana-Champaign (Accessed on 11.15.2006). [Online]. Available: http://skagit.meas.ncsu.edu/˜helena/gmslab/viz/sinter.html
25. C. E. Shannon, "The Mathematical Theory of Communication," *The Bell System Technical Journal*, vol. 27, pp. 379–423; 623–656, July, October, 1948.

26. K. L. Blackard, T. S. Rappaport, and C. W. Bostian, "Measurements and models of radio frequency impulsive noise for indoor wireless communications," *IEEE Journal on Selected Areas in Communications*, vol. 11, no. 7, pp. 991–1001, Sept. 1993.
27. R. S. Blum, R. J. Kozick, and B. M. Sadler, "An Adaptive Spatial Diversity Receiver for Non–Gaussian Interference and Noise," *IEEE Transactions on Signal Processing*, vol. 47, no. 8, pp. 2100–2111, Aug. 1999.
28. T. Yücek and H. Arslan, "Noise plus interference power estimation in adaptive OFDM systems," in *Proceedings of IEEE 61st Vehicular Technology Conference (VTC 2005–Spring)*, vol. 2, Stockholm, Sweden, May 30–June 1, 2005, pp. 1278–1282.
29. T. Yücek and H. Arslan, "MMSE Noise Power and SNR Estimation for OFDM Systems," *IEEE Transactions on Vehicular Technology*, 2007, in print.
30. S. Singh, M. Woo, and C. S. Raghavendra, "Power–aware routing in mobile ad hoc networks," in *MobiCom '98: Proceedings of the 4th annual ACM/IEEE international conference on Mobile computing and networking.* New York, NY, USA: ACM Press, 1998, pp. 181–190.
31. S. Yarkan and H. Arslan, "Identification of LOS and NLOS for wireless transmission," in *Proceedings of IEEE Cognitive Radio Oriented Wireless Networks and Communications, CROWNCOM 2006*, Mykonos Island, Greece, June 07–11, 2006.
32. S. Gezici, H. Kobayashi, and H. V. Poor, "Nonparametric nonline–of–sight identification," in *Proc. IEEE 58th Vehicular Technology Conference, VTC 2003–Fall*, vol. 4, Orlando, Florida, U.S.A., Oct. 6–9, 2003, pp. 2544–2548.
33. J. Vidal and R. E. Jativa, "First arriving path detection for subscriber location in mobile communication systems," in *Proceedings of IEEE International Conference on Acoustics, Speech, and Signal Processing ICASSP'02*, vol. 3, Orlando, Florida, U.S.A., May 13–17, 2002, pp. III-2733–III-2736.
34. J. Borras, P. Hatrack, and N. B. Mandayam, "Decision theoretic framework for NLOS identification," in *Proc. IEEE 48th Vehicular Technology Conference, VTC 1998*, vol. 2, Ottawa, Canada, May 18–21, 1998, pp. 1583–1587.
35. S. Venkatraman and J. C. Jr., "Statistical approach to non–line–of–sight BS identification," in *Proc. The 5th International Symposium on Wireless Personal Multimedia Communications*, vol. 1, Honolulu, Hawaii, U.S.A., Oct. 27–30, 2002, pp. 296–300.
36. A. Lakhzouri, E. Simona, L. Ridha, and H. M. Renforse, "Extended Kalman filter channel estimation for line-of-sight detection in WCDMA mobile positioning," *EURASIP Journal on Applied Signal Processing*, no. 13, pp. 1268–1278, Dec. 2003.
37. S. Al-Jazzar and J. Caffery Jr., "New algorithms for NLOS identification," in *Proc. The 14th IST Mobile and Wireless Communications Summit*, Dresden, Germany, June 19–23, 2005.
38. Bernard L. Malone III, "Wireless Search and Rescue: Concepts for Improved Capabilities," *Bell Labs Technical Journal*, vol. 9, no. 2, pp. 37–49, Summer 2004.
39. J. Mitola III, "Cognitive radio an integrated agent architecture for software defined radio," Ph.D. dissertation, KTH Royal Institute of Technology, Stockholm, Sweden, May 8, 2000.

9

Spectrum Sensing for Cognitive Radio Applications

Hüseyin Arslan[1] and Tevfik Yücek[2]

[1] University of South Florida, Tampa, FL, USA arslan@eng.usf.edu
[2] University of South Florida, Tampa, FL, USA yucek@eng.usf.edu

9.1 Introduction

The need for higher data rates is increasing as a result of the transition from voice-only communication to wireless multimedia and web type of applications. Given the limitations of natural frequency spectrum, it becomes obvious that current static frequency allocation schemes cannot accommodate these requirements of increasing number of higher data rate devices. As a result, innovative techniques that can offer new ways of exploiting the available spectrum are needed. *Cognitive radio* arises to be a tempting solution to spectral crowding problem by introducing the opportunistic usage of frequency bands that are not heavily occupied by licensed users [1]. While there is no agreement on the formal definition of cognitive radio as of now, the concept has evolved recently to include various meanings in several contexts [2]. One main aspect of cognitive radio is related to autonomously exploiting locally unused spectrum to provide new paths to spectrum access. Other aspects include interoperability across several networks; roaming across borders while being able to stay in compliance with local regulations; adapting the system, transmission, and reception parameters without user intervention; and having the ability to understand and follow actions and choices taken by their users to learn how to become more responsive over time.

One of the most important components of cognitive radio concept is the ability to measure, sense, learn, and be aware of the parameters related to the radio channel characteristics, availability of spectrum and power, interference and noise temperature, radio's operating environment, user requirements and applications, available networks (infrastructures) and nodes, local policies and other operating restrictions. In cognitive radio terminology, *primary users* can be defined as the users who have higher priority or legacy rights on the usage of a specific part of the spectrum. On the other hand, *secondary users*, which have lower priority, exploit this spectrum in such a way that they do not cause interference to primary users. Therefore, secondary users need to have cognitive radio capabilities, such as sensing the spectrum reliably to check

H. Arslan (Ed.), Cognitive Radio, Software Defined Radio, and Adaptive Wireless Systems, 263–289.

whether it is being used by a primary user, and to change the radio parameters to exploit the unused part of the spectrum.

Being the focus of this chapter, spectrum sensing by far is the most important task among others for the establishment of cognitive radio. Spectrum sensing includes awareness about the interference temperature and existence of primary users. As an alternative to spectrum sensing, geolocation and database or beacons[1] can be used for determining the current status of the spectrum usage [3,4]. In this chapter, we focus on spectrum sensing performed by cognitive radios because of its broader application areas while referring other methods as needed. Although spectrum sensing is traditionally understood as measuring the spectral content, or measuring the interference temperature over the spectrum; when the ultimate cognitive radio is considered, it is a more general term that involves obtaining the spectrum usage characteristics across multiple dimensions such as time, space, frequency, and code. It also involves determining what type of signals are occupying the spectrum (including the modulation, waveform, bandwidth, carrier frequency, etc.). However, this requires more powerful signal analysis techniques with additional computational complexity.

Various aspects of spectrum sensing task are illustrated in Figure 9.1. The goal of this chapter is to point out several aspects of spectrum sensing as shown in this figure. These aspects will be discussed in the rest of this chapter. We start by explaining some challenges associated with spectrum sensing in Section 9.2. Section 9.3 explains the main spectrum sensing methods. Cooperative sensing concept and its various forms are introduced in Section 9.4, followed by a discussion of external sensing algorithms in Section 9.5. Statistical modeling of network traffic and utilization of these models for prediction of primary user behavior is studied in Section 9.6. Section 9.7 explains the factors on deciding the frequency of spectrum sensing. Hardware perspective of sensing problem is discussed in Section 9.8. We introduce the multi-dimensional spectrum sensing concept in Section 9.9. Finally, sensing features of some current wireless standards are explained in Section 9.10 and our conclusions are given in Section 9.11.

9.2 Challenges

Before getting into the details of spectrum sensing techniques, some challenges associated with the spectrum sensing for cognitive radio is given in this section.

Hardware Requirements

Spectrum sensing for cognitive radio applications requires high sampling rate, high resolution Analog to Digital Converter (ADCs) with large dynamic range,

[1] When beacons are used, the transmitted information can be occupancy of a spectrum as well as other advanced features such as channel quality.

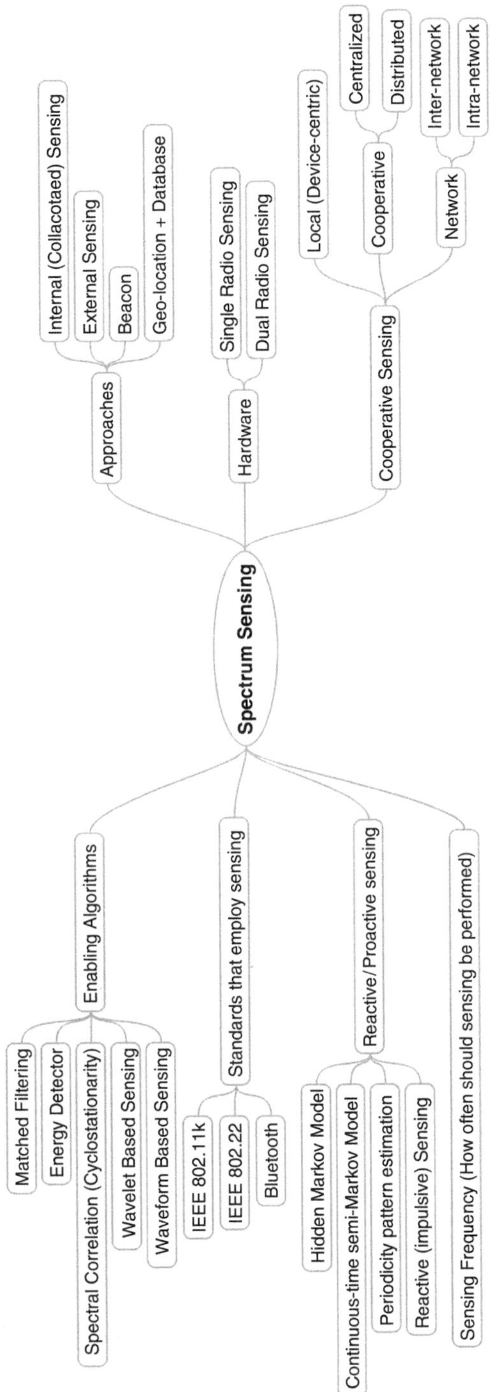

Fig. 9.1. Various aspects of spectrum sensing for cognitive radio.

multiple analog front end circuitry, and high speed signal processors. Estimating the noise variance or interference temperature over transmission of desired narrowband signals is not new. Such noise variance estimation techniques have been popularly used for optimal receiver designs like channel estimation, soft information generation, as well as for improved hand-off, power control, and channel allocation techniques. The noise/interference estimation problem is easier for these purposes as receivers are tuned to receive signals that are transmitted over a desired bandwidth. Moreover, receivers are capable of processing the narrowband baseband signals with reasonably low complexity and low power processors. However, in cognitive radio, terminals are required to process transmission over a much wider band for sensing any opportunity.

Hidden Primary User Problem

Hidden primary user problem is similar to the hidden node problem in Carrier Sense Multiple Accessing (CSMA). This problem can be caused by many factors including severe multipath fading or shadowing that secondary users observe while scanning primary users' transmissions. Figure 9.2 shows an illustration of hidden node problem. Here, cognitive radio device causes unwanted interference to the primary user (receiver) as the primary transmitters signal could not be detected because of the positioning of devices in space.

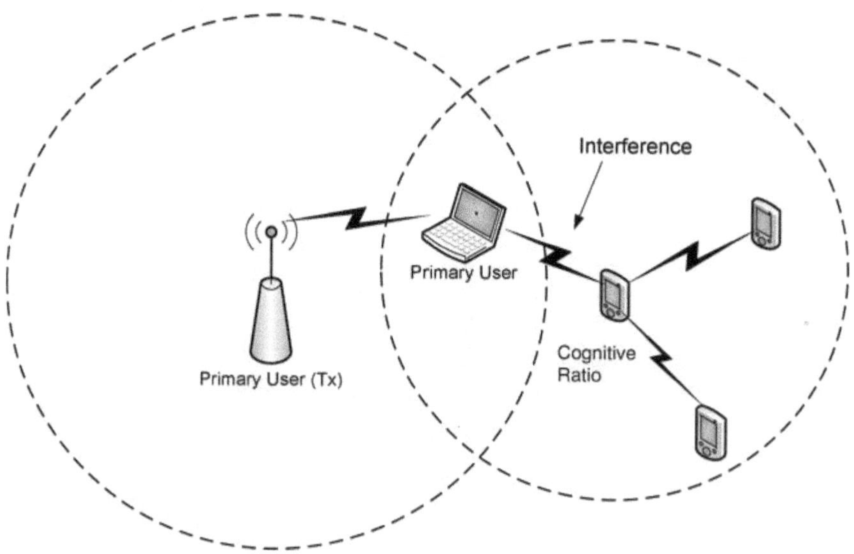

Fig. 9.2. Illustration of hidden primary user problem in cognitive radio systems.

Spread Spectrum Primary Users

Primary users that use frequency hopping or spread spectrum signaling, where the power of the primary user signal is distributed over a wider frequency even though the actual information bandwidth is much narrower, are difficult to detect. Especially, frequency hopping-based signaling creates significant problems regarding to spectrum sensing. This problem can be partially avoided if the hopping pattern is known and perfect synchronization to the signal can be achieved.

Sensing Time

Primary users can claim their frequency bands anytime while cognitive radio is operating at that band. In order to prevent interference to and from primary license owners, cognitive radio should be able to identify the presence of primary users as quickly as possible and should vacate the band immediately. Hence, sensing method should be able to identify the presence of primary user within a certain duration. This requirement possesses a limit on the performance of sensing algorithm and create a challenge for cognitive radio design.

Other Challenges

Some other challenges that need to be considered while designing effective spectrum sensing algorithm include implementation complexity, presence of multiple secondary users, coherence times, multipath and shadowing, cooperation, competition, robustness, heterogeneous propagation losses, and power consumption.

9.3 Spectrum Sensing Methods for Cognitive Radio

The present literature for spectrum sensing is still in its early stages of development. A number of different methods are proposed for identifying the presence of signal transmission. In some approaches, characteristics of the identified transmission are detected for deciding the signal transmission as well as identifying the signal type. In this section, some of the most common spectrum sensing techniques in the cognitive radio literature are explained.

9.3.1 Matched Filtering

Matched filtering is known as the optimum method for detection of primary users when the transmitted signal is known [5]. The main advantage of matched filtering is the short time[2] to achieve a certain probability of false

[2] The required number of samples grows as $O(1/SNR)$ for a target probability of false alarm or miss detection at low SNRs [6].

alarm or probability of miss detection [6] as compared to other methods that are discussed in this section. However, matched filtering requires the cognitive radio to demodulate received signals. Hence, it requires perfect knowledge of the primary users signaling features such as bandwidth, operating frequency, modulation type and order, pulse shaping, frame format, etc. Moreover, since cognitive radio needs receivers for all signal types, implementation complexity of sensing unit is impractically large [7]. Another disadvantage is large power consumption as various receiver algorithms need to be executed for detection.

9.3.2 Waveform-Based Sensing

Known patterns are usually utilized in wireless systems to assist synchronization or for other purposes. Such patterns include preambles, midambles, regularly transmitted pilot patterns, spreading sequences, etc. In the presence of a known pattern, sensing can be performed by correlating the received signal with a known copy of itself [8,9]. This method is only applicable to systems with known signal patterns, and it is termed as waveform-based sensing. In [8], it is shown that waveform-based sensing outperforms energy detector-based sensing in reliability and convergence time. Furthermore, it is shown that the performance of the sensing algorithm increases as the length of the known signal pattern increases. As one of the methods for analyzing the Wireless Local Area Network (WLAN) channel usage characteristics, packet preambles of IEEE 802.11b [10] signals are exploited in [11,12]. Measurement results presented in [13] show that waveform-based sensing requires short measurements time, however, it is susceptible to synchronization errors.

Let us assume that the received signal has the following simple form:

$$y(n) = s(n) + w(n), \tag{9.1}$$

where $s(n)$ is the signal to be detected, $w(n)$ is the Additive White Gaussian Noise (AWGN) sample, and n is the sample index. Note that $s(n) = 0$ when there is no transmission by primary user. The waveform-based sensing metric[3] can be obtained as [8]

$$M = \mathrm{Re}\left[\sum_{n=1}^{N} y(n)s^*(n)\right], \tag{9.2}$$

where N is the length of known pattern. In the absence of the primary user, the metric value becomes

$$M = \mathrm{Re}\left[\sum_{n=1}^{N} w(n)s^*(n)\right]. \tag{9.3}$$

[3] In this chapter, time-domain sampling is explained as an example. Modified versions of the method explained in this chapter can be used in frequency domain as well. Likewise, the method given in this chapter can be modified depending on the available pattern.

Similarly, in the presence of a primary user's signal, the sensing metric becomes

$$M = \sum_{n=1}^{N} |s(n)|^2 + \mathrm{Re}\left[\sum_{n=1}^{N} w(n)s^*(n)\right] . \tag{9.4}$$

The decision on the presence of a primary user signal can be made by comparing the decision metric M against a fixed threshold λ_W. This is equivalent to distinguishing between the following two hypotheses:

$$\mathcal{H}_0 : y(n) = w(n), \tag{9.5}$$

$$\mathcal{H}_1 : y(n) = s(n) + w(n). \tag{9.6}$$

The performance of the detection algorithm can be summarized with two probabilities: probability of detection P_D and probability of false alarm P_F. P_D is the probability of detecting a signal on the considered frequency when it truly is present, thus large detection probability is desired. It can be formulated as

$$P_D = \Pr\left(M > \lambda_W | \mathcal{H}_1\right), \tag{9.7}$$

where λ_W is the threshold value. P_F is the probability that the test incorrectly decides that the considered frequency is occupied when it actually is not, and it can be written as

$$P_F = \Pr\left(M > \lambda_W | \mathcal{H}_0\right). \tag{9.8}$$

P_F should be kept as small as possible. The decision threshold λ_W can be selected for finding an optimum balance between P_D and P_F. However, this requires the knowledge of noise and detected signal powers. The noise power can be estimated, but the signal power is difficult to estimate as it changes depending on ongoing transmission characteristics and the distance between the cognitive radio and primary user. In practice, the threshold is chosen to obtain a certain false alarm rate. Hence, the knowledge of noise variance is enough for selection of a threshold.

9.3.3 Cyclostationarity-Based Sensing

Cyclostationarity feature detection is a method for detecting primary user transmissions by exploiting the cyclostationarity features of the received signals [7, 14–19]. Cyclostationary features are caused by the periodicity in the signal or in its statistics like mean and autocorrelation. Instead of Power Spectral Density (PSD), cyclic correlation function is used for detecting signals present in a given spectrum. The cyclostationarity-based detection algorithms can differentiate noise from primary users' signals. This is a result of the fact that noise is wide-sense stationary Wide-Sense Stationary (WSS) with no correlation while modulated signals are cyclostationary with spectral correlation due to the redundancy of signal periodicities [16].

The Cyclic Spectral Density (CSD) function of received signal (9.1) can be calculated as [20]

$$S(f,\alpha) = \sum_{\tau=-\infty}^{\infty} R_y^{\alpha}(\tau) e^{-j2\pi f \tau}, \tag{9.9}$$

where

$$R_y^{\alpha}(\tau) = E\left[y(n+\tau)y^*(n-\tau)e^{j2\pi\alpha n}\right] \tag{9.10}$$

is the Cyclic Autocorrelation Function (CAF), and α is the cyclic frequency. The CSD function outputs peak values when the cyclic frequency is equal to the fundamental frequencies of transmitted signal $x(n)$. Cyclic frequencies can be assumed to be known [14,19] or they can be extracted and used as features for identifying transmitted signals [17].

9.3.4 Energy Detector-Based Sensing

Energy detector-based approaches, also known as radiometry or periodogram, are the most common ways of spectrum sensing because of their low computational and implementation complexities [7–9,11,12,18,22–28,28–30]. Moreover, they are more generic as receivers do not need any knowledge on the primary users' signals. The signal is detected by comparing the output of energy detector with a threshold which depends on the noise floor [31]. Some of the challenges with energy detector-based sensing include selection of the threshold for detecting primary users, inability to differentiate interference from primary users and noise, and poor performance under low Signal-to-Noise-Ratio (SNR) values [8]. Moreover, the energy detector does not work efficiently for detecting spread spectrum signals [7].

Using the same model given in (9.1), decision metric for energy detector can be written as

$$M = \sum_{n=0}^{N} |y(n)|^2. \tag{9.11}$$

The white noise can be modeled as a zero-mean Gaussian random variable with variance σ_w^2, i.e. $w(n) = \mathcal{N}(0, \sigma_w^2)$. For a simplified analysis, let us model the signal term as a zero-mean Gaussian variable as well,[4] i.e. $s(n) = \mathcal{N}(0, \sigma_s^2)$. Because of these assumptions, the decision metric M follows chi-square distribution with $2N$ degrees freedom χ_{2N}^2 and hence, it can be modeled as

$$M = \begin{cases} \frac{\sigma_w^2}{2}\chi_{2N}^2 & \mathcal{H}_0, \\ \frac{\sigma_w^2+\sigma_s^2}{2}\chi_{2N}^2 & \mathcal{H}_1. \end{cases} \tag{9.12}$$

For energy detector, the probabilities P_F and P_D can be calculated as [22][5]

$$P_F = 1 - \Gamma\left(L_f L_t, \frac{\lambda_E}{\sigma_w^2}\right), \tag{9.13}$$

$$P_D = 1 - \Gamma\left(L_f L_t, \frac{\lambda_E}{\sigma_w^2 + \sigma_s^2}\right), \tag{9.14}$$

[4] In fact, the model for $s(n)$ is more complicated as fading should also be considered.

[5] Please note that the notation used in [22] is slightly different. Moreover, the noise power is normalized before it is fed into the threshold device in [22].

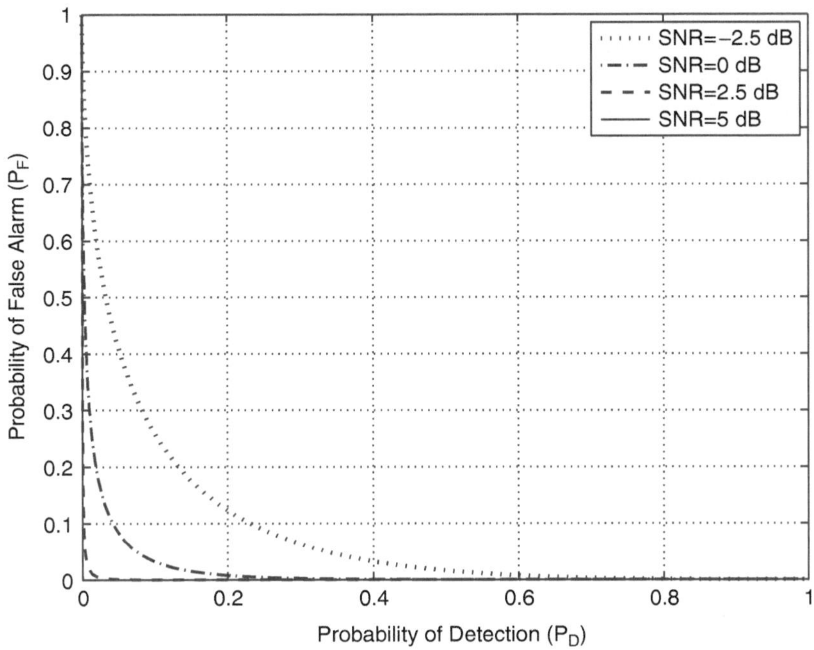

Fig. 9.3. ROC curves for energy detector-based spectrum sensing under different SNR values.

where λ_E is the decision threshold, and $\Gamma(a, x)$ is the incomplete gamma function as given in [32] (see Equation 6.5.1). Figure 9.3 shows the Receiver Operating Characteristics (ROCs) for different SNR values. SNR is defined as the ratio of the primary user's signal power to noise power, i.e. SNR $= \sigma_s^2/\sigma_w^2$. The averaging size is set to 15 in this figure, $N = 15$. As this figure clearly shows, the performance of the threshold detector increases at high SNR values.

The threshold used in energy detector-based sensing algorithms depends on the noise variance. Consequently, small noise power estimation errors causes significant performance loss [33]. As a solution to this problem, in [34], noise level is estimated dynamically by applying a reduced-rank eigenvalue decomposition to incoming signal's autocorrelation. Then, the estimated value is used to choose the threshold for satisfying a constant false alarm rate.

Measurement results are analyzed in [11,12] using energy detector to identify the idle and busy periods of WLAN channels. Energy level for each Global System for Mobile (GSM) slot is measured and compared in [25] for identifying the idle slots for exploitation. The sensing task in this work is different in the sense that the cognitive radio has to be synchronized to the primary user network and the sensing time is limited to slot duration. A similar approach is used in [35] as well for opportunistic exploitation of unused cellular slots. In [26], power at the output of Fast Fourier Transform (FFT) of incoming signal is compared with a threshold value in order to identify the number of used TV channels. FFT is performed on the data sampled at 45 kHz around the centered TV carrier frequency for each TV channel. The performance of energy detector-based sensing over various fading channels is investigated in [22]. Closed-form expressions for probability of detection under AWGN and fading (Rayleigh, Nakagami, and Ricean) channels are derived. Average probability of detection for energy detector-based sensing algorithms under log-normal shadowing and Rayleigh fading channels is derived in [8]. It is observed that the performance of energy-detector degrades considerably under Rayleigh fading. Forward methods based on energy measurements are studied for unknown noise power scenarios in [37]. The proposed method adaptively estimates the noise level, hence suitable for practical cases where noise variance is not known.

9.3.5 Radio Identification

A better knowledge about the spectrum characteristics can be obtained by identifying the transmission technology used by primary users. Such an identification enables cognitive radio with a higher dimensional knowledge as well as providing higher accuracy [30]. For example, assume that the primary users technology is identified as a Bluetooth signal. Cognitive radio can use this information for extracting some useful information in space dimension as the range of Bluetooth signal is known to be around 10 m.[6] Furthermore, cognitive radio may want to communicate with the identified communication systems in some applications. For radio identification, feature extraction and classification techniques are used in the context of European Transparent Ubiquitous Terminal (TRUST) project [38]. The goal is to identify the presence of some known transmission technologies and achieve communication through them. The two main tasks are Initial Mode Identification (IMI) and Alternative Mode Monitoring (AMM). In IMI, the cognitive device searches for a possible transmission mode (network) following the power on. AMM is the task of monitoring other modes while cognitive device is having communication in a certain mode. Some of the proposed methods for blind radio identification are shown in Table 9.1. Several features are extracted from the received signal and they are used for selecting the most probable primary user technology by employing various classification methods.

9.3.6 Other Sensing Methods

Other alternative spectrum sensing methods include multitaper spectral estimation, wavelet transform-based estimation, Hough transform, and time-frequency analysis. Multitaper spectrum estimation is proposed in [43]. The proposed algorithm is shown to be an approximation to maximum likelihood PSD estimator, and for wideband signals, it is nearly optimal. Although the

[6] Please see Section 9.9 for more examples.

Table 9.1. Blind radio identification algorithms.

Article	Used Features	Classification Method
Mehta [39], Vardoulias [40]	Amount of energy detected, its distribution across the spectrum, and its correlation with some predefined functions (Briefly mentioned, not explained in detail)	–
Palicot [41]	Channel bandwidth and its shape: this feature is found to be the most discriminating parameter using tables and cross-tables, i.e. by comparing with other parameters	Radial Basis Function (RBF) neural networks
Gandetto [42]	The standard deviation of the instantaneous frequency and the maximum duration of a signal (time–frequency analysis)	Feed Forward Back-Propagation Neural Networks (FFBPNNs) and Support Vector Machines (SVMs) with RBF
Fehske [17]	Spectral Correlation Density (SCD) and Spectral Coherence Function (SCF)	Multilayer Linear Perception Network (MLPN) neural networks
Oner [14]	Spectral Correlation Density (SCD) and Spectral Coherence Function (SCF)	Statistical tests for identifying the presence of cyclostationarity

complexity of this method is less than the maximum likelihood estimator, it is still computationally demanding. Random Hough transform of received signal is used in [44] for identifying the presence of radar pulses in the operating channels of IEEE 802.11 systems. This method can be used to detect any type of signals with periodic patterns as well. In [45], wavelets are used for detecting edges in the PSD of a wideband channel. Once the edges, which correspond to transitions from occupied band to empty band or vice versa, are detected, the power within bands between two edges are estimated. Using this information and edge positions, the PSD can be characterized as occupied or empty in a binary fashion. The assumptions made in [45], however, need to be relaxed for building a practical sensing algorithm.

9.4 Cooperative Sensing

The estimation of traffic in a specific geographic area can be done locally (by one cognitive radio only) or information from different cognitive radios can be combined. In the literature, cooperation is discussed as a solution to problems that arise in spectrum sensing due to noise uncertainty, fading, and shadowing. Cooperative sensing decreases the probability of mis-detections and the probability of false alarms considerably. In addition, cooperation can solve the hidden primary user problem and can decrease sensing time [13, 27, 28].

The interference to primary users caused by cognitive radio devices employing spectrum access mechanisms based on simple Listen-Before-Talk (LBT) scheme is investigated in [29] via analysis and computer simulations. Results show that even simple local sensing can be used to explore the unused spectrum without causing interference to existing users. On the other hand, it is shown analytically and through numerical results that collaborative sensing provides significantly higher spectrum capacity gains than local sensing. The fact that cognitive radio acts without any knowledge about the location of the primary users in local sensing degrades the performance.

The challenges of cooperative sensing include developing efficient information sharing algorithms and increased complexity [46]. The advantages and disadvantages of local and cooperative (or collaborative) sensing methods are tabulated in Table 9.2.

In cooperative sensing architectures, the control channel can be implemented using different methodologies. These include a dedicated band, unlicensed band such as ISM, and underlay Ultra Wideband (UWB) system [47]. Depending on the system requirements, one of these methods can be selected. The shared information can be soft or hard decisions made by each cognitive device [48]. Furthermore, various techniques for combining sensing results can be employed. The performances of Equal Gain-Combining (EGC), Selection Combining (SC), and Switch and Stay Combining (SSC) are investigated in [22] for energy detector-based spectrum sensing under Rayleigh fading. The EGC method is found to have a gain of approximately two

Table 9.2. Local versus cooperative sensing.

Sensing Method	Advantages	Disadvantages
Non-cooperative sensing (Local sensing)	Computational & implementation simplicity	Hidden node problem Multipath and shadowing
Cooperative sensing	Higher accuracy (close to optimal) Reduced sensing time [27] Shadowing effect and hidden node problems can be prevented	Complexity (complexity of sensor, complexity of within-system cooperation, complexity of among-system cooperation) Traffic overhead The need for a control channel

orders of magnitude while SC and SSC having one order of magnitude gain. As far as the networking is concerned, the coordination algorithm should have reduced protocol overhead and it should be robust to changes and failures in the network. Moreover, the coordination algorithm should introduce minimum amount of delay.

Cooperative sensing can be implemented in two fashions: centralized or distributed [49]. These two methods will be explained in the following sections.

Centralized Sensing

In centralized sensing, a central unit collects sensing information from cognitive devices, identifies the available spectrum, and broadcasts this information to other cognitive radios or directly controls the cognitive radio traffic.

The hard (binary) sensing results are gathered at a central place which is known as Access Point (AP) in [28]. The goal is to mitigate the fading effects of the channel and increase detection performance. Resulting detection and false alarm rates are given in [50] for the sensing algorithm used in [28]. In [48], the sensing results are combined in a central node, termed as master node, for detecting TV channels. Hard and soft information combining methods are investigated for reducing the probability of missed opportunity. The results presented in [28, 48] show that soft information-combining outperforms hard information-combining method in terms of the probability of missed opportunity.

Distributed Sensing

In the case of distributed sensing, cognitive nodes share information among each other but they make their own decisions as to which part of the spectrum they can use. Distributed sensing is more advantageous in the sense that there is no need for a backbone infrastructure.

An incremental gossiping approach termed as GUESS (Gossiping Updates for Efficient Spectrum Sensing) is proposed in [51] for performing efficient coordination between cognitive radios in distributed collaborative sensing. The proposed algorithm is shown to have low-complexity with reduced protocol overhead. The GUESS algorithm has fast convergence and robust to network changes as it does not require a setup phase to generate the clusters. Incremental aggregation and randomized gossiping algorithms are also studied in [51] for efficient coordination within a cognitive radio network. A distributed collaboration algorithm is proposed in [28]. The collaboration is performed between two secondary users. The user closer to primary transmitter, which has a better change of detecting the primary user transmission, cooperates with a far away user. An algorithm for pairing secondary users without a centralized mechanism is also proposed. A distributed sensing method is proposed in [8] where secondary users share their sensing information among themselves. Only final decisions are shared in order to minimize the network overhead due to collaboration. A secondary user receives decisions from other users and decides $\mathcal{H}_1$ if any of the received decisions plus its own is $\mathcal{H}_1$, a fusion rule known as OR-rule. The results presented in [8] clearly show the performance improvements achieved through collaborative sensing.

9.5 External Sensing

Another technique for obtaining spectrum information is external sensing. In external sensing, an external agent performs the sensing and broadcasts the channel occupancy information to cognitive radios. External sensing algorithms solve some problems associated with the internal sensing, which is termed as collocated sensing in [18]. The main advantages are overcoming hidden primary user problem as well as the uncertainty due to shadowing and fading. Furthermore, as the cognitive radios do not spend time for sensing, spectrum efficiency is increased. The sensing network does not need to be mobile and not necessarily powered by batteries. Hence, power consumption problem of internal sensing can also be addressed.

A sensor node detector architecture is used in [52]. The presence of passive receivers, viz. television receivers, is detected by measuring the Local Oscillator (LO) power leakage. Once a receiver and the channel is detected, the sensor node notifies cognitive radios in the region of passive primary user via a control channel. Similar to [52], a sensor network-based sensing architecture is proposed in [18]. A dedicated network composed of only spectrum sensing units is used to sense the spectrum continuously or periodically. The results are communicated to a sink (central) node which further processes the sensing data and shares the information about the spectrum occupancy in the sensed area with opportunistic radios. These opportunistic radios use the information obtained from sensing network for selecting the bands (and time durations) of their data transmissions. The sensing results can also be shared via a pilot channel similar to Network Access And Connectivity Channel (NACCH) [53]. External sensing is one of the methods proposed for identifying primary users in IEEE 802.22 standard as well (see Section 9.10).

9.6 Statistical Approaches and Prediction

For minimizing interference to primary users while making the most out of the opportunities, cognitive radios should keep track of the variations in spectrum and should make predictions. Stemming from the fact that cognitive radio senses the spectrum steadily and has the ability of learning, the history of the spectrum usage information can be used for predicting the future profile of the spectrum. Towards this goal, knowledge about currently active devices or prediction algorithms based on statistical analysis can be used.

Channel access patterns of primary users are identified and used for predicting spectrum usage in [54]. Assuming a TDMA transmission, periodicity pattern of channel occupancy is extracted using cyclostationary detection. This parameter is then used to forecast the channel idle probability for a given channel. Furthermore, [54] proposes to use Hidden Markov Models (HMMs) in order to model the channel usage patterns of primary users. A multivariate time series approach is taken in [55] to be able to learn the primary user characteristics and predict the future occupancy of neighboring channels. A binary scheme (*empty* or *occupied*) is used to reduce the complexity and storage requirements as shown in Figure 9.4. It is noted in [11] that the statistical model of primary users behavior should be kept simple enough to be able to design optimal higher order protocols. On the other hand, it will be useless if the primary user's behavior could not be predicted well. In order to strike a balance between complexity and effectiveness, continuous-time semi-Markov process model is used to describe the statistical characteristics of WLAN channels that can be used by cognitive radio to predict transmission opportunities. The investigation of VoIP and FTP-type traffic scenarios for semi-Markov model is performed in [12]. Pareto, phase-type (hyper-Erlang) and mixture distributions are used for fitting to the empirical data. Statistics of spectrum availability is employed in [24] for dynamically selecting the operating frequency, i.e. for identifying the spectrum holes. The statistics of the spectral occupancy of a bin (FFT output) is assumed to be at least piecewise stationary over the time at which they are observed in order to guarantee that these statistics are still reliable when a spectrum access request is received. Using the statistics, the likelihood that the spectral opportunity will remain available for at least the requested time duration is calculated for each bin. Then, these likelihood values are used to identify the range of frequencies which can be used for transmission.

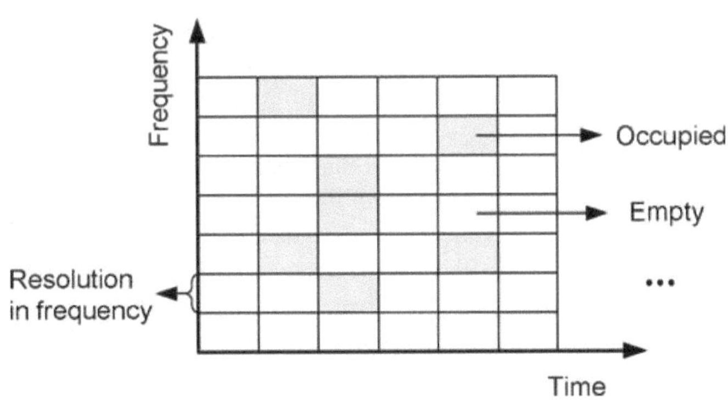

Fig. 9.4. Binary scheme used for modeling spectrum occupation in [55].

9.7 Sensing Frequency

Sensing frequency, i.e. how often cognitive radio should perform spectrum sensing, is a design parameter that needs to be chosen carefully. The optimum value depends on the capabilities of cognitive radio itself and temporal characteristics of primary users in the environment. If the status of primary users are known to change slowly, sensing frequency can be relaxed. A good example for such a scenario is detection of TV channels. The presence of a TV station usually do not change frequently in a geographical area unless a new station starts broadcasting or an existing station goes offline. Another factor that affects the sensing frequency is the interference tolerance of primary license owners. For example, when the cognitive radio is exploiting opportunities in public safety bands, sensing should be done as frequently as possible in order to prevent any interference. Cognitive radio should immediately vacate the band if it is needed by public safety units. In the IEEE 802.22 draft standard (see Section 9.10), the sensing period is defined as 30 seconds. In addition to these, the channel detection time, channel move time and some other timing related parameters are also defined [56].

9.8 Hardware Requirements and Approaches

In this section, several aspects of spectrum sensing from hardware perspective are investigated. As explained before, one of the main challenges lies on the requirements of high sampling rate, high resolution ADCs with large dynamic range. This requirement is a result of the need for a wideband sensing. Cognitive radio should be able to capture and analyze a relatively large band for identifying spectrum opportunities. Moreover, high speed processing units (Digital Signal Processors (DSPs) or Field Programmable Gate Arrays (FPGAs)) are needed for performing computationally demanding signal processing tasks with relatively low delay.

Sensing can be performed via two different architectures: single-radio and dual-radio [18, 40]. In the single-radio architecture, only a specific time slot is allocated for spectrum sensing. As a result of this, only a certain accuracy can be guaranteed for spectrum sensing results. Moreover, the spectrum efficiency is decreased as some portion of the available time slot is used for sensing instead of data transmission. The obvious advantage of single-radio architecture is its simplicity and lower cost. In the dual-radio sensing architecture, one radio chain is dedicated for data transmission and reception while the other chain is dedicated for spectrum monitoring. The drawback of such an approach is the increased power consumption and hardware cost. Note that only one antenna would be sufficient for both chains as suggested in [40]. A comparison of advantages and disadvantages of single and dual-radio architectures is given in Table 9.3. In conclusion, one might prefer one architecture over the other depending on the available resources, and performance and/or data rate requirements.

Table 9.3. Comparison of single-radio and dual-radio sensing algorithms.

	Single-Radio Architecture	**Double-Radio Architecture**
Advantages	Simplicity Lower cost	Higher spectrum efficiency Better sensing accuracy
Disadvantages	Lower spectrum efficiency Poor sensing accuracy	Higher cost Higher power consumption Higher complexity

9.9 Multi-dimensional Spectrum Awareness

The definition of opportunity determines the ways of measuring and exploiting the spectrum space. The conventional definition of the spectrum opportunity which is often referred as "*band of frequencies that are not being used by the primary user of that band at a particular time in a particular geographic area*" [57] only exploits three dimensions of the spectrum space: frequency, time, and space. The problems stated in the previous section also relates to sensing the spectrum in these three dimensions. However, there are other dimensions that need to be explored further for spectrum opportunity. For example, the code dimension of the spectrum space has not been explored well in the literature. Therefore, the conventional spectrum sensing algorithms do not know how to deal with signals that use spread spectrum, time or frequency hopping codes. As a result, these types of signals constitute a major problem in sensing the spectrum. If the code dimension is interpreted as part of the spectrum space, this problem can be avoided, and new opportunities for spectrum usage can be created. Naturally, this will bring about other new challenges for detection and estimation of this new opportunity. Similarly, the angle dimension has not been exploited well enough for spectrum opportunity. It is assumed that the primary users and/or the secondary users are transmitting in all the directions. However, with the recent advances in multi-antenna technologies, e.g. beam forming, multiple users can be multiplexed into the same channel at the same time in the same geographical area. In other words, an additional dimension of spectral space can be created as opportunity. This will also create new opportunities for spectral estimation, where not only the frequency spectrum but also the angle of arrivals might need to be estimated. With these new dimensions, sensing only the frequency spectrum usage falls short. The radio space with the introduced dimensions can be defined as "*a theoretical hyperspace occupied by radio signals, which has dimensions of location, angle-of-arrival, frequency, time, and possibly others*" [58]. This hyperspace is called electrospace, transmission hyperspace, radio spectrum space, or simply spectrum space by various authors, and it can be used to describe how radio environment can be shared among multiple (primary and/or secondary) systems [59]. Various dimensions of this space and the corresponding measurement/sensing requirements are summarized in Table 9.4 along

Table 9.4. Multi-dimensional radio spectrum space and transmission opportunities.

Dimension	What needs to be sensed?	Comments	Illustrations
Frequency	Opportunity in the frequency domain.	Availability in part of the frequency spectrum. The available spectrum is divided into narrower chunks of bands. Spectrum opportunity in this dimension means that all the bands are not used simultaneously at the same time, i.e. some bands might be available for opportunistic usage.	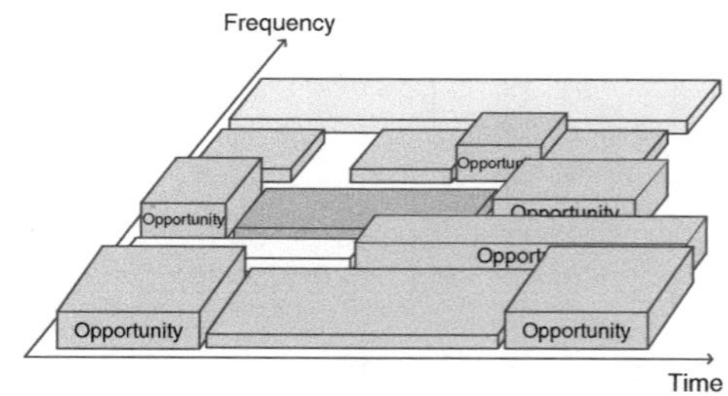
Time	Opportunity of a specific band in time.	This involves the availability of a specific part of the spectrum in time. In other words, the band is not continuously used. There will be times where it will be available for opportunistic usage.	
Geographical space	Location (latitude, longitude, and elevation) and distance of primary users.	The spectrum can be available in some parts of the geographical area while it is occupied in some other parts at a given time. This takes advantage of the propagation loss (path loss) in space. These measurements can be avoided by simply looking at the interference temperature. No interference means no primary user transmission in a local area. However, one needs to be careful because of hidden terminal problem.	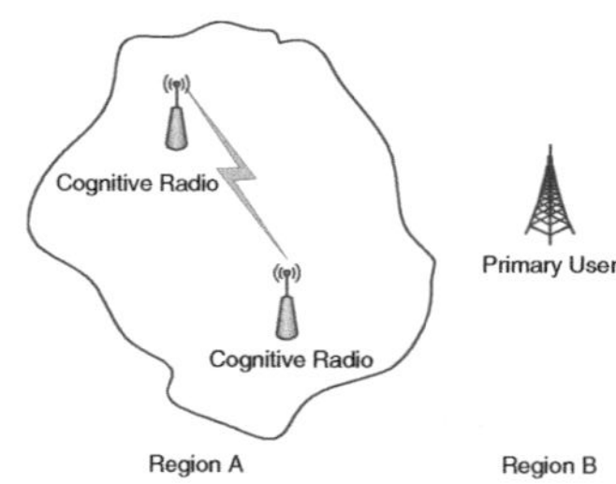

Continued on next page.

Dimension	What needs to be sensed?	Comments	Illustrations
Code	The spreading code, time hopping (TH), or frequency hopping (FH) sequences used by the primary users. Also, the timing information is needed so that secondary users can synchronize their transmissions w.r.t. primary users. The synchronization estimation can be avoided with long and random code usage. However, partial interference in this case is unavoidable.	The spectrum over a wideband might be used at a given time through spread spectrum or frequency hopping. This does not mean that there is no availability over this band. Simultaneous transmission without interfering with primary users would be possible in code domain with an orthogonal code with respect to codes that primary users are using. This requires the opportunity in code domain, i.e. not only detecting the usage of the spectrum, but also determining the used codes, and possibly multipath parameters as well.	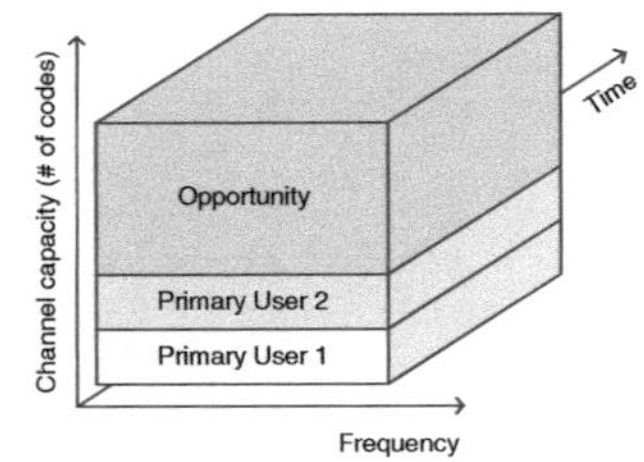
Angle	Directions of primary users' beam (azimuth and elevation angle) and locations of primary users.	Along with the knowledge of the location/position or direction of primary users, spectrum opportunities in angle dimension can be created. For example, if a primary user is transmitting in a specific direction, the secondary user can transmit in other directions without creating interference on the primary user.	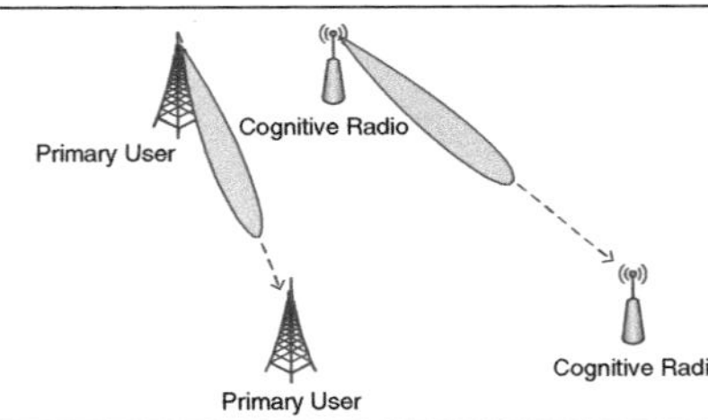
Signal	Signal polarization and waveforms of primary users.	Primary users and secondary users might be transmitting a waveform at a specific band for a given time in a geographical area in all the directions but secondary users can exploit the signal dimension to transmit an orthogonal waveform so that it does not create interference with primary users. This requires not only spectrum estimation but also waveform identification.	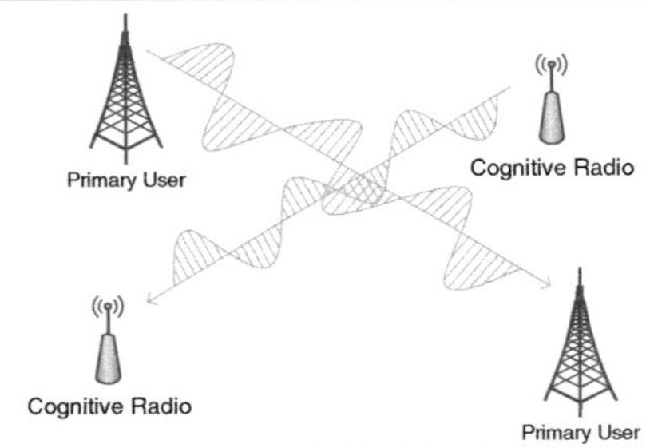

with some representative pictures. Each dimension has its own parameters that should be sensed for a complete spectrum awareness as indicated in the Table.

It is of crucial importance to define such an n-dimensional space for spectrum sensing. Spectrum sensing should include the process of identifying occupancy in all dimensions of the spectrum space and finding spectrum holes, or more precisely spectrum space holes. For example, a certain frequency can be occupied for a given time, but it might be empty in another time. Hence, temporal dimension is as important as frequency dimension. This example can be extended to the other dimensions of spectrum space given in Table 9.4. As a result of this requirement, advanced spectrum sensing algorithms that offer awareness in multiple dimensions of the spectrum space should be developed.

9.10 Spectrum Sensing in Current Wireless Standards

Recently developed wireless standards has started to include cognitive features. Even though it is difficult to expect a wireless standard that is based on wideband spectrum sensing and opportunistic exploitation of spectrum, the trend is in this direction. In this section, wireless technologies that require some sort of spectrum sensing for adaptation or for Dynamic Frequency Selection (DFS) will be discussed. However, the spectrum knowledge can also be used to initiate advanced receiver algorithms such as adaptive interference cancellation [60].

9.10.1 IEEE 802.11k

A proposed extension to IEEE 802.11 specification is IEEE 802.11k which defines several types of measurements [61]. Some of the measurements include channel load report, noise histogram report and station statistic report. The noise histogram report provides methods to measure interference levels that display all non-802.11 energy on a channel as received by the subscriber unit. The access point (AP) collects channel information from each mobile unit and makes its own measurements. This data is then used by the AP to regulate access to a given channel.

The sensing (or measurement) information is used to improve the traffic distribution within a network as well. WLAN devices usually connects to the AP that has the strongest signal level. Sometimes, such an arrangement might not be the optimum and can cause overloading on one AP and underutilization of others. In 802.11k, when an AP with the strongest signal power is loaded to its full capacity, new subscriber units are assigned to one of the underutilized APs. Despite the fact that the received signal level is weaker, the overall system throughput is better thanks to more efficient utilization of network resources.

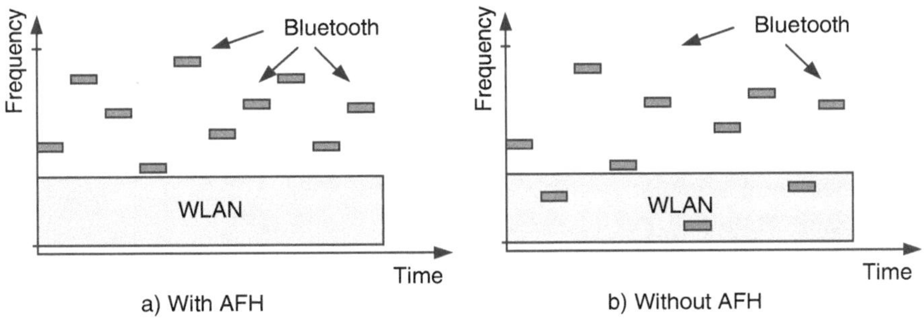

Fig. 9.5. Bluetooth transmission with and without adaptive frequency hopping (AFH). AFH prevents collusions between WLAN and Bluetooth transmissions.

9.10.2 Bluetooth

A new feature, namely Adaptive Frequency Hopping (AFH), is introduced to Bluetooth standard to reduce interference between wireless technologies sharing the 2.4 GHz unlicensed radio spectrum [62]. In this band IEEE 802.11b/g devices, cordless telephones, microwave ovens use the same wireless frequencies as Bluetooth. AFH identifies the transmissions in the ISM band and avoids their frequencies. Hence, narrow-band interference can be avoided and better Bit-Error-Rate (BER) performance can be achieved as well as reducing the transmit power. Figure 9.5 shows an illustrative Bluetooth transmission with and without AFH. By employing AFH, collusions with Wireless Local Area Network (WLAN) signals are avoided in this example.

AFH requires a sensing algorithm for determining whether there are other devices present in the ISM band and whether or not to avoid them. The sensing algorithm is based on statistics gathered to determine which channels are occupied and which channels are not occupied. Channel statistics can be packet-error rate, BER, Received Signal Strength Indicator (RSSI), Carrier-To-Interference Noise Ratio (CINR) or other metrics [63]. The statistics are used to classify the channel as *good*, *bad*, or *unknown* [62].

9.10.3 IEEE 802.22

IEEE 802.22 standard is known as *cognitive radio standard* because of the cognitive features that it has. The standard is still in the development stage. One of the most distinctive feature of 802.22 standard is its sensing requirements [40]. IEEE 802.22-based wireless rural area network (WRAN) devices sense the TV channels and identify transmission opportunities.

The sensing is envisioned to be based on two stages: fast and fine sensing [56]. In the fast sensing stage, a fast sensing algorithm is employed, e.g. energy detector. The fine sensing stage is initiated based on the fast sensing results. Fine sensing involves a more detailed sensing where more powerful methods are used. Several techniques that have been proposed and included in the draft standard include energy detection, waveform-based sensing (PN511 or PN63 sequence detection and/or segment sync detection), cyclostationary feature detection, and matched filtering. A Base Station (BS) can distribute the sensing load among Subscriber Stations (SSs). The results are returned to BS which uses these results for managing the transmissions. Hence, it is a practical example of centralized collaborative sensing explained in Section 9.4.

Another approach for managing the spectrum in IEEE 802.22 devices is based on a centralized method for available spectrum discovery. The BSs would be equipped with a Global Positioning System (GPS) receiver which would allow its position to be reported. The location information would then be used to obtain the information about available TV channels through a central server. For low-power devices[7] operating in the TV bands, external sensing is proposed as an alternative technique. These devices periodically transmit beacons with a higher power level. These beacons are monitored by IEEE 802.22 devices to detect the presence of such low-power devices which are otherwise difficult to detect due to the low-power transmission.

9.11 Conclusions

Spectrum is a very valuable resource in wireless communication systems, and it has been a focal point for research and development efforts over the last several decades. Cognitive radio, which is one of the efforts in utilization of the available spectrum more efficiently through opportunistic spectrum usage, become an exciting and promising concept. One of the important elements of the cognitive radio is sensing the available spectrum opportunities. In this chapter, various aspects of spectrum sensing task is explained in detail. Several sensing methods are studied and collaborative sensing is considered as a solution to some common problems in spectrum sensing. Hardware aspects of spectrum sensing and pro-active approaches are given and sensing methods employed in current wireless systems are discussed. Furthermore, the spectrum opportunity and spectrum sensing concepts are re-evaluated by considering different dimensions of the spectrum space. The new interpretation of spectrum space will create new opportunities and challenges for spectrum sensing while it will solve some of the traditional problems. Estimating real levels of usage of the spectrum in multiple dimensions including time, frequency, space, angle, and code; identifying for opportunities in multiple dimensions including prediction into the future using past information and making reasoning can be considered some of these challenges for future research.

References

1. J. Mitola and G. Q., Maguire "Cognitive radio: making software radios more personal," *IEEE Personal Commun. Mag.*, vol. 6, no. 4, pp. 13–18, Aug. 1999.

[7] These devices include wireless microphone, wireless camera, etc.

2. J. O. Neel, "Analysis and design of cognitive radio networks and distributed radio resource management algorithms," Ph.D. dissertation, Virginia Polytechnic Institute and State University, Sept. 2006.
3. Federal Communications Commission, "In the matter of unlicensed operation in the TV broadcast bands," Notice of proposed rule making ET Docket No. 04-186 (FCC 04-113), May 2004.
4. M. Marcus, "Unlicensed cognitive sharing of TV spectrum: the controversy at the federal communications commission," *IEEE Commun. Mag.*, vol. 43, no. 5, pp. 24–25, 2005.
5. J. G. Proakis, *Digital Communications*, 4th ed. McGraw-Hill Higher Eduation, 2001.
6. R. Tandra and A. Sahai, "Fundamental limits on detection in low SNR under noise uncertainty," in *Proc. IEEE Int. Conf. Wireless Networks, Commun. and Mobile Computing*, vol. 1, Maui, HI, June 2005, pp. 464–469.
7. D. Cabric, S. Mishra, and R. Brodersen, "Implementation issues in spectrum sensing for cognitive radios," in *Proc. Asilomar Conf. on Signals, Systems and Computers*, vol. 1, Pacific Grove, California, Nov. 2004, pp. 772–776.
8. H. Tang, "Some physical layer issues of wide-band cognitive radio systems," in *Proc. IEEE Int. Symposium on New Frontiers in Dynamic Spectrum Access Networks*, Baltimore, Maryland, Nov. 2005, pp. 151–159.
9. A. Sahai, R. Tandra, S. M. Mishra, and N. Hoven, "Fundamental design tradeoffs in cognitive radio systems," in *Proc. of Int. Workshop on Technology and Policy for Accessing Spectrum*, Aug. 2006.
10. *Supplement to IEEE standard for information technology telecommunications and information exchange between systems – local and metropolitan area networks – specific requirements. Part 11: wireless LAN Medium Access Control (MAC) and Physical Layer (PHY) specifications: high-speed physical layer externsion in the 2.4 GHz band*, The Institute of Electrical and Electronics Engineering, Inc. Std. IEEE 802.11b, Sept. 1999.
11. S. Geirhofer, L. Tong, and B. Sadler, "A measurement-based model for dynamic spectrum access in WLAN channels," in *Proc. IEEE Military Commun. Conf.*, Washington, DC, Oct. 2006.
12. S. Geirhofer, B. Sadler, and L. Tong, "Dynamic spectrum access in WLAN channels: Empirical model and its stochastic analysis," in *Proc. of Int. Workshop on Technology and Policy for Accessing Spectrum*, Boston, Massachusetts, Aug. 2006.
13. D. Cabric, A. Tkachenko, and R. Brodersen, "Spectrum sensing measurements of pilot, energy, and collaborative detection," in *Proc. IEEE Military Commun. Conf.*, Washington, DC, Oct. 2006, pp. 1–7.
14. M. Oner and F. Jondral, "Cyclostationarity based air interface recognition for software radio systems," in *Proc. IEEE Radio and Wireless Conf.*, Atlanta, Georgia, Sept. 2004, pp. 263–266.
15. M. Oner and F. Jondral, "Cyclostationarity-based methods for the extraction of the channel allocation information in a spectrum pooling system," in *Proc. IEEE Radio and Wireless Conf.*, Atlanta, Georgia, Sept. 2004, pp. 279–282.
16. D. Cabric and R. W. Brodersen, "Physical layer design issues unique to cognitive radio systems," in *Proc. IEEE Int. Symposium on Personal, Indoor and Mobile Radio Commun.*, Berlin, Germany, Sept. 2005.
17. A. Fehske, J. Gaeddert, and J. Reed, "A new approach to signal classification using spectral correlation and neural networks," in *Proc. IEEE Int. Symposium*

on New Frontiers in Dynamic Spectrum Access Networks, Baltimore, Maryland, Nov. 2005, pp. 144–150.
18. S. Shankar, C. Cordeiro, and K. Challapali, "Spectrum agile radios: utilization and sensing architectures," in *Proc. IEEE Int. Symposium on New Frontiers in Dynamic Spectrum Access Networks*, Baltimore, Maryland, Nov. 2005, pp. 160–169.
19. M. Ghozzi, F. Marx, M. Dohler, and J. Palicot, "Cyclostationarity-based test for detection of vacant frequency bands," in *Proc. IEEE Int. Conf. Cognitive Radio Oriented Wireless Networks and Commun. (Crowncom)*, Mykonos, Greece, June 2006.
20. W. Gardner, "Exploitation of spectral redundancy in cyclostationary signals," *IEEE Signal Processing Mag.*, vol. 8, no. 2, pp. 14–36, 1991.
21. T. Weiss, J. Hillenbrand, and F. Jondral, "A diversity approach for the detection of idle spectral resources in spectrum pooling systems," in *Proc. of the 48th Int. Scientific Colloquium*, Ilmenau, Germany, Sept. 2003, pp. 37–38.
22. F. Digham, M. Alouini, and M. Simon, "On the energy detection of unknown signals over fading channels," in *Proc. IEEE Int. Conf. Commun.*, vol. 5, Seattle, Washington, May 2003, pp. 3575–3579.
23. M. Wylie-Green, "Dynamic spectrum sensing by multiband OFDM radio for interference mitigation," in *Proc. IEEE Int. Symposium on New Frontiers in Dynamic Spectrum Access Networks*, Baltimore, Maryland, Nov. 2005, pp. 619–625.
24. S. Jones and N. Wang, "An experiment for sensing-based opportunistic spectrum access in CSMA/CA networks," in *Proc. IEEE Int. Symposium on New Frontiers in Dynamic Spectrum Access Networks*, Baltimore, Maryland, Nov. 2005, pp. 593–596.
25. P. Papadimitratos, S. Sankaranarayanan, and A. Mishra, "A bandwidth sharing approach to improve licensed spectrum utilization," *IEEE Commun. Mag.*, vol. 43, no. 12, pp. 10–14, Dec. 2005.
26. A. Leu, K. Steadman, M. McHenry, and J. Bates, "Ultra sensitive TV detector measurements," in *Proc. IEEE Int. Symposium on New Frontiers in Dynamic Spectrum Access Networks*, Baltimore, Maryland, Nov. 2005, pp. 30–36.
27. G. Ganesan and Y. Li, "Agility improvement through cooperative diversity in cognitive radio," in *Proc. IEEE Global Telecomm. Conf. (Globecom)*, vol. 5, St. Louis, Missouri, Nov./Dec. 2005, pp. 2505–2509.
28. G. Ganesan and Y. Li, "Cooperative spectrum sensing in cognitive radio networks," in *Proc. IEEE Int. Symposium on New Frontiers in Dynamic Spectrum Access Networks*, Baltimore, Maryland, Nov. 2005, pp. 137–143.
29. A. Leu, M. McHenry, and B. Mark, "Modeling and analysis of interference in listen-before-talk spectrum access schemes," *Int. Journal of Network Management*, vol. 16, pp. 131–147, 2006.
30. T. Yücek and H. Arslan, "Spectrum characterization for opportunistic cognitive radio systems," in *Proc. IEEE Military Commun. Conf.*, Washington, DC, Oct. 2006.
31. H. Urkowitz, "Energy detection of unknown deterministic signals," *Proc. IEEE*, vol. 55, pp. 523–531, Apr. 1967.
32. M. Abramowitz and I. Stegun, Eds., *Handbook of Mathematical Functions With Formulas, Graphs, and Mathematical Tables*, 9th ed. National Bureau of Standards, 1970.

33. A. Sahai, N. Hoven, and R. Tandra, "Some fundamental limits on cognitive radio," in *Proc. Allerton Conf. on Commun., Control, and Computing*, Monticello, Illinois, Oct. 2004.
34. M. Olivieri, G. Barnett, A. Lackpour, and A. Davis, "A scalable dynamic spectrum allocation system with interference mitigation for teams of spectrally agile software defined radios," in *Proc. IEEE Int. Symposium on New Frontiers in Dynamic Spectrum Access Networks*, Baltimore, Maryland, Nov. 2005, pp. 170–179.
35. S. Lal and A. Mishra, "A look ahead scheme for adaptive spectrum utilization," in *Proc. IEEE Radio and Wireless Conf.*, Boston, Massachusetts, Aug. 2003, pp. 83–86.
36. A. Ghasemi and E. Sousa, "Collaborative spectrum sensing for opportunistic access in fading environments," in *Proc. IEEE Int. Symposium on New Frontiers in Dynamic Spectrum Access Networks*, Baltimore, Maryland, Nov. 2005, pp. 131–136.
37. J. Lehtomaki, J. Vartiainen, M. Juntti, and H. Saarnisaari, "Spectrum sensing-with forward methods," in *Proc. IEEE Military Commun. Conf.*, Washington, DC, Oct. 2006, pp. 1–7.
38. T. Farnham, G. Clemo, R. Haines, E. Seidel, A. Benamar, S. Billington, N. Greco, N. Drew, T. Le, B. Arram, and P. Mangold, "IST-TRUST: A perspective on the reconfiguration of future mobile terminals using software download," in *Proc. IEEE Int. Symposium on Personal, Indoor and Mobile Radio Commun.*, London, UK, Sept. 2000, pp. 1054–1059.
39. M. Mehta, N. Drew, G. Vardoulias, N. Greco, and C. Niedermeier, "Reconfigurable terminals: an overview of architectural solutions," *IEEE Commun. Mag.*, vol. 39, no. 8, pp. 82– 89, 2001.
40. G. Vardoulias, J. Faroughi-Esfahani, G. Clemo, and R. Haines, "Blind radio access technology discovery and monitoring for softwaredefined radio communication systems: problems and techniques," in *Proc. Int. Conf. 3G Mobile Communication Technologies*, 2001, pp. 306–310.
41. J. Palicot and C. Roland, "A new concept for wireless reconfigurable receivers," *IEEE Commun. Mag.*, vol. 41, no. 7, pp. 124–132, 2003.
42. M. Gandetto, M. Guainazzo, and C. S. Regazzoni, "Use of time-frequency analysis and neural networks for mode identification in a wireless software-defined radio approach," *EURASIP Journal on Applied Signal Processing*, vol. 2004, pp. 1778–1790, 2004.
43. S. Haykin, "Cognitive radio: brain-empowered wireless communications," *IEEE J. Select. Areas Commun.*, vol. 3, no. 2, pp. 201–220, Feb. 2005.
44. K. Challapali, S. Mangold, and Z. Zhong, "Spectrum agile radio: Detecting spectrum opportunities," in *Proc. Int. Symposium on Advanced Radio Technologies*, Boulder, CO, Mar. 2004.
45. Z. Tian and G. B. Giannakis, "A wavelet approach to wideband spectrum sensing for cognitive radios," in *Proc. IEEE Int. Conf. Cognitive Radio Oriented Wireless Networks and Commun. (Crowncom)*, Mykonos, Greece, June 2006.
46. T. Weiss, J. Hillenbrand, A. Krohn, and F. Jondral, "Efficient signaling of spectral resources in spectrum pooling systems," in *Proc. IEEE Symposium on Commun. and Veh. Technol.*, Eindhoven, Netherlands, Nov. 2003.
47. D. Čabrić, S. Mishra, D. Willkomm, R. Brodersen, and A. Wolisz, "A cognitive radio approach for usage of virtual unlicensed spectrum," in *Proc. IST Mobile and Wireless Communications Summit*, Dresden, Germany, June 2005.

48. E. Visotsky, S. Kuffner, and R. Peterson, "On collaborative detection of TV transmissions in support of dynamic spectrum sharing," in *Proc. IEEE Int. Symposium on New Frontiers in Dynamic Spectrum Access Networks*, Baltimore, Maryland, Nov. 2005, pp. 338–345.
49. I. F. Akyildiz, W. Y. Lee, M. C. Vuran, and S. Mohanty, "NeXt generation / dynamic spectrum access / cognitive radio wireless networks: A survey," *Computer Networks Journal (Elsevier)*, Sept. 2006.
50. J. Hillenbrand, T. Weiss, and F. Jondral, "Calculation of detection and false alarm probabilities in spectrum pooling systems," *IEEE Commun. Lett.*, vol. 9, no. 4, pp. 349–351, Apr. 2005.
51. N. Ahmed, D. Hadaller, and S. Keshav, "GUESS: gossiping updates for efficient spectrum sensing," in *Proc. International workshop on Decentralized resource sharing in mobile computing and networking*, Los Angeles, California, 2006, pp. 12–17.
52. B. Wild and K. Ramchandran, "Detecting primary receivers for cognitive radio applications," in *Proc. IEEE Int. Symposium on New Frontiers in Dynamic Spectrum Access Networks*, Baltimore, Maryland, Nov. 2005, pp. 124–130.
53. D. Hunold, A. Barreto, G. Fettweis, and M. Mecking, "Concept for universal access and connectivity in mobile radionetworks," in *Proc. IEEE Int. Symposium on Personal, Indoor and Mobile Radio Commun.*, vol. 2, London, UK, Sept. 2000, pp. 847–851.
54. T. Clancy and B. Walker, "Predictive dynamic spectrum access," in *Proc. SDR Forum Technical Conference*, Orlando, Florida, Nov. 2006.
55. S. Yarkan and H. Arslan, "Binary time series approach to spectrum prediction for cognitive radio," in *Proc. IEEE Veh. Technol. Conf.*, Baltimore, Maryland, Sept./Oct. 2007.
56. C. Cordeiro, K. Challapali, and D. Birru, "IEEE 802.22: An introduction to the first wireless standard based on cognitive radios," *Journal of communications*, vol. 1, no. 1, Apr. 2006.
57. P. Kolodzy et al., "Next generation communications: Kickoff meeting," in *Proc. DARPA*, Oct. 2001.
58. R. Matheson, "The electrospace model as a frequency management tool," in *Int. Symposium On Advanced Radio Technologies*, Boulder, CO, Mar. 2003, pp. 126–132.
59. W. D. Horne, "Adaptive spectrum access: Using the full spectrum space," in *Proc. Annual Telecommunications Policy Research Conf.*, Arlington, Virginia, Oct. 2003.
60. J. Andrews, "Interference cancellation for cellular systems: a contemporary overview," *IEEE Wireless Commun. Mag.*, vol. 12, no. 2, pp. 19–29, 2005.
61. *Draft Supplement to STANDARD FOR Telecommunications and Information Exchange Between Systems – LAN/MAN Specific Requirements – Part 11: Wireless Medium Access Control (MAC) and physical layer (PHY) specifications: Specification for Radio Resource Measurement*, The Institute of Electrical and Electronics Engineering, Inc. Std. IEEE 802.11k/D0.7, Oct. 2003.
62. *Specification of the Bluetooth system, Master Table of Contents & Compliance Requirements*, Bluetooth, SIG Std. Bluetooth Standard, Nov. 2004.
63. N. Golmie, N. Chevrollier, and O. Rebala, "Bluetooth and WLAN coexistence: challenges and solutions," *IEEE Wireless Commun. Mag.*, vol. 10, no. 6, pp. 22–29, Dec. 2003.

64. *Standard for Wireless Regional Area Networks (WRAN) – Specific requirements – Part 22: Cognitive Wireless RAN Medium Access Control (MAC) and Physical Layer (PHY) Specifications: Policies and procedures for operation in the TV Bands*, The Institute of Electrical and Electronics Engineering, Inc. Std. IEEE 802.22.

10

Location Information Management Systems for Cognitive Wireless Networks

Hüseyin Arslan[1] and Hasari Celebi[2]

[1] University of South Florida, Tampa, FL, USA, arslan@eng.usf.edu
[2] University of South Florida, Tampa, FL, USA, hcelebi@eng.usf.edu

10.1 Introduction

Location information has been traditionally used for the positioning systems to estimate and track the location of a target device or object. The tremendous growth in the number of mobile users initiates the development of location-based services, which are mainly based on the positioning systems. Moreover, the demands on the higher Quality of Service (QoS) such as global mobility and seamless connectivity from the users as well as wireless network operators motivate to exploit the utilization of location information in the wireless networks. Recently, it has been recognized that location-based services are not the only applications, where the location information can be used, but also it can be utilized to solve some other issues in the wireless networks. The applications based on the utilization of location information are folded under four categories: location-based services, network optimization, transceiver algorithm development and optimization, and environment characterization. For instance, location-assisted handover mechanism, routing, drop call management, and adaptive coverage systems are some examples of network optimization.

Increasing utilization of location information for different applications in the wireless networks and ever-growing number of mobile users require incorporating a location information hierarchy into network structure. Although some of the existing wireless network structures have a miniature location information management system, cognitive wireless networks [1–3] are promising systems that comprehensive location information management systems (location awareness engine) can be realized. Location information management system in cognitive wireless networks is expected to provide location awareness capability to the cognitive radios and networks. Moreover, it can provide global mobility and seamless connectivity to large number of mobile cognitive radios and handle tremendous amount of real-time multimedia traffic. Basically, high quality location-based services can be realized and aforementioned location-based applications can be achieved in cognitive wireless

H. Arslan (Ed.), Cognitive Radio, Software Defined Radio, and Adaptive Wireless Systems, 291–323.

networks with the introduction of a comprehensive location information management system to network architecture.

In this chapter, a brief taxonomy of cognitive wireless networks based on cognition, collaboration, and node diversity criteria is presented. A system model for the location information management in cognitive wireless networks that is responsible for obtaining and utilizing location information for different applications is proposed. Classification of location information along with an overview of cognitive and legacy location estimation and sensing algorithms are provided. Fundamentals of mobility models and location tracking in cognitive wireless networks are discussed. Moreover, great amount of location-assisted applications for cognitive wireless networks are presented in details. Privacy concerns due to the utilization and tracking of user location are addressed and potential solutions are discussed followed by the concluding remarks.

10.2 Cognitive Wireless Network Model

A model for the location information management system in cognitive wireless networks is illustrated in Figure 10.1. According to this model, the measurement and/or sensing devices are used to obtain data from the operational environments. The acquired data are sent to location information management system for post-processing, which is embedded in central cognitive engine of the network and/or the cognitive engine of cognitive radio node. Location estimation and/or sensing algorithms process the data to determine location information. Since different location estimation and/or sensing methods provide the estimated and/or sensed location information in different coordinate systems, cognitive engine needs a coordinate system converter to manage the transition between different coordinate systems. Finally, location information management system utilizes location information for different applications such as location-based services (i.e. positioning and tracking), network optimization, transceiver algorithm optimization, and environment characterization. Furthermore, location information management system has a mechanism to handle mobility and tracking tasks as shown in Figure 10.1. The details of each of these blocks are discussed in later sections. Cognition, collaboration, and node diversity are the main criteria that are used to classify cognitive wireless networks in this chapter. These networks consist of two major parts, which are central cognitive engine and cognitive nodes. One can think of a central cognitive engine in cognitive wireless networks like a mobile switching center with more intelligence in the cellular networks. We consider both cognitive radio Base Station (BS) and Mobile Station (MS) as a node in this chapter for the sake of discussion.

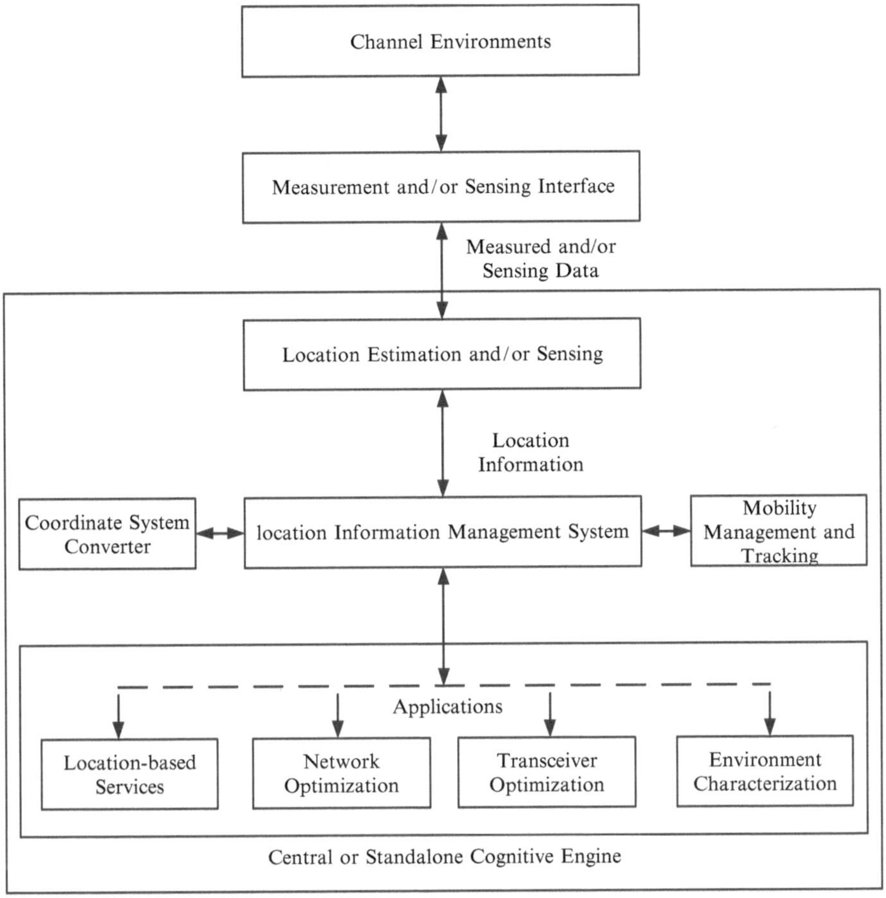

Fig. 10.1. A simplified system model for location information management in cognitive wireless networks.

10.2.1 Cognition

Partitioning cognition features between central cognitive engine of network and the cognitive engine of nodes plays key role to classify cognitive wireless networks. In theory, various types of cognitive wireless networks can be developed between two extreme cognition limits, which are absolute centralized and distributed (decentralized) cognition [4]. In the former limit, central cognitive engine has the full cognition capabilities whereas the nodes possess limited cognition capabilities. Basically, central cognitive engine can be considered as the brain and the nodes are the members. On the other hand, the structure is totally opposite in the absolute distributed cognition case, where the full cognition capability is embedded into the cognitive engine of each node. Note that this type of network will always have a central cognitive engine to maintain network organization regardless of the cognition level of cognitive radio node. Indeed, the nodes will have more cognition capabilities as the technology advances. Mankind is a good example to prove this idea. Although the human being is the most intelligent creature in the universe, there exist organizations or collective works in every aspects of the life. All of the organizations have a head, which can be either permanent or interim, to manage relationships between the members of organization. On the other hand, ad hoc cognitive radio network [5] can be considered as a network type between these two cognition limits.

10.2.2 Collaboration

Cognitive wireless networks can be classified into three types from the perspective of collaboration within the network [6, 7]. Collaboration can be between nodes and central cognitive engine of the network and/or between different nodes. Since there is a natural collaboration between central cognitive engine and cognitive nodes, the following classification is based on collaboration of the nodes. The first type is called as *cooperative network* in which all the nodes agree on performing the predefined (e.g. cooperative spectrum sensing) or ideally random tasks collectively. For instance, nodes can collaborate on estimation of location of the licensed users.

Non-cooperative network is the second type of network in which there are no collaborations between nodes. For example, all the nodes retrieve their location information from the Global Positioning System (GPS) satellites. On the other hand, if a group of nodes do and the rest of them do not agree to collaborate, then this forms the third type of the network that is called as *partially cooperative network*. For example, non-cooperative nodes can obtain their location information from the GPS satellites whereas the rest of them can estimate their location information by collaborating among themselves. Ideally, cognitive wireless networks can have the capability to transit from one type to another type dynamically depending on the collaboration of the nodes. It is important to note that cognitive radio nodes and cognitive wireless networks can have a capability to cooperate with the satellites such as in the case of Assisted-Global Positioning System (A-GPS).

10.2.3 Node Diversity

Based on the node diversity criteria, cognitive wireless networks can be grouped under two main categories: *pure cognitive wireless networks* and *mixed cognitive and non-cognitive wireless networks*. As the names imply, the former consists of only cognitive radio nodes and the latter composed of both cognitive and non-cognitive radio nodes. Pure cognitive wireless networks can be divided further into two subgroups, which are *homogeneous* and *heterogeneous networks*. For a given geographical area or cell, all cognitive radio nodes are identical (they use same waveform) in the homogeneous networks.

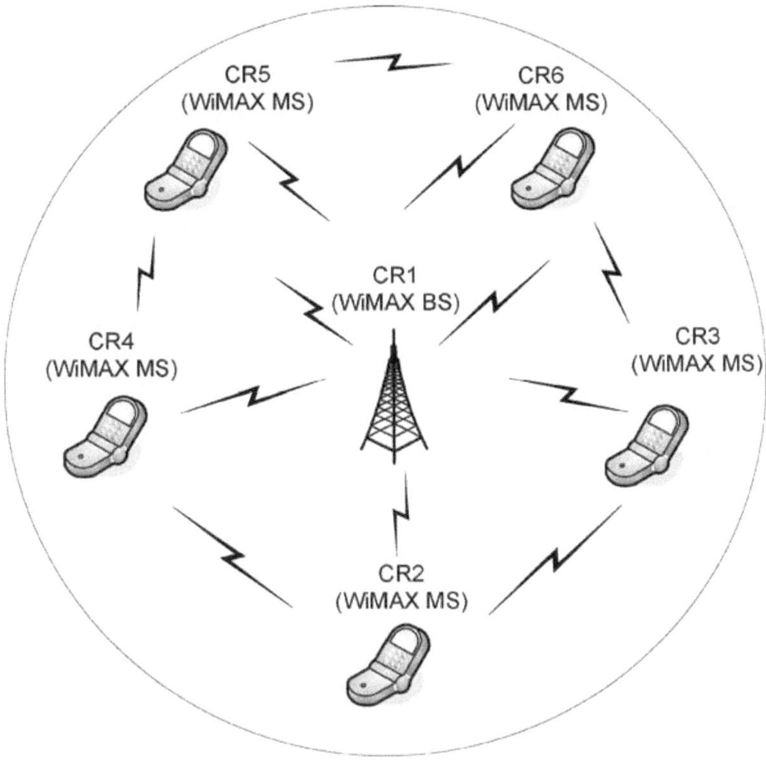

Fig. 10.2. Illustration of the homogenous pure cognitive wireless networks. Text in the parentheses shows the instantaneous waveform of each cognitive radio node. CR stands for cognitive radio node.

For instance, a network consisting of only cognitive Worldwide Interoperability for Microwave Access (WiMAX) nodes is an example for the homogeneous networks, which is illustrated in Figure 10.2. On the other hand, if the given geographical area or cell consists of mixture of different cognitive radio nodes (i.e. WiMAX base stations, Ultrawideband (UWB) nodes), this type of network is called as heterogeneous network [8]. In this type of network, a mobile device can roam across the cell borders of the other networks and inter-operate with the other wireless devices. A representative example of mixed cognitive and non-cognitive wireless networks is illustrated in Figure 10.3. Notice that in this type of networks interoperability is an issue. Although cognitive radios can communicate among themselves and with non-cognitive radios, non-cognitive radio nodes may not be able to communicate among themselves. Location information of mobile cognitive radio nodes play important roles to achieve global mobility and seamless connectivity in cognitive wireless networks. For instance, mobile cognitive radio node can switch from WLAN to WiMAX networks as the user leaves home by tracking the location of the user. Note

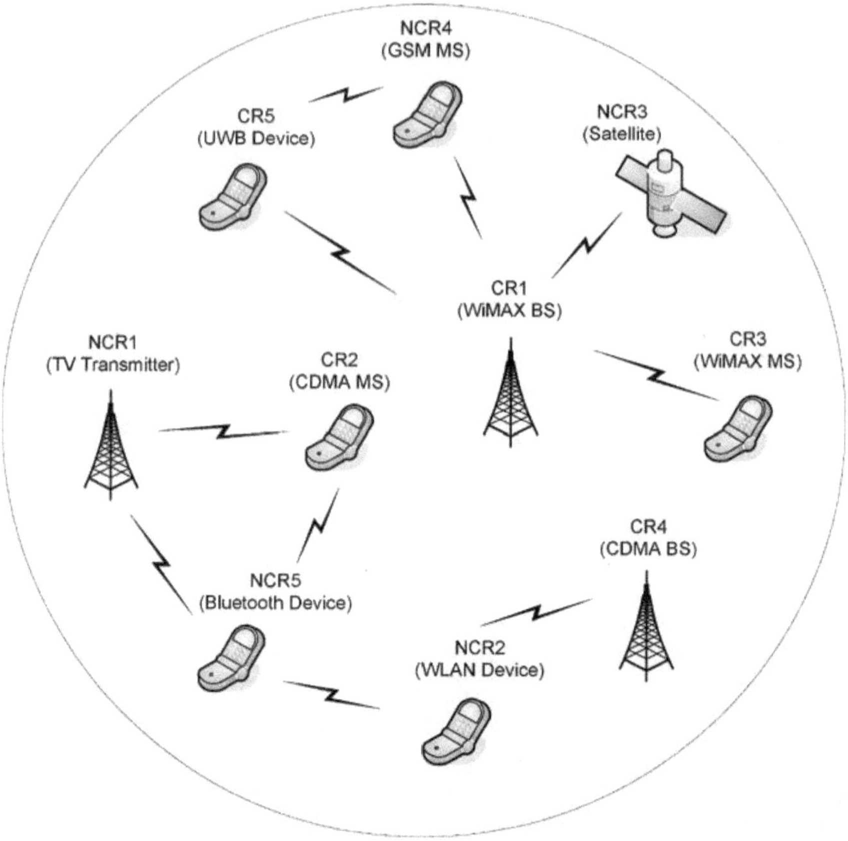

Fig. 10.3. Illustration of the mixed cognitive and non-cognitive wireless networks. NCR stands for non-cognitive radio nodes.

that, in reality, the heterogeneity of cognitive wireless networks can change dynamically as well.

10.3 Location Estimation and Sensing

10.3.1 Location Information

The type of location information that needs to be estimated or sensed plays an important role to determine the complexity of positioning systems. Therefore, classification of location information is needed for cognitive radio technologies. As a result, a taxonomy of location information for cognitive radios are shown in Figure 10.4.

In this chapter, the place occupied by a designated user, device or mainly object is referred as *location*. The object can be physical or virtual and consequently location information of the object can be either physical or virtual

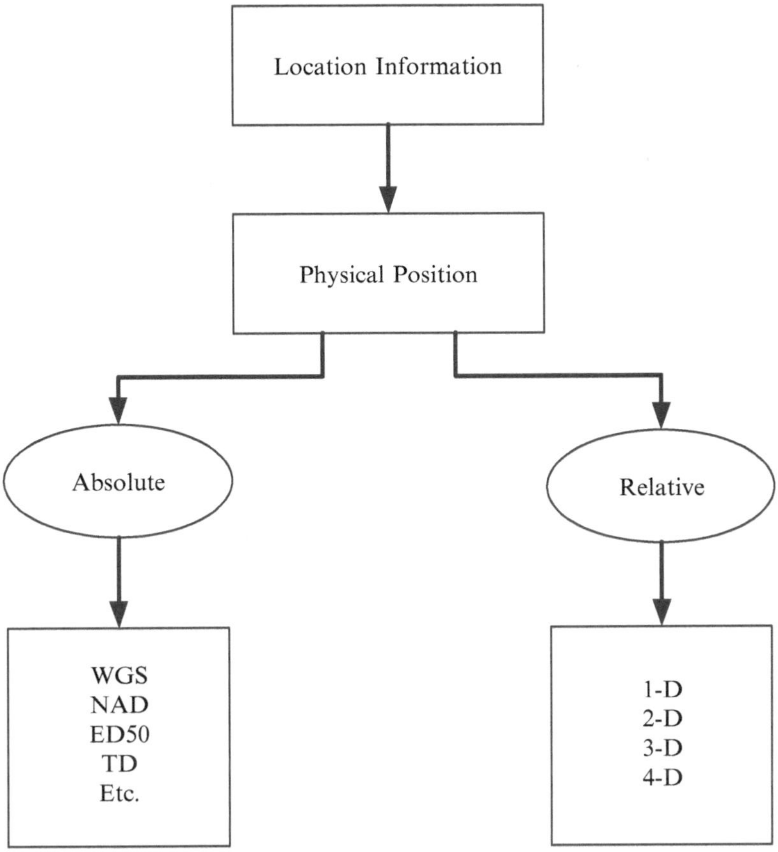

Fig. 10.4. A taxonomy of location information.

position [9]. The *position* term is defined as the coordinates of a single point in space that represents the location of an object. The *physical position* of an object is obvious as the name implies. On the other hand, the *virtual position* is defined as the invisible position (e.g. Internet Protocol (IP) address), which is relative to some known entity whose physical location may or may not be precisely known [9]. This type of location information is commonly encountered in the wired networks such as World Wide Web (WWW) access networks. For instance, a cognitive radio can log into a remote computer (or device) through Internet on the other side of the earth, but the geographic position of that computer may not be known precisely. The position of such device is referred as *virtual position.* However, cognitive radio can retrieve the physical position of the remote computer from the virtual position along with some additional information. Therefore, virtual position is considered as a form of physical position in this chapter. Mapping virtual position to the corresponding physical position already exists in the Internet domain.

Extracting physical position of a remote device from its virtual position information can be useful for cognitive radios. Such information can be used to develop efficient location-assisted routing protocol. Physical position of an object can be either absolute or relative. The *absolute position* is referred to the complete coordinate knowledge of an object. On the other hand, the *relative position* is defined as the position of an object (e.g. cognitive radio device) relative to another or neighbor objects that do or do not know their absolute positions [10]. Note that a cognitive radio device can estimate its absolute position using its relative position along with the absolute position of the reference device that is used during relative positioning. Absolute position estimation techniques are more mature and widely used compared to relative position estimation methods. As a result, cognitive radio device can switch between absolute and relative position estimation methods depending on the accuracy requirements. We refer to [10] for the details on relative positioning techniques and the details on the absolute position estimation and sensing techniques will be provided in the later sections.

The absolute and relative position of a cognitive radio can be quantified using coordinate systems. There are numerous global, continental, and country-specific reference coordinate systems for absolute position of an object such as North American Datum (NAD), European Datum (ED50), Tokyo Datum (TD), and Earth Centered Fixed (ECF), World Geodetic Systems (WGS). Each of these reference coordinate systems have various revisions. Among these, WGS-84 is a well-known standard reference coordinate system, which is also currently being used by GPS. To achieve interoperability between these different reference coordinate systems, cognitive radios can have coordinate systems converter which is shown in Figure 10.1. For instance, cognitive radios can employ standard Molodensky datum conversion algorithm [11].

Relative position information can be classified under four groups of reference coordinate systems [8]:

- *1-dimensional (1-D)*: It provides the location of a cognitive radio in a single axis (x or y or z). For instance, the distance between a transmitter and a receiver (or two cognitive radios) is a *1-D* location information. This information can be used for the ranging and network authorization purposes in cognitive wireless networks, which is also currently being used by many wireless networks. Time parameter can be added to this type of location information.
- *2-D*: It provides the position of a cognitive radio in a plane (i.e. (x, y)). This type of location information is also estimated by some of the existing wireless communications systems. Time parameter can be added to this type of location information, too.
- *3-D*: It provides the location of a cognitive radio in three dimensions (x, y, z). For instance, cognitive wireless networks can have a capability to estimate the *3-D* location of a cognitive radio node.

- *4-D*: It provides the location of a cognitive radio in four dimensions, which are *x, y, z,* and *time* parameters. This type of information currently is estimated by the GPS receivers for the geolocation applications. For instance, cognitive radio can estimate its position in the *4-D* format by having a built-in GPS receiver [12].

Notice that the accuracy of reference coordinate system model along with the resolution of positioning technique that are employed can affect the performance of location-based applications.

10.3.2 Location Estimation vs. Sensing

Recent advances in sensing technologies such as cognitive vision systems [13] show the feasibility of designing cognitive radios with the sensing capabilities. Cognitive vision systems can convert video signals even into a natural language text describing the acquired scene state. In the positioning context, scene analysis method [14, 15] is a promising location sensing technique for the cognitive radios. Hence, the location of a device or object can be determined by the estimation and/or sensing techniques. The former techniques require certain amount of measurements using the measurement devices such as antenna signals whereas the latter ones require sensing devices such as video camera to observe or acquire information from the environment. Moreover, the measurement and sensing methods differ in the consequent data post-processing steps as well, which the former uses the location estimation algorithms whereas the latter uses the location sensing algorithms.

10.3.3 Location Estimation Approaches

Location estimation methods can be broadly folded under three groups, which are range-based, range-free, and pattern matching-based schemes. These schemes are discussed in the following sections.

Range-Based Schemes

In these schemes, precise distance and angle of arrival of signal are measured to estimate the location information of the target device. Time Of Arrival (TOA), Receive Signal Strength Indicator (RSSI), and Angle Of Arrival (AOA) statistics of the received signal are used for the estimation of location information.

Range-free Schemes

In this approach, the range or angle of arrival information is not used as the name implies. This is one of the simplest form of positioning in which coarse positioning accuracy is provided. There are three main approaches in this category, which are Hop-count-based (e.g. DV-Hop), Centroid, and Area-based schemes.

Pattern Matching-Based Schemes

This approach is based on probabilistic models that describe the dependency of observed signal properties on the location of the terminal, and the motion of the terminal. The models are used to estimate the terminals location when signal measurements are available. The methodology for pattern matching-based approach is given as follows. The first step is *calibration* that is the process of obtaining a model of the signal properties at various locations. Different signal properties can be used such as signal strength and power delay profile. Furthermore, different probabilistic model for the specified signal properties can be employed. For instance, Bayes and Hidden Markov are well-known models that can be used for this purpose. For the sake of discussion, we use Bayes rule [16] in this section. In the sequel, the target area is divided into a number of grid ($i \times j$) depending on the accuracy requirements. Each location l is marked with the corresponding coordinates (e.g. x_i, y_j) as shown in Figure 10.5. Measurement data are collected from each location l to form observation vector o. Consequently, probability distribution of o for given any location l ($P(o|l)$) is obtained in the calibration phase. In the next step, which is the location estimation is performed using the following Bayes rule:

$$P(l|o) = \frac{P(o|l)P(l)}{P(o)} , \qquad (10.1)$$

where $P(l)$ is the prior probability of being at location l and this distribution can be used to incorporate background information such as user profiles. $P(o)$ is the normalizing constant for the case of relative probabilities or probability ratios. The posterior distribution $P(l|o)$ are used to select an optimal location estimator. A simple estimator can be to select the location with maximum $P(l|o)$ for given observation vector. Note that pattern matching-based

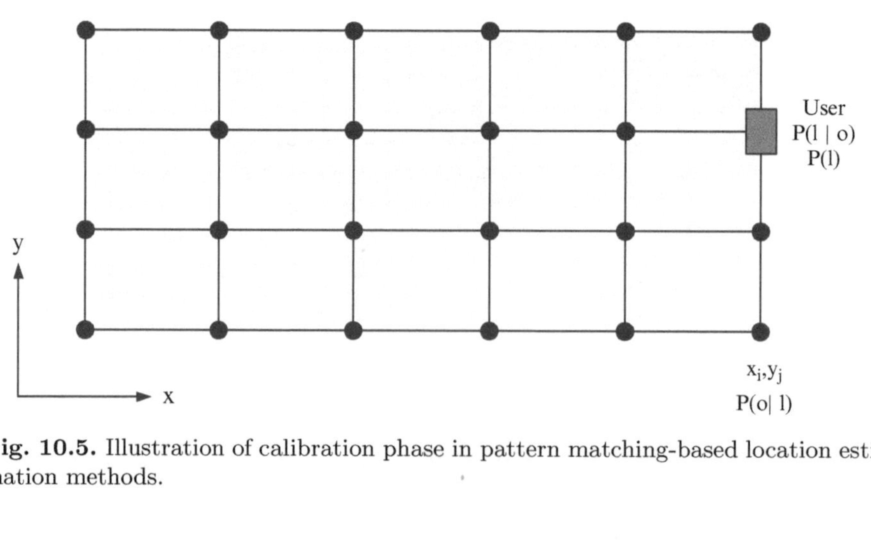

Fig. 10.5. Illustration of calibration phase in pattern matching-based location estimation methods.

approaches are useful for not only location estimation but also for active learning, tracking with history, and error estimation. We refer to [16] for further details on the pattern matching-based methods. The measurement errors and noise can result in the incompatibility of the distance and angle estimates and there is not a structured solution to deal with such problem. This is one of the main drawbacks of the range-based approach. However, this approach is computationally more efficient than the pattern matching-based approach [16]. Cognitive wireless networks can have a capability to support both approaches and employ them depending on the optimization criteria. Legacy range-based and pattern matching-based positioning techniques do not provide the required cognition capability that a cognitive radio demands. Even the GPS that is being considered for some of the next generation wireless systems as a positioning technique cannot provide the cognition capability and flexibility. Furthermore, the GPS is not a solution that can satisfy low-cost, high-accuracy, low-size, and low-power design requirements [17], which cognitive radio may require. However, some deficiencies of the GPS such as low-accuracy and indoor inoperability are addressed with the recent advances in this technology such as the A-GPS [18] and Indoor GPS [19]. For instance, the A-GPS technology can provide a ranging accuracy of 10 m or less [14]. In spite of the advances in the GPS technology, it needs more enhancements to be considered as a practical solution for cognitive radio positioning. Another alternative solution is UWB positioning, which has the capability to provide centimeter ranging accuracy due to the use of large bandwidth during the transmission [20]. The drawback of UWB positioning techniques is that they can provide such a high-precision positioning within only a short range. However, UWB and GPS can be combined along with absolute (through GPS) and relative (through UWB) positioning to obtain high-precision and long-range location estimation. In [21], a hybrid distance estimation technique for a legacy positioning system that is based on TOA and signal strength methods is proposed. The technique provides flexibility to improve the accuracy using a priori distance information. However, it does not offer a capability to change the accuracy adaptively.

10.3.4 Legacy Positioning Techniques

Triangulation and proximity are two well-known legacy range-based location estimation techniques. However, each of them requires additional enhancements in order them to be considered as the candidate positioning systems for cognitive wireless networks [8, 12, 22].

Triangulation

Triangulation is one of the most recognized legacy range-based positioning techniques. The basic idea behind the triangulation is use of geometric properties of triangles to find the location of a device. Lateration and angulation are two triangulation techniques, which are discussed briefly in this section. The lateration techniques use the distance information whereas the angulation ones utilize the distance and angle information together to estimate the position of a device.

According to the lateration methods, the position of a device is estimated by measuring its distance from the position of a certain number of reference devices. The dimension of location information determines minimum number of reference devices required and the geometric relationship between them. For instance, distance measurements from three devices (multilateration) that are located in a non-collinear manner are required to estimate the position of a device in *2-D* as shown in Figure 10.6. On the other hand, *3-D* lateration requires the distance measurements from four non-coplanar devices. As a result, cognitive wireless networks can optimize the positioning algorithm by having a priori information about the dimension of the location information. For instance, the signal traffic and power consumption due to the positioning in cognitive wireless networks can be reduced. There are several lateration techniques such as TOA and RSSI to estimate the distance from the received signal.

TOA estimation technique is based on computing the travel time takes between a transmitter and receiver. For instance, the GPS utilizes the concept of one-way TOA ranging to estimate the position. By having the estimated travel

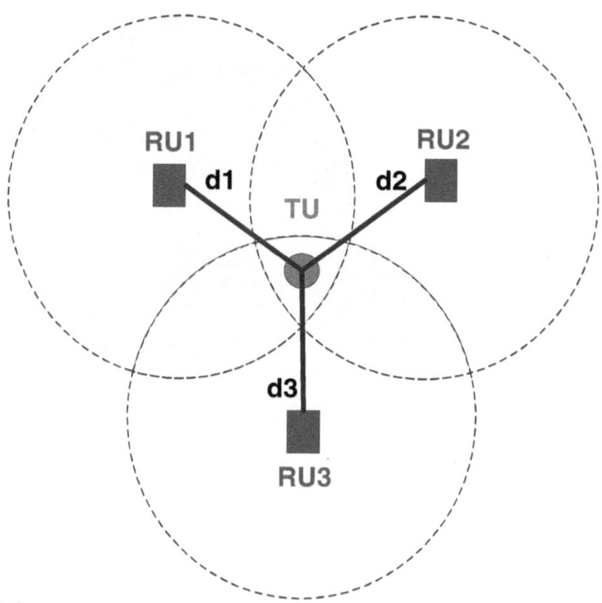

Fig. 10.6. 2-D lateration method. Three Reference Units (RU) are used to find the 2-D location of Target Unit (TU). The distances (d_1, d_2, d_3) can be measured using lateration methods.

time $\hat{\tau}$, the distance between a transmitter and receiver $\hat{d}$ can be estimated using $\hat{d} = v\hat{\tau}$, where v is the velocity of the waves used. Various waves such as electromagnetic and sound waves with different velocity can be used during the transmission by cognitive radios. For instance, the velocity of the sound wave is 344 m/s whereas the velocity of the light is approximately 3×10^8 m/s in the air. As a result, Ultrasound-based positioning systems [23] can be developed for cognitive radios as well. Therefore, cognitive radio can have a capability to use different waves for the positioning purpose depending on the environments; underground, underwater [24], and on the ground. For instance, in the case of emergency, a cognitive radio can utilize acoustic wave to find a location of a victim with cognitive radio device in underwater (due to flooding).

Angulation techniques are based on using angle and distance measurements to estimate the position of a device. The angles are measured with relative to a reference vector. The location information estimated through the angulation techniques are also in the form of different dimensions. Similar to the lateration techniques, the number of devices that need to be involved and their geometric relationships can be determined a priori by knowing the dimension of the location information. For instance, angle information from two different reference devices and the distance information between both are required to estimate the location of a device in *2-D* format as shown in Figure 10.7. On the other hand, *3-D* angulation requires an elevation information in addition to the requirements of *2-D* angulation [14].

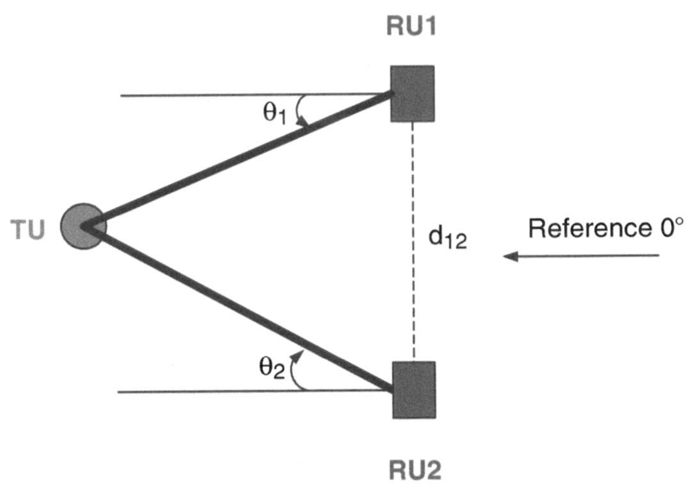

Fig. 10.7. 2-D angulation method. Two Reference Units (RU) are used to find the 2-D location of Target Unit (TU). The angles (θ_1, θ_2) and distance (d_{12}) can be measured using angulation and lateration methods, respectively.

Proximity

Proximity location estimation technique is one of the simplest forms of the positioning. The basic idea behind this type of positioning technique is to use the position of the antenna in a cell as the approximate position of a device [25]. Each cell is identified by a Location Area Identifier (LAI) in the cellular networks. The BS broadcasts the LAI and its Cell-ID to its cells. The MSs receive these messages and extract the Cell-ID, consequently, they know their Cell-IDs. The MSs also can determine its actual location information using the location information of the corresponding BS. Since the device can be anywhere in the cell, the accuracy depends on the size of cell. As a result, the accuracy can vary from a few meters (e.g. nanocell) to several kilometers (e.g. up to 30 km in large macrocell) depending on the cell type. The accuracy of the cell proximity can be improved by using additional information. For instance, in TDMA systems, BS calculates the timing delay between the MS and itself and sent this timing delay to the MS. In the sequel, the MS can estimate the distance between the BS and itself [25]. With this distance information, the MS can determine its location information relative to the location information of the BS. This method is a low-complex technique and cognitive wireless networks can use it when a coarse positioning and/or low computations are required.

10.3.5 Cognitive Positioning Techniques

Alternative to the aforementioned positioning techniques, the methods specific to cognitive wireless networks can be developed due to the deficiencies of the legacy positioning systems in terms of providing cognition features such as flexibility, adaptability, and high QoS. The cognitive positioning systems [12] and scene analysis [14] are two positioning systems specific to cognitive wireless networks that are discussed in this chapter.

Cognitive Positioning Systems

Cognitive positioning systems allow cognitive radio to adjust the positioning accuracy adaptively in both indoor and outdoor environments. A cognitive positioning system is proposed in [12], and this technique is composed of two modes, which are bandwidth determination and hybrid overlay and underlay enhanced dynamic spectrum management system. In the first mode, the cognitive positioning system determines the required effective bandwidth for a given accuracy. Adaptive TOA estimation technique that is an enhanced version of legacy TOA method is used in this mode. The required effective bandwidth is determined using the bandwidth determination equation, which is derived through Cramer-Rao Lower Bound (CRLB) for Additive White Gaussian Noise (AWGN) channels. Once the effective bandwidth is determined, the second mode that is hybrid overlay and underlay enhanced dynamic system is initiated. This is an improved dynamic spectrum management system that includes an additional set of rules in order to support positioning capability of cognitive radio. The main responsibility of the hybrid overlay and underlay enhanced dynamic is to search, find, and provide the optimum available bandwidth to cognitive positioning systems. Finally, the specified relative bandwidth is used to estimate the location using adaptive TOA technique. The details of these modes and overall technique can be found in [12].

Scene Analysis

Scene analysis is a location sensing technique that is based on using the features of a scene acquired by a sensing device (i.e. video camera) to infer the location of the observer and the objects in the environments. Not only information to be used for the positioning but also for the communication systems can be extracted using the scene analysis technique. This is one of the enabling techniques for cognitive radio to sense the environment. The acquired scene (such as a snapshot from a digital camera as shown in Figure 10.8) is simplified using basic geometric shapes for the data post-processing since analyzing a scene is a complex and difficult process. There are two popular scene analysis techniques, which are static and differential [14]. In the static method, the simplified scene is compared with the predefined datasets to draw conclusions regarding the scene. On the other hand, the location is estimated

Fig. 10.8. A snapshot of Mountain Nemrut acquired by digital camera (Courtesy of www.kahtanet.com). The objects in the scene can be represented by the geometric shape to simplify the scene. For instance, the top part of the mountains (Mountain Nemrut) can be represented by a triangle shape and one of the buildings in the scene can be represented by rectangular shape.

by keeping track of the differences between the consecutive snapshots from the scene according to the differential method. Note that one of the challenges for the scene analysis technique is the requirement of tremendous amount of digital data and image processing. For instance, if the position of the objects in the environments change there can be a need to construct new datasets to represent the objects. Another disadvantage of the scene analysis technique is that the device needs to have an access to a library for the features of environments. Legacy wireless devices do not have such capability. However, cognitive radio has potential to have such access to the predefined dataset of feature of environments stored either in central cognitive engine of the network or in the Internet.

10.3.6 Implementation Options

Depending on the choice of implementations, there are two approaches to estimate position of a device. The first approach is based on cognitive radio node, which is determining its location information. This approach is referred as *Decentrelized (Self) Positioning*. The second approach is based on cognitive wireless networks, which is determining the location information of target cognitive radio node using central cognitive engine. This approach is so-called *Centralized (Remote) Positioning*. These two approaches are discussed as follows.

Decentrelized (Self) Positioning

Cognitive radio device can estimate its own location information either independently or collaboratively using location estimation and/or sensing methods. For instance, cognitive radio can employ the scene analysis to sense its location independently. In the case of collaboration, cognitive radio can have a GPS receiver that collaborates with the satellites or it can request the distance measurements from the reference devices in cognitive wireless networks or terrestrial networks to estimate its location.

Centralized (Remote) Positioning

The location of a cognitive radio can be estimated using aforementioned positioning techniques such as distance measurements between the certain number of reference devices in cognitive wireless networks or terrestrial networks and the target cognitive radio. The measurements from the reference nodes are collected and processed by central cognitive engine of network. The position of target cognitive radio can be obtained from the geometric relationship (i.e. triangulation) between the devices that involve.

10.4 Mobility Management

Mobility management in cognitive wireless networks enables different cognitive wireless networks and legacy wireless networks to interconnect by means of a global interoperability. It can be achieved by integrating all access networks to an IP core network, which can be based on an enhanced Mobile-IP architecture. Moreover, enhanced Mobile-IP protocols support vertical handoffs in the heterogenous cognitive wireless networks to provide a seamless connectivity. For instance, a Voice over IP (VoIP) call made by cognitive radio users can switch from the wireless metropolitan area network (e.g. WiMAX MS) to the wireless personal area network (e.g. UWB or Bluetooth devices) through the wireless local area network (e.g. WLAN access point) infrastructure as they drive to work and then walk to their offices. Two important elements of the mobility management in cognitive wireless networks, mobility models, and location tracking, are discussed in this section.

10.4.1 Mobility Models

Utilization of location information in cognitive wireless networks for different applications will have a major impact on system complexity. Introduction of such additional services and applications into cognitive wireless networks will exacerbate the mobility issues. Consequently, the system capacity and implementation cost can be affected by these issues. Therefore, it is desirable to develop an accurate mobility model during the network planning phase. There are several common models that can be used to describe the mobility of mobile cognitive radio users, which are mainly derived from the human movement behavior [26].

Fluid Flow Model

As the name implies, the network traffic is modeled based on the idea of flow of a fluid, which is used to describe the macroscopic movement behavior. According to this model, the amount of traffic flowing out of a region is proportional to the population density within the region, the average velocity, and the region boundary length. Diffusion model is an advanced fluid model that characterizes the flow of traffic as a diffusion process [26].

Random-walk (Markovian-walk) Model

In this model, geographical area is divided into the grids (i.e. cells) and assuming a cognitive radio user can move from one cell to another cell without crossing the third cell. In order to model the movement of cognitive radio users in cognitive wireless networks, it is assumed that the time is slotted and a user can make at most one move during a slot. The movements of cognitive radio user is assumed to be a Markov random process and independent from one user to another in this model. According to the Markovian-walk model, the user can be in one of the three states, which are stationary (S), right-move (R), and left-move (L), during a slot. However, additional states such as back-move and cross-move can be introduced to extend the model. For instance, if a cognitive radio user is in the cell n according to the previous slot $i-1$, the next state determines the movement of the user during the slot i as follows:

- if the next state is S, then it remains in cell n,
- if the next state is R then it moves to cell $n+1$ (the cell on the right-hand side of the cell n),
- if the next state is L then it moves to cell $n-1$ (the cell on the left-hand side of the cell n).

Let assume that $X(t)$ is the state during the slot t and $\{X(t)|t=0,1,...\}$ is a Markov chain with the following transition probability function:

$$P_{i,j} = P[X(t+1) = j|X(t) = i] \ , \quad \forall t \ , \tag{10.2}$$

where $i = \{S, R, L\}$, $j = \{S, R, L\}$, and $P[\cdot]$ represents the probability function. The state diagram of the Markovian-walk model is illustrated in Figure 10.9 [26]. There are three common location reporting strategies, which are time-based, movement-based, and distance-based location update. In time-based update, each cognitive radio user reports a message containing its location update every predefined number of slots whereas this update is performed by each cognitive radio user at every completion of predefined number of movements between the cells in the movement-based strategy. On the other hand, each cognitive radio user transmits the location update message every predefined amount of distance in terms of number of cells, which is defined as the distance between the current and the last reported cells in the distance-based

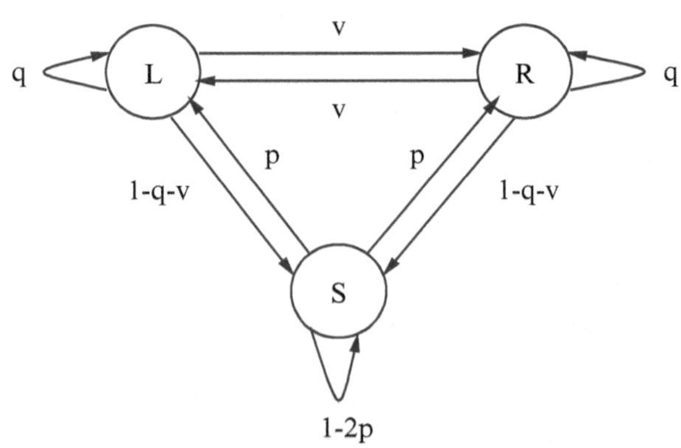

Fig. 10.9. A state diagram for the Markovian-walk model. The transition probabilities are defined as follows: $P_{R,R} = P_{L,L} = q$, $P_{L,R} = P_{R,L} = v$, $P_{S,R} = P_{S,L} = p$, $P_{L,S} = P_{R,S} = 1-q-v$, and $P_{S,S} = 1-2p$.

update [26]. Depending on the trade-offs between the complexity and performance of cognitive wireless networks, one or combination of these strategies can be utilized.

Gravity Model

Gravity models have been used to model the traffic in the regions of various sizes such as city, national, and international models. The amount of traffic T_{ij} moving from the region i to region j is formulated by

$$T_{ij} = K_{ij} D_i D_j, \tag{10.3}$$

where D_i and D_j are the population in region i and j, respectively. Moreover, K_{ij} is the parameter that describes the characteristics of the traffic movement between the region i and region j and it has to be computed for all (i, j) combinations. For instance, K_{ij} can be equal to d_{ij}^{-2}, which is analogous to the Newton gravitational law [26].

Mobility traces are useful tools that can indicate the current movement behavior of mobile cognitive radio users and are more realistic than the mobility models. Three different scales are commonly used for the mobility traces of the large population sizes and geographical areas, which are metropolitan, national, and international mobility models. Further details on the mobility models and traces can be found in [26].

10.4.2 Location Tracking

Once location information is estimated and/or sensed using the previously mentioned methods, then tracking cognitive mobile devices is another main task in cognitive wireless networks. The location of a mobile cognitive radio device can be tracked using successive range-based and pattern matching-based approaches. However, the latter approaches are more widely used than the former ones. The pattern matching-based methods are mainly based on learning process, which requires successive collection of statistics from mobile cognitive radio nodes on their movement behaviors. Location tracking process is usually modeled as a hidden Markov model and it is characterized by five elements [27]:

- *Number of hidden states N.* Hidden state variables $L = \{l_t, l_{t+1}, \dots, l_{t+N-1}\}$ are indexed by time $t + j$ and state at $t + j$ is denoted as q_{t+j}. In the context of the location tracking, the hidden states l_{t+j} correspond to the location of a cognitive radio device at time $t + j$ and the sequence of the states corresponds to the trajectory of cognitive radio device.
- *Number of distinct observation symbols per state M.* These symbols are denoted as $S = \{S_0, S_1, ..., S_{M-1}\}$. The observed variable for each hidden state is denoted as $O = \{O_t, O_{t+1}, ..., O_{t+N-1}\}$. Each O_t is assumed to be dependent only on the current location l_t.

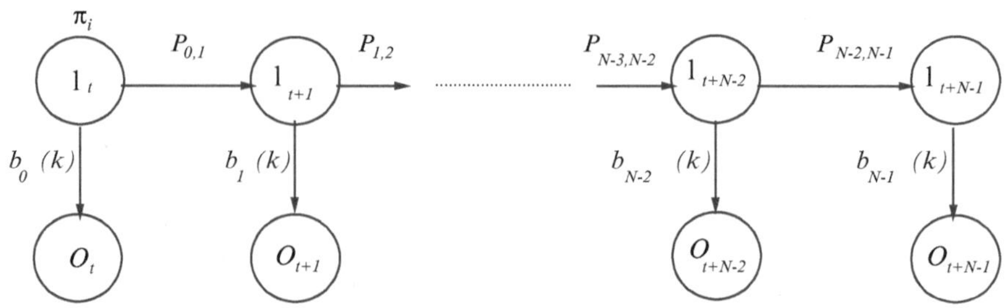

Fig. 10.10. Hidden Markov model for location tracking.

- *State transition probability* $P_{i,j}$ that is defined by (14), $i = \{0, 1, ..., N-1\}$ and $j = \{0, 1, ..., N-1\}$ in this case.
- *Observation symbol probability for each state* $b_j(k)$, which is defined as

$$b_j(k) = P[S_k \; at \; t | q_{t+j} = l_{t+j}] \;, \quad 0 \leq k \leq M-1. \tag{10.4}$$

- *Initial state probability* π_i, which is defined as

$$\pi_i = p[q_1 = l_{t+i}] \;. \tag{10.5}$$

The hidden Markov model for location tracking is illustrated in Figure 10.10. With this model, probability distribution of location of cognitive radio device at any given time can be determined for a given series of observations [16]. In [28], a location tracking algorithm based on hybrid TOA and hidden Markov models to estimate the driven route of a vehicle in the cellular networks is proposed. Similar location tracking algorithms can be developed for cognitive wireless networks as well.

10.5 Applications

10.5.1 Location-Based Services

Location-based services have been gaining an increasing popularity and they are becoming part of our daily lives. These services found applications in different areas such as for personal, commercial, governmental, and military purposes. Some representative location-based services for the personal purpose are navigation systems for sightless people, tracking systems for parents to monitor children in the residential and school areas, navigation systems to guide travelers (i.e. locate restaurants and hotels) and tourists during travels. Some of the many commercial location-based services are tracking vehicles on the roads, monitoring employees in the offices and finding the location of on-demand equipments in the industrial areas, supervising patients in the hospitals, and tracking animals in the wildlife conservation areas and farms.

For instance, one of the products of Qualcomm, which is called SnapTrack, has capability to provide some of the above location-based services [29]. Moreover, the Google Maps for mobile phone allows the users to obtain the detailed directions for their destinations and track the traffic in real-time for the selected routes [30].

Location-based services also can be used by the government agencies such as to find the location of the victims and rescue them in the case of emergency and disaster (i.e. E911 in the United States and E112 in the Europe). Moreover, location-based services can be utilized for military applications such as tracking soldiers during the combat missions. In short, countless location-based services can be developed for different purposes.

10.5.2 Network Optimization

Numerous location-aided network optimization applications can be developed such as location-assisted dynamic spectrum management, network upgrade, handover, drop call management, dynamic channel allocation, routing, power control, internetworking, and adaptive coverage system. However, we discuss the first four applications with details in this section.

Location-Assisted Dynamic Spectrum Management

Under-utilization of the licensed bands and spectral crowdness in the license-exempt bands prompt the concept of using the spectrum in an opportunistic manner. Basically, in the absence of licensed users that have the privileges to use their bands, unlicensed users can utilize these bands temporarily. This can be achieved by using overlay and/or underlay spectrum access techniques. There are several different proposals for the dynamic spectrum management in cognitive wireless networks from the academia and industry. However, we consider three schemes proposed by the Federal Communications Commission (FCC) in the United States, which are *Listen-before-talk*, *Geolocation-database*, and *Local Beacon* techniques [31]. These schemes are proposed for allowing unlicensed operation in the TV bands while preventing interference to the television (TV) reception. Moreover, a recently formed IEEE 802.22 working group for the Wireless Regional Area Networks (WRANs) has the same vision as the FCC. The main goal of this group is to establish a cognitive radio-based standard that allows unlicensed users to utilize the bands allocated to the TV broadcast services in a non-interfering fashion. It is worth to mention that low frequency analog TV bands (54–862 MHz) have some attractive features for wireless broadband services such as achieving long-distance transmission. Since the Geolocation-database and Local Beacon techniques utilize the location information, the other technique is out of scope of this section. However, it is worth to mention the main concern regarding the Listen-before-talk scheme, which is the hidden node problem [31].

According to Geolocation-database scheme, licensed users (i.e. TV transmitter) are equipped with location estimation and/or sensing device to estimate their current location information. Licensed users provide their frequency and location information to the FCC central database, which is a part of central cognitive engine. The FCC database broadcasts the available channel information along with the location information of licensed users. On the other hand, unlicensed users are also equipped with location estimation and/or sensing device to estimate their current locations. Unlicensed users cross-check their locations with the location of licensed users in order to obtain the channel information that they can use locally. Two concerns regarding this scheme are the reliability of current geolocation technologies and the performance of the FCC central database. Some TV broadcasters point out that GPS does not provide a reliable location information of a user that is located in the indoor environments. However, this concern can be addressed by use of Indoor GPS. Alternatively, licensed users and unlicensed users can utilize more reliable cognitive positioning techniques such as cognitive positioning system to obtain their current location information. The other concern is that the FCC central databases can be slow to catch up the frequency changes and they are not 100% reliable. This is a valid concern and the FCC is aware of this issue. As a result, the FCC needs to upgrade its database technology in order to be used for the spectrum sensing and allocation purposes.

Local Beacon method is pretty similar to the Geolocation-database scheme except that the database will be placed in a local cell to manage the spectrum. The details and example of local Geolocation-database approach can be found in [8]. The FCC states that short-range signal will be used in this method to broadcast the available channel information. Since the FCC does not specify the "short-range signal" term, long-range signal can be used and they can introduce interferences to the other cells. Consequently, such signals can degrade performance of this scheme significantly and even it can shut down the system.

Location-Assisted Network Upgrade

Network design is usually thought of as the initial phase of the network build out, however, it is an ongoing work that is needed to upgrade the capacity and optimize the network. From a design point of view, revenue, license, capacity, and performance are the main criteria that are used to make decisions for the network expansion. Last two criteria are considered in this section since they possess technical merits.

The current wireless network operators plan and expand their networks in a semi-computerized manner. They have a group of design engineers that are responsible for the network planning and expansion. The network expansion is determined based on the measurement data that is obtained from a specific geographical area. Such data are generally collected by driving a vehicle equipped with the measurement instruments within the target areas. The engineers map the collected data to the morphology tables, which are used to help them to predict the traffic intensity and consequently the capacity in the various areas. The engineers carefully study the tables and perform area visits regularly to achieve an optimal capacity prediction. This is a critical part in the initial design and expansion phases of the network. Over-prediction can cause an inefficient use of resources (i.e. hardware and spectrum) whereas under-prediction can affect the QoS provided by the network.

Cognitive wireless networks have a potential to perform network planning and expansion procedure automatically. In this approach, cognitive radio nodes are equipped with location estimation and/or sensing devices. Central cognitive engine can request or cognitive radio nodes can report the information that is beneficial for the network planning, expansion, and optimization along with their location information. This information can be also incidents for the drop call or deep fading. Further information on location-assisted network upgrade for cognitive wireless networks can be found in [8].

Location-Assisted Handover

Handover mechanism is an important mechanism that can affect the performance of the legacy cellular as well as cognitive wireless networks. In [32], a location-based handover algorithm is proposed for the GSM networks and the performance of the conventional handover mechanisms (i.e. RSSI-based handover) and the proposed handover algorithms are compared. The measurement results show that the location-based handover provides 30% reduction in the number of handovers compare to the RSSI-based handover for a given call. The majority of the current handover mechanisms are based on only signal strength, signal quality and cell traffic parameters due to some complexity limitations. There can be situations where the number of competing cells is large. This scenario can be, for example, a long bridge over a water surface such as a bridge on highway 60 in Tampa, Florida as shown in Figure 10.11. Here we have a long road and due to the fact that signal propagates further over the water, many cells will appear as attractive candidates for the neighbor selection. The problem here is that optimizing handover margins will not help since we want to control the handover along highway 60 only (in this example). If we make handover hard to certain sites, it will affect the handovers in the different areas in which case we have partially solved one problem, but created additional ones. For instance, the number of handover decisions can increase in the other areas and this can affect the amount of overhead in the network and the computational power that is needed.

As the QoS requirements increase, the current handover algorithms such as RSSI-based handover start to lag. Therefore, they can be replaced with more efficient handover algorithms such as location-assisted handover method, which is based on use of the location information of MSs. In [8], a location-assisted handover mechanism for cognitive wireless networks is proposed to

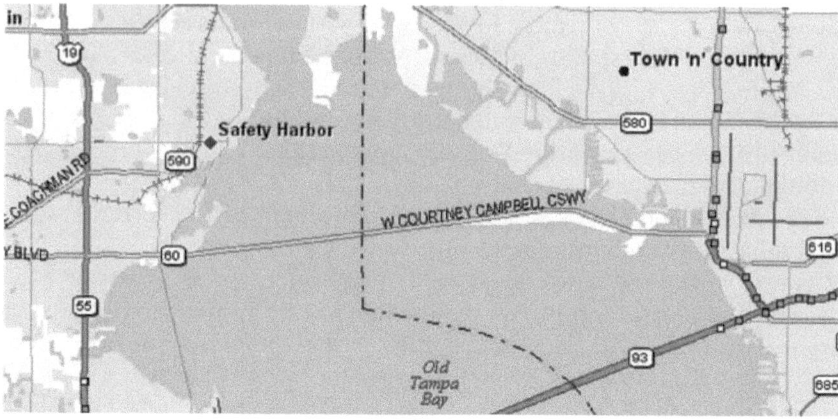

Fig. 10.11. Map for a bridge on the US Route 60 in Tampa, FL, USA. This is a real life geographical area (a bridge), where the number of competing cells are large during the handover process and it is difficult to manage handover mechanism in such areas. The location-assisted handover algorithms are promising methods to optimize the handover mechanisms in such problematic areas.

reduce the number of handovers and consequently the network load. In this algorithm, the handover zones in which all the MSs are served by the predefined candidates are defined. The MSs are equipped with location estimation and/or sensing devices to report their locations back to the MSC. Once a MS enters into a handover zone, the MS can be served by one of the predefined candidates. We refer to [8] for the details of the location-assisted handover mechanism for cognitive wireless networks.

Location-Assisted Drop Call Management

Drop call is another important issue that can affect the QoS of wireless networks. The drop call occurs when the call in progress is terminated unexpectedly due to some technical reasons. The reasons that cause the drop calls can be classified under three major categories, which are air-interface, hardware, and handover issues and the distribution of these issues for the cellular networks is shown in Figure 10.12. It has been reported that the main factor that produces the drop calls is the air-interface issues (58.33% according to the Figure 9). For instance, one of the air-interface issues that creates the drop call is the poor coverage or moving out of range of a wireless network. In other words, an active call can be lost when the users move out to the areas where the communications is unavailable, interrupted, interfered with, or jammed. Co-channel and adjacent channel interferences are some other reported reasons that can cause the drop calls. The neighbor cells using the same frequencies interfere with each other resulting in drop calls as well. Moreover, sun spots and

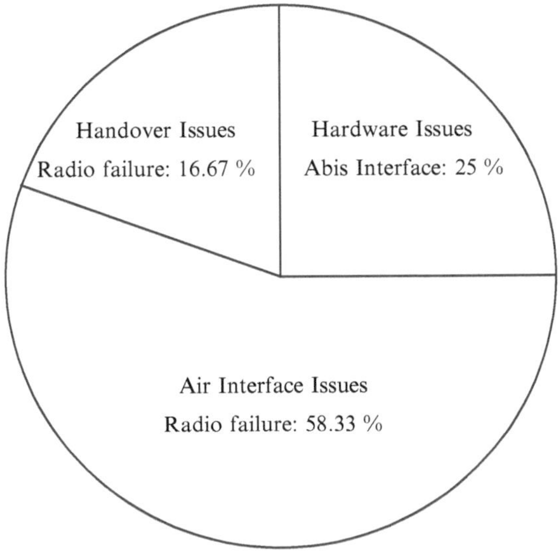

Fig. 10.12. The distribution of the factors that cause the drop calls in the cellular networks. The abis interface within the GSM architecture is the interface between the BTS (Base Transceiver Station) and BSC (Base Station Controller).

solar flares are rarely reported to be the reasons for causing the interference leading to dropped calls.

Another main category that produces the drop calls is the hardware issues such as transmission problems and faulty hardware components in the mobile phone or base station. Calls also can be lost if the mobile phone runs out of the power and consequently stops to transmit.

The drop calls can occur due to handover issues, which is the third main category. The calls can be dropped during the handover process due to the various reasons such as the imbalance traffic between the cell sites and improper network configuration. For instance, if the target cell is running out of its capacity, it can not accept the additional users requesting admission during the handover process. In such scenarios, the calls are dropped as well. Furthermore, if the mobile phone tries to handover to a target cell that is not aware of such mobile phone and the phone can not find out an alternative target cell to move in, then the call will be lost.

The drop call management in the existing cellular networks is often as follows. Anytime a customer calls to complain about drop call, a support professional first validates the complaint and then tries to get as much information as possible about the location of the incident. The tickets that are generated by the agents are sent to the network engineers, who in turn will position them on the maps for further analysis. The main drawback of this approach is that the majority of the customers do not take the time to report the problems they experience in the field and hence, a lot of information is not fed back into the drop call management system. One issue here would be how can the network know or even predict the reason and the place of the drop call. That is, during the drop call the connection is lost between the mobile phone and the network and hence it is difficult to gather intelligent information from the mobile phone at the time the drop calls happen.

The drop call management in cognitive wireless networks can be automated and consequently location-assisted drop call management systems can be developed. The reason and location information for the drop call incidents are two important parameters that affect the complexity of the location-assisted drop call management systems. Cognitive wireless networks can have a capability to determine the reason and location of the incidents. Note that the location information is not necessarily to be estimated for all the drop call incidents. Basically, the reason of the incidents determine whether the location information needs to be estimated or not. For instance, the location information is not necessary for the case of the hardware issues that cause the drop calls. Therefore, the reason of the incidents is determined initially. It is difficult to identify the reason if the several issues (i.e. running out of power and poor coverage) occur at the same time. Once, cognitive radio nodes determine the reason of the incidents, then they report back the reason along with the location information if necessary to central cognitive engine of the network for further analysis in a cooperative manner. Central cognitive engine classifies the incidents depending on their reasons. For a duration of time, the information regarding the drop calls are accumulated and central cognitive engine monitors the incidents, basically it tries to learn behavior of the incidents. Based on the accumulated information and using statistical prediction and estimation techniques, central cognitive engine can predict ahead of time the reason and location of the drop calls and consequently initiate the drop call avoidance or suppression algorithms. Numerous solutions for each reason to predict and avoid or reduce the drop calls can be developed. Alternatively, the information regarding the drop call problems can be sent to the aforementioned location-assisted network planning and expansion system to optimize the network in order to avoid or reduce the drop calls.

Cognitive radio nodes can monitor the hardware components such as battery life, the handover mechanism, and air-interface in order to predict the drop call. For instance, the battery life is monitored and if the power level goes below a predefined limit, then cognitive radio warns its user that drop call can occur within the estimated time. Moreover, cognitive radio sends this information to central cognitive engine as a feedback to let it know that the battery most likely will be the reason if the call is dropped. Let us assume that cognitive radio node has enough battery life, cognitive wireless network does not experience any air-interface issue that can lead to a drop call, and central cognitive engine monitors the handover mechanism. If the drop call occurs, then it will be reported that the drop call happens most likely (i.e. the probability can be specified) due to the handover mechanism. Optimizing the handover mechanism such as by employing the previously discussed location-assisted handover can avoid or reduce the drop calls. Similarly, assuming that cognitive radio node has enough battery life and it does not need to perform a handover, then cognitive radio can predict and avoid the drop calls (or reduce the number of drop calls) by employing the location-assisted drop call management system. For instance, if the user is moving toward out of range of the network, cognitive radio can warn the users ahead of time that the drop call can happen within the estimated time or distance. Same information is sent to central cognitive engine as a feedback. If the user responds that he/she must go that area, then different solutions can be developed to avoid the drop calls. For instance, cognitive radio can inform central cognitive engine (or serving cognitive base station) that the cell coverage needs to be extended by the estimated distance to maintain the communications. The serving cognitive base station increases its coverage towards the specified area where the user is located (i.e. by having a beamforming capability). The location of the user is tracked to adaptively change the coverage. This algorithm is so called location-assisted adaptive coverage algorithm, which is part of the location-assisted drop call management system. Note that the location-assisted drop call management system is embedded in each cognitive radio node in the case of decentralized cognitive wireless networks.

10.5.3 Transceiver Algorithm Optimization

Utilization of location information for improving and optimizing transceiver algorithms is a current research topic. We briefly discuss several location-assisted transceiver algorithms such as ranging, beamforming and interference avoidance, and link adaptation. However, more location-aided transceiver algorithm can be developed.

Location-Assisted Adaptive Beamforming and Interference Avoidance

Adaptive beamforming consists of adaptive array whose outputs are combined based on adjustable weight vector to radiate the transmit signal to the desired user while minimizing the interference to the other users [33]. There are two major challenges in the legacy beamforming systems; estimation of the coefficients of arrays from the received signal and minimization of interference that is created during the radiation of the beam. The coefficients of elements in the arrays are estimated mainly based on the direction of arrival and strength of the signal that comes from the desired user. However, the received signal is distorted by the other interference signals that comes from the undesired users as well as the background noise. It is a tedious task to estimate the element coefficients from the distorted signal. Since the direction of the desired user is precisely not known, the radiated power can interfere the neighbor users.

The beam is generated in such a way to minimize the interference to the other users. For instance, one way of doing that is to place nulls in the interfering directions [33].

The aforementioned problems can be eliminated by introducing location-assisted adaptive beamforming and interference avoidance algorithms to the cognitive radios. The first problem is eliminated since the explicit absolute location information of the desired user can be obtained easily through the positioning systems such as GPS. Since cognitive radio that generates the beam can obtain its absolute location information as well, it directs the beam to the desired user precisely. The width of the beam plays an important role in such algorithms. The optimum width of the beam needs to be determined based on two optimization criteria; the width of the beam needs to be large enough to eliminate periodic steering while avoiding the interference to the others.

Location-Assisted Link Adaptation

Link adaptation is adjusting the transmission parameters adaptively according to the condition of wireless propagation channel, which is one of the key features of cognitive radios. Link adaptation can be defined as a set of algorithms and protocols that constitute adaptive modulation and coding [34]. The parameters that are adapted are not limited to the modulation and coding. Other transmission parameters such as transmit power level and bandwidth can also be adjusted. Link adaptation techniques are powerful methods to improve energy efficiency and increase data rate in the fading channels [35,36].

One of the key metrics in the link adaptation techniques is the distance between transmitter and receiver. Hence, the link adaptation techniques in the literature mainly rely on physical metrics, which are RSSI and Signal-to-Noise-Ratio (SNR) [37]. Alternative to these metrics, the location information of the transmitter and receiver can also be used. Since cognitive radios are equipped with location estimation and/or sensing devices, the precise distance between two cognitive radios can be easily determined. Such precise distance information can be utilized to improve the performance of the link adaptation techniques.

Location-Assisted Ranging

Traditionally the ranging information between transmitter and receiver is estimated using the training sequence [38]. The time of arrival of paths, angle of the received signal, or strength of the signal are mainly utilized to obtain the ranging information. Similarly, the delay of the first arrival paths are estimated to establish synchronization between a transmitter and receiver.

Alternative to the legacy ranging algorithms, techniques that are aided by location information can be developed for cognitive wireless networks. The ranging and synchronization can be performed separately or jointly. However, joint ranging and synchronization methods [39] are desirable for cognitive radios. In the sequel, a location-assisted joint ranging and synchronization is discussed briefly. The ranging (distance) between two cognitive radios using their absolute location information can be estimated easily. The transmitting cognitive radio sends training sequence to the receiving cognitive radio to establish synchronization. Once both are synchronized, the transmitting cognitive radio transmits its absolute location information in the preamble to the receiving cognitive radio. It extracts the absolute location information of the transmitting cognitive radio from the preamble and measure the distance between itself and transmit cognitive radio. The complexity and accuracy of the location-assisted ranging algorithms can be compared to those of the legacy ranging algorithms. However, it is expected that the former will have less complexity and higher accuracy than the latter.

10.5.4 Environment Characterization

Similar to transceiver algorithm optimization, location-assisted channel environment characterization is a current research topic. Transceiver algorithms are generally developed for the specific environments. Hence, propagation channel of the target environment is modeled as the first step. Consequently, the transceiver algorithms are developed based on the channel models. There are two main approaches for modeling wireless propagation channels in an environment: 1) statistical modeling [40], 2) deterministic modeling [41]. In the statistical modeling methods, extensive field measurements are performed and the statistical parameters of the channel are extracted. Then, the channel is modeled based on these statistical parameters. On the other hand, deterministic models (also known as site-specific models), which are based on numerical calculations like ray tracing methods, strive to attain that ultimate pinpoint accuracy via computer simulations.

In order to realize the adaptation capability of cognitive radio, it needs to be aware of its operating environment, characterize it, and adapt itself accordingly. We believe that statistical modeling is more suitable than the deterministic modeling to be adopted by cognitive radios to achieve this goal. The statistical parameters of wireless channels are mainly folded under two groups, which are *large-scale statistics* and *small-scale statistics* [40]. With the aid of location information, these statistics can be estimated through the empirical models since most of them are environment-dependent parameter (e.g. rural, urban, and indoor).

In the literature, there is not any clear-cut distinction between the environmental classes. However, in the classification, there are major differences that affect the channel parameters. Indoor/outdoor separation is the prominent example of this case. For indoor environments, it is expected not to have a higher Doppler spread as compared to that for outdoor due to the mobility restrictions. On the other hand, for an outdoor environment, it is more likely to have larger delay spread as compared to that for indoor.

The main small-scale statistical parameters can be retrieved by having a priori information about the position of user and environment. As a result, location information of cognitive radio along with the rough description of operating environment are required for performing adaptation through the empirical models. The location information of cognitive radio can be obtained using one of the location estimation and/or sensing methods (e.g. GPS) discussed in earlier sections. Moreover, cognitive radio can identify the distinctive properties of the surrounding environments (e.g. topographical information) using basic topographical databases such as digital elevation models, neural networks, and Markov models. By using the location information along with rough description of the operating environment, cognitive radio can determine the type of propagation environment and employ the corresponding channel model for the adaptation.

10.6 Privacy Concerns

Extensive utilization of location information for the network optimization and adoption of location-based services in cognitive wireless networks prompt the security [42] and user privacy issues [43, 44]. Some users may not agree on the use of their instantaneous location information for the aforementioned location-assisted applications. In such cases, different options need to be available to the users. For instance, in the emergency cases such as dialing a 911 call, cognitive wireless networks should have a right to locate the position of the users. On the other hand, providing the location information to the network can be optional in the non-emergency cases. Another possible solution is that cognitive radio can have self-positioning capability (i.e. GPS receiver) and the users can have a full control to setup the frequency of reporting the location of their devices to the network. Furthermore, secure positioning techniques robust to position and distance spoofing attacks [42] can be developed for cognitive radios. With employing such potential solutions, the privacy of users can be protected. Furthermore, by having a secure network and employing encryption technologies [45] can reduce the fears of users from the illegal activity that can threat their privacies.

10.7 Conclusions

Evolution of the utilization of location information in wireless networks is provided. The need for a comprehensive location information management system for cognitive wireless networks to realize location awareness is emphasized and a location information management system model for cognitive wireless networks is presented with the details. Detailed discussion on location estimation and sensing in cognitive wireless networks is provided. Furthermore, mobility management along with the mobility models and location tracking for cognitive wireless networks are discussed. Several representative examples for location-based services and location-assisted network optimization, transceiver algorithm optimization, and environment characterization are demonstrated to support location information management system model. The user privacy concerns regarding the utilization of their location information along with some solutions are addressed. It can be concluded that location information management system is a crucial part of cognitive wireless networks and utilization of location information in cognitive wireless networks can be exploited. Moreover, it is strongly recommended to balance the utilization of location information in cognitive wireless networks and the protection of user privacy.

References

1. Y. Xing, R. Chandramouli, and S. Mangold, "Dynamic Spectrum Access in Open Spectrum Wireless Networks," *IEEE J. Selec. Commun.*, vol. 24, no. 3, pp. 626–637, March 2006.
2. N. Sai Shankar, C. Cordeiro, and K. Challapali, "Spectrum Agile Radios: Utilization and Sensing Architectures," in *First IEEE International Symposium on New Frontiers in Dynamic Spectrum Access Networks*, Baltimore, MD, USA, Nov. 2005, pp. 160–169.
3. H. Arslan and M. E. Sahin, "Cognitive UWB-OFDM: Pushing Ultrawideband Beyond Its Limit via Opportunistic Spectrum Usage," *J. Commun. Networks – Special Issue Spectrum Resource Optimization*, vol. 8, No. 2, pp. 151–157, June 2006.
4. S. Haykin, "Cognitive Radar," *IEEE Signal Processing Mag.*, vol. 23, no. 1, pp. 30–40, Jan. 2006.
5. T. Fujii and Y. Suzuki, "Ad-hoc Cognitive Radio-Development to Frequency Sharing System by Using Multi-hop Network," in *First IEEE International Symposium on New Frontiers in Dynamic Spectrum Access Networks*, Baltimore, MD, USA, Nov. 2005, pp. 589–592.
6. H. Zheng and C. Peng, "Collaboration and Fairness in Opportunistic Spectrum Access," in *Proc. IEEE International Conference on Communications*, vol. 5, Seoul, Korea, May 2005, pp. 3132–3136.
7. S. Mishra, A. Sahai, and R. Brodersen, "Cooperative Sensing among Cognitive Radios," in *Proc. IEEE International Conference on Communications*, vol. 4, Istanbul, Turkey, June 2006, pp. 1658–1663.
8. H. Celebi and H. Arslan, "Utilization of Location Information in Cognitive Wireless Networks," in *IEEE Wireless Communications Magazine – Special issue on Cognitive Wireless Networks*, August 2007.
9. J. Agre, A. Akinyemi, L. Ji, R. Masuoka, and P. Thakkar, "A Layered Architecture for Location-based Services in Wireless Ad hoc Networks," in *Proc. IEEE Aerospace Conference*, vol. 3, Big Sky, MT, USA, March 2002, pp. 1085–1097.
10. N. Patwari, A. Hero, M. Perkins, N. Correal, and R. O'Dea, "Relative Location Estimation in Wireless Sensor Networks," *IEEE Trans. Signal Processing*, vol. 51, no. 8, pp. 2137–2148, Aug 2003.

11. "Standard Molodensky Datum Conversion," 2006. [Online]. Available: http://www.colorado.edu/geography/gcraft/notes/datum/gif/molodens.gif.
12. H. Celebi and H. Arslan, "Cognitive Positioning Systems," 2007, *IEEE Trans. on Wireless Communications*, to appear, 2007.
13. H.-H. Nagel, "Steps Towards a Cognitive Vision System," *Artif. Intell. Mag.*, vol. 25, no. 2, pp. 31–50, 2004.
14. V. Zeimpekis, G. Giaglis, and G. Lekakos, "A Taxonomy of Indoor and Outdoor Positioning Techniques for Mobile Location Services," *ACM SIGecom Exchanges*, vol. 3, no. 4, pp. 19–27, 2002.
15. R. Koch, "Dynamic 3-D Scene Analysis Through Synthesis Feedback Control," *IEEE Trans. Pattern Analysis Machine Intell.*, vol. 15, no. 6, pp. 556–568, June 1993.
16. R. Ganesh, S.L. Kota, K. Pahlavan, and R. Agusti, *Emerging Location Aware Broadband Wireless Ad Hoc Networks.* Springer, 2005.
17. C. Fretzagias and M. Papadopouli, "Cooperative Location-Sensing for Wireless Networks," in *Proc. Second IEEE Annual Conference on Pervasive Computing and Communications*, Orlando, FL, USA, March 2004, pp. 121–131.
18. S. Feng and C. Law, "Assisted GPS and Its Impact on Navigation in Intelligent Transportation Systems," in *Proc. IEEE 5th International Conference on Intelligent Transportation Systems*, Singapore, Sept. 2002, pp. 926–931.
19. F. van Diggelen, "Indoor GPS Theory & Implementation," in *IEEE Position Location and Navigation Symposium*, Palm Springs, CA, USA, Apr. 2002, pp. 240–247.
20. H. Celebi, I. Guvenc, and H. Arslan, "On the Statistics of Channel Models for UWB Ranging," in *Proc. IEEE Sarnoff Symposium*, Princeton, NJ, USA, March 2006.
21. Y. Qi and H. Kobayashi, "On Relation Among Time Delay and Signal Strength Based Geolocation Methods," in *Proc. IEEE Globecom Conference*, vol. 7, San Francisco, CA, USA, Dec. 2003, pp. 4079–4083.
22. H. Celebi and H. Arslan, "Adaptive Positioning Systems for Cognitive Radios," in *IEEE Symposium on New Frontiers in Dynamic Spectrum Access Networks (DySpan)*, Dublin, Ireland, 17–20 April 2007.
23. H. Piontek, M. Seyffer, and J. Kaiser, "Improving the Accuracy of Ultrasound-Based Localisation Systems," in *1st International Workshop on Location- and Context-Awareness*, vol. 3479, Munich, Germany, May 2005.
24. V. Chandrasekhar, W. Seah, Y. Choo, and H. Ee, "Localization in Underwater Sensor Networks: Survey and Challenges," in *Proc. 1st ACM international Workshop on Underwater Networks*, Los Angeles, CA, USA, Sept. 2006, pp. 33–40.
25. E. Trevisani and A. Vitaletti, "Cell-ID Location Technique, Limits and Benefits: an Experimental Study," in *Proc. 6th IEEE Workshop on Mobile Computing Systems & Applications*, English Lake District, UK, Dec. 2004, pp. 51–60.
26. A. Mukherjee, S. Bandyopadhyay, and D. Saha, *Location Management and Routing in Wireless Networks.* Artech House Publishers, 2003.
27. L. Rabiner, "A Tutorial on Hidden Markov Models and Selected Applications Inspeech Recognition," *Proc. IEEE*, vol. 77, no. 2, pp. 257–286, Feb. 1989.
28. S. Mangold, S. Kyriazakos, and T. Aachen, "Applying Pattern Recognition Techniques Based on Hidden Markov Models for Vehicular Position Location in Cellular Networks," in *IEEE Vehic. Tech. Conference-Fall*, vol. 2, Amsterdam, The Netherlands, Sep. 1999, pp. 480–484.

29. "SnapTrack," 2006. [Online]. Available: http://www.snaptrack.com/
30. "Google Maps for Mobile Phones," 2006. [Online]. Available: http://google.com/gmm.
31. M.J. Marcus, P. Kolodzy, and A. Lippman, "Reclaiming the Vast Wasteland: Why Unlicensed Use of the White Space in the TV Bands will not Cause Interference to DTV Viewers," *New America Foundation: Wireless Future Program*, Tech. Rep., 2005.
32. H. Lin, R. Juang, and D. Lin, "Validation of an Improved Location-Based Handover Algorithm Using GSM Measurement Data," *IEEE Trans. Mobile Computing*, vol. 4, no. 5, pp. 530–536, Sept-Oct 2005.
33. T.-J. Shan and T. Kailath, "Adaptive Beamforming for Coherent Signals and Interference," *IEEE Trans. Signal Processing*, vol. 33, no. 3, pp. 527–536, June 1985.
34. S. Catreux, V. Erceg, D. Gesbert, R. Heath Jr, I. Inc, and C. San Jose, "Adaptive Modulation and MIMO Coding for Broadband Wireless Datanetworks," *IEEE Communi. Mag.*, vol. 40, no. 6, pp. 108–115, June 2002.
35. A. Goldsmith and S. Chua, "Adaptive Coded Modulation for Fading Channels," *IEEE Trans. Commun.*, vol. 46, no. 5, pp. 595–602, May 1998.
36. T. Keller and L. Hanzo, "Adaptive Modulation Techniques for Duplex OFDM Transmission," *IEEE Trans. Vehic. Tech.*, vol. 49, no. 5, pp. 1893–1906, Sep. 2000.
37. T. Braskich, N. Smavatkul, and S. Emeott, "Optimization of a Link Adaptation Algorithm for Voice Over Wireless LAN Applications," in *Proc. IEEE Wireless Communications Networking Conference*, vol. 3, New Orleans, LA, USA, March 2005, pp. 1602–1607.
38. H. Celebi and H. Arslan, "A Data-Aided Timing Acquisition for Pulse Position Modulated UWB Signals," in *Proc. IEEE International Conference on Ultra-Wideband*, Zurich, Switzerland, 2005, pp. 344–348.
39. B. Denis, J. Pierrot, and C. Abou-rjeily, "Joint Distributed Synchronization and Positioning in UWB Ad Hoc Networks Using TOA," *IEEE Trans. Microwave Theory Techniques*, vol. 54, no. 4, pp. 1896–1911, June 2006.
40. A.F. Molisch, "Ultrawideband Propagation Channels-Theory, Measurement, and Modeling," *IEEE Trans. Veh. Technol.*, vol. 54, no. 5, pp. 1528–1545, Sept. 2005.
41. T. Sarkar, Z. Ji, K. Kim, A. Medouri, and M. Salazar-Palma, "A Survey of Various Propagation Models for Mobile Communication," *IEEE Antennas Propagation Mag.*, vol. 45, no. 3, pp. 51–82, June 2003.
42. S. Capkun and J. Hubaux, "Secure Positioning in Wireless Networks," *IEEE J. Sel. Areas Commun.*, vol. 24, no. 2, pp. 221–232, Feb. 2006.
43. L. Barkhuus and A. Dey, "Location-Based Services for Mobile Telephony: A Study of Users' Privacy Concerns," *Proc. INTERACT*, vol. 3, 2003.
44. B. Rao and L. Minakakis, "Evolution of Mobile Location-Based Services," *Commun. ACM*, vol. 46, no. 12, pp. 61–65, Dec. 2003.
45. J. Hong and J. Landay, "An Architecture for Privacy-Sensitive Ubiquitous Computing," in *Proc. 2nd International Conference Mobile Systems, Applications, and Services*. ACM Press New York, NY, USA, 2004, pp. 177–189.

11

OFDM for Cognitive Radio: Merits and Challenges

Hüseyin Arslan[1], Hisham A. Mahmoud[2], and Tevfik Yücek[3]

[1] University of South Florida, Tampa, FL, USA arslan@eng.usf.edu
[2] University of South Florida, Tampa, FL, USA hmahmoud@eng.usf.edu
[3] University of South Florida, Tampa, FL, USA yucek@eng.usf.edu

11.1 Introduction

With emerging technologies and with the ever increasing number of wireless devices, the radio spectrum is becoming scarcer everyday. On the other hand, measurements show that wide ranges of the spectrum are rarely used most of the time while other bands are heavily used. However, those unused portions of the spectrum are licensed and thus cannot be utilized by users other than the license owners. Hence, there is a need for a novel technology that can benefit from these opportunities. Cognitive radio arises to be a tempting solution to spectral crowding problem by introducing the opportunistic usage of frequency bands that are not heavily occupied by licensed users [3]. It can be defined as an intelligent wireless system that is aware of its surrounding environment through sensing and measurements; a system that uses its gained experience to plan future actions and adapt to improve the overall communication quality and meet user needs. One main aspect of cognitive radio is its ability to exploit unused spectrum to provide new ways of communication. Hence, cognitive radio should have the ability to sense and be aware of its operational environment, and dynamically adjust its radio operating parameters accordingly. For cognitive radio to achieve this objective, the Physical Layer (PHY) needs to be highly flexible and adaptable. A special case of multicarrier transmission known as OFDM is one of the most widely used technologies in current wireless communications systems and it has the potential of fulfilling the aforementioned requirements of cognitive radios inherently or with minor changes. By dividing the spectrum into sub-bands that are modulated with orthogonal subcarriers, OFDM removes the need for equalizers and thus reduces the complexity of the receiver. Because of its attractive features, OFDM has been successfully used in numerous wireless technologies including Wireless Local Area Network (WLAN), Wireless Metropolitan Area Network (WMAN), and the European Digital Video Broadcasting (DVB). It is believed that OFDM will also play an important role in realizing cognitive radio concept by providing a proven, scalable,

H. Arslan (Ed.), Cognitive Radio, Software Defined Radio, and Adaptive Wireless Systems, 325–353.

and adaptive technology for air interface. In this chapter, the application of OFDM to cognitive radio is discussed. We identify the advantages of OFDM over other technologies and provide challenges associated with its application to cognitive radio.

11.2 A Basic OFDM System Model

A simplified block diagram of a basic OFDM system is given in Figure 11.1. In a multipath fading channel, due to the frequency selectivity, each subcarrier can have different attenuation. The power on some subcarriers may be significantly less than the average power because of deep fades. As a result, the overall BER may be dominated by a few subcarriers with low power levels. To reduce the degradation of system performance due to this problem, channel coding can be used prior to the modulation of bits. Channel coding can reduce the BER significantly depending on the code rate, decoder complexity, and SNR level among other factors. Interleaving is also applied to randomize the occurrence of bit errors and introduce system immunity to burst errors. Coded and interleaved data is then mapped to the constellation points to obtain data symbols. This step is represented by the modulation block of Figure 11.1. The serial data symbols are then converted to parallel data symbols which are fed to the Inverse Discrete Fourier Transform (IDFT) block to obtain the time domain OFDM symbols. Time domain samples can be written as

$$\begin{aligned} x(n) &= IDFT\{X(k)\} \\ &= \sum_{k=0}^{N-1} X(k) e^{j2\pi nk/N} \qquad 0 \leq n \leq N-1 \,, \end{aligned} \tag{11.1}$$

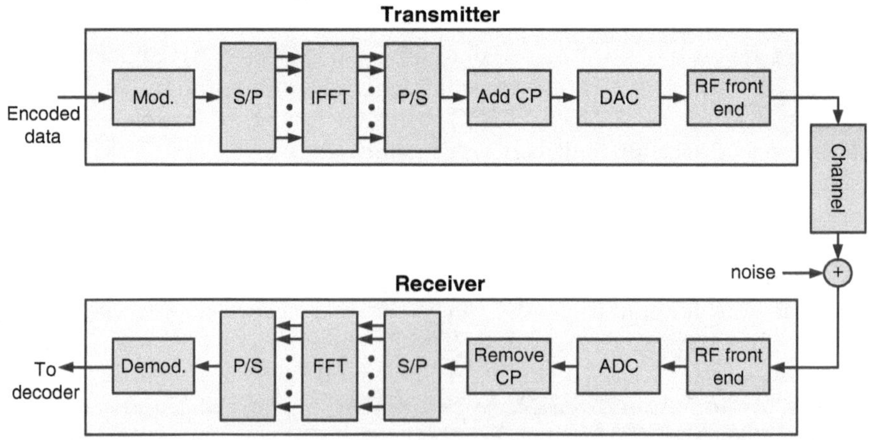

Fig. 11.1. Block diagram of a generic OFDM transceiver.

where $X(k)$ is the transmitted symbol on the kth subcarrier and N is the number of subcarriers. Time domain signal is cyclically extended to avoid residual Inter-Symbol Interference (ISI) from the previous OFDM symbols. Baseband digital signal is converted to analog signal through the Digital-to-Analog Converter (DAC) block. Then, the signal is fed to the Radio Frequency (RF) frontend. The RF frontend up-converts the signal to the RF frequencies using mixers, amplifies it using Power Amplifiers (PAs), and transmits the signal through antennas.

In the receiver side, the received signal is passed through a band-pass noise rejection filter and down-converted to baseband by the RF frontend. The Analog-to-Digital Converter (ADC) digitizes the analog signal and re-samples it. After frequency and time synchronization (which are not shown in the figure for simplicity), Cyclic Prefix (CP) is removed and the signal is transformed to frequency domain using the Discrete Fourier Transform (DFT) block. A simplified baseband model of the received symbols in frequency domain can be written as

$$Y(k) = H(k)X(k) + W(k) \ , \tag{11.2}$$

where $Y(k)$ is the received symbol on the kth subcarrier, $H(k)$ is the frequency response of the channel on the same subcarrier, and $W(k)$ is the additive noise plus interference sample which is usually assumed to be a Gaussian random variable with zero mean and variance of σ_w^2. Note that OFDM converts the convolution in time domain into multiplication in frequency domain, and hence simple one-tap frequency domain equalizers can be used to recover the transmitted symbols. After DFT, symbols are demodulated, deinterleaved, and decoded to obtain the transmitted information bits.

Figure 11.2 shows a typical OFDM waveform in frequency domain. The figure shows the orthogonal subcarriers that modulates the transmitted data. For a given bandwidth, the communication channel affects some of the design parameters of the OFDM system. The main parameters of an OFDM system

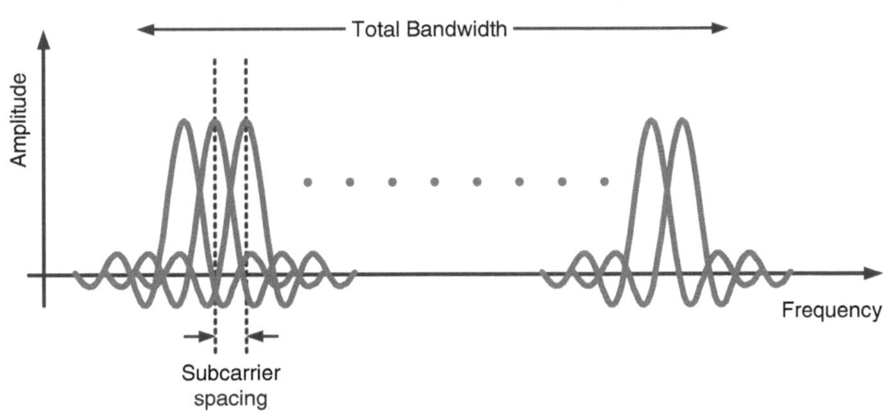

Fig. 11.2. OFDM waveform.

are the symbol time, subcarrier spacing (or consequently the number of subcarriers), and CP length.

The transmitted signal usually arrives at the intended receiver either directly (what is called Line-of-Sight (LOS) communication) or after being reflected on surfaces of buildings, cars, and other surroundings in the environment (also called as None-Line-of-Sight (NLOS)). As a result of the signal being reflected on multiple surfaces, the received signal becomes a sum of the transmitted signal with different delays and gains corresponding to the multiple paths the signal travels through. Such a channel is usually referred to as multipath channel (see Figure 11.3).

The main effects of multipath channel on the received signal are frequency-selective fading and ISI. Frequency selective fading means that the channel

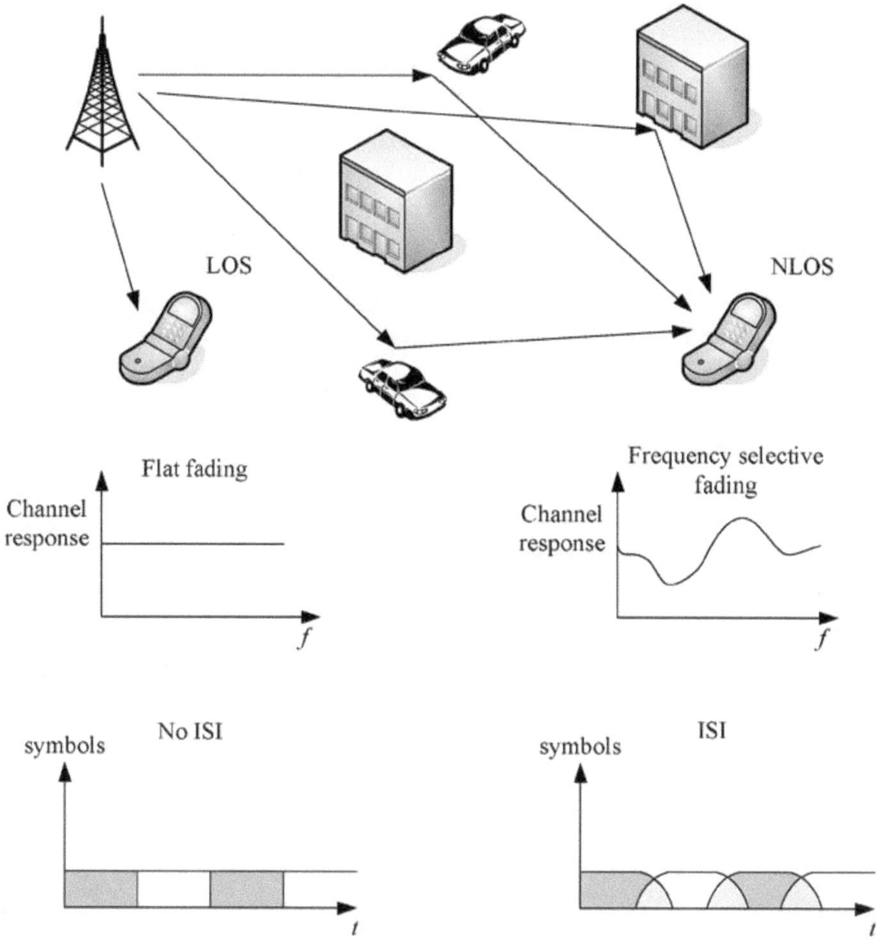

Fig. 11.3. Illustration of multipath fading.

does not affect all frequency components of the signal equally which results in distortion of the received signal. On the other hand, ISI is the interference between consequent OFDM symbols. Due to receiving multiple copies of the signal with different delays, a symbol can leak part of its energy into the following symbol. If not dealt with, both frequency selective fading and ISI can result in a significant degradation of system performance.

Channel equalizers are usually used to compensate multipath effects. Equalizers can considerably increase the system complexity as their complexity increases depending on the number of channel paths. In OFDM system, however, the need for equalizers can be avoided by careful system design. To avoid ISI, symbols duration are extended by adding a guard band to the beginning of each symbol in what is known as CP. If we define the delay spread (or multipath spread) of the channel as the delay between the first and last received paths over the channel, the CP should be longer than that delay. On the other hand, frequency selective fading is avoided by decreasing the subcarrier spacing or consequently increasing the number of subcarriers. We define the channel coherence bandwidth as the bandwidth over which the channel could be considered flat. Since ODFM signal can be considered as group of narrow band signals, by increasing the number of subcarriers, the bandwidth of each subcarrier (subcarrier spacing) becomes narrower. By choosing the subcarrier spacing to be less than the coherence bandwidth of the channel, each subcarrier is going to be affected by a flat channel and thus no channel equalization is needed.

Another channel effect that should be considered in OFDM system design is mobility. For fixed communication systems, the channel can be considered constant over time. However, if either transmitter or receiver is mobile, the channel is going to vary over time resulting in fast fading of the received signal. Coherence time of the channel is defined as the time over which the channel is considered constant. To avoid fast fading effect, OFDM symbol time is chosen to be shorter than the coherence time of the channel. In the frequency domain, mobility results in a frequency spread of the signal which dependents on the operating frequency and the relative speed between the transmitter and receiver, also know as Doppler spread [2]. Doppler spread of OFDM signals results in Inter-Carrier Interference (ICI) which can be reduced by increasing the subcarrier spacing.

In conclusion, while increasing the symbol time reduces ISI effect, shorter symbol time is desirable to avoid fast fading of the signal. And while decreasing subcarrier spacing reduces ICI, narrower subcarrier spacing helps avoiding frequency selectivity. As a matter of fact, there exists an optimum value of these parameters that should be used to improve the system performance [3].

In this section, a single user system model is represented, where the available channel is used by a single user. Note that OFDM by itself is not a multiaccess technique. However, it can be combined with existing multiple accessing methods to allow multiple users to access the available channel. Some of the most common multiple access techniques that can be employed by OFDM systems are TDMA, Carrier Sense Multiple Accessing (CSMA) [4], Frequency Division Multiple Accessing (FDMA), and CDMA based schemes [5]. In addition, a mix of TDMA and FDMA known as Orthogonal Frequency Division Multiple Access (OFDMA) [6] is also possible. Note that in the above system model, the interference from other users and other technologies (like co-channel interference, adjacent channel interference, narrow band interference, etc.) are all folded into the noise term for the sake of simplicity. However, in reality, when the received signal is impaired by an interferer, a more accurate model needs to be used, where the noise term will be colored effecting each carrier differently.

11.3 OFDM-Based Cognitive Radio

Application of OFDM to cognitive radio brings about new aspects and challenges to system design. The cognitive OFDM conceptual model considered in this chapter is shown in Figure 11.4[1]. The cognitive engine is responsible for making the intelligent decisions and configuring the radio and PHY parameters. The spectral opportunities are identified by the decision unit based on the information from policy engine as well as local and network spectrum sensing data.

The policy engine provides information to the cognitive engine concerning the current policies to be considered depending on the system location. This will ensure that the cognitive radio will not use illegal waveforms or breach any policies. On the other hand, the local spectrum sensing unit process the spectrum information and identify licensed users accessing the spectrum, their signal specifications such as the their bandwidth and power level, and detect spectrum opportunities that can be exploited by cognitive radio.

Once the required information is available, the decision unit can make a conclusion on the best course of action for the system. The decision includes choosing the appropriate channel coding, modulation, operation frequencies, and bandwidth. At this stage, OFDM technology gets the upper hand over other similar transmission technologies with its adaptive features and great flexibility. By only changing the configuration parameters of OFDM (see Table 11.1 for some example parameters) and radio, the cognitive system can communicate with various radio access technologies in the environment, or it can optimize the transmission depending on the environmental characteristics.

The radio circuit is divided into a digital part (digital IF, ADC, and DAC) and an analog part (software tunable analog radio). Both parts are reconfigurable by the cognitive engine to increase the flexibility of the system. This includes controlling the operating frequency, bandwidth, filters, and mixers. Even antenna parameters (e.g. number of antennas, beam forming) can be configured to improve the system performance.

[1] Some OFDM functions are skipped or simplified in order to keep the figure simple.

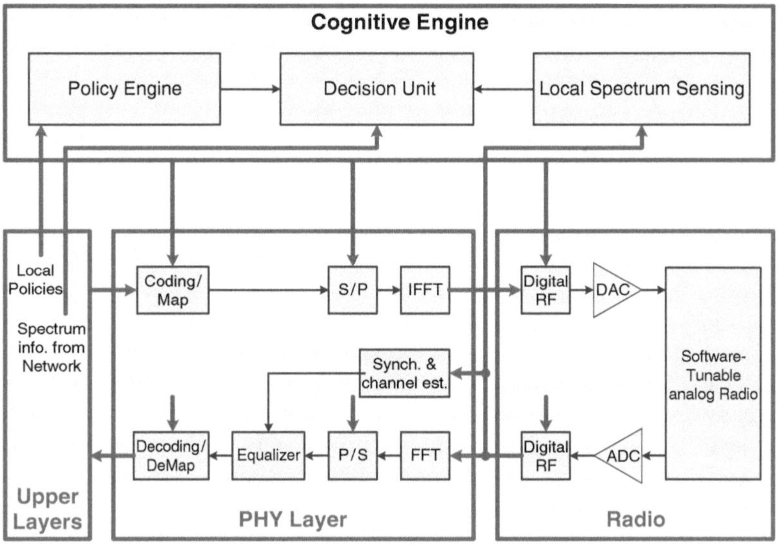

Fig. 11.4. OFDM-based Cognitive radio system block diagram. All of the layers can interact with the Cognitive engine. OFDM parameters and radio are configured by the Cognitive engine.

Table 11.1. OFDM-based wireless standards.

Standard	**IEEE 802.11(a/g)**	**IEEE 802.16(d/e)**	**IEEE 802.22**	**DVB-T**
FFT size	64	128, 256, 512, 1024, 2048	1024, 2048, 4096	2048, 8192
CP size	1/4	1/4, 1/8, 1/16, 1/32	Variable	1/4, 1/8, 1/16, 1/32
Bit per symbol	1, 2, 4, 6	1, 2, 4, 6	2, 4, 6	2, 4, 6
Pilots	4	Variable	96, 192, 384	62, 245
Bandwidth (MHz)	20	1.75 to 20	6, 7, 8	8
Multiple accessing	CSMA	OFDMA /TDMA	OFDMA /TDMA	N/A

11.4 Why OFDM is a Good Fit for Cognitive Radio

The underlying sensing and spectrum shaping capabilities together with flexibility and adaptiveness make OFDM probably the best candidate for cognitive radio systems. In the following, we present some of the requirements for cognitive radios and explain how OFDM can fulfill these requirements. A summary of these requirements and OFDM's strength in addressing them are presented in Table 11.2.

11.4.1 Spectrum Sensing and Awareness

Cognitive radio should be able to scan the spectrum and measure different channel characteristics such as power availability, interference, and noise temperature [7]. In addition, the system should be able to identify different users' signals in the spectrum and also identify if they are either licensed or rental users. These abilities allow cognitive radio system to identify unused parts of the spectrum and spectral opportunities.

However, since for a rental system it is important not to interfere with other licensed systems using the spectrum, other measures should be taken to guarantee an interference-free communication between rental users. One approach is to share the spectrum sensing information between multiple cognitive radio devices to decrease or even eliminate the probability of interference with licensed users [8]. On the other hand, more sophisticated algorithms can be used for sensing the spectrum.

Table 11.2. OFDM cognitive radio

Cognitive radio requirements	**OFDM's strength**
Spectrum sensing	Inherent FFT operation of OFDM eases spectrum sensing in frequency domain.
Efficient spectrum utilization	Waveform can be easily shaped by simply turning off some subcarriers, where primary users exist.
Adaptation/Scalability	OFDM systems can be adapted to different transmission environments and available resources. Some parameters include: FFT size, subcarrier spacing, CP size, modulation, coding, subcarrier powers.
Advanced antenna techniques	Multiple-Input Multiple-Output (MIMO) techniques are commonly used with OFDM mainly because of the reduced equalizer complexity. OFDM also supports smart antennas.
Interoperability	With WLAN (IEEE 802.11), WMAN (IEEE 802.16). WRAN (IEEE 802.22), WPAN (IEEE 802.15.3a) all using OFDM as their PHY techniques, interoperability becomes easier compared to other technologies.
Multiple accessing and spectral allocation	Support for multiuser access is already inherited in the system design by assigning groups of subcarriers to different users (OFDMA).
NBI immunity	NBI affect only some subcarriers in OFDM systems. These subcarriers can be simply turned off.

While the efficiency of the spectrum sensing and analyzing process is important for a successful implementation of cognitive radio, the processing time can be even more important. The periodicity of spectrum sensing should be short enough to allow for detection of new spectrum opportunities and, at the same time, to detect licensed users accessing the previously-identified-as-unused parts of the spectrum. On the other hand, if spectrum sensing is done so frequently, the overhead of sharing such information will increase reducing the spectrum efficiency of the whole system not to mention the increase in system complexity. In OFDM systems, conversion from time domain to frequency domain is achieved inherently by using DFT. Hence, all the points in the time–frequency grid can be scanned without any extra hardware and computation because of the hardware reuse of Fast Fourier Transform (FFT) cores. Using the time–frequency grid, the selection of bins that are available for exploitation (spectrum holes) can be carried out using simple hypothesis testing [9]. The DFT outputs can be filtered across time and frequency dimensions to reduce the uncertainty in detection as well [10]. Note that the resolution of the frequency grid is dependent on subcarrier spacing.

11.4.2 Spectrum Shaping

After a cognitive radio system scans the spectrum and identifies active licensed users and available opportunities, comes the next step: spectrum shaping. Theoretically, it is desired to allow the cognitive users to freely use available unused portions of the spectrum.

Cognitive users should be able to flexibly shape the transmitted signal spectrum. It is desired to have control over waveform parameters such as the signal bandwidth, power level, center frequency, and most of all a flexible spectrum mask. OFDM systems can provide such flexibility due to the unique nature of OFDM signaling. By disabling a set of subcarriers, the spectrum of OFDM signals can be adaptively shaped to fit into the required spectrum mask. Assuming the spectrum mask is already known to the cognitive radio system, choosing the disabled subcarriers is a relatively simple process [11].

The main parameters of an OFDM system that can be used to shape the signal spectrum are number of subcarriers, subcarrier's power, and pulse-shaping filters. Increasing the number of subcarriers for a fixed bandwidth allows the OFDM system to have a higher resolution in the frequency domain. However, this results in increasing the complexity of the FFT operations and thus increasing the overall system complexity.

Subcarrier power can be used to shape the signal into the desired mask. One reason to assign subcarriers different powers is to better fit into the channel response [12]. For example, subcarriers with higher SNR values can be assigned lower power than those with lower SNR to improve the overall system BER. Another reason is to reduce the adjacent channel interference from an OFDM system by reducing the power assigned to edge subcarriers.

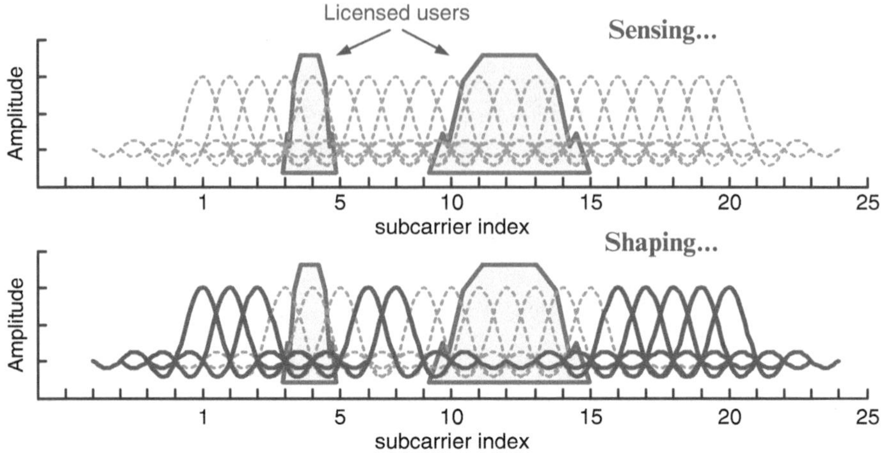

Fig. 11.5. Spectrum sensing and shaping using OFDM.

An example of spectrum sensing and shaping procedures in OFDM-based cognitive radio systems is illustrated in Figure 11.5. Two licensed users are detected using the output of FFT block, and subcarriers that can cause interference to licensed user are turned off. The transmitter then uses the free parts of the spectrum for signal transmission. In addition, pulse-shaping filters can also be used to reduce the interference to adjacent bands. More discussions on reducing interference is introduced in Section 11.5.5.

11.4.3 Adapting to Environment

Adaptivity is one of the key requirements for a cognitive radio [13]. By combining the measured information (awareness) with the knowledge of current system abilities and limitations, cognitive radio can adapt waveforms to interoperate with other friendly communication devices, choose the most appropriate communications channel or network for transmission, allocate appropriate frequency to transmit in an open area of spectrum, adapt the waveform to compensate for channel fading, and null an interfering signal [14].

OFDM offers a great flexibility in this regard as the number of parameters for adaptation is quite large [15]. The transmission parameters that can be changed based on the spectrum awareness include bandwidth, FFT size, filters, windows, modulation, transmit power, and active subcarriers used for transmission. Moreover, the parameters that can be adapted depending on the characteristics of the environment in order to optimize the transmission include cyclic prefix size, coding rate/type, modulation type, interleaving method, pilot patterns, preambles/midambles, duplexing method.

The adaptivity in OFDM systems can be divided into two groups [16]: algorithm-selection level adaptivity and algorithm-parameter level adaptivity.

In classical wireless systems, usually parameters of the algorithms, e.g. coding rate, have been adapted in order to optimize the transmission. However, in cognitive OFDM systems, algorithm type, e.g. channel coding method, may also be adapted in order to achieve interoperability with other systems and/or to further optimize the transmission. To achieve such adaptivity, a fully configurable hardware platform would be needed.

11.4.4 Advanced Antenna Techniques

Advanced antenna techniques are not necessarily required for cognitive radios. However, they are desirable as they will provide better spectral efficiency which is the primary motivation for cognitive radio. Smart antennas and MIMO systems can be used to exploit the spatial dimension of spectrum space (e.g. through beam forming) to improve the efficiency. In essence multi-antenna systems can help to find spectral opportunities in the spatial domain and can help to exploit these opportunities in full. The use of MIMO techniques offers several important advantages including spatial degree of freedom, increased spectral efficiency and diversity [17]. These advantages can be used to increase the spectrum utilization of the overall system. Furthermore, beamforming, diversity combining, and space-time equalization can also be applied to cognitive OFDM systems. Another application of adaptive antenna techniques is the reduction of the interference in OFDM systems [18].

MIMO systems commonly employ OFDM as their transmission technique because of the simple diversity combination and equalization, particularly at high data rates. In MIMO–OFDM, the channel response becomes a matrix. Since each tone can be equalized independently, the complexity of space–time equalizers is avoided and signals can be processed using relatively straightforward matrix algebra. Moreover, the advantages of OFDM in multipath are preserved in MIMO–OFDM system as frequency selectivity caused by multipath increases the capacity.

11.4.5 Multiple Accessing and Spectral Allocation

The resources available to a cognitive system often have to be shared among users. Several techniques can be used to achieve such tasks. OFDM supports the well-known multiple accessing techniques such as TDMA, FDMA, and CSMA. Moreover, CDMA can also be used together with OFDM for multiplexing different users, in which case the transmission is known as Multicarrier Code Division Multiple Access (MC–CDMA) or Multicarrier Direct Spread Code Division Multiple Access (DS–CDMA) [5].

OFDMA, a special case of FDMA, has gained tremendous attention recently with its usage in fixed and mobile Worldwide Interoperability for Microwave Access (WiMAX) [19]. In OFDMA, subcarriers are grouped into sets each of which is assigned to a different user. Interleaved, randomized, or clustered assignment schemes can be used for this purpose. Hence, it offers very flexible multiple accessing and spectral allocation capability for cognitive radios without any extra complexity or hardware. The allocation of subcarriers can be tailored according to the spectrum availability. The flexibility and support of OFDM systems for various multiple accessing enables the interoperability and increases the adoption of cognitive radio as well.

11.4.6 Interoperability

Another desirable feature of cognitive radio is interoperability. Interoperability can be defined as *the ability of two or more systems or components to exchange information and to use the information that has been exchanged* [20]. Since cognitive radio systems have to deal with licensed users as well as other cognitive users, the ability to detect and encode existing users' signals can expedite the adoption and improve the performance of cognitive radio systems. Furthermore, some recent unfortunate events manifested the importance of interoperability in terms of wireless communications for the first responders [21,22]. For interoperability problems, cognitive radio can improve the disaster relief operations by developing the coordination between first responders [23, 24]. For such tasks, OFDM is one of the best candidates as OFDM signaling has been successfully used in various technologies. Systems based on OFDM include 802.11a and 802.11g Wireless LAN standards, Digital Audio Broadcasting (DAB), DVB, and WiMAX. Figure 11.6 shows some of the OFDM-based wireless technologies according to communication range. As shown in this figure, OFDM has been used in both short range and long range communication systems. Hence, a cognitive radio system employing OFDM can communicate with systems using other OFDM-based technologies with much ease. Only the knowledge of the signal parameters used by the desired users will be needed (see Table 11.1). However, for such task to be successful, the cognitive radio system should be built around a software-defined radio architecture. In addition, the cognitive radio system should have all the standard-related information required to decode other signals, such as the data and pilot mapping to the frequency subcarriers, frame structure and the coding type and rate.

11.5 Challenges to Cognitive OFDM Systems

As an intelligent system with features such as awareness, adaptability and learning, cognitive radio represents the future of wireless systems with the promise of offering solutions to various communication problems as outlined in the previous section. However, with this new technology, new challenges appear, raising interesting research topics. In the considered OFDM-based cognitive radio systems, the challenges can be grouped into three categories as illustrated in Figure 11.7. The first category includes the challenges that are unique to classical OFDM systems including Peak-to-Average Power Ratio (PAPR),

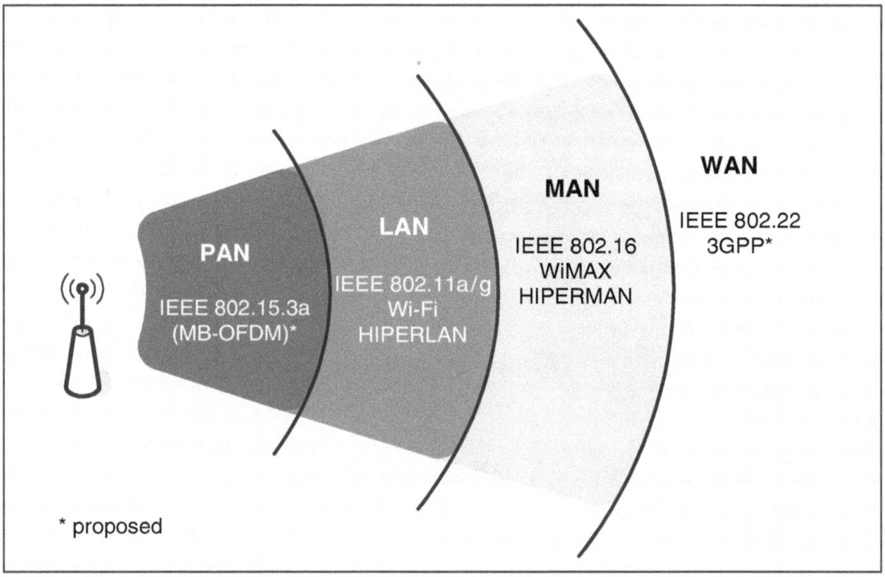

Fig. 11.6. OFDM-based wireless technologies.

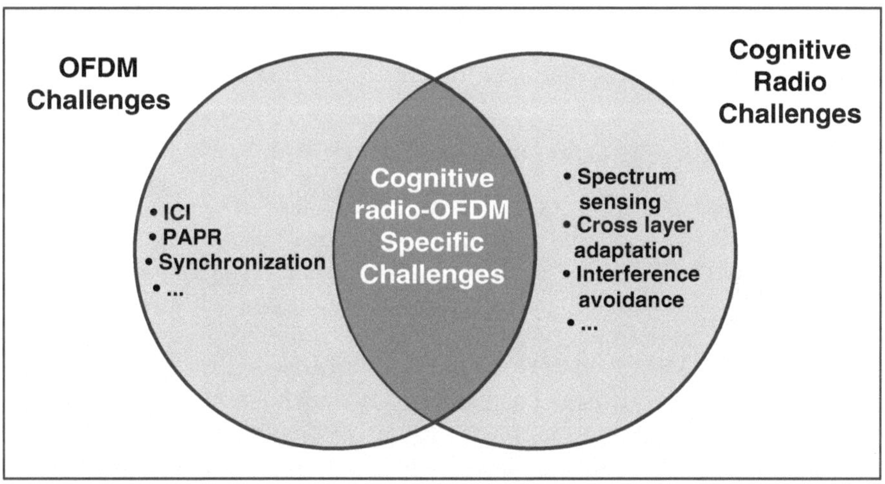

Fig. 11.7. Research challenges in cognitive radio and OFDM.

sensitivity to frequency offset and phase noise, synchronization, etc. The second category includes problems faced by all cognitive radios such as spectrum sensing, cross layer adaptation, and interference avoidance. Our main focus in this chapter is on the third category: challenges that arise when OFDM technique is employed by cognitive radio systems. In the following, some challenges and approaches for solving them are given.

11.5.1 Spectrum Shaping

One of the main challenges in OFDM cognitive radio systems is spectrum shaping. In OFDM-based systems, spectrum shaping means determining the subcarriers to be used by the OFDM system while keeping the interference to and from primary users at a negligible level. Once spectrum sensing information is acquired, this knowledge should be utilized to select the subcarriers to be used by the secondary/cognitive users. This problem is addressed in [25] by using energy detectors over each subcarrier. Moreover, a detection criterion is used to determine used subcarriers. Spectrum sensing is directly related to the sensing problem for spectrum hole identification. However, cognitive radio might prefer to skip some opportunities depending on the power and network traffic requirements.

11.5.2 Effective Pruning Algorithm Design

Once the subcarriers to be used are determined, there might be many subcarriers that are deactivated. In such a case, the efficiency of FFT algorithms can be increased and/or execution time can be decreased by removing operations on input values which are zero, a process known as pruning. Designing effective pruning algorithms is important for cognitive OFDM systems for achieving higher performance. Specific implementation of pruning technique for CR-OFDM systems is discussed in [26].

11.5.3 Signaling the Transmission Parameters

OFDM system can adjust its waveform by turning off some subcarriers in order to exploit the available spectrum holes (see Figure 11.5). The receiver, however, should be informed about subcarriers that are deactivated and that are to be used. Signaling of this information should be performed carefully in order to prevent interference to primary users while keeping the bandwidth loss at minimum. Detection of those unused subcarriers can also be performed blindly. However, to the best knowledge of the authors, no work in this area has been done yet. One method to reduce the overhead due to signaling is proposed in [27]. The activation/deactivation of subcarriers is performed over a block of subcarriers instead of each individual subcarrier. Hence, the signaling overhead can be reduced by a factor of each block's size. Moreover, depending on the channel quality and available resources, parameters like FFT size, CP size, etc. can be changed and this information should also be conveyed to the receiver.

11.5.4 Synchronization

Synchronization is another important issue that needs to be addressed in OFDM system design. With the introduction of cognitive radio, new aspects are introduced to the problem. The NBI, which can interfere with the preamble, is one of the problems [28]. Furthermore, the incomplete subcarrier set might be an issue for preambles, and pilots might fall into unused subcarriers if used. Moreover, if multiple user accessing is employed, the subcarriers can be assigned to different users. To keep the orthogonality between subcarriers and avoid interference, all users should be synchronized to the receiver. In [28], it is shown that longer preambles are needed in CR-OFDM systems as compared to conventional systems. Moreover, new preamble structures are introduced and their performance for time and frequency synchronization is investigated.

11.5.5 Mutual Interference

Mutual interference should be carefully considered when designing cognitive radio systems. The side lobes of modulated OFDM subcarriers are known to be large as shown in Figure 11.8. As a result, there will be power leakage from used subcarriers to nulled subcarriers which causes interference to the licensed users. Various methods are proposed in the literature to reduce this leakage and to enable co-existence of cognitive-OFDM systems with primary license owner systems. One method is to make the *sinc* function (see Figure 11.8) decay faster by windowing the time domain OFDM samples [29]. Similar techniques have already been investigated to reduce ICI and out-of-band radiation in OFDM systems [30, 31]. In [29], a raised-cosine window is applied. By changing the roll-off factor of the raised-cosine window, interference reduction of upto 6 dB has been achieved. Figures 11.9 and 11.10 show

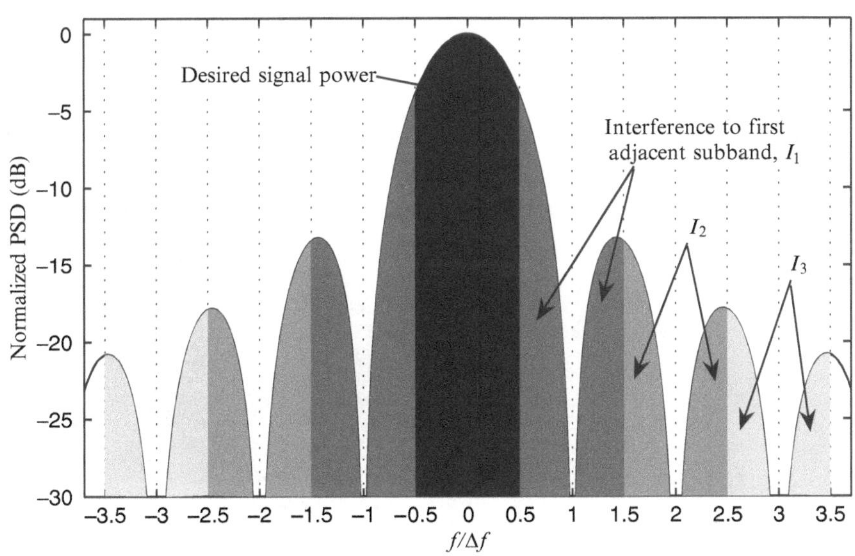

Fig. 11.8. Power spectrum density of a single OFDM subcarrier.

the raised-cosine window shape for different roll-off values and corresponding power spectral densities. The drawback of this method is the reduction of system throughput due to the temporal extension of time domain signal to maintain orthogonality. Another method for reducing the interference is to adaptively deactivate the subcarriers that are adjacent to the subcarriers occupied by licensed users [29]. This way the interference can be greatly reduced as most of the interference comes from the neighboring subcarriers. However, the obvious disadvantage of this method is the reduction of spectral efficiency. Instead of deactivating the neighboring subcarriers, their values

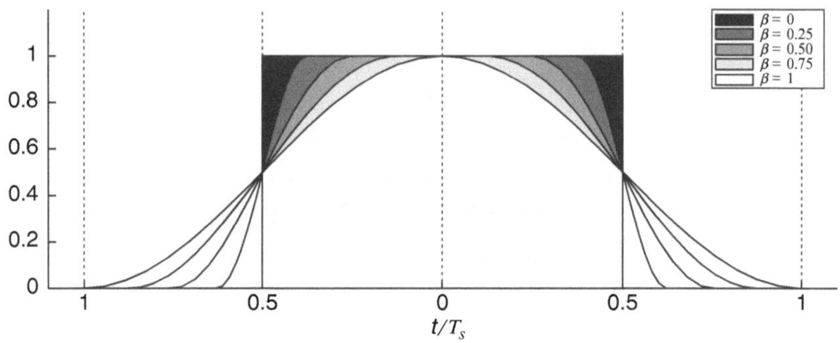

Fig. 11.9. Raised cosine windowing with different roll-off (β) values.

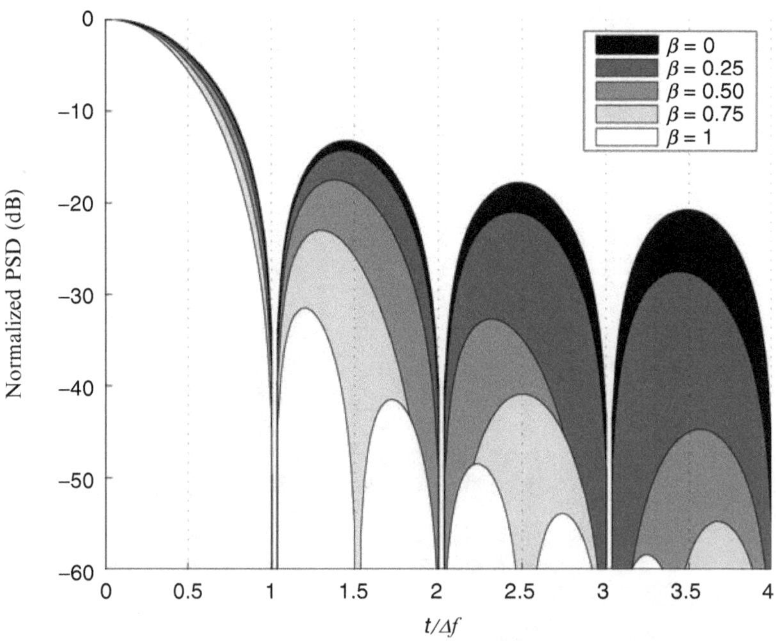

Fig. 11.10. Roll-off effect on the PSD of a single OFDM subcarrier.

can be determined actively in order to cancel the interference in the deactivated bands. This technique is proposed in [32, 33] and referred to as active interference cancellation and cancellation carriers, respectively. It is shown that the performance can be improved, however, determination of the values for cancellation subcarriers is complex as it requires optimization. One last method for reducing the interference to and from the narrowband primary users is subcarrier weighting [34, 35]. In this method, the subcarrier weights are determined in such a way that the sidelobes of the transmission signal are minimized according to an optimization algorithm which allows several optimization constraints. This way, more than 10 dB reduction in the sidelobes of OFDM signal can be achieved. Note that subcarrier weighting requires constant envelope modulation such as BPSK or QPSK. Moreover, the receiver does not need to know the weighting sequence as the phase information is not changed.

In addition to the aforementioned challenges, there are other issues for practical implementation of OFDM cognitive radio systems. While cognitive radio is such a promising technology, more research is needed to build a practical system with affordable complexity.

11.6 Multi-band OFDM

In previous sections, only Single-Band OFDM (SB-OFDM) systems were considered. In this section, using OFDM signaling over multiple bands is discussed along with the advantages and motivations of using Multi-band OFDM (MB-OFDM). The proposition of MB-OFDM for UWB is addressed. However, the discussion of MB-OFDM is not limited to one application but rather it is presented from a broader point of view.

For systems utilizing wide bands of the spectrum, multi-band signaling approach –where the total bandwidth is divided into smaller bands– can prove to be more advantageous over using single band signaling. While using a single band simplifies the system design, processing a wide band signal requires building highly complex RF circuitry for signal transmission/reception. In addition, high speed ADCs are required to sample and digitize the signal. Moreover, higher complexity channel equalizers are also needed to capture sufficient multipath signal energy for further processing. On the other hand, multi-band signaling relaxes the requirements on system hardware as smaller portions of the spectrum are processed at a time. Furthermore, dividing the spectrum into smaller bands allows for better spectrum allocation. The system can drop some of the available bands to achieve other goals (e.g. avoid interference, save power, allow for multiuser accessing).

For OFDM-based cognitive radio, the question becomes when to use multi-band and when to use single band. Given a certain scanned spectrum shape, choosing the number of bands depends on various parameters. Required throughput, hardware limitations, computational complexity, number of spectrum holes and their bandwidths, and interference level are examples of what could affect cognitive system's choice. We further illustrate the importance of multi-band signaling with the next example.

Consider the scenario, where a cognitive radio senses the spectrum and finds the results shown in Figure 11.11. Two spectrum holes are detected with 10 and 15 MHz bandwidth. One of the spectrum holes contains narrow band interference. The detected vacancies in the spectrum are 1 GHz apart. In such scenario, if the system chooses to use SB-OFDM, then the bandwidth of the OFDM signal is going to be 1.025 GHz. A signal with such bandwidth requires ADCs with very high sampling rate capability. In addition, a large number of subcarriers is required to guarantee that subcarriers can better fit into the spectrum holes as well as to keep the subcarrier's channel relatively flat. Unfortunately, large number of subcarriers results in a more complex FFT operation. Moreover, in order to avoid the narrow band interference in the 15 MHz band, the system drops two out of three subcarriers that are used to fill the spectrum hole, resulting in a low spectrum efficiency. By using MB-OFDM, the spectrum holes can be filled with two OFDM signals with 15 and 10 MHz bandwidths. Hence, ADCs with practical sampling rates can now be used. A large number of subcarriers is not necessary for either OFDM signals in that case. In addition, the system has more control over the signal spectrum in each band due to the small subcarrier spacing. Hence, avoiding narrow band interference is done with lower spectrum loss. However, the system is

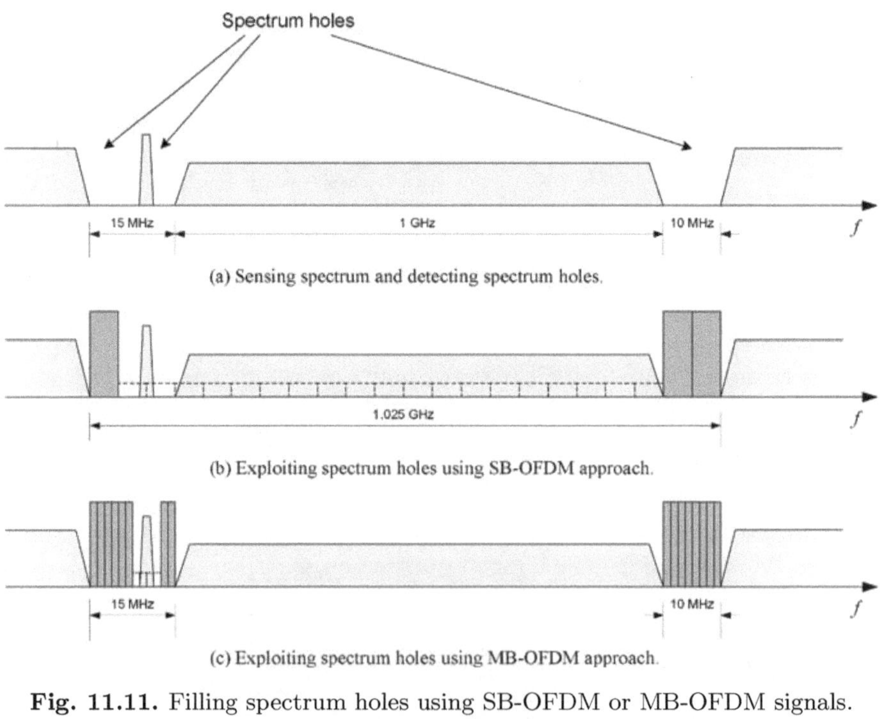

Fig. 11.11. Filling spectrum holes using SB-OFDM or MB-OFDM signals.

now processing two OFDM signals. While sampling frequency is reduced in MB-OFDM case, the system is performing receiver algorithms (e.g. synchronization, channel estimation, and equalization) separately for each band.

Another example is shown in Figure 11.12. In this example a SB-OFDM could be a better choice for the system rather than using five bands in a MB-OFDM scheme. The improvement of spectrum efficiency introduced by using MB-OFDM may not be as significant as the increase in system complexity.

It is worth mentioning that MB-OFDM is also employed in some UWB systems. Instead of using a single band UWB signal, the spectrum is divided into sub-bands (with approximately 500 MHz bandwidth each) and OFDM signals are used to transmit data over each band [36, 37]. Figure 11.13 shows an illustration of the UWB signal in frequency domain. However, while UWB is one of the applications of MB-OFDM, it is only limited to a specific scenario where all sub-bands have almost equal size and OFDM signals used in sub-bands are identical in many other parameters such as the CP and the subcarrier spacing.

In summary, to choose between SB-OFDM and MB-OFDM, the system needs to identify some system and environment parameters. Note that MB-OFDM is only effective if spectrum holes are far apart or if the covered bandwidth is relatively wide. MB-OFDM can lessen the strain on the required ADC sampling rate, the memory size, and speed. On the other hand, the complexity of the baseband signal processing increases as the system needs to carry out receiver algorithms over multiple OFDM signals. In this section, an overview with some discussions is provided on the use of multi-band signaling

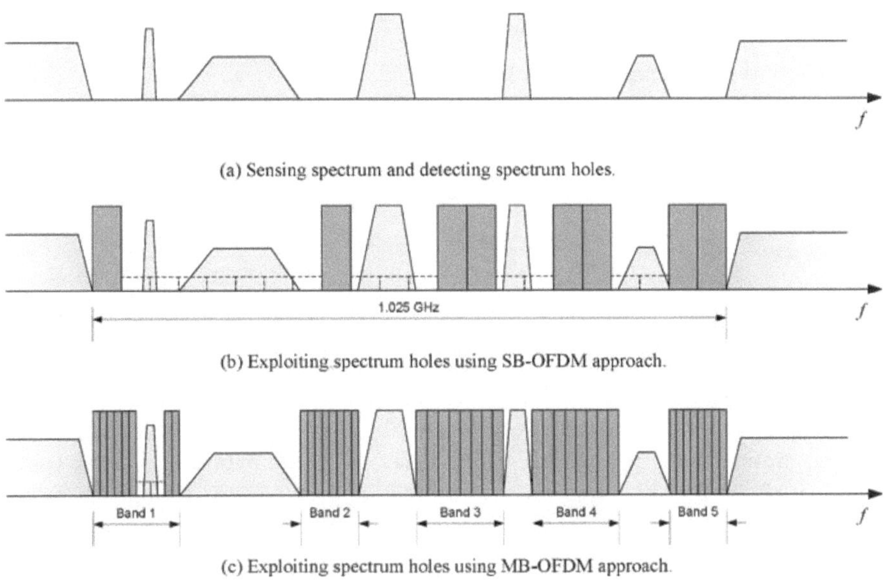

Fig. 11.12. Filling spectrum holes using SB-OFDM or MB-OFDM signals.

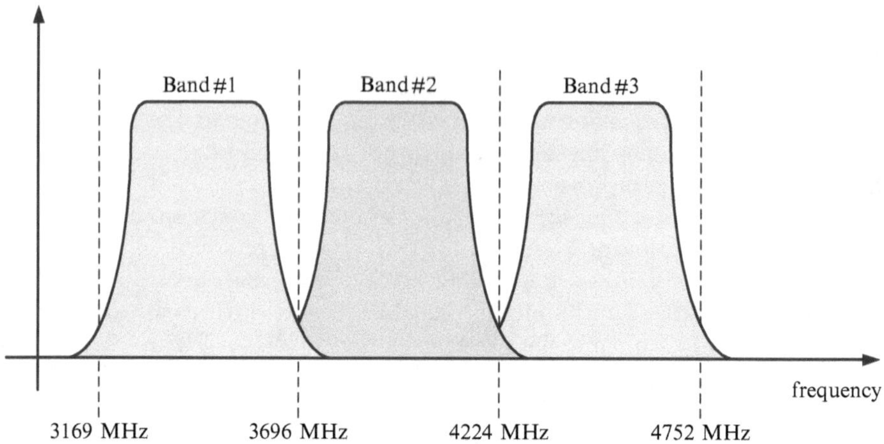

Fig. 11.13. Sub-bands of MB-OFDM-based UWB systems in frequency domain [36].

vs. single band. However, in practicality, more parameters are considered when designing the system and algorithms that are capable of identifying the best signaling choice are needed. Hence, more research is required on this topic.

11.7 A Step Toward Cognitive-OFDM: Standards and Technologies

As cognitive radio concept is attracting more interest everyday, recently developed standards are considering more cognitive features. Dynamic frequency allocation, Transmit Power Control (TPC), and spectrum sensing are just a few examples of features that are included in some of the current standards. These can be considered as a step toward the future implementation of a cognitive radio. It is worth to note that most of these standards consider OFDM as their choice for transmission technology (see Figure 11.6).

11.7.1 WiMAX – IEEE 802.16

One of the technologies that is getting a fair amount of interest lately – in both academia and industry – is WiMAX. The first WiMAX standard, IEEE 802.16a [38], operates in the 10–66 GHz range. In this frequency range, only LOS communication is possible. The standard later evolved in 2004 to the IEEE 802.16-2004 [39]; also know as IEEE 802.16d. The IEEE 802.16-2004 standard supported the operation in the 2–11 GHz range allowing for a NLOS communications. It provides point to multipoint access for fixed subscribers. The IEEE 802.16e-2005 [19] standard updates and extends this standard to allow for mobile subscriber stations traveling at vehicular speeds. A scalable version of OFDMA is introduced improving the overall system performance. While IEEE's role is to develop standards for the PHY and MAC layers, WiMAX forum ensures compatibility and interoperability between vendors' equipments through its certification process.

OFDMA PHY mode is probably the most interesting mode supported by WiMAX. Using OFDM signaling, OFDMA PHY mode enables a WiMAX Base Station (BS) to support multiple fixed or mobile users at the same time. In this mode, a BS system utilizes the available channel by dividing the available subcarriers into subchannels. A number of subchannel grouped with a number of OFDMA symbols constitutes a slot. A slot is defined as the minimum data allocation unit [39]. The slot definition shows that the system resources are being shared between users in two domains. The first dimension is frequency which is represented by the number of subchannels in each slot. The second dimension is time which is represented by number of OFDMA symbols. Figure 11.14 shows the OFDMA signal structure used in WiMAX. Note that this figure is only for illustration purposes and thus the number of subcarriers or the slot size does not reflect the actual numbers used by the standard.

In WiMAX-based systems, users can be assigned different bandwidths, time durations, transmit power levels, and modulation orders (see Table 11.1) based on various parameters such as user Carrier-to-Interference Noise Ratio (CINR), Received Signal Strength Indicator (RSSI), or the available bandwidth. Moreover, OFDMA PHY offers multiple FFT sizes, CP sizes, and pilot allocation schemes. The FFT size can be selected as 128, 256, 512, 1024, or 2048 depending on the transmission bandwidth.[2] Similarly, the CP length can be set to 1/4, 1/8, 1/16, and 1/32 time the OFDM symbol length. The CP size can be changed depending on the various environmental characteristics. With all these adaptive features, WiMAX has the ability to adapt to various channel conditions and communication scenarios. Indeed, a WiMAX BS measures the available channel and received signal parameters, makes a plan on what the most appropriate settings for communication with current subscribers (with certain goals in mind such as maximum throughput, Quality of Service (QoS).) is, and executes this plan.

WiMAX standard is very rich in terms of advanced antenna techniques as well. Table 11.3 shows the MIMO features available in the mobile-WiMAX standard, IEEE 802.16E-2005 [19]. Although these antenna techniques are not required, they are well suited to cognitive radio and useful for achieving high data rates.

The amendment to IEEE 802.16 standard IEEE 802.16h, which is currently being developed, introduces cognitive features to WiMAX. The goal is to achieve coexistence of WiMAX devices in unlicensed bands. Furthermore, methods for coexistence with primary users are also developed.

---

[2] This is known as scalable OFDMA. Various FFT sizes are used to keep the subcarrier spacing constant for different transmission bandwidths.

Table 11.3. Advanced antenna features of WiMAX.

Techniques	Details	Advantages
Adaptive Antenna Systems (AAS) beamforming	The BS uses multiple antennas to form the beams in the direction of a subcarrier.	Extended range and increased capacity thanks to lower interference
Space-Time Coding (STC)	Transmit diversity such as Alamouti code is used.	Increase in system gain through spatial diversity and reduced fade margin
Spatial Multiplexing (SM)	Independent and separately encoded data signals are transmitted over multiple antennas.	Increase in capacity (higher data rates)
Collaborative SM	Two uplink users can transmit collaboratively in the same slot as if two streams are spatially multiplexed from two antennas of the same user.	Increased coverage and throughput
Antenna selection	Any combination of antennas are selected (on–off type of selection of group of antennas from the available antennas) based on the channel feedback.	Efficient use of available power
Antenna grouping	The BS can group multiple antennas for different carriers in different ways based on the feedback from BSs. For example, if we have 3 Tx antennas, the BS can group the first two antennas in some carriers, and the last 2 antennas in some other carriers.	Maximum diversity / capacity gain
MIMO precoding	The antenna elements are weighted with a matrix before mapping them to transmit antennas based on the feedback from subscriber stations. This scheme is similar to a water-pouring algorithm.	Increased capacity gain
STC sub-packet combining	In the initial transmission, the packets are transmitted in a full MIMO spatial multiplexing mode (no diversity) If the data cannot be decoded correctly (CRC did not check), then the packets are sent in full STC mode (full transmit diversity mode). The receiver combines the initial data and the later data for better detection.	Provides incremental redundancy
Frequency Hopping Diversity Coding (FHDC)	This scheme (as for STC) transmits two complex symbols using the multiple input single output channel.	
Adaptive MIMO Switch (AMC)	STC or SM is selected adaptively to adopt channel conditions.	Optimum spectral efficiency is achieved

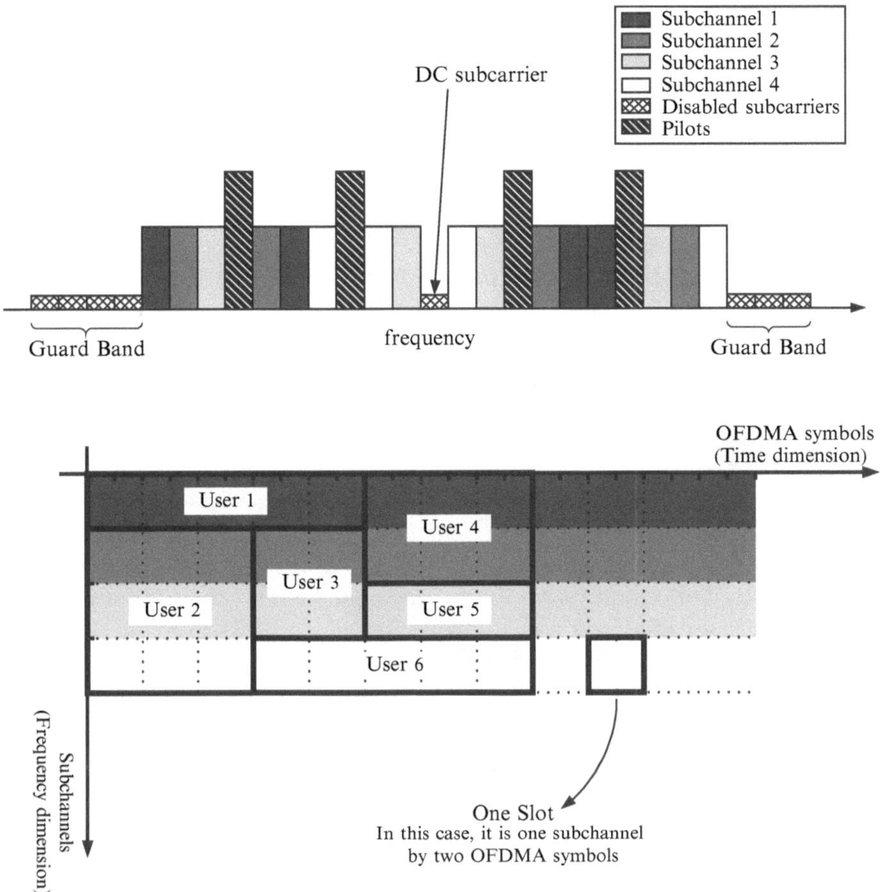

Fig. 11.14. Illustration of OFDMA signal structure used in WiMAX.

11.7.2 IEEE 802.22

Another standard that is being developed recently is the IEEE 802.22 standard [40]. IEEE 802.22 is designed to operate within the TV bands. However, IEEE 802.22-based systems should avoid interfering with incumbent signals. Systems should be able to sense the channel, detect incumbent signals within spectrum, and adjust their operating frequency and transmit power accordingly [41]. With features such as channel sensing, licensed users detection, Dynamic Frequency Selection (DFS), and TPC, IEEE 802.22 can be considered as a cognitive radio standard. Even though this standard is not finalized yet, it is anticipated that it will be based on OFDM transmission as well.

The main difference between WiMAX and the IEEE 802.22 standard is that the later is mainly targeting rural and remote areas, since it is reusing TV bands. However, this gives 802.22 the upper hand in terms of range.

A coverage area of up to 100 km is possible making it the first Wireless Rural Area Network (WRAN) standard. Current specifications can establish a coverage area of up to 33 km.

The 802.22 standard is designed for a fixed point-to-multipoint communication topology, where the BS acts as the master mandating all the operation parameters of users with its cell. And while the users (slaves) can share sensing information with the BS through distributed sensing, it is up to the BS to change a user transmit power, modulation, coding or operating frequency. In such topology, it is the responsibility of the service provider to ensure that users' signals is causing no interference to the incumbent signals by within the coverage area. This is a crucial issue for the coexistence of the standard with the already existing TV services.

Another challenge in designing the 802.22 standard is the initialization of new users who desire to communicate with the BS. Unlike current wireless technologies, the frequency and time duration of the initialization channel is not predefined. In other words, initial users will have to scan parts (if not all) of the TV bands to find the current BS operating frequency and time. In addition, users should be able to differentiate between incumbent signals and the BS signal. This could prove to be very challenging especially if the BS is operating over a combination of multiple frequency bands.

A discussion of the aforementioned challenges and more issues related to the design of the 802.22 standard can be found in [41].

11.7.3 IEEE 802.11

The well-known WLAN standard, IEEE 802.11a/g [4,42], is amended to have cognitive features with IEEE 802.11h and IEEE 802.11k standards. IEEE 802.11h [43] is designed to allow estimation of channel characteristics and DFS, which is the ability to switch to different physical RF channels for transmit and receive activity based on channel measurements. In addition, TPC is incorporated as well, providing the system with more control over the signal range and interference level. The purpose of the IEEE 802.11h standard is to allow WLAN systems to share the 5 GHz spectrum with primary users (e.g. military radar systems).

The DFS detects devices using the same radio channel and the system switches to other radio channels if necessary avoiding interference with other existing primary users. WLAN station reports a list of channels that it can support to an access point. When it's necessary to switch to a new channel, the access point uses this data to determine the best channel. On the other hand, TPC is used to reduce the interference from stations to other devices by controlling the power level of the transmitted signal. In addition, TPC is used to manage the power consumption of wireless devices and the range between access points and wireless devices.

On the other hand, IEEE 802.11k standard is proposed for radio resource management. The aim is to improve the traffic distribution in a WLAN. The standard defines a list of several radio parameters to be estimated by the system. While this list is limited and aimed for IEEE 802.11 standards, it is further enhanced compared to earlier standards. WLAN devices can be upgraded to support the new standard, since it is designed to be implemented in software. The standard allows access points to collect data from clients regarding access points they can detect and their signal power. After the collected data is analyzed, access points within range of a client are ordered into a list according to their signal strength, services, and encryption types supported by the client. This list is called *site report*. The access points provide the clients with the site report and thus improve the roaming decisions and increase the overall network throughput.

The access point could gather informations from clients about the RF channel. For example, the access point could request the client to measure the channel noise level, or to provide the access point with informations regarding the traffic load on the channel and the time duration over which the channel is occupied. Using this information, the access point can make a decision whether a certain channel is being crowded or if the channel contains high level of noise/interference.

Other features including tracking of hidden nodes and sharing clients statistics are included in the standard. By applying both 802.11h and 802.11k standards to current 802.11-based WLAN systems, the performance and efficiency of wireless networking can be improved significantly. Adding cognitive features such as channel sensing and estimation, statistics distribution, DFS, TPC to WLAN devices will soon be possible. It is important to remember that 802.11 standards mainly use OFDM making it the signaling of choice for future technologies. Figure 11.15 shows an illustration of the discussed current and future technologies and standards.

11.8 Conclusion

Cognitive radio is an exciting and promising effort for solving the spectrum crowding problem. On the other hand, OFDM technique is used in many wireless systems and proven as a reliable and effective transmission method. OFDM can be used for realizing cognitive radio concept because of its inherent capabilities that are discussed in detail in this chapter. By employing OFDM transmission in cognitive radio systems; adaptive, aware and flexible systems that can interoperate with current technologies can be realized. However, the identified challenges need to be studied further in order to find solid solutions. The adoption of OFDM in cognitive radios may happen in two ways: current wireless technologies might evolve to have more and more cognitive features or new systems might be developed that has full cognitive features. In either case, we foresee that OFDM will be the dominant PHY technology for cognitive radio.

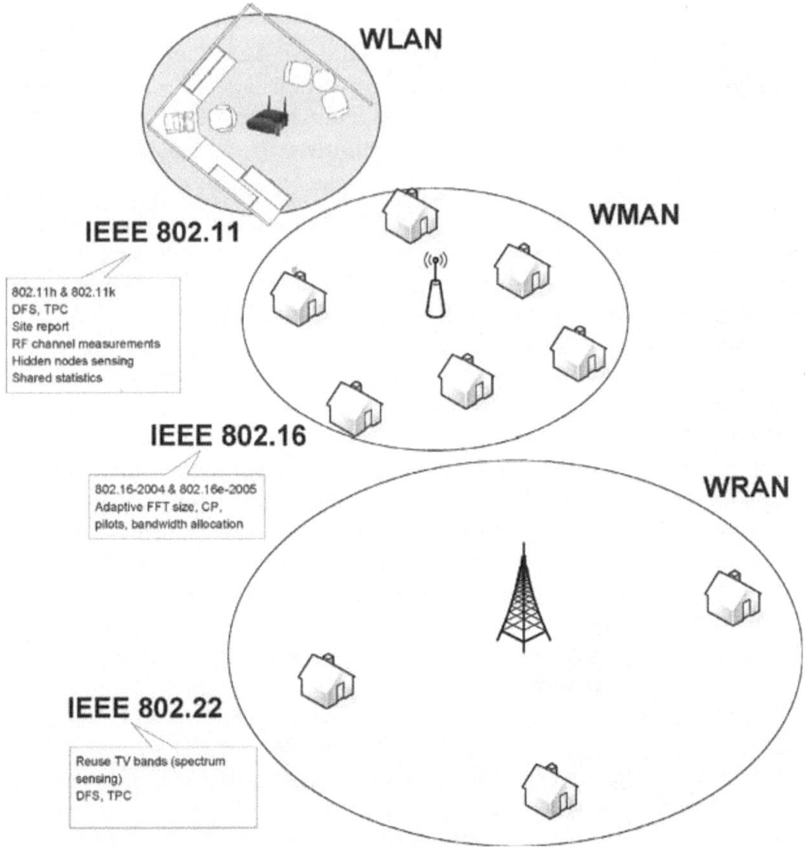

Fig. 11.15. Standards and technologies developments.

References

1. J. Mitola and G. Q. Maquire Jr., "Cognitive radio: making software radios more personal," *IEEE Personal Commun.*, vol. 6, no. 4, pp. 13–18, Aug. 1999.
2. P. Robertson and S. Kaiser, "The effect of Doppler spreads in OFDM(A) mobile radio systems," in *Proc. IEEE Veh. Technol. Conf.*, vol. 1, Sept. 1999, pp. 329–333.
3. D. Harvatin and R. Ziemer, "Orthogonal frequency division multiplexing performance in delay and doppler spread channels," in *Proc. IEEE Veh. Technol. Conf.*, vol. 3, May 1997.
4. *Supplement to IEEE standard for information technology telecommunications and information exchange between systems – local and metropolitan area networks – specific requirements. Part II: wireless LAN Medium Access Control (MAC) and Physical Layer (PHY) specifications: high-speed physical layer in the 5 GHz band*, The Institute of Electrical and Electronics Engineering, Inc. Std. IEEE 802.11a, Sept. 1999.

5. S. Hara and R. Prasad, "Overview of multicarrier CDMA," *IEEE Commun. Mag.*, vol. 35, no. 12, pp. 126–133, Dec. 1997.
6. H. Yaghoobi, "Scalable OFDMA physical layer in IEEE 802.16 WirelessMAN," *Intel Technol. J.*, vol. 8, no. 3, pp. 201–212, Aug. 2004.
7. T. Clancy and W. Arbaugh, "Measuring interference temperature," in *Virginia Tech Wireless Personal Communications Symposium*, June 2006.
8. A. Ghasemi and E. Sousa, "Collaborative spectrum sensing for opportunistic access in fading environments," in *Proc. IEEE Int. Symposium on New Frontiers in Dynamic Spectrum Access Networks*, Baltimore, MD, Nov. 2005, pp. 131–136.
9. M. Wylie-Green, "Dynamic spectrum sensing by multiband OFDM radio for interference mitigation," in *First IEEE International Symposium on DySPAN 2005*, 2005, pp. 619–625.
10. T. Yücek and H. Arslan, "MMSE noise power and SNR estimation for OFDM systems," in *Proc. IEEE Sarnoff Symposium*, Princeton, NJ, Mar. 2006.
11. T.A. Weiss and F.K. Jondral, "Spectrum pooling: an innovative strategy for the enhancement of spectrum efficiency," *IEEE Commun. Mag.*, vol. 42, no. 3, pp. 8–14, Mar. 2004.
12. F. Tufvesson, M. Faulkner, and T. Maseng, "Pre-compensation for Rayleigh fading channels in time division duplex OFDM systems," *Wireless Personal Commun.*, vol. 16, no. 1, pp. 21–33, 2001.
13. S. Haykin, "Cognitive radio: brain-empowered wireless communications," *IEEE J. Select. Areas Commun.*, vol. 3, no. 2, pp. 201–220, Feb. 2005.
14. M. Uhm, "Making the adaptivity of SDR and cognitive radio affordable, going beyond flexibility to adaptivity in FPGAs," *Xcell DSP Mag.*, pp. 25–27, May 2006.
15. H. Arslan and T. Yücek, "Adaptation of Wireless Mobile Multi-carrier Systems". *Adaptation Techniques in Wireless Multimedia Networks.* Nova Science Publishers, 2006.
16. Q. Zhang, F. Hoeksema, A. Kokkeler, and G. Smit, "Towards Cognitive Radio For Emergency Networks" *Mobile Multimedia: Communication Engineering Perspective.* Nova Science Publishers, 2006 .
17. S. Haykin and M. Moher, *Modern Wireless Communications.* New York, NY: Prentice-Hall, 2004.
18. Y. Li and N. Sollenberger, "Interference suppression in OFDM systems using adaptive antenna arrays," in *Proc. IEEE Global Telecommunications Conf. (Globecom)*, vol. 1, Sydney, Austalia, Nov. 1998, pp. 213–218.
19. *IEEE Standard for Local and metropolitan area networks Part 16: Air Interface for Fixed and Mobile Broadband Wireless Access Systems Amendment 2: Physical and Medium Access Control Layers for Combined Fixed and Mobile Operation in Licensed Bands and Corrigendum 1*, IEEE Std 802.16e-2005 and IEEE Std 802.16-2004/Cor 1-2005 (Amendment and Corrigendum to IEEE Std 802.16-2004) Std., 2006.
20. Institute of Electrical and Electronics Engineers, "IEEE standard computer dictionary: A compilation of IEEE standard computer glossaries," New York, NY, 1990.
21. A. Meissner, T. Luckenbach, T. Risse, T. Kirste, and H. Kirchner, "Design challenges for an integrated disaster management communication and information system," in *Proc. 1st IEEE Workshop on Disaster Recovery Networks (DIREN '02) [co-located with IEEE INFOCOM 2002]*, New York City, June 24, 2002.

22. National Task Force on Interoperability, "Why Can't We Talk? Working Together To Bridge the Communications Gap To Save Lives," Feb. 2003.
23. K.J. Imel and J.W. Hart, *Understanding Wireless Communications in Public Safety: A Guidebook to Technology, Issues, Planning, and Management*, 2nd ed. National Institute of Justice, Jan. 2003.
24. SDR Forum, "Software Defined Radio Technology for Public Safety," SDR Forum, Apr. 14, 2006, approved Document SDRF-06-A-0001-V0.00.
25. T. Weiss, J. Hillenbrand, and F. Jondral, "A diversity approach for the detection of idle spectral resources in spectrum pooling systems," in *Proc. 48th Int. Scientific Colloquium*, Ilmenau, Germany, Sept. 2003, pp. 37–38.
26. R. Rajbanshi, A.M. Wyglinski, and G.J. Minden, "An efficient implementation of NC-OFDM transceivers for cognitive radios," in *1st International Conference on Cognitive Radio Oriented Wireless Networks and Communications*, Mykonos Island, Greece, June 2006.
27. A. M. Wyglinski, "Effects of bit allocation on non-contiguous multicarrier-based cognitive radio transceivers," in *Proc. IEEE Veh. Technol. Conf.*, Sept. 2006.
28. T. Weiss, A. Krohn, F. Capar, I. Martoyo, and F. Jondral, "Synchronization algorithms and preamble concepts for spectrum pooling systems," *IST Mobile & Wireless Telecommunications Summit*, June 2003.
29. T. Weiss, J. Hillenbrand, A. Krohn, and F. Jondral, "Mutual interference in OFDM-based spectrum pooling systems," in *Proc. IEEE Veh. Technol. Conf.*, vol. 4, Milan, Italy, May 2004, pp. 1873–1877.
30. C. Muschallik, "Improving an OFDM reception using an adaptive Nyquist windowing," *IEEE Trans. Consumer Electron.*, vol. 42, no. 3, pp. 259–269, Aug. 1996.
31. S. H. Muller-Weinfurtner, "Optimum Nyquist windowing in OFDM receivers," *IEEE Trans. Commun.*, vol. 49, no. 3, pp. 417–420, Mar. 2001.
32. H. Yamaguchi, "Active interference cancellation technique for MB-OFDM cognitive radio," in *Proc. IEEE European Microwave Conf.*, vol. 2, Amsterdam, the Netherlands, Oct. 2004, pp. 1105–1108.
33. S. Brandes, I. Cosovic, and M. Schnell, "Sidelobe suppression in OFDM systems by insertion of cancellation carriers," in *Proc. IEEE Veh. Technol. Conf.*, vol. 1, Dallas, TX, Sept. 2005, pp. 152–156.
34. I. Cosovic, S. Brandes, and M. Schnell, "A technique for sidelobe suppression in OFDM systems," in *Proc. IEEE Global Telecommunications Conf. (Globecom)*, vol. 1, St. Louis, MO, Nov./Dec. 2005, pp. 204–208.
35. I. Cosovic, S. Brandes, and M. Schnell, "Subcarrier weighting: a method for sidelobe suppression in OFDM systems," *IEEE Commun. Lett.*, vol. 10, no. 6, pp. 444–446, June 2006.
36. A. Batra, J. Balakrishnan, G. Aiello, J. Foerster, and A. Dabak, "Design of a multiband OFDM system for realistic UWB channel environments," *IEEE Trans. Microw. Theory Tech.*, vol. 52, no. 9, pp. 2123–2138, 2004.
37. V. Somayazulu, J. Foerster, and S. Roy, "Design challenges for very high data rate UWB systems," in *Proc. Asilomar on Systems, Signals, and Computation Conf.*, Pacific Grove, CA, 2002, pp. 717–721.
38. *Broadband Wireless Access: IEEE MAN standard*, IEEE LAN/MAN standards committee 802.16a, 2003.
39. *IEEE Standard for Local and Metropolitan area networks Part 16: Air Interface for Fixed Broadband Wireless Access Systems*, The Institute of Electrical and Electronics Engineering, Inc. Std. IEEE 802.16-2004, 2004.

40. "IEEE 802.22 working group on wireless regional area networks." [Online]. Available: http://www.ieee802.org/22/
41. C. Cordeiro, K. Challapali, D. Birru, and S. Shankar, "IEEE 802.22: the first worldwide wireless standard based on cognitive radios," in *Proc. IEEE Int. Symposium on New Frontiers in Dynamic Spectrum Access Networks*, Baltimore, Maryland, Nov. 8–11 2005.
42. *IEEE standard for information technology – telecommunications and information exchange between systems – local and metropolitan area networks – specific requirements Part II: wireless LAN medium access control (MAC) and physical layer (PHY) specifications*, IEEE Std 802.11g-2003 (Amendment to IEEE Std 802.11, 1999 Edn. (Reaff 2003) as amended by IEEE Stds 802.11a-1999, 802.11b-1999, 802.11b-1999/Cor 1-2001, and 802.11d-2001) Std., 2003.
43. *IEEE Std. 802.11h - 2003*, IEEE Std 802.11h-2003 (Amendment to IEEE Std 802.11, 1999 Edn. (Reaff 2003)) Std., 2003.

12

UWB Cognitive Radio

Hüseyin Arslan[1] and Mustafa E. Şahin[2]

[1] University of South Florida, Tampa, FL, USA arslan@eng.usf.edu
[2] University of South Florida, Tampa, FL, USA mesahin@eng.usf.edu

12.1 Introduction

Cognitive radio, which is a recent concept introduced by Mitola [1], has been attracting significant interest and has the potential of shaping the future of wireless communication systems. The basic idea of the concept can be stated as employing smart wireless devices that are furnished with awareness, sensing, learning, and adaptation capabilities in order to utilize the available radio resources as efficiently as possible.

Since the concept of cognitive radio is still at the stage of being developed, there is no consensus on what kind of wireless technologies to employ for realizing it. There are a number of requirements a wireless system has to satisfy in order to be considered a suitable candidate for cognitive radio. These requirements include causing no interference to licensed systems, having an adjustable pulse shape, bandwidth, and transmit power, supporting various throughputs, providing adaptive multiple access, and ensuring the security of information. When the wireless systems that are potential candidates for cognitive radio are considered, Ultra Wideband (UWB) seems to be one of the tempting choices because it has an inherent potential to fulfill some of the key cognitive radio requirements. Therefore, the focus of this chapter will be on UWB and its suitability for cognitive radio. However, it is worth to stress that a cognitive radio might have the ability to synthesize and process various waveforms and wireless technologies. In this chapter, it is not claimed that a cognitive wireless system with only the UWB capability can handle all the requirements of the ideal cognitive radio. Instead, we would like to show how UWB can accommodate some of the major requirements and can be incorporated to cognitive radio devices as an access technology.

Although the term ultra wideband can be used to describe any wireless system that has a bandwidth greater than 500 MHz or a fractional bandwidth[1]

[1] Fractional Bandwidth $= 2 \cdot \frac{F_H - F_L}{F_H + F_L}$, where F_H and F_L are the upper and lower edge frequencies, respectively.

H. Arslan (Ed.), Cognitive Radio, Software Defined Radio, and Adaptive Wireless Systems, 355–381.

greater than 0.2, the number of commonly proposed technologies to implement UWB is limited to two. These two technologies are the Orthogonal Frequency Division Multiplexing based UWB (UWB-OFDM) and the impulse radio based UWB (IR-UWB). Throughout this chapter, the impulse radio option will be considered as the means of realizing ultra wideband. The OFDM technology (in a broader sense) is discussed in detail in Chapter 11.

Ultra wideband systems have been attracting an intense attention from both the industry and academic world since 2002, when the United States Federal Communications Commission (FCC) released a spectral mask officially allowing the unlicensed usage of UWB. Under the current regulation, UWB is a promising technology for future short and medium range wireless communication networks with a variety of throughput options including very high data rates. UWB's most significant property is that it can coexist in the same temporal, spatial, and spectral domains with other licensed/unlicensed radios because it is an underlay system. Other tempting features of UWB include that it has a multi-dimensional flexibility involving adaptable pulse shape, bandwidth, data rate, and transmit power. On top of these, UWB has a low-power consumption, and it allows significantly low complexity transceivers leading to a limited system cost. It also has an advanced multipath resolution capability. Another very important feature of UWB is providing secure communications. It is very hard to detect UWB transmission as its power spectrum is embedded into the noise floor. This feature introduces very secure transmission in addition to other possible higher layer encryption techniques.

The attractiveness of UWB for cognitive radio is not limited to the inherent attributes of this technology. UWB offers some exceptional uses that can add a number of extra intellective features to cognitive systems. These special uses are brought by the high multipath resolution property, which enables UWB to act as an accurate radar, ranging, and positioning system. Examples of specific UWB features include sensing the physical environment to enable situation awareness, providing geographical location information, and specifying the mobile communication parameters when one or more parties of communication are nomadic.

The fundamental ideas that constitute the cognitive radio concept are not restricted to optimizing the spectrum usage. Cognitive radio's awareness about temporary conditions and needs of the user, for example, is also a very significant issue. Regarding providing user awareness, UWB has a very distinguished feature. The extremely fine multipath resolution of UWB can even make it possible to image different organs and tissues inside the body of the cognitive radio user, and more surprisingly, to find the densities of certain substances in the content of the user's blood. Therefore, this UWB feature can be utilized to determine the biomedical conditions of the cognitive radio user to optimize his/her making use of the cognitive radio. Similarly, other user contextual awareness information can be obtained using UWB. For example, the fine resolution capability can tell where exactly the user is, i.e. the UWB can capture images of the user environment (like a camera) and can tell how far the objects are from the user, or what type of objects are around the user, etc. This is especially important for safety, security, and rescue operations. If a victim is stuck under a rubble after an earthquake, and if the rescue team is able to detect the victim, but the victim is not able to describe the condition and the environment, the radio with UWB capability could do this automatically.

Owing to all its distinctive properties mentioned, in this chapter, impulse radio based UWB is considered as one of the enabling/access technologies of cognitive radio. In the remainder of the chapter, the suitability of IR-UWB for being employed in cognitive communications is proven by two means. First, a number of one-to-one matches between the cognitive radio requirements and IR-UWB properties are revealed; and second, some supplementary uses of IR-UWB that can be utilized by cognitive radio systems are proposed, and their feasibility is demonstrated.

The flow of the chapter is as follows. In Section 12.2, the implementation of impulse radio is discussed. Section 12.3 addresses a number of IR-UWB properties that are tempting for cognitive radio. In Section 12.4, various approaches that aim either at realizing cognitive radio via IR-UWB or at making use of UWB to enhance cognitive radio systems are investigated. Finally, Section 12.5 concludes the chapter by summarizing the ideas conveyed throughout the chapter.

12.2 Fundamentals of Impulse Radio Ultra Wideband

Impulse radio based implementation of UWB is carried out by transmitting extremely short, low-power pulses that are on the order of nanoseconds [2, 3]. According to the current FCC regulations in the USA, UWB systems are allowed to operate in the 3.1–10.6 GHz band without a license requirement. However, the transmit power of these systems is strictly limited. As shown in Figure 12.1, both in indoors and outdoors, UWB systems are not permitted

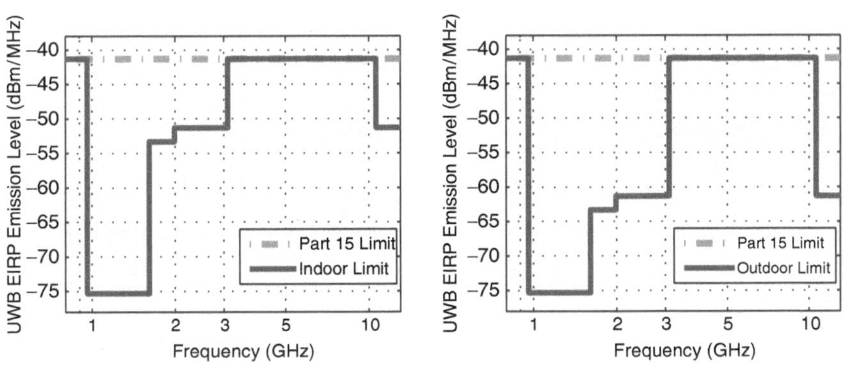

Fig. 12.1. FCC indoors and outdoors UWB masks.

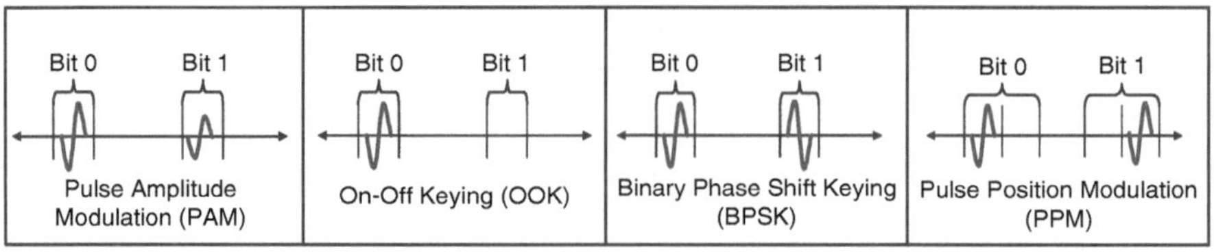

Fig. 12.2. Various modulation options for IR-UWB systems.

to transmit more than −42 dBm/MHz in the specified band. This limitation ensures that the UWB systems do not affect the licensed operators that use various frequency bands inside the UWB band. However, it should also be kept in mind that it is not unlikely that revisions can be made in the UWB related FCC regulations, especially regarding the transmit power limits. In the near future, if the UWB radios are provided with cognitive properties that allow them to sense the spectrum to determine the occupancy of their target bands and to ensure not interfering with licensed users, it is possible that regulatory agencies may consider to offer more freedom to UWB.

Impulse radio UWB is advantageous in that it enables to employ various types of modulations, including On–Off Keying (OOK), Pulse Amplitude Modulation (PAM), Pulse Shape Modulation (PSM), Pulse Interval Modulation (PIM), Pulse Position Modulation (PPM), and Phase Shift Keying (PSK) [4]. Some of these modulation types are illustrated in Figure 12.2. This variety of modulation choices adds another dimension to the adaptive properties of UWB. For example, in case of limited available transmit power, a UWB transceiver may choose binary PSK to obtain power efficiency. If the spectral spikes that occur due to the periodicity of the transmitted pulses cause a problem, again, BPSK may be a desirable solution. If there is a concern about the complexity and cost of the transceiver, on the other hand, then PAM, PPM, and OOK become preferable over BPSK. When M-ary modulation is applied, PPM can be favored over PAM because of its better performance, as long as the decreased data rate in PPM due to the longer symbol duration is acceptable. In [5], a pseudo-random modulation index is proposed for PPM to mitigate the effect of cross-modulation interference. It is demonstrated that the randomness in the pulse positions can increase the robustness of UWB signal reception. This approach can be utilized for adding another dimension of adaptivity to PPM by determining the modulation index in an adaptive manner according to the channel conditions. All these examples indicate the importance of UWB's being able to select the type of modulation in an adaptive manner according to the system design requirements and available resources.

In Figure 12.3, the building blocks of UWB signals are shown as chips, frames, symbols, and blocks. For multi-user access, IR-UWB systems employ Time Hopping (TH) codes that are specific to each user [6]. In Figure 12.3, a Time Hopping Ultra Wideband (TH-UWB) system is demonstrated. In the

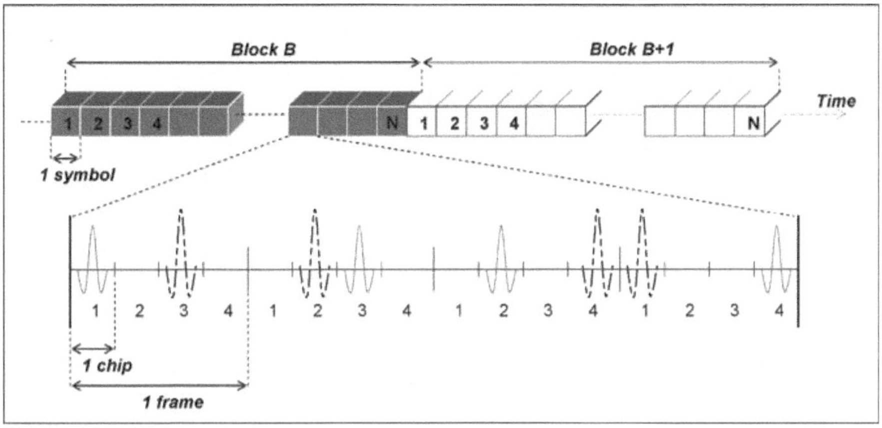

Fig. 12.3. Impulse radio based time hopping UWB.

illustrated scenario, each information carrying symbol is transmitted with four pulses. Pulses occupy a location in the time frame based on the specific Pseudo-random Noise (PN) code assigned for each user. Two different codes (for two different users) and the corresponding pulse locations are shown in the figure. Note that these two codes are orthogonal (i.e. they do not overlap with each other). This simple scenario indicates how conveniently UWB can provide access to multiple users. The multi-user parameters can be adaptively modified according to the change in number of users. To enable more users to communicate, for example, the UWB system can increase the number of chips in each frame at the expense of decreasing each user's data rate. Besides this, as shown in the illustrated scenario, by assigning user specific PN codes the UWB system ensures additional security for each user.

The receiver types that can be utilized for IR-UWB communications are also various and include coherent receivers (such as Rake and correlator receivers) as well as non-coherent ones such as energy detector and transmitted reference receivers, which can be seen in Figure 12.4. In wireless communications, the choice of the receiver generally depends on hardware complexity and cost, throughput requirement, and targeted Quality-of-Service (QoS). In the case of UWB, however, also channel conditions become one of the primary factors. The power delay profile of the UWB channel can be various depending on the physical environment. Channel impulse response examples for CM1 and CM4 [7] are illustrated in Figure 12.5. The first channel response reflects short distance (up to 4 m), Line-of-Sight (LOS) channel conditions, whereas the second one is a long distance (10 m and above), non-LOS channel. The main difference between these two channels is that the energy is spread over a much longer time frame because of the relatively much higher number of multipaths in the latter one. In UWB signal reception, coherent receivers are preferred due to their high power efficiency, which they owe to

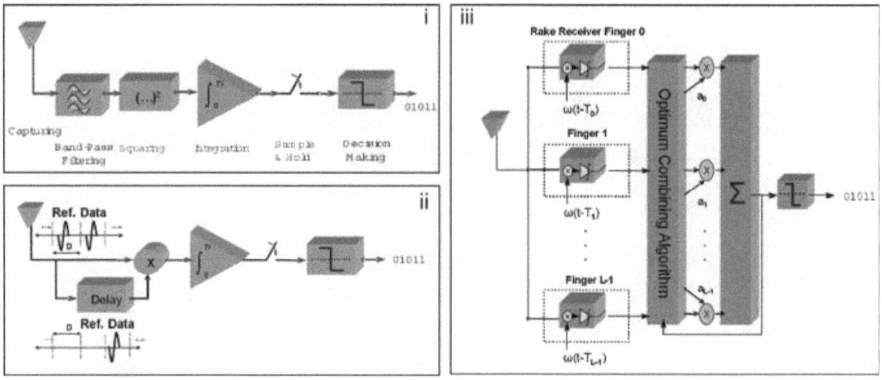

Fig. 12.4. Various receiver options for UWB signal reception. i – Energy detector. ii – Transmit reference receiver. iii – Rake receiver.

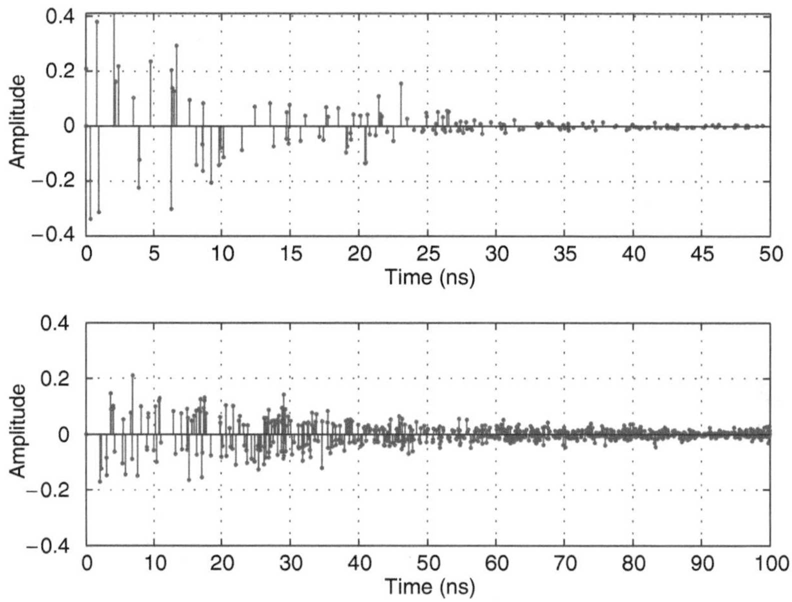

Fig. 12.5. Channel impulse responses for channel model 1 and channel model 4 in [7]. (Note that the time frame in the second response is twice as long as the first one.)

their advanced ability to collect multipath energy. However, implementation of such receivers requires estimation of *a priori* channel information regarding the timing, fading coefficient, and the pulse shape for each individual channel tap. Coherent signal reception also mandates a high sampling rate and an accurate synchronization. Therefore, coherent receivers have a high complexity, and may be feasible only when the number of channel taps is limited, or the targeted reception quality is high regardless of the hardware cost. On the other hand, non-coherent receivers have less stringent *a priori* information requirements and can be implemented with lower complexity. For example, in transmitted reference (TR) UWB systems, a reference pulse that includes the channel information is transmitted along with the information bearing pulse. In the receiver part, these two pulses are correlated with each other, and this way, the need for estimating the channel parameters is eliminated. Non-coherent receivers can be desirable in cases where the complexity and cost are the limiting factors, and QoS is not a primary concern. In short, the variety of employable receivers make UWB a tempting choice under various system design considerations and physical conditions.

Along with the flexibility in modulation methods and receiver types, IR-UWB also offers a variety of options regarding the shapes of the transmitted pulses. Employing various pulse shapes provides the UWB system with a spectrum shaping capability that allows it to adapt to changing spectral conditions. Along with the primary pulse shaping options, which are the Gaussian pulses [8] and the prolate spheroidal wavelet functions [9, 10], raised cosine windowing based pulse shaping (analyzed in Section 12.4) can be considered for impulse radio. Other various analog and digital methods to implement pulse shaping for impulse radio can be found (among others) in [11–16].

UWB has a considerable resistance against the Multi-user Access Interference (MAI), which is investigated in detail in [17–21]. The IR-UWB systems are immune to not only MAI, but also against Narrowband Interference (NBI), which is caused by the licensed and unlicensed systems that exist in the frequency band occupied by the UWB system [22–24].

Besides being a communication system, IR-UWB is a precise radar technology as well as a highly accurate ranging and positioning system. These extra features are owed to the fact that IR-UWB systems have an excellent multipath resolving capability because of the extremely wide frequency band that they occupy. The UWB system can be used as a very precise radar because of the following reason. Short pilot pulses transmitted by a UWB transmitter are reflected, scattered, and refracted by the objects in the physical environment. The resolving capability allows the receiver part of the UWB system to distinguish between the arriving signals from objects that are very close to each other. This way, the UWB system can detect the physical details of objects like a high resolution radar. The ability to separate the signals in time that reach at the receiver through different paths can be utilized to determine the duration that is needed for the signals to travel the distance between the transmitter and receiver. This feature can be utilized to make an accurate range estimation. By combining the range estimates obtained by various UWB transceivers, a position estimation can be achieved that has an inaccuracy limited to a couple of centimeters.

12.3 Cognitive Radio Requirements vs. IR-UWB Features

One of the main goals targeted with cognitive radio is to utilize the existing radio resources in the most efficient way. To ensure the optimum utilization, cognitive radio requires a number of conditions to be satisfied. A wireless system that is a potential candidate for cognitive radio applications is expected to fulfill some of these conditions.

The primary cognitive radio requirements include

- negligible interference to licensed systems and ability to avoid/cancel interference (such as NBI, MAI),
- capability to adapt itself to various link qualities (link adaptation capability),
- ability to sense and measure critical parameters about the environment, channel, etc. using the received signal (the internal/self measurement and sensing capability),
- ability to exploit variety of spectral opportunity (related to bandwidth adaptation), and
- flexible pulse shape and bandwidth, adaptable data rate, transmit power, information security, and limited cost.

At this point, if the main properties of impulse radio based UWB are considered, it is seen that there is a strong match between what the cognitive radio requires and what IR-UWB offers. In the following, the primary features of IR-UWB will be investigated from the point of satisfying the requirements of cognitive radio.

12.3.1 Limited Interference to Licensed Systems

Cognitive radios aim at an opportunistic usage of frequency bands that are owned by their licensed users. Since the licensed users will not necessarily be willing to share the spectra that they have paid for with unlicensed users, it is the responsibility of the cognitive radio system to make sure that its existence is not felt by the licensed user. Hence, one of the most significant requirements of cognitive radio is that the interference caused by cognitive devices to licensed users remains at a negligible level.

Offering the possibility of being implemented both in *underlay* and *overlay* modes, IR-UWB has a significant potential for fulfilling the limited interference requirement of cognitive radio. The difference between the two modes of impulse radio operation is the amount of transmitted power. In the underlay mode, IR-UWB has a considerably restricted power, which is spread over a wide frequency band. In this mode, it complies with the corresponding regulations of the FCC in the USA. When an IR-UWB system is operating in the underlay mode, it is quite unlikely that any coexisting licensed systems are affected from it. On top of this, underlay IR-UWB can employ various narrowband interference avoidance methods such as pulse shaping [25], time-hopping code adjustment [26], and antenna design [27] to abstain interfering with licensed systems.[2] Therefore, from the point of licensed narrowband systems that reside inside the UWB spectrum, underlay IR-UWB is a *stealth* way of wireless communication because its signals are undetectable.

In the overlay mode, on the other hand, the transmitted power can be as high as the upper limits for electromagnetic emission set by the regulatory agencies. The overlay mode impulse radio is only applicable if the UWB transmitter ensures that the targeted spectrum is completely free of signals of other systems, i.e. there are no other licensed or unlicensed systems utilizing the same band, and, of course, if the regulations allow this mode of operation. If these conditions are met, the transmitted UWB power can be increased to a certain level that is comparable to the power of licensed systems. UWB can also operate simultaneously in both underlay and overlay modes. Depending on the spectrum opportunities, the signaling and the spectrum of the transmitted signal can be shaped in such a way that part of the spectrum is occupied in an underlay mode, and some other parts are occupied in an overlay mode. Shaping the spectrum in such a way is possible with a single simultaneous transmission. Apparently, in either mode of operation, IR-UWB causes negligible interference to other communication systems, if it does at all. This special feature of impulse radio makes it very tempting for the realization of cognitive radio.

12.3.2 Flexibility in Pulseshape/Bandwidth (Dynamic Spectrum)

One of the main features of the cognitive radio concept is that the targeted frequency spectrum is scanned periodically in order to check its availability for opportunistic usage. According to the results of this spectrum scan, the bands that will be utilized for cognitive communication are determined. Since at different times and locations the available bands can vary, cognitive radio is expected to have a high flexibility in determining the spectrum it occupies. This requires the cognitive radio to be capable of modifying its transmitted pulse in such a way that the spectrum of the pulse fits into the available bands as accurately as possible. Therefore, cognitive radios have to employ a system that has an advanced capability of adapting the pulse shape to changing conditions.

Flexible spectrum shaping is a part of IR-UWB's nature. Since the IR-UWB communication is basically realized via the transmission of short pulses, varying the duration or the form of the pulses directly alters the occupied spectrum. Various pulse shaping options were mentioned in Section 12.2. An IR-UWB transmitter can select one of these pulse shapes, and the occupied spectrum will change according to this selection. This attribute of IR-UWB

[2] For a detailed discussion of narrowband interference avoidance and cancelation methods in UWB systems, the reader can be referred to [22].

is excellent from the point of satisfying the dynamic spectrum requirement of cognitive radio.

12.3.3 Dynamically Adjustable Data Rate and Quality-of-Service

Cognitive radio aims at using licensed frequency bands opportunistically by utilizing them when they are not being accessed by their licensed users. Although this sounds like free and unlimited communications in the first glance (which is partially true), it should be kept in mind that the continuity of cognitive radio communications depends on the availability of unused bands. Any increase in the utilization of the bands by the licensed systems directly results in narrowed freedom for the cognitive radio, which can force it to decrease its data rate and quality-of-service, or even to terminate its communication. Therefore, any system that is a candidate for being employed by cognitive radio should have the capability of dynamically adjusting the throughput. Such systems may also be expected to provide a solution for the cases when the available bandwidth is so limited that the communication cannot be continued (i.e. for dropped calls).

An impulse radio based UWB system is able to make abrupt changes in its data rate. According to the simple scenario illustrated in the first figure in Table 12.1, if the amount of available band decreases, the only thing impulse radio has to do is to elongate the transmitted pulses such that less bandwidth is occupied, and the data rate is decreased. If there is more band to use, it can respond by doing the opposite. Therefore, it satisfactorily meets the adjustable data rate necessity of cognitive radio. The adjustment of the occupied bandwidth may not always be that simple. The available bands can be anywhere in the frequency spectrum, and their widths can vary. This may require the IR-UWB transmitters to include hardware capable of transmitting and receiving various types of UWB pulses with different spectral features.

In some cases, the interference from UWB systems to licensed users may become unusually effective. This is most likely to occur when the UWB transceiver is physically very close to the licensed system's transceiver. In such a case, the interference can be mitigated by decreasing the *duty cycle* of the UWB system rather than modifying its bandwidth. The duty cycle can be lowered by decreasing the number of pulses per frame, which, again, corresponds to decreasing the data rate. A lowered duty cycle leads to less coexistence in time of these two systems. Therefore, the increase in the licensed system's Bit-Error-Rate (BER) due to the UWB interference can be diminished by decreasing the duty cycle.

On top of its flexible data rate property, IR-UWB provides an exceptional solution regarding the dropped calls. As mentioned earlier, IR-UWB can be performed both in underlay and overlay modes. Assuming that the normal operation mode is overlay, in cases when it becomes impossible to perpetuate the communication, IR-UWB can switch to the underlay mode (as illustrated in Table 12.1). Since the licensed systems are not affected by the impulse radio

Table 12.1. Adaptive features of impulse radio UWB systems.

Flexible spectrum shaping Impulse Radio UWB is implemented via the transmission of short pulses. Therefore, varying the duration and form of the pulses directly alters the occupied spectrum. This makes flexible spectrum shaping a part of IR-UWB's nature. The figure illustrates the adjustment of the occupied bandwidth by modifying the pulse duration.	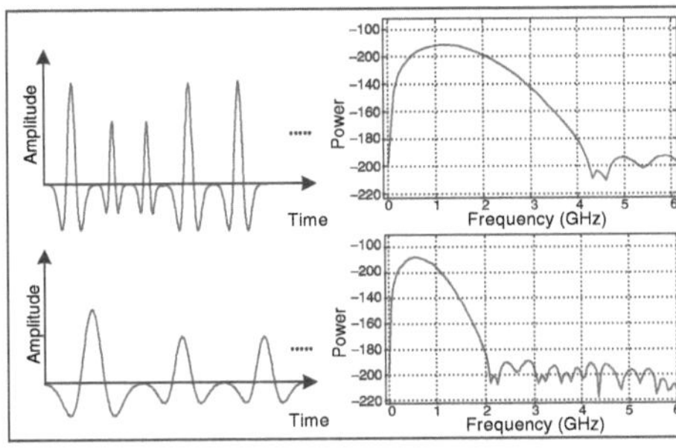
Adjustable Data Rate/Processing Gain In case of decreasing communication link quality, IR-UWB boosts the processing gain at the expense of lowering the datarate. This way, the Qos is maintained. The trade-off between the processing gain and the data rate is demonstrated in the figure. This feature can be especially useful if the link quality is not high due to NLOS conditions or increased distance between the transmitter and receiver.	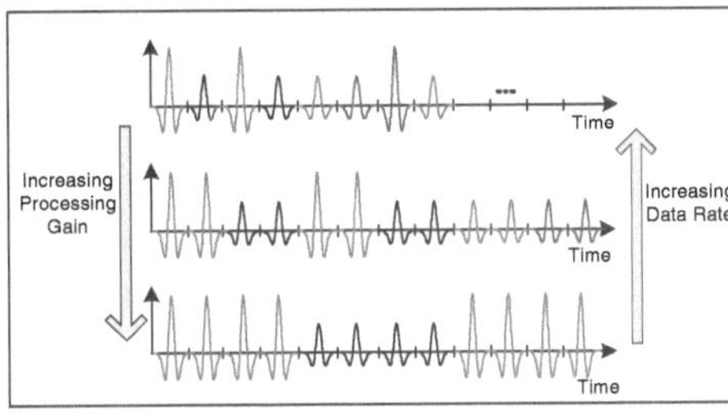
Switching between overlay and underlay Modes IR-UWB can be performed both in underlay and overlay modes. If the normal operation mode is overlay, in cases when the number of licensed users increases, IR-UWB can switch to the underlay mode as illustrated in the figure. Since the licensed systems are not affected when it is in the underaly mode, the IR-UWB communication link can be maintained, and thus, any discontinuity in the communication can be prevented.	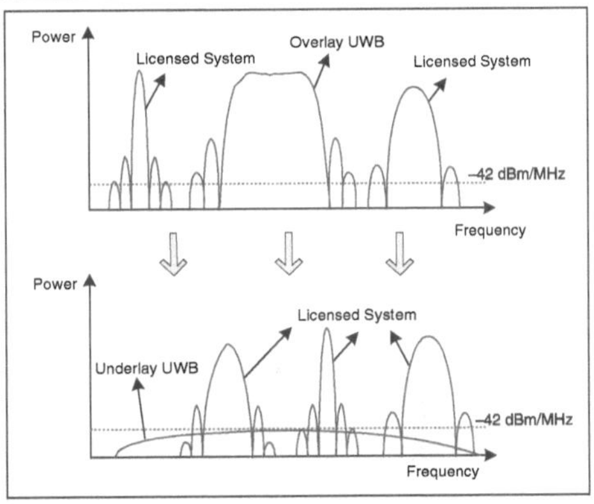

when it is in the underlay mode, this gives the IR-UWB the opportunity to maintain the communication link even though it is at a low quality.

If the quality of the communication link between the transmitter and receiver decreases due to any reason, IR-UWB can maintain the QoS by lowering the data rate and boosting the processing gain. The trade-off between the processing gain and the data rate is illustrated in Table 12.1. This feature can be especially useful if the link quality is not high due to non-existence of a LOS path or increased distance between the transmitter and receiver. The dynamic QoS adjustment feature of IR-UWB systems can be of vital usefulness in certain cognitive radio applications that cannot tolerate any discontinuity in the communication.

12.3.4 Adaptable Transmit Power

The existence of licensed systems and other unlicensed users is not the only limitation regarding the secondary usage of spectrum. The spectral masks that are imposed by the regulatory agencies are also determinative in spectrum usage in that they set a limit to the transmit power of wireless systems. These masks may have been set to regulate the usage of spectrum both in underlay and overlay modes. Even though cognitive radio is conceptualized to make use of the spectrum to the maximum extent, it cannot ignore these power limitations. Taking into account that the spectral masks can be totally different at different locations (and even at different times in the same location), a cognitive radio system may have to modify its transmitted power on a frequent basis.

IR-UWB offers a satisfactory solution to the adaptable transmit power requirement of cognitive radio. Since the impulse radio is based on the transmission of separate pulses, adapting the total transmit power is as simple and convenient as modifying the power of a single pulse. Based on the pulse shaping method employed, adaptation of pulse power can be accomplished by changing the pulse amplitude as well as by using different order derivatives of a given basis pulse. By adapting its transmit power, impulse radio can comply with any set of spectral rules mandated upon the cognitive radio system.

12.3.5 Adaptive Multiple Access

The cognitive radio concept includes free utilization of frequency bands that are temporarily not used by their licensed owners. Since this sounds to be a very tempting promise, it is not hard to imagine that there will be a number of users willing to make use of the same opportunity bands at the same time. Therefore, cognitive radio networks should be able to provide access to multiple users simultaneously and in a robust manner.

During the operation of a cognitive radio, changes may occur in the overall spectrum occupancy, or the signal quality observed by each user can fluctuate because of various factors. These changes may require the cognitive radio to modify its multiple access parameters accordingly. For example, if the amount of opportunity bands declines because of increasing number of licensed users, the cognitive radio is forced either to decrease the number of cognitive users or to lower the upper data rate it offers to each user. Another example can be a cognitive network where the users are mobile. Since the positions of users are changing continuously, their received signal strengths may vary considerably in time. This would require the cognitive system to keep track of each user's signal quality and adapt the transmission parameters accordingly. Therefore, cognitive radio systems have to employ a technology that is capable of adaptively changing its multiple access parameters.

IR-UWB is very flexible in terms of multiple access. By changing the number of chips in a frame, the number of multiple users can be determined.

The duty cycle in a frame can be adjusted according to the state of spectral occupancy. More importantly, the data rate and the processing gain, as well as the transmit power level and even the type of modulation can be adjusted for each user separately according to the user's distance, temporary fading conditions, and noise level [28]. This allows the IR-UWB system to optimize the service it provides to its users both from throughput and signal quality perspectives. Therefore, also from the point of adaptive multiple access, IR-UWB proves to be a proper candidate for cognitive radio applications.

12.3.6 Information Security

The primary objectives targeted with cognitive radio include preserving the privacy of information. This can be a vital necessity especially in military communications.

In many communication systems, the security of information is guaranteed by using an excessive amount of encryption. Encryption schemes, which are only known by the transmitter and the targeted receiver, prevent unwanted users to eavesdrop on the communication. In some other systems, information security is an inherent attribute of the physical layer employed. Direct Sequence Spread Spectrum (DSSS) systems and frequency hopping systems, which can be considered as examples for such technologies, enable secure communication by assigning each user a specific spreading code or a hopping sequence. This way a number of users can utilize the same frequency band or the same chunk of frequencies without listening to each other.

Impulse radio UWB is also one of the systems that have information security in their nature. As it was pointed out in the discussion regarding the UWB interference on other systems, underlay mode UWB is a stealth communication technology because for unwanted users it is impossible to detect even the existence of the UWB signals. Therefore, underlay UWB is a highly secure means of exchanging information. Overlay mode UWB, on the other hand, can also be considered a safe communication method. In overlay UWB, if time hopping is employed for multiple access, different UWB users are assigned different time hopping sequences. If multiple accessing is enabled by direct sequencing, each user has a specific spreading code. Therefore, the reception of a user's information is only possible if the exact knowledge about the user's access code is available. Hence, ensuring the privacy of multiple accessing codes ensures the security of information in UWB communications. Apparently, UWB is a secure way of communicating in both its underlay and overlay modes. Therefore, UWB proves to be a strong candidate for cognitive radio applications also from the point of guaranteeing the information security.

12.3.7 Limited Cost

One of the primary conditions for long-lasting market success of any wireless technology is an affordable cost. This economical condition applies to all components of a radio system including the infrastructure expenses, base station and transceiver costs, and service charges. Being a future wireless concept, cognitive radio also targets at a low cost for each of its components. This is necessary for the system to be able to reflect the profit earned by using the spectrum in an opportunistic way (rather than purchasing a license) to its subscribers.

Impulse radio UWB signals, which are basically simple pulses, can be generated and processed by inexpensive analog transmitter and receiver circuitries (unless the pulses need to be generated by digital means for a precise spectrum shaping, which may increase the cost of the circuitry). The RF front-end required to send and capture the UWB signals are also quite uncomplicated and they are available for reasonable prices. Therefore, impulse radio UWB communication can be accomplished by employing very low cost transmitter and receivers. This property of impulse radio makes it very attractive for cognitive radio, which aims at limited infrastructure and transceiver costs.

12.4 Merging Impulse Radio with Cognitive Radio

As it is pointed out throughout the previous section, the impulse radio UWB is highly competent in satisfying the basic requirements of the cognitive radio concept. Therefore, merging IR-UWB into the cognitive radio could be very instrumental for the successful penetration of cognitive radio into the wireless world.

Combining UWB with cognitive radio can be mainly accomplished in two ways. The first one is performing the practical implementation of cognitive radio directly via impulse radio. The second way is to supplement cognitive radio with IR-UWB through various methods. In the following, these two options will be investigated in detail.

12.4.1 Implementing Cognitive Radio via Impulse Radio

Owing to its very tempting adaptive features addressed in Section 12.3, impulse radio UWB can be considered a means of realizing cognitive radio. Since the pulses transmitted by IR-UWB are very influential on the spectrum occupied, a careful selection of the set of pulses to be utilized is essential for an opportunistic spectrum usage. In the remainder of this section, an IR-UWB implementation method that employs pulses shaped by raised cosine (or root raised cosine) windows is going to be provided as a case study.

Raised cosine (and root raised cosine) windows, which can be exemplified as in Figure 12.6a (and b), can be an attractive means for generating special pulses that

- have sharp fall-offs and suppressed side lobes in the frequency domain,
- are limited both in time and bandwidth, and
- have a pulse width and bandwidth that can be controlled simultaneously.

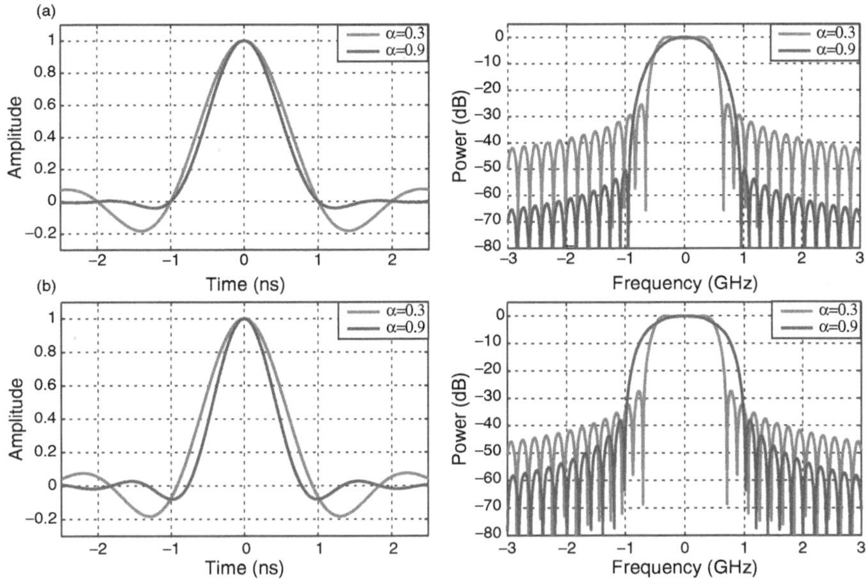

Fig. 12.6. Various pulse envelopes and their spectra a – Raised cosine windows with roll-off factors $\alpha = 0.3$ and $\alpha = 0.9$. b – Root raised cosine windows with roll-off factors $\alpha = 0.3$ and $\alpha = 0.9$.

Therefore, a method that uses pulses generated by means of raised cosine windowing can lead to a very efficient spectrum usage. In this method, it is assumed that detailed information about the spectrum opportunities, which are licensed bands temporarily not being used by their owners, is available through the spectrum sensing performed. The first step is to determine the center frequencies and bandwidths of each spectrum opportunity. Cognitive radio devices are expected to have the facility to allow users to load information and statistics, which can be recalled and processed when needed. Hence, it can be conveniently assumed that information regarding the temporal and spectral properties of a number of raised cosine windows is available in cognitive radio's memory. Therefore, in the second step of implementation, cognitive radio selects the raised cosine windows that are most suitable for each of the spectrum opportunities determined. It should be noted that filling a higher percentage of a white space requires a higher roll-off raised cosine window, and this corresponds to a longer symbol in time. An elongated time symbol either leads to inter-symbol interference or forces the system to set a lower throughput. Therefore, cognitive radio determines the window to be used according to the amount of available bandwidth and the data rate aimed. The selected windows are mixed with locally generated (analog or digital) cosine signals yielding

$$\phi_i(t) = \cos(2\pi f_{c_i} t) \cdot r_i(t), \tag{12.1}$$

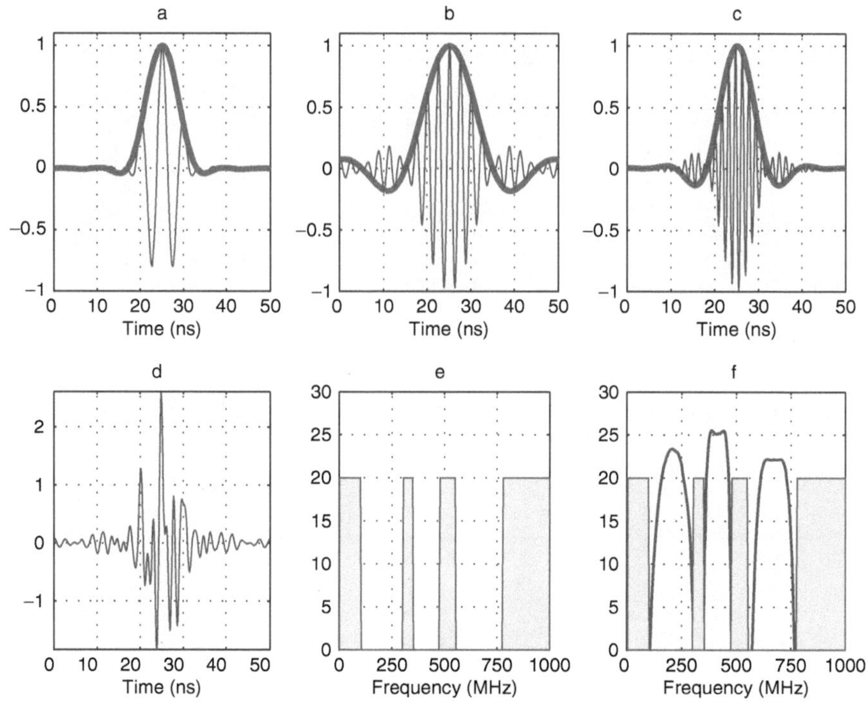

Fig. 12.7. a,b,c – Separate pulses obtained via raised cosine windowing that fit into different opportunities. d – Sum of the separate pulses. e – Classification of frequency bands as "occupied" or "opportunity". f – Spectrum of the designed pulse filling the opportunities.

where f_{c_i} stand for the center frequencies of the spectrum opportunities, and $r_i(t)$ denote the windows selected. $\phi_i(t)$ can be exemplified as in Figure 12.7a,b, and c, which are generated using raised cosine windows with roll-off coefficients 0.9, 0.3 and 0.5, respectively. Each of these pulses is filling one of the spectrum opportunities in Figure 12.7e. The final pulse shape (demonstrated in Figure 12.7d) is obtained by taking the sum of all these separate pulses

$$p(t) = \sum_{i=1}^{N} \phi_i(t) , \tag{12.2}$$

and this combined pulse fills the targeted opportunities as shown in Figure 12.7f.

As it has been demonstrated with the raised cosine windowing based implementation method, impulse radio based UWB technologies can be a convenient way of realizing cognitive radio. However, it has to be kept in mind that under the current FCC regulations in the USA, the overlay mode of UWB is not permitted, and it is reasonable to consider that similar spectral limitations will be mandated upon UWB systems in other regions of the world, as well. Under these conditions, it should be stated that IR-UWB can be *the technology* for cognitive radio only when the regulatory agencies are convinced that IR-UWB devices are equipped with advanced sensing and spectrum shaping capabilities that leave zero possibility to interfering with licensed users.

12.4.2 Various Ways of Supplementing CR with IR-UWB

Due to the spectrum regulations of today, which prohibit employing UWB systems in the overlay mode, IR-UWB based implementation of cognitive radio might not become a reality in the near future. However, besides being a strong candidate for practical cognitive radio implementation, IR-UWB can be considered as a supplement to cognitive radio systems that are realized by means of other wireless technologies. Therefore, it can be concluded that this way or the other, UWB will be an inseparable part of cognitive radio applications.

IR-UWB can offer various kinds of support to cognitive radios. These include sharing the spectrum sensing information via IR-UWB, locating the cognitive nodes in a cognitive network by means of IR-UWB, sensing the physical environment/channel with IR-Radar, and sensing the cognitive user's physical conditions. In the following, each of these supplementary uses of IR-UWB will be discussed in detail.

Spectrum Sensing Information Exchange in Cognitive Networks

In order to be able to opportunistically utilize the available licensed frequency bands, cognitive radio systems periodically scan their target spectrum and detect the spectrum opportunities. In cognitive radio communications, it is mandatory that all parties of communication agree on the spectral opportunities to be utilized. Therefore, it is a major issue for a cognitive radio device to share the spectrum sensing information with other cognitive devices. In some works in the literature, it is considered to have an allocated control channel to transmit this information [29]. In some other works, it is proposed to have a centralized controller that gathers this information, decides for spectrum availability, and allocates distinct bands to different cognitive users [30, 31]. An alternative to these methods is to transmit spectrum sensing results via low-power IR-UWB signaling that complies with the FCC regulations [32]. Since this transmission will be accomplished in an underlay manner, it can be done simultaneously with the real data communication without affecting it regardless of the wireless technology employed to realize the cognitive radio itself. Considering the relatively low throughput needed to transmit the sensing information as well as the low cost transceiver requirement, it turns out to be a proper option to use an uncomplicated non-coherent receiver such as an

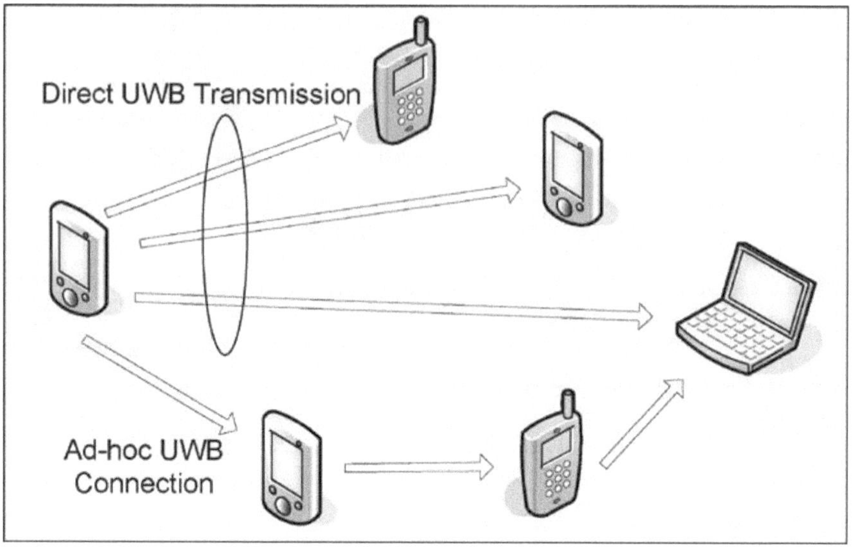

Fig. 12.8. Sharing the spectrum sensing information via IR-UWB signaling in a cognitive network.

energy detector, and to employ on–off keying (OOK) modulation.[3] By using a proper mapping scheme (from sensing information to binary codewords), coding, and OOK modulation, spectrum information can be conveniently shared between the two parties of cognitive communication.

A cognitive network (see Figure 12.8) can be realized by allowing its nodes to communicate with each other using UWB to exchange spectrum information. One of the aims of cognitive radio is to increase the range of communication as much as possible, and at the first glance, UWB signaling may not seem to be very appropriate for this purpose because of the limited range of underlay UWB. The answer to this question can be obtained by looking at the bit-error-rate expression for OOK modulated UWB signals. This BER expression can be stated as

$$BER = Q\left(\sqrt{\frac{N_s A E_p}{2N_0}}\right), \tag{12.3}$$

where N_s is the number of pulses per symbol, A is the pulse amplitude, E_p is the normalized pulse energy, and the Additive White Gaussian Noise (AWGN) has a double sided spectrum of $\frac{N_0}{2}$. In this expression, it is seen that increasing the number of pulses per symbol results in lower BER. Increasing N_s requires a repeated transmission of data, i.e. processing gain. By applying the necessary

[3] The implementation issues regarding the OOK based energy detector receivers such as estimating the optimal threshold and determining the optimum integration interval can be found in [33].

amount of processing gain, it can be made possible that the farthest nodes in a cognitive network can share the spectrum sensing information. Although this comes at the expense of lowered throughput, it is not a limiting factor in this case because a quite low data rate is enough to transmit the spectrum sensing information. By enabling all the nodes in a cognitive network to talk to each other via UWB, there is no need

- either to allocate a separate channel for sharing the sensing information,
- or to employ a centralized controller that collects this information, processes it, and sends it to the cognitive users in the network.

Spectrum information can also be shared in an ad hoc multi-hopping scheme that uses UWB as it is demonstrated in a simple way in Figure 12.8. This way, long range transmission is not needed. Multiple nodes collaboratively share the information and route this information to other nodes using low power, low cost UWB technology. In essence, a UWB based sensor network with the collaboration of multiple radios are formed.

The sensing information received from all the other nodes in the network can be combined in each node, and pulse design can be done according to the common white spaces. This way, the range of cognitive communications can be extended to the coverage area of a medium-sized network. Increasing the network size results in an increased probability of overlapping with licensed systems. This fact sets a practical limit to the size of the cognitive network, because continuing to enlarge the network, the common available spectra become less and less, and after some point their amount becomes insufficient to ensure the minimum quality-of-service. For the details of how the common white bands are going to be shared by the cognitive nodes in the network, the reader can be referred to [34] and [35].

Sharing Various Kinds of Supportive Data via UWB

Cognitive communications do not necessarily have to be based on indoor networks that are mostly made up of personal mobile devices. Cognitive networks can be established to enable military communications or communications at extreme outdoor conditions such as natural disasters.

In outdoors, the factors affecting the cognitive communications will be quite different than in indoor scenarios. The weather and temperature conditions are among these factors. The range of a wireless device used outdoors can considerably decrease when there is a thick fog in the air, for example. Therefore, a moisture level sensor may be very instrumental. To determine rain and snow, a high resolution camera combined with a Digital Signal Processor (DSP) that can recognize the type of precipitation, can be utilized. A temperature sensor embedded in the device can help to adjust the temperature dependent hardware parameters and to verify that the weather condition is detected correctly.

It is known that wireless systems operating at lower frequencies have better penetration capabilities. If a cognitive network has information about the weather conditions, it can adjust its communication frequency accordingly. In a foggy environment, for example, the spectral opportunities at lower frequencies can be preferred, or the transmission power can be increased to maintain the communication range targeted under normal conditions.

It is apparent that it would not be feasible to embed the numerous sensors or a digital camera and DSP to each outdoor communication device. However, it would be enough that each of these sensing capabilities exists in only one of the nodes in a cognitive network, if these nodes were able to communicate the corresponding data to the other nodes. Ultra wideband can fulfill a vital role for this purpose. As in the case of spectrum sensing information exchange, any type of data required for outdoor communications can be shared between the cognitive nodes via ultra wideband simultaneously with the real data.

Locating the Cognitive Nodes via IR-UWB

Owing to the extremely wide band they occupy, IR-UWB systems have an advanced multipath resolving capability. This desirable feature enables these systems to be considered as a means of highly accurate (centimeter range) positioning besides being communication systems [36]. Because of this reason, IR-UWB is the primary candidate for the IEEE 802.11.4a standardization group, which aims at determining a new physical layer for very low power, low data rate communications with a special emphasis on accurate location finding.

Details of positioning systems for cognitive radio are discussed in Chapter 10. Here, they will be summarized for the sake of the completeness of this chapter. Positioning, in general terms, can be done by estimating either the Angle Of Arrival (AOA), the Signal Strength (SS), or the Time Of Arrival (TOA).

In AOA based positioning, two reference nodes determine the angle at which the signal from the target node is arriving, and the location is found by combining the information from the reference nodes. AOA estimation is not suitable for IR-UWB positioning mainly because of the very high number of multipath components. SS based positioning depends on finding the distance to the target node by measuring the energy received from that node. This method requires to employ at least three reference nodes (to implement triangulation). This method is inferior to the TOA estimation based technique in the case of UWB positioning because it requires an accurate channel estimation and it does not take the advantage of the very wide band occupied. The position of the target node can also be determined by finding the TOA of multipath components of the received signal. The wide bandwidth allows the IR-UWB receivers to resolve virtually all the multipath components and to determine the arrival time of each of them. Since these times of arrival provide an information about the delay that is caused by the propagation in the channel, the physical distance between the IR-UWB transmitter and receiver can be estimated accordingly [37].

The positioning capability can make IR-UWB systems an excellent supplement for small-sized cognitive networks. Since such networks aim at not interfering with other radios in their physical environment, it can be very beneficial for them to be able to determine the locations of the nodes in the network closely.

Having information about the precise locations of the nodes in a cognitive network, accurate and high efficiency beamforming [38] can be achieved towards the direction of the target nodes. Also, spatial nulls can be generated towards undesired receivers/signal sources to avoid interference. Beamforming can be accomplished by planar antenna arrays, which can be put onto very small areas for high frequency systems (such as 60 GHz radios), and these arrays can be employed even by wireless nodes that are smaller than a hand palm in size.

The accurate positioning capability of IR-UWB can also be utilized to determine the transmit power adaptively. Using the positioning data, the distance between the transmitting and receiving nodes can be found, and based on the distance information the radiated power can be set. Such an implementation would not only optimize the power consumption but also help to ensure the link quality between the distant nodes.

Another nice utilization of the positioning capability can be tracking cognitive nodes or devices that are mobile. Updating the corresponding positioning information in a frequent manner, a cognitive node can be tracked in space. This way, any communication link directed to it would not be lost although its location is changing continuously.

Examples of using the positioning feature to augment the cognitive communication quality can be increased. Many examples of other advantages of positioning/geolocationing in cognitive radio are given in detail in Chapter 10. The interested readers are referred to this chapter. All these examples lead to the idea that IR-UWB can leverage cognitive radio networks by providing a very strong support through its accurate positioning capability.

Cognitive Node Identification via UWB

In cognitive communications, every cognitive device might be given a serial number when being manufactured. Since this number will be unique to each cognitive device, it would allow the device to be identified. Cognitive devices can transmit their serial numbers steadily via ultra wideband. This way, cognitive devices can recognize each other, and secure cognitive communications can be established between targeted cognitive nodes. Other useful data can accompany this serial number, as well. These include

1. A beacon that indicates if the device is currently communicating or it is in the stand-by mode. This information can be extremely useful because

it would allow spectrum allocations to be done accordingly. If a cognitive node can inform the other nodes about its existence even if it is not in the active mode, some bandwidth can be allocated to this sleeping node. This way, the highest possible number of nodes in a cognitive network can utilize a given band at the same time.

2. Some short data that identify the device by providing information about the type of modulation and coding it uses, its bandwidth, and its carrier frequency. This kind of data can enable different types of cognitive radios recognize each other very easily. Currently, a significant concern about the cognitive radio communications is how to supply the advanced cognitive devices with spectral sensing capabilities that allow these devices not just to detect the existence of other systems, but also to determine the type of modulation and coding they are using. Since this technology requires quite complex algorithms and very costly hardware, the feasibility of such advanced cognitive devices becomes questionable. However, if the cognitive devices provide information about their transmitted signal properties via UWB, there may be no need for an advanced sensing algorithm, and this would be a big step toward the realization of the cognitive radio concept.

Providing Awareness via Impulse Radar

Among the various impulse radio applications, impulse radar is one of the oldest, and it has been used especially for military purposes [39,40]. Practical implementations of impulse radar have been addressed in [41–46]. As in the case of the other IR-UWB applications mentioned so far, impulse radar can improve cognitive communications from a number of aspects when combined with cognitive radio systems.

A. Sensing the Physical Environment/Channel

One of the potential uses of impulse radar can be to determine objects and walls in the indoor environments. Determining the objects can yield a rough estimation of the directions of multipath components, which can improve the channel estimation. Determining the walls, on the other hand, yields information about the physical borders of an indoor network, which may be very useful when establishing a cognitive network.

In mobile applications, impulse radar can allow to estimate the speed of the mobile users. Also, it can enable a cognitive mobile device to measure its own speed. Such a capability would result in being able to estimate the Doppler spread and the channel coherence time, which are important parameters to know in mobile communications.

Impulse radar can also be used to detect the movement of human beings in the wireless channel, which can be very effective on the link quality between cognitive nodes especially for extremely high frequency systems such as the 60 GHz radios [47].

B. Sensing the Cognitive User's Physical Conditions

Cognitive personal communication devices of the future are expected to be much more user aware than they are today. In commonly used sense, user awareness includes having information about the user's preferences about the electronic device such as language, visual interface, and user menu choices. User awareness also requires having statistical data regarding user's habits, which can be exemplified as the average duration of communication or frequently contacted wireless devices and frequently used services.

In cognitive radios, the user awareness is expected to be improved much beyond the usual features. For example, cognitive radio can be made aware of the physical and biomedical conditions of the user to enhance the quality of communications in a different dimension. This kind of awareness can especially improve the quality of lives of users with certain illnesses and conditions. If specific indicators of these health problems in the user's body can be monitored, the communication of the user can be adjusted according to his/her current health status. If an emergency case occurs during the communication, the communication can be interrupted, and an emergency center can be alarmed by the cognitive radio.

Impulse radio radar offers a unique feature to cognitive radio devices to support them with physical condition awareness. Owing to its extremely high multipath resolution capability, impulse radar can non-invasively look at various tissues and organs in human body, and it can even find the densities of certain substances in user's blood. The possible related capabilities of impulse radar can be itemized as follows [48]:

- monitoring the contractions of heart and, this way, determining the heart rate [49],
- detecting chest movements to realize respiratory monitoring,
- imaging the inside of the chest (the lungs),
- imaging the heart (for cardiography purposes), and
- determining the densities of certain substances in user's blood such as glucose, urea, and oxygen [50].

Along with the listed uses, imaging capability of impulse radar can be utilized to investigate the vocal tract of the user to accomplish speaker identification/verification.

The capabilities of impulse radar that are not directly related to cognitive communications but may provide indirect information about the cognitive radio users are as follows:

- imaging to help diagnosing diseases such as breast cancer and brain tumor,
- fetus imaging in pregnant women and determining the heart and breath rate of the baby inside the mothers body,
- monitoring Obstructive Sleep Apnoea (OSA).

12.5 Conclusions

In this chapter, the attractiveness of the impulse radio UWB technology for purposes of implementing cognitive radio is investigated from two main approaches. The first one considers IR-UWB as a direct means of practical cognitive radio realization. Under this approach, the UWB features such as negligible interference to licensed systems, dynamically adjustable bandwidth and data rate, and adaptive transmit power and multiple access are discussed emphasizing their closeness to the cognitive radio requirements. IR-UWB is shown to be a proper candidate for implementing the cognitive radio. The concern regarding IR-UWB's being the enabling technology of cognitive radio is that the overlay mode operation of UWB is currently not allowed by regulatory agencies. Therefore, this option may have to be deferred until it is proven that licensed systems can co-exist with specifically designed overlay UWB systems that have advanced sensing and spectrum shaping capabilities.

The second approach considers IR-UWB as a source of supplementary uses for cognitive radio. Among the numerous uses that will enhance cognitive communications, some significant ones such as sharing the spectrum sensing information and other useful data via IR-UWB, locating the cognitive nodes using IR-UWB, identifying cognitive nodes via UWB, and providing awareness via impulse radar are addressed in this chapter.

It should be emphasized that even in the case that the impulse radio UWB is not accepted as the means of implementing the cognitive radio, its supplementary uses are so beneficial that IR-UWB cannot be separated from cognitive radio systems of future.

References

1. J. Mitola, "Cognitive radio for flexible mobile multimedia communications," in *Proc. Mobile Multimedia Commun. (MoMuC '99)*, Nov. 1999, pp. 3–10.
2. M. Z. Win and R. A. Scholtz, "Impulse radio: How it works," *IEEE Commun. Lett.*, vol. 2, no. 2, pp. 36–38, Feb. 1998.
3. M. Win and R. A. Scholtz, "Ultra-wide bandwidth time-hopping spread-spectrum impulse radio for wireless multiple-access communications," *IEEE Trans. Commun.*, vol. 48, no. 4, pp. 679–689, Apr. 2000.
4. I. Guvenc and H. Arslan, "On the modulation options for UWB systems," in *Proc. IEEE Military Commun. Conf.*, vol. 2, Boston, MA, Oct. 2003, pp. 892–897.
5. H. Arslan, "Cross-modulation Interference Reduction for Pulse-Position Modulated UWB Signals," in *Proc. IEEE Vehic. Technol. Conf.*, Montreal, Canada, Sep. 2006, pp. 1–5.
6. R. Scholtz, "Multiple access with time-hopping impulse modulation," in *Proc. IEEE Military Commun. Conf.*, vol. 2, Boston, MA, Oct. 1993, pp. 447–450.
7. J. Foerster, "IEEE P802.15 working group for wireless personal area networks (WPANs), channel modeling sub-committee report – final," Mar. 2003. [Online]. Available: http://www.ieee802.org/15/pub/2003/Mar03/.

8. G. Lu, P. Spasojevic, and L. Greenstein, "Antenna and pulse designs for meeting UWB spectrum density requirements," in *Proc. IEEE Ultrawideband Syst. Technol. (UWBST)*, Reston, VA, Nov. 2003, pp. 162–166.
9. R. Dilmaghani, M. Ghavami, B. Allen, and H. Aghvami, "Novel UWB pulse shaping using prolate spheroidal wave functions," in *Proc. IEEE Int. Symp. on Personal, Indoor and Mobile Radio Commun. PIMRC 2003*, vol. 1, Beijing, China, Sept. 2003, pp. 602–606.
10. K. Wallace, B. Parr, B. Cho, and Z. Ding, "Performance analysis of a spectrally compliant ultra-wideband pulse design," *IEEE Trans. on Wireless Commun.*, vol. 4, no. 5, pp. 2172–2181, Sept. 2005.
11. X. Wu, Z. Tian, T. Davidson, and G. Giannakis, "Optimal waveform design for UWB radios," *IEEE Trans. on Signal Proc.*, vol. 54, no. 6, pp. 2009–2021, June 2006.
12. Y.-P. Nakache and A. Molisch, "Spectral shaping of UWB signals for time-hopping impulse radio," *IEEE J. Select. Areas Commun.*, vol. 24, no. 4, pp. 738–744, Apr. 2006.
13. Y. Wu, A. Molisch, S.-Y. Kung, and J. Zhang, "Impulse radio pulse shaping for ultra-wide bandwidth (UWB) systems," in *Proc. IEEE Int. Symp. on Personal, Indoor and Mobile Radio Commun. PIMRC 2003*, vol. 1, Beijing, China, Sep. 2003, pp. 877–881.
14. A. Taha and K. Chugg, "On designing the optimal template waveform for UWB impulse radio in the presence of multipath," in *Proc. IEEE Ultrawideband Syst. Technol. (UWBST)*, Baltimore, MD, May 2002, pp. 41–45.
15. M. G. Di Benedetto and L. De Nardis, "Tuning UWB signals by pulse shaping," *Special Issue on Signal Proc. in UWB Commun., Eurasip Journal on Signal Proc.,* Elsevier Publishers, 2005.
16. S. Gezici, Z. Sahinoglu, H. Kobayashi, and H. Poor, "Ultra-wideband impulse radio systems with multiple pulse types," *IEEE J. Select. Areas Commun.*, vol. 24, no. 4, pp. 892–898, Apr. 2006.
17. M. Win and R. Scholtz, "Ultra-wide bandwidth time-hopping spread-spectrum impulse radio forwireless multiple-access communications," *Communications, IEEE Transactions on*, vol. 48, no. 4, pp. 679–689, 2000.
18. R. Tesi, M. Hamalainen, J. Iinatti, J. Oppermann, and V. Hovinen, "On the multi-user interference study for ultra wideband communication systems in AWGN and modified Saleh-Valenzuela channel," in *Proc. Ultra Wideband Systems, 2004. Joint with Conference on Ultrawideband Systems and Technologies. Joint UWBST & IWUWBS. 2004 International Workshop on*, May 2004, pp. 91–95.
19. L. Zhao and A. Haimovich, "Performance of ultra-wideband communications in the presence of interference," *Selected Areas in Communications, IEEE Journal on*, vol. 20, no. 9, pp. 1684–1691, 2002.
20. S. Gezici, H. Kobayashi, H. Poor, and A. Molisch, "Performance evaluation of impulse radio UWB systems with pulse-based polarity randomization," *IEEE Trans. Signal Processing*, vol. 53, no. 7, pp. 2537–2549, July 2005.
21. J. Foerster, "The performance of a direct-sequence spread ultrawideband system in the presence of multipath, narrowband interference, and multiuser interference," in *Proc. Ultra Wideband Systems and Technologies, 2002. Digest of Papers. 2002 IEEE Conference on*, 2002, pp. 87–91.

22. H. Arslan and M. E. Sahin, *Ultra Wideband Wireless Communication.* Hoboken, NJ: John Wiley & Sons, Inc., Sep. 2006, ch. Narrowband Interference Issues in Ultrawideband Systems.
23. L. Piazzo and F. Ameli, "Performance analysis for impulse radio and direct-sequence impulse radio in narrowband interference," *IEEE Trans. Commun.*, vol. 53, no. 9, pp. 1571–1580, Sep. 2005.
24. I. Bergel, E. Fishler, and H. Messer, "Narrowband interference mitigation in impulse radio," *IEEE Trans. Commun.*, vol. 53, no. 8, pp. 1278–1282, Aug. 2005.
25. A. Taha and K. Chugg, "A theoretical study on the effects of interference UWB multiple access impulse radio," in *Proc. IEEE Asilomar Conf. on Signals, Syst., Computers*, vol. 1, Pacific Grove, CA, Nov. 2002, pp. 728–732.
26. L. Piazzo and J. Romme, "Spectrum control by means of the TH code in UWB systems," in *Vehic. Technol. Conf.*, vol. 3, Apr. 2003, pp. 1649–1653.
27. H. Schantz, G. Wolenec, and E. Myszka, "Frequency notched UWB antennas," in *IEEE Conf. Ultrawideband Syst. Technol. (UWBST)*, vol. 3, Nov. 2003, pp. 214–218.
28. I. Guvenc, H. Arslan, S. Gezici, and H. Kobayashi, "Adaptation of multiple access parameters in time hopping UWB cluster based wireless sensor networks," in *Proc. IEEE Int. Conf. on Mobile Ad-hoc and Sensor Systems (MASS)*, vol. 1, 2004, pp. 235–244.
29. X. Jing and D. Raychaudhuri, "Spectrum co-existence of IEEE 802.11b and 802.16a networks using the CSCC etiquette protocol," in *Proc. IEEE Int. Symp. Dynamic Spectrum Access Networks (DySPAN) 2005*, Baltimore, MD, Nov. 2005, pp. 243–250.
30. J. Hillenbrand, T. Weiss, and F. Jondral, "Calculation of detection and false alarm probabilities in spectrum pooling systems," *IEEE Comm. Letters*, vol. 9, no. 4, pp. 349–351, April 2005.
31. G. Ganesan and Y. Li, "Cooperative spectrum sensing in cognitive radio networks," in *Proc. IEEE Int. Symp. Dynamic Spectrum Access Networks (DySPAN) 2005*, Baltimore, MD, Nov. 2005, pp. 137–143.
32. M. E. Sahin and H. Arslan, "System design for cognitive radio communications," in *Proc. Cognitive Radio Oriented Wireless Networks and Commun. (CrownCom)*, Mykonos Island, Greece, June 2006, pp. 1–5.
33. M. E. Sahin, I. Guvenc, and H. Arslan, "Optimization of energy detector receivers for UWB systems," in *Proc. IEEE Vehic. Technol. Conf.*, vol. 2, Stockholm, Sweden, May 2005, pp. 1386–1390.
34. H. Zheng and L. Cao, "Device-centric spectrum management," in *Proc. IEEE Int. Symp. Dynamic Spectrum Access Networks (DySPAN) 2005*, Baltimore, MD, Nov. 2005, pp. 56–65.
35. R. Thomas, L. DaSilva, and A. MacKenzie, "Cognitive networks," in *Proc. IEEE Int. Symp. Dynamic Spectrum Access Networks (DySPAN) 2005*, Baltimore, MD, Nov. 2005, pp. 352–360.
36. S. Gezici, Z. Tian, G. Giannakis, H. Kobayashi, A. Molisch, H. Poor, and Z. Sahinoglu, "Localization via ultra-wideband radios: a look at positioning aspects for future sensor networks," *IEEE Signal Proc. Mag.*, vol. 22, no. 4, pp. 70–84, July 2005.
37. P. Cheong, A. Rabbachin, J.-P. Montillet, K. Yu, and I. Oppermann, "Synchronization, TOA and position estimation for low-complexity LDR UWB devices,"

in *Proc. IEEE Int. Conf. on Ultra-Wideband*, Zurich, Switzerland, Sep. 2005, pp. 480–484.
38. S. K. Yong, M. E. Sahin, and Y. H. Kim, "On the effects of misalignment and angular spread on the beamforming performance," in *Proc. IEEE Consumer Commun. and Networking Conf. (CCNC)*, Las Vegas, NV, Jan. 2007.
39. R. Fontana, "Recent system applications of short-pulse ultra-wideband (UWB) technology," *IEEE Trans. Microwave Theory Tech.*, vol. 52, no. 9, pp. 2087–2104, Sep. 2004.
40. I. Immoreev and J. Taylor, "Ultrawideband radar special features & terminology," *IEEE Aerospace Electron. Syst. Mag.*, vol. 20, no. 5, pp. 13–15, Mar. 2005.
41. M. Mahfouz, A. Fathy, Y. Yang, E. Ali, and A. Badawi, "See-through-wall imaging using ultra wideband pulse systems," in *Proc. 34th Applied Imagery Pattern Recognition Workshop*, Washington, DC, Oct. 2005.
42. R. J. Fontana, J. F. Larrick, and J. E. Cade, "A low cost ultra wideband system for UAV communications and high resolution radar applications," in *Precision Strike Technology Symp.*, Baltimore, MD, Oct. 1997.
43. R. J. Fontana, J. C. Koppier, and R. W. T. Mulloy, "A low cost ultra wideband (UWB) radar altimeter," in *Proc. AUVSI'96*, Orlando, FL, July 1996, pp. 297–301.
44. I. Immoreev, S. Samkov, and T.-H. Tao, "Short-distance ultra wideband radars," *IEEE Aerospace Electron. Syst. Mag.*, vol. 20, no. 6, pp. 9–14, June 2005.
45. D. Sostanovsky and A. Boryssenko, "Experimental UWB sensing & communication system," *IEEE Aerospace Electron. Syst. Mag.*, vol. 21, no. 2, pp. 27–29, 2006.
46. A. Yarovoy, L. Ligthart, J. Matuzas, and B. Levitas, "UWB radar for human being detection," *IEEE Aerospace Electron. Syst. Mag.*, vol. 21, no. 3, pp. 10–14, Mar. 2006.
47. S. Collonge, G. Zaharia, and G. Zein, "Influence of the human activity on wide-band characteristics of the 60 GHz indoor radio channel," *IEEE Trans. Wireless Commun.*, vol. 3, no. 6, pp. 2389–2406, Nov. 2004.
48. E. Staderini, "UWB radars in medicine," *IEEE Aerospace Electron. Syst. Mag.*, vol. 17, no. 1, pp. 13–18, Jan. 2002.
49. M. Chia, S. Leong, C. Sim, and K. Chan, "Through-wall UWB radar operating within FCC's mask for sensing heart beat and breathing rate," in *Proc. European Radar Conf. (EURAD 2005)*, Oct. 2005, pp. 267–270.
50. C. Y. Lee and C. Toumazou, "Ultra-low power UWB for real time biomedical wireless sensing," in *Proc. IEEE Int. Symp. on Circuits and Syst. (ISCAS 2005)*, vol. 1, May 2005, pp. 57–60.

13

Applications of Cognitive Radio

Hüseyin Arslan[1] and Sadia Ahmed[2]

[1] University of South Florida, Tampa, FL, USA, arslan@eng.usf.edu
[2] University of South Florida, Tampa, FL, USA, ahmed3@eng.usf.edu

13.1 Introduction

Technology is futile without its application. One of the major driving forces of a new technology is the combination of man's endless thirst for knowledge, unequal intuitive mind, and limitless ambition to better his life. The other important factor consists of user demand, user necessity, and increasing user desire to have it all in one single device.

Wireless communication involves accessing and utilizing air spectrum for communication purpose. More and more devices may fall under the umbrella of "wireless communication technology" in the near and distant future. For example, every personal computer may soon be connected to its keyboard and mouse via the radio medium. This in turn can eliminate hardware manufacturing cost, manual labor for connection, and hardware pollution of the discarded parts. Wireless technology thus opens up an endless market of wireless communication applications.

"Cognitive Radio" (CR) in the wireless world is relatively new and comprises of two magical words that encompass everything! It is the radio with intelligence [1]; capable of using its radio ability in the most optimum manner by interacting with surrounding environment, learning about the environment, and using the learned knowledge to improve communication. The traditional radio on the other hand lacks cognitive intelligence; uses fixed spectrum, special purpose hardware dedicated to that spectrum, and its ability is confined to manufacturer driven specifications.

In wireless communication today, spectrum partitioning is overcrowded due to continuous rise of numerous applications. However, the majority of these spectrum partitioning often remain unused during off-peak hours. CR takes advantage of these spectrum holes by intelligently identifying and using them, and releasing when required by the primary users. CR intelligence is not only limited to unused spectrum usage but can also improve

H. Arslan (Ed.), Cognitive Radio, Software Defined Radio, and Adaptive Wireless Systems, 383–420.

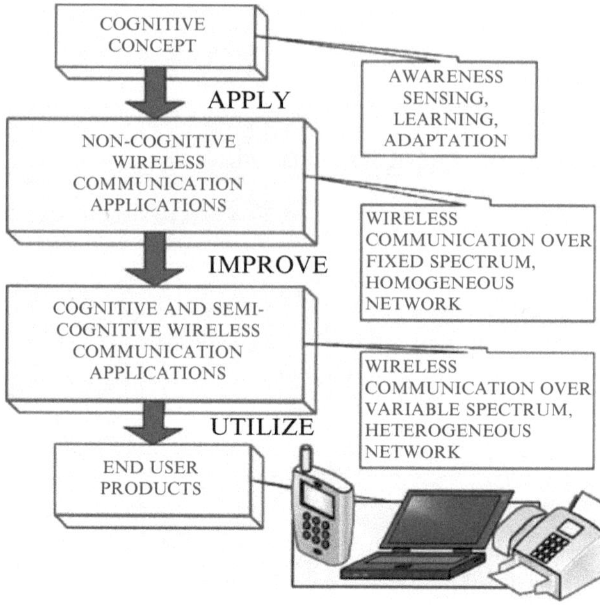

Fig. 13.1. Cognitive concept in wireless communication.

any communication that fully or partially utilizes the wireless medium. Figure 13.1 illustrates a generic flow from cognitive concept to the final end user products. It is impossible and unnecessary to include every current and future envisioned CR application within this chapter. Therefore, the major ones are presented here.

This chapter is organized as follows. Section 13.2 addresses various wireless communication applications that utilize cognitive radio. Section 13.3 presents their major implementation challenges and Section 13.4 presents the concluding remarks.

13.2 Cognitive Wireless Communication Applications

Cognitive radio comes with multitudes of features. The majority of these attributes are "Awareness", "Sensing", "Learning", and "Adaptation". They are addressed in detail in other chapters while their applications are presented in this chapter.

CR applications can be broadly categorized into those, which contribute to the full or partial improvement of system performance in the existing wireless communications; and into those, which can be beneficial in establishing new wireless communication. This section will address both of these categories. Figure 13.2 represents a high level classification of possible CR applications.

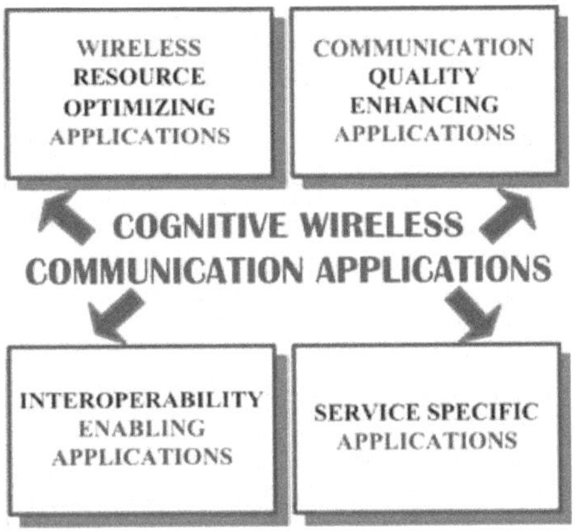

Fig. 13.2. Classification of cognitive radio applications.

This classification is conducted for better understanding and organization of each application and their relationship with "cognitive concept". The first group consists of CR applications contributing to the optimization of various wireless resources. The second group represents CR applications contributing to the improvement in communication quality. The third and forth categories reflect CR applications concerning interoperability and service/end user products, respectively.

In general, many of these applications across groups can benefit from each other, their territories may overlap, and they can (individually or combined) work together to improve wireless communication. The applications in the following sub-sections are presented as generic overview without going into each implementation detail. Although, resource optimization and quality enhancement applications are grouped separately, they are closely related and are addressed under the same section as follows.

13.2.1 Resource Optimization and Quality Enhancing Applications

Wireless resources are limited. This scarcity in resources demands careful consideration and planning in their usage. The conventional radio resources encompass spectrum, hardware/software, network infrastructure, and power. Cognitive intelligence adds another dimension to these namely "knowledge". Figure 13.3 illustrates the major wireless resources that can be optimized by CR applications. These applications can result in communication quality improvement as well. Figure 13.4 illustrates generic communication among various wireless devices, the quality of which can be greatly improved by CR applications.

Table 13.1 summarizes the various resources and their respective applications. It is beneficial to address each resource related applications separately. Although, one or more resources can work together in the optimization

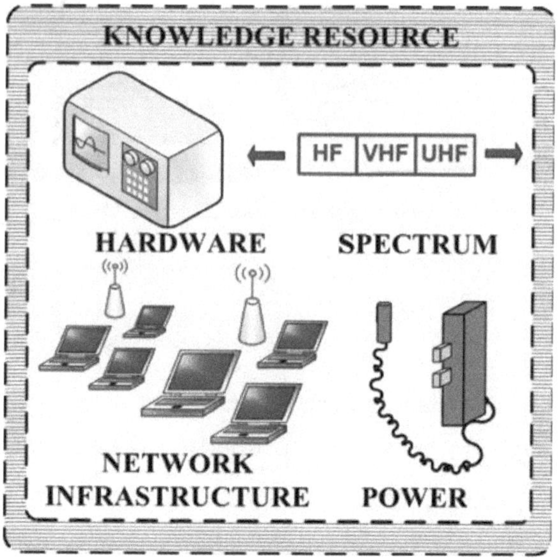

Fig. 13.3. Wireless resources.

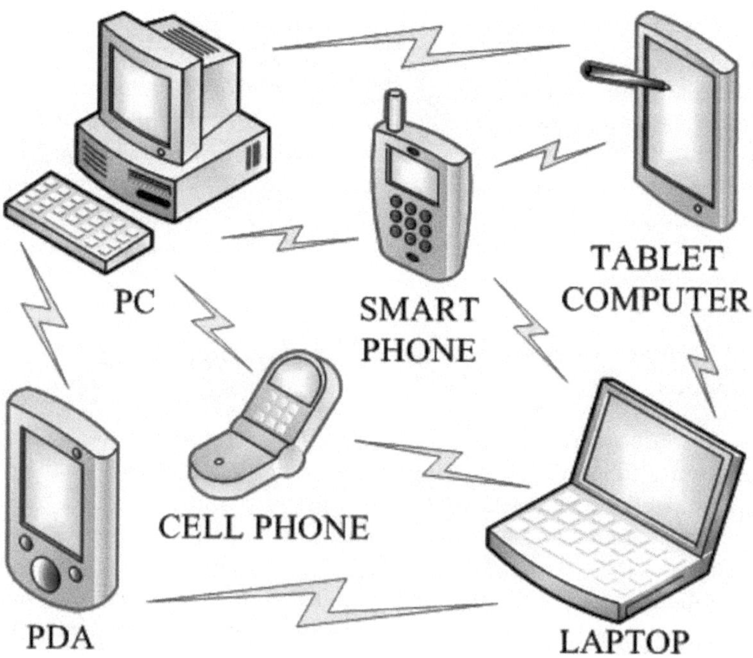

Fig. 13.4. Wireless communication among wireless devices.

Table 13.1. CR applications in wireless communication.

Wireless resources	Brief description of related CR applications
Knowledge	Awareness of available knowledge and sensing their changes Utilization of knowledge
Hardware/software	Software upgradeable hardware Hardware/software operation over diverse networks and spectrums Self diagnosis and repair of hardware/software
Power	Prioritizing tasks according to power level Power control according to channel and mobile position Power control according to audio and visual activity
Spectrum	Detection, accessing and management of available spectrum
Network	Cross layer optimization Autonomous selection of optimum route and topology Network security against intruders/hackers Self diagnosis and repair of network components

and/or quality enhancement process within the confinement of a single CR application.

Knowledge

"Knowledge" is a significant resource exploited by cognitive radio to improve wireless communication. Knowledge comprises of information on the individual entities within a communication link. Configuration and capability parameters of hardware/software, network infrastructure, power, and spectrum can represent knowledge. Measured and predicted values of channel related information such as link quality, Signal-to-noise Ratio (SNR), and channel fading parameters might also be considered as knowledge. Similarly, information about user habits, location, and bio-metric information such as thumb print, voice, and retina print can be considered as knowledge as well.

The traditional wireless systems exploited some of these items in the recent years through various adaptation techniques. But, cognitive concept has the potential to bring revolutionary changes in knowledge utilization. The major tasks to use knowledge can include cognitive awareness of various parameters, tracking their changes, and employing appropriate action. These parameters can be provided to the cognitive radio via look-up tables, through the download process from a central/semi-central database, and through a software upgrade process. Some parameters can be measured in real or non-real time as well.

Knowledge utilization tasks can be better understood through some examples as below:

- Cognitive awareness of channel fading characteristics, channel noise and/or interference level, and error correction capability can allow appropriate selection of Forward Error Correction (FEC) scheme, interleaving length, modulation order, and type, etc. This in turn, can reduce the effect of interference, and/or improve signal quality, and/or provide high data rate transmission.
- Knowledge on selectivity in multi-dimensional channel space can lead to adjusting various parameters like channel equalizer taps (in frequency selectivity dimension) [2] as illustrated in Figure 13.5, or channel tracking memory (in time selectivity dimension), or multi-antenna loading parameters (in spatial selectivity dimension).
- Knowledge on transmitted information type such as voice, data, text, video, audio or images can help determine the most appropriate source coding technique.
- Network knowledge on available routes, available bridging nodes, and protocol status of nodes can result in reducing network delay; minimizing network usage, and providing uninterrupted communication over heterogeneous network components.
- Spectrum knowledge beyond the traditional concept of fixed spectrum can lead to transmission over variable spectrums and variable dimensions of channel space. This in turn can provide ubiquitous connectivity and higher achievable data rate as well.

Cognitive radio may also deduce more knowledge from the available information. One such case can involve deriving channel characteristics [3] from the available Global Positioning System (GPS) information. It can subsequently reduce transmission of channel related information while increasing data rate.

The knowledge utilization tasks discussed so far can be carried out on individual nodes as well as over multiple nodes in a collaborative manner.

Hardware and Software

Hardware is the physical entity that makes transmission/reception a reality. For example, two cell phone handsets at the two ends of a communication represent partial hardware for that specific connection. The term "Hardware/Software" represents the hardware/software resources of individual nodes within a link. CR enables intelligent utilization of this resource to improve communication performance.

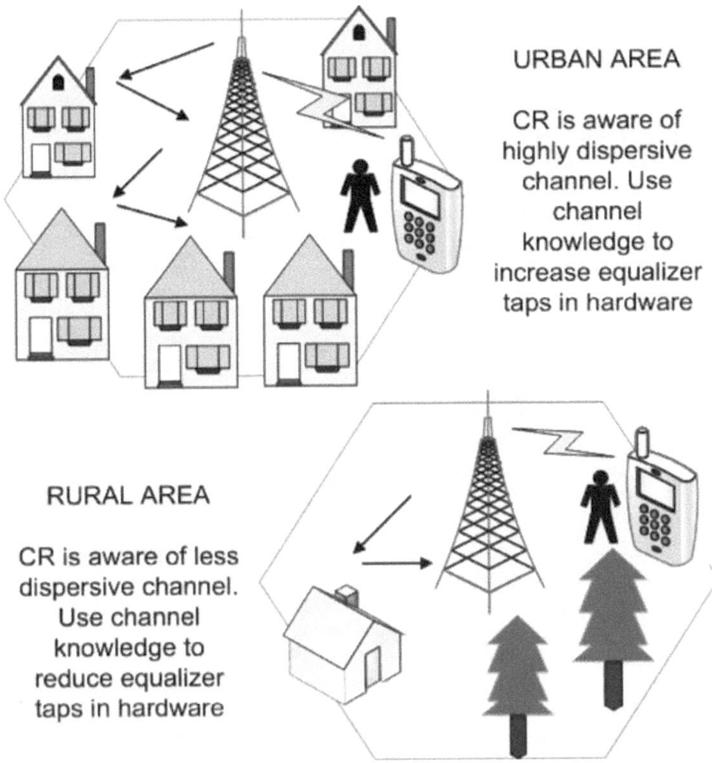

Fig. 13.5. Knowledge effect on other resources.

One of the significant Software Defined Radio (SDR) based CR applications includes partial or full upgrade of hardware through partial and/or full alteration of software. This in turn can

- reduce hardware replacement need;
- lower hardware manufacturing, replacement, and labor cost;
- provide availability for more applications;
- reduce hardware complexity; and
- eliminate redundant hardware chain.

With extensive deployment of CR, each node may also be able to download software upgrades from neighboring nodes. They may collaborate with each other to share hardware/software resources as well.

The other important SDR-based CR applications can include

- Configuration of hardware to operate over any network. For example, a cell phone operating on a Global System for Mobile Communications (GSM) network can establish connection over a Wireless Local Area Network (WLAN) in emergency or other situations. Figure 13.6 illustrates such a scenario.

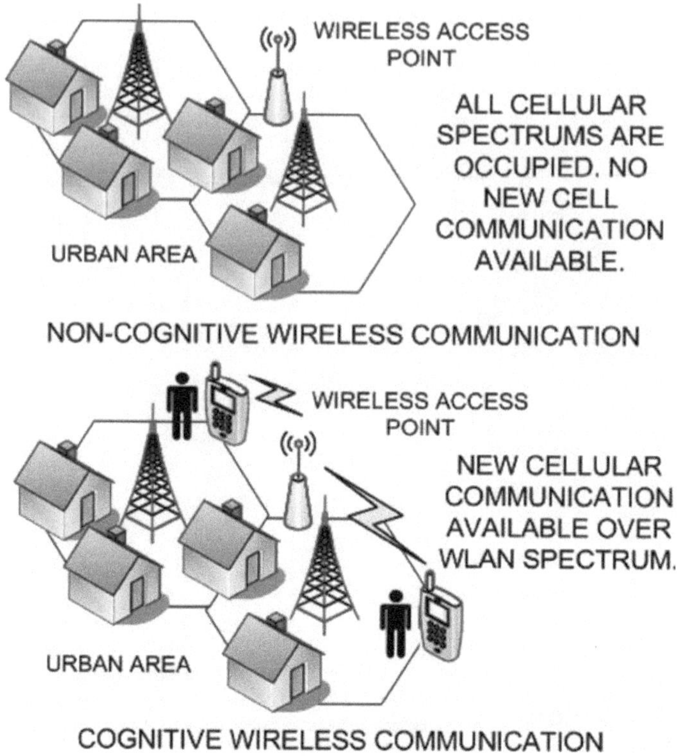

Fig. 13.6. Communication over diverse spectrum and network.

- Configuration of hardware to operate over any spectrum by generating appropriate waveform [4]. Once a new or more suitable spectrum is sensed and accessed, CR can adjust its signaling parameters (such as signal bandwidth, frequency, modulation, and coding) according to selected spectrum conditions. This can result in switching onto a new spectrum with improved signal quality, better channel fading characteristics, and reduced interference effects.

Among non-SDR based CR applications, self diagnosis and heal of hardware is a significant one, where cognitive radio enables autonomous measurement of hardware usage, detection of damage, and repairing of faulty parts of individual nodes. As an outcome, hardware downtime is minimized and usage is optimized.

Power

Power had been a never ending battle since the birth of wireless communication. Optimizing power is important for sustained wireless connectivity over long period of time. At the same time sufficient power level is necessary for maintaining the required link quality within the coverage area.

Cognitive intelligence can bring drastic improvements in the power situation in wireless communications. The major CR applications involving power can be represented as

- Adaptive power control. Power adjustment according to link quality has already been implemented in various adaptive wireless systems. However, these adjustments are mostly rule driven and limited to a certain degree. The goal of CR applications go beyond these limits to adjust power to any level for any link quality. Appropriate power control can also ensure satisfying Federal Communication Commission (FCC) spectrum mask requirement and interference control [5].
- Optimize power by executing auto shut off; deactivating low priority jobs (such as continuous spectral monitoring and collaborative sensing) or notifying user when power level reaches a defined emergency threshold. This can save sufficient power for high priority transmission such as a "911" call.
- Allow transmission only upon an audio or visual change of scene. For example, with a baby monitor set up, CR transmits baby information only when the baby cries or moves.
- In ad hoc or multi hop networks CR enables the routing protocol to be aware of available power of other nodes and the routing mechanism is adopted accordingly.

Spectrum

A whole chapter (Chapter 9) in this book is dedicated for spectrum resource and for the description of multi-dimensional spectrum space. Therefore, here, we briefly mention spectrum resource and refer the reader, to Chapter 9 for the details.

The overcrowded spectrum partitioning impose additional usage of available spectrum in alternate dimensions and instances (such as different times of day). Cognitive intelligence brought forth a breakthrough in this "additional usage" of spectrum.

Spectral resources outside the domain of fixed bands can envelop

- any licensed bands that are in use such as GSM bands in cellular;
- any licensed bands that are not in use such as spectrum holes in unused public safety and commercial TV bands [6];
- any unlicensed bands in use such as Industrial, Scientific and Medical (ISM) band;
- any unlicensed bands not in use such as 60 GHz bands; and
- any licensed bands that are auctioned such as Very High Frequency (VHF) TV bands [7].

Utilizing these additional spectral resources can be tricky. For example, accessing licensed or unlicensed bands while they are in use need to be carried out in alternate dimension (such as space, time, subcarrier, spreading code, transmission power, polarization [8]), with controlled power [5] or using multiple accessing schemes. Utilizing licensed bands that are not in use needs to be readily vacated when the licensed user returns.

Cognitive radio enables various spectrum utilization tasks to realize "additional usage" of spectrum. The major spectral tasks are presented as below:

- **Dynamic Spectrum Sensing:** Cognitive radio, with its sensing ability, allow detection of available spectral resources. The spectrum sensing techniques employed by CR can include matched filter detection, energy detection, cyclostationary feature detection [9, 10], and interference-specific detection [10]. These techniques are presented in detail in Chapter 9.
- **Dynamic Spectrum Access:** Sensing of available band, is not enough unless the detected spectrum is accessed and utilized efficiently. CR can dynamically select available spectrum and intelligently adapt its required parameters to operate on it [1]. The spectrum sharing [11] or spectrum access can be implemented using centralized, distributed, co-operative, and non-cooperative schemes [10].
  In the presence of multiple spectrum holes, CR can intelligently access the more suitable ones for present transmission needs.
- **Dynamic Spectrum Management:** Dynamic spectrum sensing and accessing both require cognitive coordination, planning, management, and characterization of available spectrum for intelligent decision making. This decision process can be dynamic to use the spectrum opportunity before it is occupied again.
  The major spectrum management tasks performed by CR can incorporate the following:
  – ***Rules:*** Identifying policy, etiquette, and other rules applicable to the identified spectrum and intelligently handling those rules.
  – ***Duration and release:*** Estimating the duration of vacant spectrum. Several users sharing an Radio Frequency (RF) interface can have different usage durations. Spectrum release can ideally occur in stages as each service completes its communications task.
  – ***Prioritize:*** The available accessed spectrum can be demanded by more than one nodes simultaneously. Cognitive radio can perform intelligent prioritizing of their spectrum access.
  – ***Coordination and synchronization:*** Cognitive spectrum management can also account for the destination(s)'s estimated receive capabilities and provision a reliable means of source–destination co-ordination and spectrum-usage synchronization.
  – ***Interference:*** Identifying existing active interference-tolerant and intolerant services who desire the available spectrum.
  – ***Performance metric:*** Spectrum management can include cognitive decision making to determine if transmission over the new available spectrum would indeed improve the communication. This can be achieved

through metric measurement and assessment of cognitive wireless communication performance.

In brief, Dynamic Spectrum Management (DSM) can allow consumers, military, government, and civil groups an improved means of information conveyance, cognition capabilities and knowledge about the locations, path-distances, trajectories, available radio resources, and desired Quality of Service (QoS) [8]. As a result, an optimal solution can be obtained for each scenario.

Spectrum sensing, detection, and management tasks can also be carried out over multiple nodes collaboratively [12]. However, research and implementation using collaboration need to be investigated further.

As an outcome of spectrum optimization, CR can allow shifting into a new spectrum with low interference and better SNR quality. CR ability to operate over diverse spectrums can provide ubiquitous connectivity, higher spectral, and user capacity and higher data rate. It may be possible to establish communication at any time and anywhere over any dimension (time, frequency, space, etc.) as long as spectrum is available. Therefore, cognitive radio may possibly eliminate wireless communication failure due to spectrum shortage alone.

Network

Network can include all hardware/software resources belonging to source, destination, and intermediate entities engaged and/or will engage in a wireless communication link. Optimizing network within a specific link can depend on individual and/or multiple participating nodes.

The network tasks carried out by CR over individual nodes are addressed below. However, their effects can influence the overall network performance.

- **Cross Layer Optimization:** Cross layer optimization involves optimizing the functions of multiple protocol stacks jointly. CR can monitor its protocol stacks intelligently and adapt its protocol composition depending on application, network, and other needs. The details of the cross-layer optimization is given in the following chapter.
- **Network Learning:** Network learning includes utilizing network knowledge to reconfigure network parameters. This can, in turn, improve network capacity, handling of multiple applications, and prioritizing applications at times of network congestion.

There are several collaborative network tasks performed over multiple nodes and are addressed as follows:

- **Optimum Route:** Cognitive intelligence can identify an optimum route and cleverly re-configure network parameters to use it; resulting in optimization of network resources and minimization of network delay and cost. Figure 13.7 illustrates an intelligent route selection for WLAN.

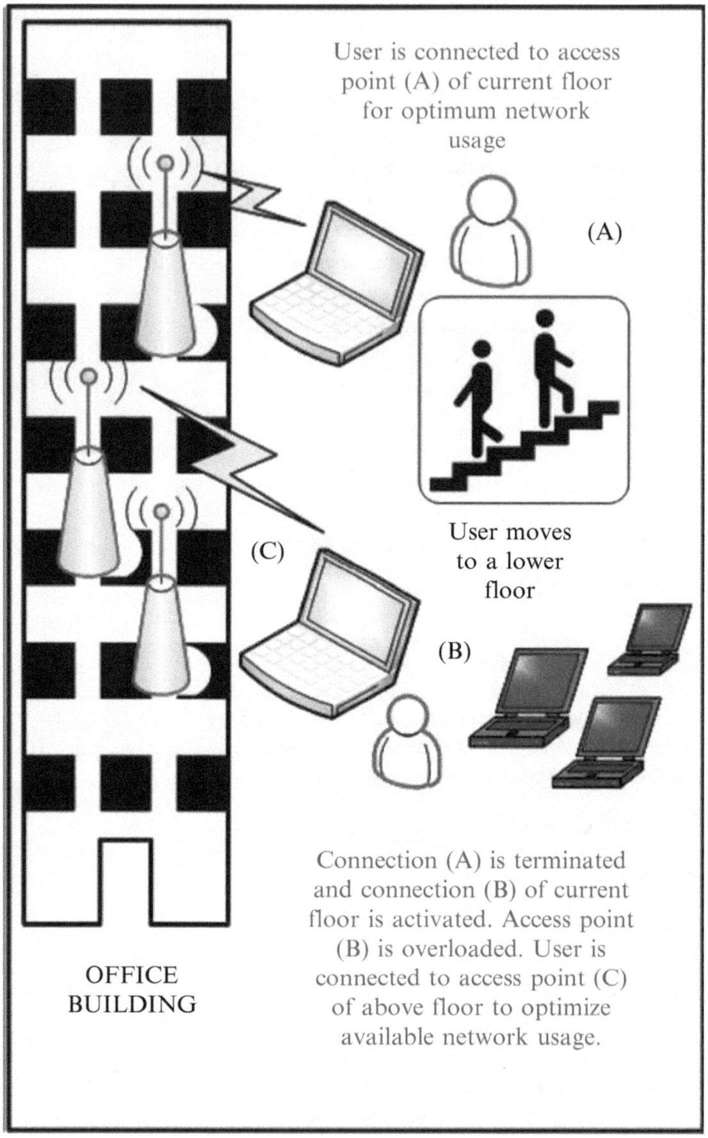

Fig. 13.7. Network optimization.

- **Topology Change:** Topology change involves autonomous change of network topology to establish and maintain communication between source and destination.

Centralized topology can include central control and peripheral nodes to be static or mobile [13]. In the event of any node falling out of range, the central control can use other fixed or mobile cognitive nodes to communicate with the distant node in a relay fashion. In a distributed environment, with the joining/leaving of a node, cognitive radio can re-arrange its network between source and destination. It can choose the appropriate leader within a domain of nodes, or sub-nets based on its location, available radio resources, and behavior patterns. Cognitive devices can act as bridging nodes between domains of ad-hoc networks or between domains of ad-hoc and wired network. Thus transforming the conventional networks into more robust and time-variant topologies.
Depending on the availability of central control, cost, delay, and other network factors, cognitive radio can establish centralized topology or a distributed one to optimize the network resources in hand.
- **Network Security:** The greater availability of network access makes it more vulnerable to intruders/hackers [14–16]. These intruders can invade the network and cause temporary or permanent damage [17]. As a consequence, the network communication may be temporarily or permanently interrupted. Cognitive radio can provide high level of network security against such catastrophic effect and minimize network disruption.
- **Self Diagnosis/Heal of Network:** One of the important network tasks performed by CR is diagnosing its network components and link. This can involve autonomous measurement of network usage and network damage. CR can thus detect faulty parts and also provide repairing without human intervention. This in turn can minimize network disruption, downtime, and provide uninterrupted communication between source and destination.
- **Software Upgrade of Network:** Software upgradability involves upgrading of network components through software modification. It can eliminate the need for recurring replacement of network hardware when required by a new link or by a new technology.

The CR applications presented so far utilize one or more wireless resources. Outside this concentration on individual resource related applications, there are several wireless topics that require attention. These topics are presented separately in the next discussions.

Interference

Interference is an important topic in wireless communication that prevents unlimited expansion of communication domain in terms of user, network, spectrum, space, time, and geographical coverage capacity. Cognitive radio helps mitigate interference issues and provide a powerful reduced interference or interference free communication system. The interference related applications can be grouped as follows:

- **Self Interference:** The radio medium imposes interference on the transmitted signal due to multipath fading characteristics of the wireless channel. Depending on the operating environment, symbol duration, mobility, and the technology used, these channel induced interference can include Inter Symbol Interference (ISI) and/or Inter-carrier Interference (ICI). CR can dynamically sense frequency and time selectivity of channel and adjust its own transmitter/receiver parameters against or ICI. Figure 13.5 briefly illustrates the CR ability to adjust equalization complexity [2] depending on channel characteristics.
- **Other User's Interference:** In a multiple user environment, the signal of other simultaneous users can cause undesired interference on the desired user signal. Multiple users may reside at the same instance of a specific dimension such as at the same time, on the same frequency band, or in the same spatial range. Examples of this type of interference and the corresponding CR applications can be
  – ***Co-Channel Interference (CCI):*** In cellular type systems, even though the users are spatially separated in different cells, the simultaneous user operation on the same frequency can cause CCI. The amount of CCI can depend on the user location within each cell. Adaptive beamforming, adaptive channel allocation, and adaptive power control according to varying transmitter–receiver distance [5] can reduce undesired user interference.
  – ***Adjacent Channel Interference (ACI):*** Spectral overlapping among adjacent channel users can create ACI. On the other hand, this same spectral overlapping greatly increases spectral efficiency. This trade-off between interference and efficiency can be optimized with the help of cognitive radio through adaptive channel allocation and by other means.
  – ***Narrow Band Interference (NBI):*** Broadband channel users (like in Ultra Wideband (UWB)) can be heavily effected by narrow band interferer where interference affects only part of the desired transmission spectrum. CR can intelligently identify the interfering user and avoid transmission over that band.
- **Interference in Spectrum Sharing:** In a spectrum sharing environment, detection of primary user can be tricky [18]. Primary user detection needs to depend on more than one parameter such as SNR values, channel fading parameters, spectrum usage statistics, and geographical location [1]. Cognitive radio can intelligently determine the presence of primary user and avoid any possible interference on primary from secondary or vice versa. In order to achieve this, RF environment monitoring [19] need to be deployed that will detect spectrum holes, primary user status, and dynamically adapt CR operating parameters.

CR can also detect primary presence in a collaborative manner by sharing the detection task among other nodes [12].

Link Adaptation

The link in wireless communication can include a transmitter, a receiver, and the wireless channel. The factors influencing a link can be radio environment, mobility, data rate, available spectrum, network, and individual application requirements. Cognitive radio employs various techniques to handle these factors and maintain desired link quality. Even though a few of these applications were mentioned in previous discussions, it is beneficial to emphasize them individually under link adaptation.

- **Adaptive Source Coding:** CR can utilize knowledge on information type such as voice, text, video, image, and emergency tone; QoS for each application, and the available transmission bandwidth [20] to choose appropriate source coding technique for transmission. The source coding techniques can include adaptive selection of the coding type and coding rate [21].
- **Adaptive Channel Coding:** Cognitive awareness of the channel can optimally decide the appropriate channel coding type, coding rate, interleaving length, and type [21]. CR can also dynamically adjust these parameters to adapt to changes in channel characteristics. This in turn, can autonomously adjust number of transmitted bits to increase or decrease redundancy, and maintain desired link quality. For example, if a strong FEC and interleaving is provided for robustness against the worst channel impairments for all times, the data rate, latency, and spectral efficiency will be sacrificed at times of improved link quality. Therefore, it is essential to adapt the coding rate based on the link quality.
- **Adaptive Modulation:** Cognitive utilization of channel quality, desired data rate, available channel bandwidth, and desired link quality can lead to the optimum choice of modulation/demodulation type, order, and format [21]. This can result in improved BER, data rate, communication robustness, and range. For example, Higher Order Modulations (HOM) allow higher number of bits to be transmitted for a given symbol rate. On the other hand, HOM is less power efficient, since it requires higher energy to maintain a given BER. Therefore, HOM should be used only when the link quality is high.
  A few scenarios are addressed below in order to emphasize modulation and coding adaptation according to link quality. Upon measuring the link quality regularly, the most appropriate modulation and coding scheme can be assigned for the next transmission interval. In incremental redundancy scheme, information is first transmitted with high coding rate. This results in a high bit rate provided that the receiver is capable of decoding the received information. However, if decoding fails with such a high

rate, additional coded bits (redundancy) needs to be sent for successful decoding of transmitted bits. At the same time, sending extra coded bits reduces the resulting bit rate incrementally and adds extra delay. Therefore, the initial code rate and modulation for the incremental redundancy scheme is based on measurements of the link quality instead of starting at an arbitrary rate. As a result, the combination of incremental redundancy with adaptive initial code rate lowers delays, lowers memory requirements, and at the same time achieves high data rates. The different initial code rates can be obtained by puncturing a different number of bits from a common convolution code (for example, a rate of 1/3 or 1/2). Incremental redundancy operation is enabled by puncturing a different set of bits each time a block is retransmitted, whereby the code rate is gradually decreased toward the mother code rate for every new transmission of the block.

Adaptive modulation and coding attracted many new generation wireless standards as an option to increase the data rates. In conjunction with the advanced receiver algorithms that reduce the required SINR, a better link quality can be exploited to increase the data rates further. The combination of adaptive modulation with multi-antenna transmitter and Multiple Input Multiple Output (MIMO) schemes based on the feedbacks related to channel estimates, channel quality, channel correlation etc. is one of the interesting research areas that exploits the information on link quality and channel selectivity. Based on the channel feedback information, the modulation type on multi-antenna transmitters can be varied. In a similar way, adapting the source coding with the channel coding and/or modulation is another interesting area of focus utilizing link adaptation. For example, Adaptive Multi Rate (AMR) codec allows to change the compression rate of speech depending on the link quality, as in GSM AMR. For weak link conditions, where heavy FEC is required, AMR has the ability to decrease the codec rate (more speech compression) to allocate more bits for FEC.

- **Adaptive Carrier Frequency Selection:** CR can sense the available spectrum and accordingly adapt its carrier frequency and cell assignment as in a cellular service [21].

  The introduction of cellular technology led to efficient usage of finite spectrum through a concept called frequency reuse. However, the capacity of cellular systems is interference limited, dominated by CCI and ACI. Early cellular systems aimed to avoid these major interference effects by designing the system for the worst case interference conditions and fixed channel allocation. This is often achieved by employing lower frequency re-use and by allowing enough carrier spacing between adjacent channels where both techniques sacrificed on spectral efficiency. Dynamic channel assignments relative to current interference, propagation, and traffic conditions is an improvement over fixed channel allocation that substantially increases spectral efficiency. In traditional cellular system designs, the channel frequencies corresponding to cells is fixed limiting each cell to use only a set of frequencies. Even in the presence of other cells with available unused frequencies the cells that are in need of frequencies (fully loaded cell) cannot use those unused frequencies belonging to other cells. On the other hand, in dynamic channel allocation, all the channels belong to a global pool and the channels are assigned according to a cost function that considers the CCI and ACI into account. As a result, for non-uniform traffic conditions, the available channels can be used more efficiently.

- **Adaptive RF Component:** CR can manipulate its RF components such as antenna hardware to optimize transmission [21]. This in turn can reduce information loss, optimum power usage, and reduced interference effects. Some of the antenna manipulation techniques can include multiple antenna usage, and adaptive beam forming. An example of multiple antenna CR application is MIMO where more than one antenna are used on transmission and reception to capture path diversity and improve communication.
- **Technology Specific Adaptation:** CR can allow individual technology specific adaptation. CR enables adaptive selection of cyclic prefix size, Fast Fourier Transform (FFT) size, and number of carriers in Orthogonal Frequency Division Multiplexing (OFDM), and adaptive selection of duty cycle and number of pulses per bit in impulse radio based UWB technology [21].
- **Other Adaptive Parameters:** In Hierarchical Cellular Systems (HCS), CR can allow adaptive cell assignments. The use of HCS has become a major component in 3G mobile systems such as UMTS and IMT-2000. In an HCS, various cell sizes can be deployed with small cell clusters overlaid by larger cells. Microcells increase capacity within a coverage area, but radio resource management becomes more difficult. Within microcells, the number of hand-offs per cell increases by an order of magnitude but the time available to make a hand-off is decreased. The HCS handles the microcell situations by assigning cells to the mobiles depending on their speed or doppler spread estimates. For example, in a two layer environment, low speed mobiles can be assigned to microcells, whereas, high speed mobiles can be assigned to macrocells. This way, the macrocell/microcell overlay architecture provides a balance between maximizing the capacity per unit area and minimizing the number of hand-offs. As a result, the risk of call dropping is reduced and handover delays is lowered, switching load is reduced, and QoS is increased. The HCS can extend to more than two layers. For example, pico-cellular layers can also be included in multi-layer HCS. Similarly, communication satellite beams can be overlaying all the terrestrial layers at the highest hierarchical level. Recently, the dynamic allocation and multi-tiered design strategies are further generalized to take power control, cell hand-off, traffic classes (like multi-media), and user priorities into account. Link adaptation schemes can also be combined with adaptive resource allocation. For example,

adaptive modulation (and coding) can be combined with dynamic channel allocation. Similarly, adaptive modulation (and coding) can be combined with handover algorithms to introduce more intelligent handover strategies. All these developments require more sophisticated adaptation of the network, and they are based on many parameter measurements.
In a multi-hop network, CR can adaptively choose routing path algorithms. Adaptation of clustering parameters for clustering based routing and network topology, scheduling algorithms, adaptive frequency hopping, and adaptive generation of spreading code can be implemented to improve communication quality as well [21]. CR also allows adapting to user preferences, user locations, and user habits [22].

13.2.2 Interoperability

"Interoperability" is a powerful tool in CR domain that enables intelligent wireless communication across any boundary and over any dimension. This ubiquitous connectivity can define the ultimate adaptive system over heterogeneous networks, varied spectrums, diverse geographical boundaries, and over different communication policy and regulations.

Achieving the ideal interoperable system may still be a long way from reality. Some of the cognitive applications that can set the way are CR ability to operate over legacy systems, intelligent policy management methods, utilizing network knowledge, cross layer optimization, and various multi-antenna configurations. Cognitive traits of frequency agility and protocol independence can allow CR to build SDR platforms that are potentially capable of solving radio and system interoperability problems. These platforms can provide seamless system operation in a highly fragmented and in a multi-terminal/multi-frequency heterogeneous communication environments.

An immediate implementation of interoperability at present can be the public safety first responders and military applications [13]. The role of interoperable devices, services, networks, and spectrums are extremely critical for these applications. In other sectors, such as in consumer applications, CRs can offer interoperability across licensed, unlicensed, and semi-licensed spectrum services and over diverse networks. Figure 13.8 represents a coalesce of interoperable devices, services, and networks for a visual understanding of interoperable wireless system.

13.2.3 End User Product/Service Specific Cognitive Wireless Communication Applications

The lowest tier in the CR application hierarchy comprises of service specific/end user products such as the cell phone, laptop, and fax machine illustrated in Figure 13.1. These services and products utilize the benefits of cognitive influence in wireless communication applications. It may be impossible to capture all present and future products within this chapter and

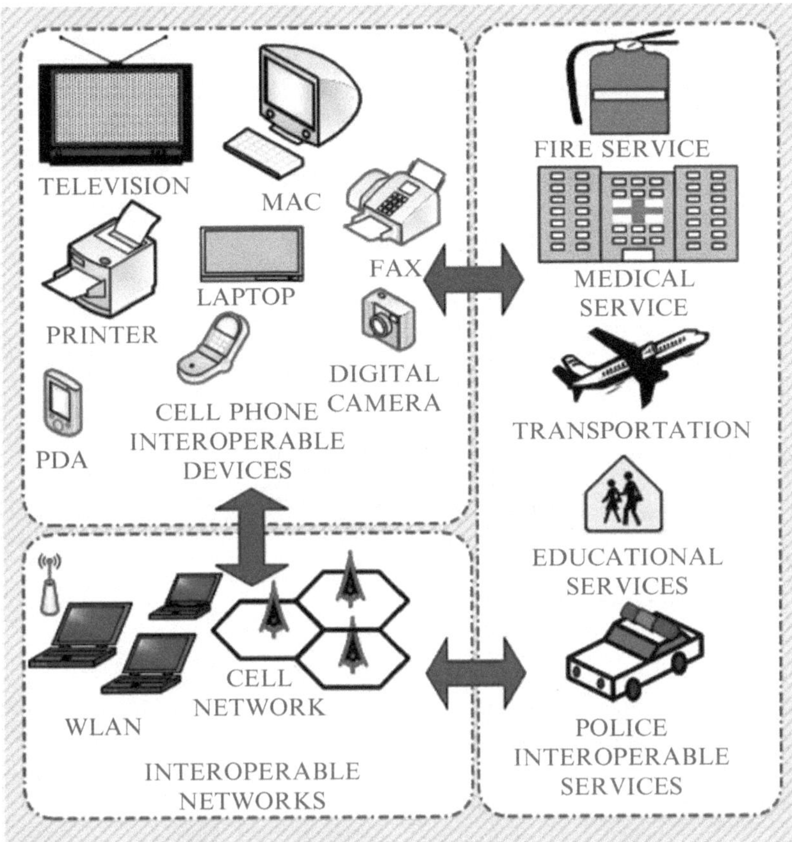

Fig. 13.8. Interoperable wireless system.

therefore, some of the major ones are described here. Figure 13.9 represents a high level classification of the service specific/end user CR applications and their usage in everyday lives. Even though these applications are grouped separately, they may overlap with each other across boundaries.

Private Sector (Personal Level)

- **Home and Family Environment:** Technology is an integral part of modern life. Spending quality time with the family has regained its importance. As a result, the differentiating border between home and office is disappearing. For example, a corporate employee may wish to engage himself in an office conference on his laptop before leaving home. The connection can be established over a WLAN transmitting real time multimedia information. At the same time, other neighbors can access WLAN for

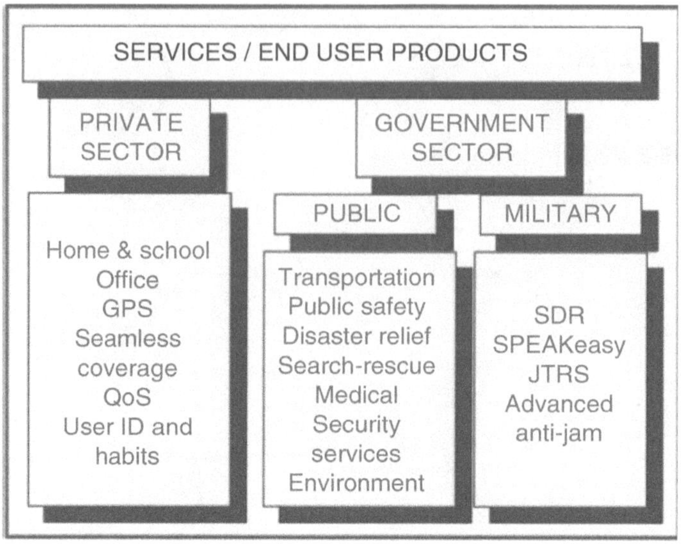

Fig. 13.9. Service specific application classification.

similar or other reasons. To avoid this rush hour congestion CR can identify available spectrum holes and utilize it. This available spectrum can belong to public safety services, such as the fire and police departments. If the access points are in less populated areas, the unused spectrum can consist of a TV band. The cognitive ability can provide appropriate bandwidth for desired information transmission [20] and choose appropriate modulation, coding techniques to provide the best wireless communication using the spectrum hole. Under the same circumstances a traditional radio may be unable to operate during radio access congestion and the person may altogether sacrifice his morning conference.

- **School Environment:** When a small child attends school, safety becomes a major concern. A short range tracking device between home and school or within school can be placed on the child's body where the central control can reside at home or at the teacher's office. Cognitive tracking node can intelligently capture a child's location information. Whenever the child is out of the designated premises, the cognitive sensor can immediately convey the message to the base. Then the parents or the teacher may intervene. The child tracker can also be equipped with an emergency channel that can activate according to behavioral changes such as crying or screaming.
  In a home schooling scenario, students can watch the classes live or recorded from their homes. Cognitive tutoring [23] can enable tutoring of a student without human intervention.

- **Office Environment:** In an office environment, the CR can prioritize a radio network connection according to pre-set priority status. For example, a conference among the CEOs can receive the highest priority in network connection and available spectrum occupancy. Then the next level of priority is set forth and so on. The priorities can be set at times of network initialization and can be updated on a regular basis within a central control. Each wireless work station can communicate with the central control and download the priority table. The network and spectrum access can vary in time, location, space or according to the employee rank.
- **Awareness in Spatial Dimension:** Determining the location or terrain surrounding a communication link or a mobile path can improve communication. CR can utilize RF signals, network information, device/system resources, sensed parameter/values, and GPS information to geolocate a system or a device. Once the location is determined, the information can be conveyed to the central control to retrieve location specific channel information. CR can then use that information to improve communication. The details of location based services and application of these in cognitive radio are discussed in detail in Chapter 10.
- **Handoff:** Efficient handoff mechanism is a must for uninterrupted communication and to avoid scenarios such as a call drop in cellular services. Handoff can be classified into predictive and unpredictive handoffs. Predictive handoff involves determining possible handoff scenario in advance and take appropriate action for efficient handling. A mobile cognitive device with its GPS capability can select either a preferred cell within an existing network type, or select an entirely different network based upon available options in the mobile's current path.
  Unpredictive handoff can involve unpredictive behavior in the possible handoff scenario. Cognitive radio can employ intelligent mechanisms to handle such scenario. One example of unpredictive behavior can be applicable to commercial markets where seamless handoff performance is critical for user satisfaction.
- **Roaming Across Borders:** CR can allow the user to roam across borders with variable policy and etiguette rules and can negotiate with several providers to establish connection which has the lowest available cost.
- **QoS and QoS Management:** Quality of service varies from one application to another. Cognitive functions to satisfy specific QoS can include prioritizing connection such as dropping streaming audio if there is an incoming call; optimizing transmission time such as executing uploads/downloads in low priority batches until a low cost and high bandwidth connection is available; selection of appropriate channel bandwidth that matches endpoint codecs and allowing adaptive compression to balance application and voice bandwidth usage. CR can also intelligently degrade the supplied services in accordance with link quality changes [8] and battery power reduction.

- **Communicating with Existing Non-Cognitive Services:** One of the wonderful traits of cognitive radio is to operate over any network and over any available services even if the network and the service are non-cognitive. At times of need, CR can scan the environment for available networks and possible services and choose the appropriate network/service dynamically [24].
- **Man–Machine Interface:** One of the first cognitive concepts proposed by Dr. Mitola [25] is the autonomous interaction between a communication device and its user. Any human such as a cell phone user, any animal such as a bird carrying a sensor, any plant such as a rare tree in the Amazon jungle or any other wireless node can fall under the definition of "user" within a communication link. Some of the major CR applications that can directly involve the user are stated below.
  – ***User authentication:*** Cognitive radio can be aware of each user's unique identification. In case of man this user authentication can include voice, thumb print, DNA, and eye retina recognition. In case of animals it can be DNA or species specific identification such as stripe orientation in Zebras. In the initialization phase, cognitive learning can allow the specific wireless node to learn about its user's unique ID [25].
  – ***User emotions:*** CR can identify its user's emotions such as anger, happiness, sadness, and fear based on voice stress level and execute appropriate action. For example, if a man is scared, CR can detect fear and autonomously dial "911".
  – ***Noise cancellation:*** The cognitive transmitter and the receiver both may tackle channel impairments to produce desired signal quality at the receiver. Yet, the received speech clarity cannot be guaranteed to the user's ears due to surrounding environment noise near the ears. CR can detect the environment noise around the receiver and apply noise cancellation techniques dynamically to adapt, adjust and maintain desired signal quality.
  – ***User habits:*** During initialization phase, cognitive radio can learn about its user's periodic habits such as the mostly dialed number on a cellphone, the mostly visited Internet sites and the mostly traveled geographical locations. CR can utilize knowledge about its user habits to make judicious decision in radio communication. Upon sensing any variations to these habits, CR can learn the new habit and adapt to it. Figure 13.10 illustrates CR application, where the environment around user's daily traveling route is incorporated into communication to provide the desired signal quality.
  It is still a long way from reaching complete user habit adaptation. As a step toward that, the present man–machine interface can be simplified to reduce the expanded number of options that require Input/Output (I/O) action between a communication device and its user [22]. Some of the autonomous CR tasks related to user habits can encompass learning and adapting to a user's common errors and restructuring the I/O sequence

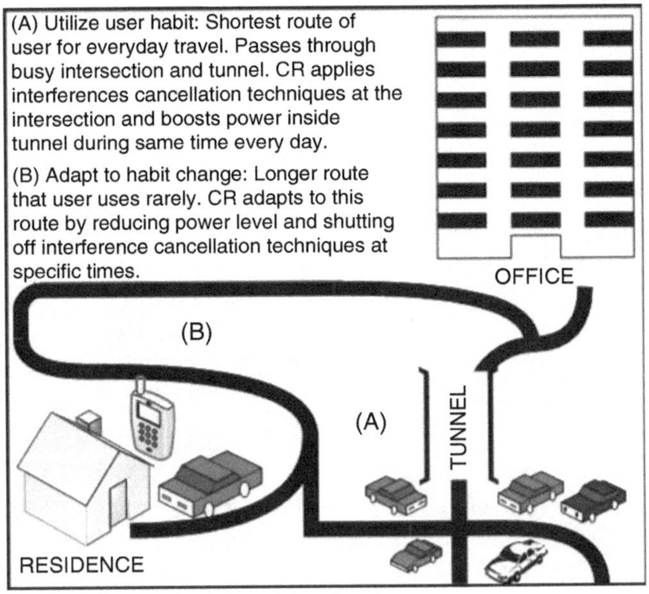

Fig. 13.10. Cognitive adaptation to user habits.

to correct those. CR can notify the user of any high priority messages autonomously and filter out some of the unnecessary ones after intelligent sorting. The notification can take place by audio/vibrate, textual and visual methods. For example, upon learning the average call time between the current user and the other end, CR can decide to notify the user of a forecasted call drop based upon current trajectory and known network conditions.

– ***User perception:*** CR intelligence can sense the perception level of a user on the receiving end. CR can intelligently decide on repeating the transmission if the user is incapable of deducing the received information. This can autonomously improve communication between two users.

Government Sector (Public: Protection, Safety, Security, and Disaster Relief)

The application of CR in public safety and disaster situations can bring revolutionary changes [26, 27]. Cognitive radio enables creation and maintenance of communication over diverse networks and spectrums. This in turn, enables an uninterrupted communication link during public protection and disaster relief scenarios [28].

In a disaster situation, the private wireless networks such as a cellular network can be inoperable and the public safety spectrum can be overcrowded due to numerous emergency connections. Under such situation, cognitive radio

Table 13.2. Some general problems encountered in disaster environments and their effects.

Problem	Effect
Damage to infrastructure	Communication blackout Capacity degradation Batteries running out
Overwhelming network load	Escalation of Probability of Blocking Escalation of interference around the incident
Interoperability	Slow rescue operations Inefficient disaster recovery
Drastic changes in environment	Malfunctioning wireless devices Malfunctioning applications (such as localization)

can utilize available licensed/unlicensed spectrum holes and heterogeneous network components to create and maintain temporary emergency connection. As an example, CR can establish a communication link over GSM band using WLAN access points.

The fruits of cognitive wireless communication can benefit the public security, safety and disaster scenarios, perhaps the most. Therefore, this type of CR application deserves special emphasis and is presented in detail within this section. Table 13.2 presents some common problems that arise during and/or after a disaster situation and how they effect wireless communication.

The causes of communication failure in a disaster situation can be categorized into two, both of which, can be tackled by cognitive intelligence. These are presented below.

– ***Impaired wireless communication infrastructure:*** Wireless communication infrastructure can be divided into two categories based on their functionalities: 1) Power system that feeds the communication network and 2) Network elements that haul the communication traffic. Generally, disasters create damage to both categories and that damage causes similar consequences. Although the damages to 1) and 2) have similar consequences, it is appropriate to consider each of them individually for the sake of clarity.

Power failure, which is caused by a damage to the power system, can be divided coarsely into two sub-categories as (A) major and (B) minor (or regional) failures. In (A), the communication suddenly drops out encompassing a vast geographical area probably including some major communication nodes.[1] In (B), the failure does not cover as vast a geographical area as (A),

[1] On August 14, 2003, the North–East part of the USA experienced one of the biggest power outages of all times. It is reported that several parts of New York City went through landline communication loss because of three of the central offices of one phone company experienced power outages. Cellular phones failed as well since some of the repeaters were not equipped with power backup. The

and therefore, the impact on the communication is limited. As a result of (A) and (B), when a disaster occurs, power failure causes the devices connecting the terminals and network to cut off, and may end up with drastic capacity drops and serious communication blackouts [30–33].

Similar to the damages to power systems, breaking down of infrastructure that hauls the communication cause serious problems, such as drastic capacity drops or total loss of communications [34, 35]. On September 11, 2001, along with the collapse of the towers, the radio repeaters situated on top of the towers were also destroyed. Besides the repeaters, the debris and the impact of the collapse further damaged other communication infrastructures such as fiber-optics near the vicinity of the disaster area [36, 37]. Although communications can be established via other routes at times of partial network damage, excessive amount of traffic can lead to an overloaded network. Consequently, the system will respond by dropping the calls and/or denying the accesses, which in turn, will cause a drastic decrease in the capacity. This domino effect was observed on September 11, 2001, as well. The people who wanted to reach their loved ones inside or around the incident, created network access bursts, which gave rise to an overwhelmed network and congestion [37].

In connection with the damage to infrastructure, power shortage in batteries constitute another serious problem. The disaster victims carrying wireless devices run into battery problem due to (i) inaccessible electric outlets, (ii) a power outage and/or (iii) inability to withstand long hours of search and rescue operations.

In the event of communication infrastructure failure, the knowledge of call blocking rate can be exploited by cognitive radios [38, p. 441]. The cognition capabilities of CR, can overcome the failures caused by damage to infrastructures, to some extent. For instance, when the infrastructure is demolished, cognitive radios try to search for other cognitive radios to construct an ad hoc network in order to reach the devices that are out-of-coverage area. Thus, when the rescue teams approach the disaster area, the newly created ad hoc network will enable detection of nodes that were undetected by the rescue teams [39].

– ***Unfavorable changes in the propagation channel:*** In addition to infrastructure failure, there is a serious indirect impact of the disasters on wireless communications causing wireless environment and conditions to change drastically. Extraordinary alteration in the environment, such as collapsing of an office into a rubble with piled up thick layer of concrete and steel, and environmental conditions such as rain and dense dust particles in air (due to collapse), will greatly complicate the radio transmission. This stems from the fact that radio propagation characteristics in those environments

---

rest, which had batteries, dried out in four hours. In addition, the phone systems were overloaded with excessive amount of calls. Furthermore, due to the massive call volume, 911 system of the city failed, too [29].

and conditions are very different than those in the regular ones [34,39,40]. As reported in [34,39–41], due to the presence of pile of steel and concrete in the debris, locating the people inside the rubble becomes hard.[2] Furthermore, first responders reported loss of wireless connections during the rescue operation in harsh environments in the aftermath of some of the disasters [34, 41, 43]. In such environments and conditions, the unique propagation characteristics may lead to the failure of communication that would generally work fine under normal conditions. On September 11, 2001, it was reported that there were thick clouds of dust particles in the air due to the collapses. Moreover, during the recovery stage, it rained occasionally. It is known that precipitation (such as rain, snow, and fog) increases the signal attenuation and degrades QoS [44,45].

In disaster areas, it is seen that, detecting the presence of the signals coming out of or getting into a rubble is extremely difficult. There are several factors that affect the signal propagation through rubble. The structure of the collapsed building, flame, smoke, and water coming from sprinklers (due to the fire) are some of the factors affecting the radio propagation. In an aftermath of a hurricane and/or flood, the radio propagation channel may vary even more, imposing a wireless communication through a partial or a full water channel. Thus, depending on the specific disaster situation, the wireless propagation environment will vary widely and can cause adverse affects on communication resulting in degradation or a total communication failure.

Of course, it is not practical to manufacture a device containing every possible technology, channel model, and other capabilities inside. It is wiser to have a generic hardware that can be controlled and directed according to changing needs. Obviously, CR is the answer. If CR is equipped with the relevant wireless channel models and can classify the corresponding environment with its sensing ability, then a simple adjustment of SDR parameters will lead to selection of the appropriate channel model from its repository. As a result, the wireless connection failures can be minimized [3, 19, 39, 48]. Moreover, cognitive radio can consider the signal penetration characteristics [38, p. 437]. This is very important, especially in disaster environments [49], where CR can switch between the bands to make sure that the "help signal" is traversed.

There are numerous crucial scenarios that fall under the public protection, security, and/or disaster categories. Cognitive radio can bring radical improvements to the current handling capability of these public safety and disaster situations. It is impossible to include all of them here and therefore, a few of the major ones are addressed below:

- **Emergency Management or Disaster Recovery:** According to Federal Emergency Management Agency (FEMA), a disaster can be defined as the occurrence of one of the situations in the following list: (a) Chemical Emergencies, (b) Dam Failure, (c) Earthquake, (d) Fire or Wildfire,

[2] For localization methods of victims through sensor networks, Ref. [42] can be referred.

(e) Flood, (f) Hazardous Material, (g) Heat, (h) Hurricane, (i) Land Slide, (j) Nuclear Power Plant Emergency, (k) Terrorism, (l) Thunderstorm, (m) Tornado, (n) Tsunami, (o) Volcano, and (p) Winter Storm. However, disasters are not only limited to these. A plane, train, or a vehicle crash can also be considered as disasters from the context perspective. Regardless of their impact, in these situations, the communications is of vital importance to carry out emergency responses and disaster relief operations.
In the recent years, it is proved from experience that the communication systems, more often than not, fail to meet people's needs when they are needed the most. The reasons can be due to partial or full breakdown of communication infrastructures, extreme environmental condition changes, and due to other factors such as existence of overwhelming interference and inability of communication among heterogeneous components. Cognitive radio can resolve these problems in disaster situations. For instance, various nodes can work as a relay to establish communication when there is a power outage or a loss of central control. Under normal circumstances, this type of ad hoc relay (such as a mesh connectivity) may not exist. CR can adapt its operating parameters to provide an ad hoc connection at times of need.

- **Fire Services:** Fire service personnel play a vital role in many disaster scenarios. Structure fire fighting, wild land fire fighting, search and rescue, lifesaving, damage control, salvage operations, decontamination operations, and safety management through a restricted area (due to fire related disaster) can be a few of them. Communication requirements will vary depending on the environment and the specific scenario. In general, fire fighters work under severe conditions of extremely high temperature and high humidity. Wireless equipments are required to function under such adverse conditions that force communication over atypical channel characteristics.
For instance, in case of wild land fires, wireless equipments need to cover a wide range of landscapes and autonomously reconfigure with respect to the available communication link. Fire fightings are long run operations that enforce efficient energy/battery usage. Wild land fires can spread very rapidly and change direction with the changing physical conditions (such as winds). The agile characteristic requires immediate action and perfect coordination among the fire fighting team members. Video/audio conferencing and fast data transfer along with video streaming capabilities of wireless systems can help monitor the on-site operations from the control center, instruct, and limit the damage. The environmental and personnel data, such as the temperature, wind speed and direction, hazardous chemical levels at the direction of fire fighters, and number of fire fighters are important for wild land fires. Similarly, structure and equipment data, such as 3D models of the building, fire hydrant locations, occupant information, engine temperatures, and water pressures are important for structure fire fighting. These information need to be transferred to the

fire fighters and the rescue crew in a reliable and efficient way through the adverse wireless channel (due to wind, heat and humidity). Cognitive radio can be of great assistance to eliminate most or all of the above problems in fire fighting.

- **Search and Rescue:** In a typical search and rescue scenario the person in distress can inform about his location by creating visual signs such as a smoke or a fire from a flare gun. Since the wireless coverage is scarce in remote areas, it may be impossible for the distressed person to inform the rescuer using a wireless device. As the search party nears the rescue area, the cognitive radios of the distressed and the rescuer can establish communication without any central control. The GPS capability of CR can come in handy to appropriately detect the position of the rescued person. At the same time, if a special channel over available spectrum hole is used for short range signaling, this channel can work like a beacon for the distressed person. The rescue team can home onto this beacon and find the lost individual.
- **Mining:** CR can help minimize mine accidents in the future by maintaining continued wireless communication between the people inside and the people outside of the mine. In such situation, CR can choose appropriate waveform and apply other techniques to establish a clear signal between the adverse environment within the mine and the outside world.
- **Crime Prevention:** The crime prevention sectors of the government, such as the police and the Federal Bureau of Investigation (FBI), can greatly benefit from utilizing CR. The operation of these departments demand portable coverage throughout the community with transparent and high-speed access to a myriad of public safety files for criminal identification and investigation. This information may include files on stolen articles (such as firearms, automobiles, trucks, tracks and industrial machinery), suspect fingerprints, photos, mug shots, iris scans, property descriptions, and criminal history files. Besides the information itself, it is also crucial to transfer the data securely so that unauthorized users do not get or corrupt the transferred information. Current wireless communication applications are incapable of providing such sophisticated services and cognitive radio is the answer.

Government Sector (Public: Others)

- **Traffic Control:** Traffic is a big problem especially during rush hours in the mornings and in the evenings. Under such situations, the local traffic control can transmit the congested traffic location, the predicted traffic delay and an alternate route to the mobile user. Cognitive intelligence can be applied on traffic signals themselves as well to determine how long the red or green signal may remain on depending on the traffic volume in each direction. This traffic information can be gathered at each signal location by cognitive sensors. The appropriate decision can be taken locally or via a central control. Balloon born repeaters can be utilized to convey traffic information or images to the central control for broadcast as well.

- **Medical Applications:** The application of cognitive radio can bring improvements in areas of medical and bio-medical engineering. In a hospital environment, a new born baby needs to be identified with its mother. If the hospital personnel conducts incorrect visual or manual identification of any infant, there is always a chance of the wrong baby associating with the wrong mother. In order to avoid this unfortunate situation, a wireless tag can be provided on the infant's body that responds to the base tag around the mother's wrist. In an extreme case of a wireless tag mix up, finger prints or even retina imprints can be utilized to relate each baby with its mother. Cognitive radio tag can inform the mother intelligently if the baby is carried outside of a designated premises such as outside of the hospital baby ward.
  In case of adult patients, each may be provided with a personal cognitive ID tag. A central control within the same hospital unit can utilize, store, and update each patient information and keep track of their changes. The cognitive tag can record the vital signs of the patient and intelligently inform the respective authority if an abnormality is detected. Long range tags on the patient's body can monitor his readings and forward the information to the doctor while the patient is away from the hospital.
  There are several other special medical applications of CR that need attention. A few of these are addressed here.
  - ***Emergency medical services:*** Emergency medical services are provided to the public, primarily in two phases, first, by the mobile medical personnel and equipment (such as an ambulance or other motorized vehicles) at the field and then, within a controlled environment with enhanced medical assistance at the medical institutions (such as metropolitan hospitals). Modern advances in the medical fields provide emergency services the ability to serve advanced level of medical assistance. The mobile medical personnel wireless systems can help transfer information of the patient to the controlled environment. This information transmission requires sufficient bandwidth and may include video and sound that are gathered by the medical equipments on the mobile unit. It is of vital importance to transmit the information in a fast and reliable way so that specific medical preparations can be made at the controlled environment before the patient reaches there. Secondly, if an urgent intervention of any medical procedure is required, personnel at the mobile unit may conduct it with the help of specialists located at the controlled environment. It is also important that the operation of sensitive medical equipments is not affected or corrupted during the information transfer of wireless devices. Cognitive radio can efficiently assist the emergency medical services, ease the tasks of medical personnel and save human lives.
  - ***Biomedical engineering:*** Biomedical engineering can involve insertion/attachment of electronic devices in a human body to monitor

different functions of the body and convey the monitored information when needed. Cognitive radio can enable intelligent detection of abnormal tissues or blood cells within a human body and notify the doctor. This can play crucial role in saving human lives.
– ***Assisting blind people:*** Cognitive radio can play an important role in becoming the eyes of a blind person. Dogs and white sticks are commonly utilized to assist a blind man. Cognitive radio can replace these guides with intelligence that can carry out tasks such as finding safe areas for travel, safe time to cross the road and sketch the path from home to shop, work, and other places. Depending on the daily activities of the blind person, CR can adapt itself and provide a true guide. These generic habits and daily activities can be applicable to other blind people as well [50] and CR can customize to the needs of each individual blind user.

- **Environmental Applications:**
  – ***Weather forecast:*** Sensors and sensor networks have been employed in detecting weather parameters such as temperature, wind speed, air pressure, and humidity for over a long time. If these sensors are equipped with cognitive ability, they can communicate with each other without any user intervention. This way the sensors out in the field can detect, collect and share information among themselves for optimum performance. When the required amount of data is collected, the data can be passed to the central control by the sensor closest to the station for optimum power usage, optimum network usage, and minimized delay.
  – ***Research on the behaviors of endangered species:*** The ever increasing human population and their need for new developments are causing rapid destruction of wild animal habitats. As a result many animal species are listed under "endangered". Environmental scientists are therefore, conducting research projects to understand the behavior and life cycle of such animals. Wireless technology, especially, cognitive sensors can play an extensive role in tracking the life of such species. For example, conducting research on animal behavior within adverse terrains such as the rain forest, may be cumbersome and seldom impossible. Cognitive radio can be utilized to track an animal from its birth to all the way through its migration and mating to death to collect information intelligently and convey to the scientists.
  – ***Air pollution control:*** Detecting impurities in air is a major priority to protect human health. Cognitive radio can intelligently assess impurities in air, retrieve various pollution related data and notify authority when the limit crosses a pre-defined safety threshold. CR may intelligently adjust the threshold as well depending on various other factors without user intervention.
  – ***Global warming:*** Global Warming is once again an intensely discussed topic in the modern world. With environment pollution in the rise from industry, transportation and other pollutions, our planet is in the verge of major natural and geographical changes. Environmental scientists

principal concern is the melting of polar ice caps resulting in ocean level rise. This can in turn diminish the entire existence of various wildlife. To evaluate this catastrophic situation and to estimate prevention measures, tracking of such weather pattern and geological changes is necessary. This can be carried out using sophisticated and self steered devices like cognitive radio.

Government Sector (Military)

Among all the applications of CR, military is perhaps, one of the most significant areas, where different cognitive radio aspects have been deployed for over some time now. The SDR, SPEAKeasy, Joint Tactical Radio System (JTRS), and jamming and anti-jamming are some of the most important technologies in the military fields, that utilize the CR concepts.

In the simplest form, SDR can be defined as the radio that is controlled and/or implemented by software. The benefits of SDR have been recognized by the military to resolve their communication deficiencies (for example, deficiencies arising from inoperability among legacy systems) during wartime activities. Some of these defficiencies were addressed in the SPEAKeasy project, which is an advanced implementation of the SDR. The efforts in SPEAKeasy led to the development of JTRS project. Present situation in military imposes many different radio systems to be manually carried, maintained, and operated. JTRS can reduce the total communication equipments weight and footprint by eliminating multiple systems. Some of the significant features of JTRS include reduced maintenance costs and complexity due to the elimination of excessive equipments, making military communications to be less vulnerable to enemy intercept and jamming owing to improved spectrum usage over a wide range of frequencies, and ease of implementation due to its compatibility with current legacy systems. The SDR, SPEAKeasy, and JTRS concepts and technologies are presented in detail in Chapter 4, and are therefore, not elaborated further here. The following discussion will address the jamming and anti-jamming applications of cognitive radio.

Wireless communication has been the critical factor in military combat scenarios. The ability to communicate and destroy the communication of the enemy is very important for modern armies. There has been a strong need to deploy jammers properly in the battle fields so that the communication of the enemy is disrupted. Deploying an effective jammer needs the awareness of the enemy waveform so that the jamming signal will be properly positioned to destroy the communication of the enemy while having energy efficient jammers. On the other hand, there is also strong need to protect our own communication against the enemy jammers. This requires identification and position of jamming signal and shaping our own communication signal so that the jamming signal's effect will be minimal. For example, employing narrowband interference (or jamming interference) cancellation, deploying adaptive frequency hoping, etc. require various information (as well as statistics) about the interfering/jamming signal. The present tools and techniques for jammer placement and for protection of our own signal against the enemy jammers, are substantially limited. Both of these functions require radio awareness in the battle field and capability to identify the waveforms (signal and/or jammer) of the enemy transmission. Blind signal identification and extraction of the parameters of any arbitrary radio signal captured over the air is extremely important for the success of both goals.

Cognitive radio can coordinate a series of devices to stimulate a physical action to remove jamming and adjust communication waveforms to avoid jams that are intentionally applied in the military markets. The CR can identify the signal type (for example, multi-carrier or single carrier signal, frequency hoping or not, Code Division Multiple Access type or not, wideband or narrowband, etc.), signal modulation (such as Quadrature Amplitude Modulation, OFDM, Adaptive versus fixed modulation, etc.), occupied bandwidth, carrier frequency, number of signals over the selected band, signal statistics (such as temporal, spatial, and frequency), and estimate the geolocation of the radio source (enemy jammers or signals) from the captured enemy signal.

13.3 Cognitive Application Challenges

Cognitive dominance in the wireless world is boundless. At the same time, CR associated challenges impose high level of complexity and concern. For example, usage of unused spectrum (public or private) requires identifying the spectrum hole. Primary user absence may be misguiding if SNR is the only determining factor for spectrum availability [9]. Any large or small scale fading can cause a dip in the signal strength that may lead to a wrong conclusion. Primary detection may be more reliable through collaboration. On the other end, secondary user needs to vacate spectrum as soon as the primary user returns. Identifying the exact time of transition and handling back of spectrum can also be challenging.

Spectrum management tasks such as evaluation of new spectrum suitability for present transmission needs and selection of appropriate band from multiple available ones can be complex. Secondary user may access spectrum in individual chunks at different time/locations and transmit signal over those separate chunks.

In the presence of multiple concurrent users over the same band (such as multiple users on ISM band), power level of each requires control to mitigate undesired interference. The encryption and encoding techniques must also ensure privacy and protection against each other in such cases [17].

Pricing for the secondary spectrum usage is a topic of big discussion as well [51, 52]. The licensed spectrum services are provided to the consumers through a pre-established pricing mechanism. In order to avoid unnecessary exploitation of secondary spectrum and to eliminate unfairness to the licensed owner, some means of pricing mechanism needs to exist for secondary usage.

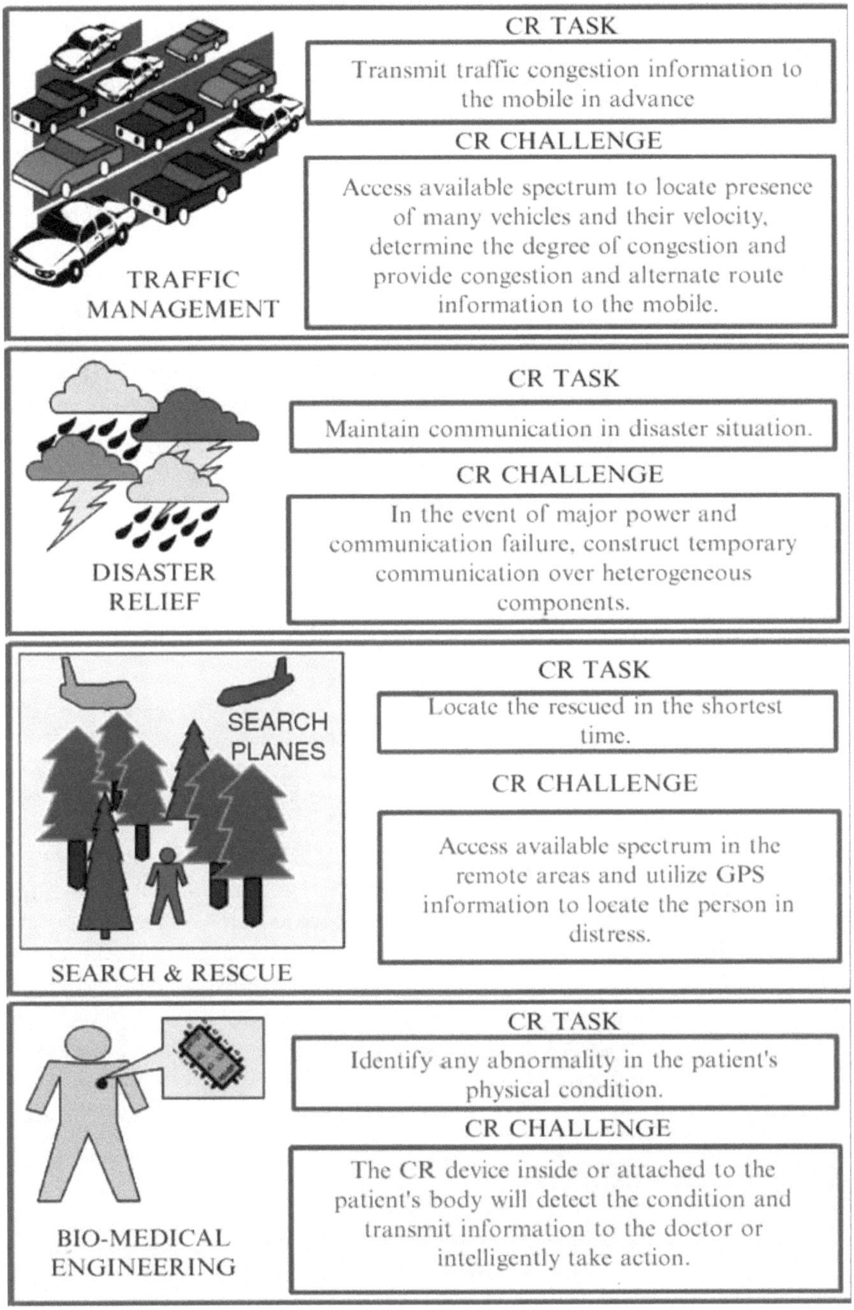

Fig. 13.11. Service specific applications and their challenges.

On the network side, there are many issues to consider as well. Identification and maintenance of optimum route; efficient network queue management; establishing appropriate control channel among heterogeneous network components and cross layer optimization can all pose great degree of challenges in CR deployment. Mobility of secondary spectrum users can create additional challenges for dynamic network configuration [13].

The hardware complexity rises with the increase in cognitive intelligence implementation. Therefore, present and future hardware capability and complexity can pose a great level of difficulty to the realization of CR applications. SDR is perhaps the first step toward resolving these issues. With the implementation of ubiquitous connectivity across geographical locations, CR devices may need to comply dynamically with policy and etiquette variations across geographical boundaries. Figure 13.11 represents a group of major present and future envisioned CR applications, their related CR tasks and associated challenges for a brief illustration.

As cognitive radio takes over, our lives may be blessed with extraordinary comfort, convenience, and ease in wireless communication. Perhaps, it is important to note that usage of such large number and variety of radio devices may cause short and long term harmful effects on health. More medical research may need to be carried out in the present and future to determine these health related issues.

13.4 Conclusion

The "cognitive" concept is still in its infancy. Therefore, cognitive wireless communication applications are still far from reaching its potential impact through wide implementation. This chapter provided an overview of potential present and future cognitive radio applications and their multidimensional usage to facilitate wireless communication. Each application may further be investigated and concentrated research in each area may be conducted in near and distant future. This may in turn open up other applications or possibilities to benefit our human lives.

Acknowledgment

The authors would like to thank Serhan Yarkan, PHD student, USF and Ali Gorcin, PHD student, USF, for their contribution on the usage of cognitive radio in public safety, security, and rescue operations.

References

1. S. Haykin, "Cognitive radio: Brain-empowered wireless communication," *IEEE J. Select. Areas Commun.*, vol. 23, no. 2, pp. 201–220, Feb. 2005.
2. J. Mitola III and G. Maguire Jr, "Cognitive radio: making software radios more personal," in *Proc. IEEE Personal Commun. [see also IEEE Wireless Commun.]*, vol. 6, no. 4, Aug. 1999, pp. 13–18.
3. T. Rondeau, C. Rieser, T. Gallagher, and C. Bostian, "Online modeling of wireless channels with hidden Markov models and channel impulse responses for cognitive radios," in *Proc. IEEE Int. Microwave Symp. Digest (MTT-S)*, vol. 2, Fort worth, TX, USA, Jun. 2004, pp. 739–742.
4. V. Chakravarthy, A. Shaw, M. Temple, and J. Stephens, "Cognitive radio-an adaptive waveform with spectral sharing capability," in *Proc. IEEE Wireless Commun. and Networking Conf. (WCNC)*, vol. 2, New Orleans, LA, USA, Mar. 2005, pp. 724–729.
5. N. Hoven and A. Sahai, "Power Scaling for Cognitive Radio," in *Proc. IEEE Int. Conf. Wireless Networks, Commun. and Mob. Computing (WIRELESSCOM)*, vol. 1, Maui, Hawaii, USA, Jun. 2005, pp. 250–255.
6. M. Marcus, "Unlicensed cognitive sharing of TV spectrum: the controversy at the Federal Commun. Commission," *IEEE Communications Magazine*, vol. 43, no. 5, pp. 24–25, May 2005.
7. S. Ellingson, "Spectral occupancy at VHF: implications for frequency-agile cognitive radios," in *Proc. IEEE Vehicular Technol. Conf. (VTC)*, vol. 2, Dallas, TX, USA, Sep. 2005, pp. 1379–1382.
8. L. Berlemann and B. Walke, "Spectrum load smoothing for optimized spectrum utilization-rationale and algorithm," in *Proc. IEEE Wireless Commun. and Networking Conf. (WCNC)*, vol. 2, New Orleans, LA, USA, Mar. 2005, pp. 735–740.
9. D. Cabric, S. Mishra, and R. Brodersen, "Implementation issues in spectrum sensing for cognitive radios," in *Proc. IEEE Asilomar Con. on Signals, Syst. and Computers*, Pacific Grove, CA, USA, Oct. 2004, pp. 772–776.
10. I. Akyildiz, W. Lee, M. Vuran, and S. Mohanty, "Next generation/dynamic spectrum access/cognitive radio wireless networks: a survey," *Computer Networks*, vol. 50, no. 13, pp. 2127–2159, Sep. 2006.
11. D. Cabric, I. O'Donnell, M. Chen, and R. Brodersen, "Spectrum sharing radios," *IEEE Circuits and Systems Magazine*, vol. 6, no. 2, pp. 30–45, 2006.
12. G. Ganesan and Y. Li, "Cooperative spectrum sensing in cognitive radio networks," Baltimore, MD, USA, pp. 137–143, Nov. 2005.
13. L. DaSilva, S. Midkiff, J. Park, G. Hadjichristofi, N. Davis, and K. Phanse, "Network mobility and protocol interoperability in ad hoc networks," *IEEE Commun. Magazine*, vol. 42, no. 11, pp. 88–96, Nov. 2004.
14. R. Molva, D. Samfat, and G. Tsudik, "Authentication of mobile users," *IEEE Network*, vol. 8, no. 2, pp. 26–34, Mar./Apr. 1994.
15. M. Borsc and H. Shinde, "Wireless security & privacy," in *Proc. IEEE Int. Conf. Personal Wireless Commun. (ICPWC)*, New Delhi, India, Jan. 2005, pp. 424–428.
16. S. Russell, "Wireless network security for users," in *Proc. IEEE Int. Conf. Information Technology: Coding and Computing*, Las Vegas, NV, USA, Apr. 2001, pp. 172–177.

17. S. Gultchev, C. Mitchell, K. Moessner, and R. Tafazolli, "Securing reconfigurable terminals – mechanisms and protocols," in *Proc. Personal, Indoor and Mobile Radio Communications*, vol. 2, Pavilhao atlantico, Lisboa, Portugal, Sep. 2002, pp. 804–808.
18. B. Wild and K. Ramchandran, "Detecting primary receivers for cognitive radio applications," in *Proc. IEEE Int. Symp. New Frontiers in Dynamic Spectrum Access Networks (DySPAN)*, Baltimore, MD, USA, Nov. 2005, pp. 124–130.
19. J. Laskar, R. Mukhopadhyay, Y. Hur, C. Lee, and K. Lim, "Reconfigurable RFICs and modules for cognitive radio," *Topical Meeting Silicon Monolithic Integrated Circuits in RF Syst. Digest of Papers*, pp. 283–286, 2005.
20. J. Mitola III, "Cognitive Radio for Flexible Mobile Multimedia Commun." in *Proc. IEEE Mob. Multimedia Commun. (MoMuC)*, San diego, CA, USA, Nov. 1999, pp. 3–10.
21. H. Arslan, *Signal Processing for Mob. Commun. Handbook, M. Ibnkahla, Ed. Boca Raton, FL: CRC.* CRC Press, 2004, ch. 28, Adaptation Techniques and the Enabling Parameter Estimation Algorithms for Wireless Commun. Syst., pp. **28**-1–**28**-26.
22. T. Costlow, "Cognitive radios will adapt to users," *IEEE Intelligent Syst., IEEE [see also IEEE Intelligent Syst. and Their Applications]*, vol. 18, no. 3, pp. 4–7, May/Jun. 2003.
23. V. Aleven, O. Popescu, and K. Koedinger, "Towards tutorial dialog to support self-explanation: Adding natural language understanding to a cognitive tutor," in *Proc. Artificial Intelligence in Education: (AI-ED) in the Wired and Wireless Future*, San Antonio, TX, USA, May 2001, pp. 246–255.
24. G. Lee, P. Faratin, S. Bauer, and J. Wroclawski, "A user-guided cognitive agent for network service selection in pervasive computing environments," in *Proc. IEEE Conf. Pervasive Computing and Commun. (PerCom)*, Orlando, FL, USA, Mar. 2004, pp. 219–228.
25. J. Mitola III, "Cognitive INFOSEC," in *Proc. IEEE Int. Microwave Symp. Digest (MTT-S)*, vol. 2, Philadelphia, PA, USA, Jun. 2003, pp. 1051–1054.
26. J. Krikke and B. Alfonsi, "In the news," *IEEE Intelligent Systems*, vol. 21, no. 2, pp. 4–7, Mar./Apr. 2006.
27. D. Maldonado, B. Le, A. Hugine, T. Rondeau, and C. Bostian, "Cognitive radio applications to dynamic spectrum allocation: a discussion and an illustrative example," in *Proc. IEEE Int. Symp. New Frontiers in Dynamic Spectrum Access Networks (DySPAN)*, Baltimore, MD, USA, Nov. 2005, pp. 597–600.
28. K. Lorincz, D. Malan, T. Fulford-Jones, A. Nawoj, A. Clavel, V. Shnayder, G. Mainland, M. Welsh, and S. Moulton, "Sensor networks for emergency response: challenges and opportunities," in *Proc. IEEE Pervasive Computing*, vol. 3, no. 4, Linz/Vienna, Austria, Apr. 2004, pp. 16–23.
29. A. J. DeBlasio, T. J. Regan, M. E. Zirker, K. Lovejoy, and K. Fichter, "Learning from the 2003 blackout," *Public Roads*, vol. 68, no. 2, pp. 22–37, 2004.
30. S. Masuda, M. Morita, and S. Nakagawa, "The signal control system with radio in a disaster," in *Proc. IEEE/IEEJ/JSAI International Conference on Intelligent Transportation Systems.* IEEE, 1999, pp. 594–598.
31. J. S. Griswold, T. L. Lightle, and J. G. Lovelady, "Hurricane Hugo: effect on state government communications," *IEEE Communications Magazine*, vol. 28, no. 6, pp. 12–17, Jun. 1990.

32. D. of Homeland Security, "The federal response to hurricane katrina – lessons learned," Department of Homeland Security, Feb. 2006. [Online]. Available: http://www.whitehouse.gov/reports/katrina--lessons--learned.pdf
33. R. C. Coile, "The role of amateur radio in providing emergency electronic communication for disaster management," *Disaster Prevention and Management*, vol. 6, no. 3, pp. 176–185, Aug. 1997.
34. WERT, "September 11, 2001 New York City World Trade Center Terrorist Attack," Wireless Emergency Response Team WERT, Tech. Rep., Oct. 2001.
35. H. Foltsa and N. C. System, "Standards initiatives for Emergency Telecommunications Service (ETS)," *IEEE Communications Magazine*, vol. 40, no. 7, pp. 102–107, Jul. 2002.
36. J. Eisenberg and C. Partridge, "The Internet under crisis conditions: Learning from September 11," *ACM Computer Communication Review*, vol. 33, no. 2, 2003.
37. M. F. Myers, *Beyond September 11th: An Account of Post-Disaster Research.* Natural Hazards Research and Applications Information Center, University of Colorado, 2003, ch. Telecommunications, pp. 58–59.
38. J. M. III, "Cognitive Radio for Flexible Mobile Multimedia Communications," *Mobile Networks and Applications*, vol. 6, no. 5, pp. 435–441, 2001.
39. Bernard L. Malone III, "Wireless Search and Rescue: Concepts for Improved Capabilities," *Bell Labs Technical Journal*, vol. 9, no. 2, pp. 37–49, Summer 2004.
40. H. Sanson and M. Mitsuji, "Localization for emergency sensor networks," in *Proc. The 7th International Conference on Advanced Communication Technology (ICACT)*, vol. 2, Feb. 2005, pp. 982–987.
41. T. H. Kean and L. H. Hamilton, *The 9/11 Commission Report.* National Commission on Terrorist Attacks upon the United States, 2004.
42. T. He, C. Huang, B. M. Blum, J. A. Stankovic, and T. Abdelzaher, "Range-free localization schemes for large scale sensor networks," in *MobiCom '03: Proceedings of the 9th Annual International Conference on Mobile Computing and Networking.* New York, NY, USA: ACM Press, 2003, pp. 81–95.
43. SAFECOM. Library of emergency response wireless interoperability issues. Department of Homeland Security. Visited on 10.10.2006. [Online]. Available: http://www.safecomprogram.gov/SAFECOM/library/
44. F. S. Solheim, J. Vivekanandan, R. H. Ware, and C. Rocken, "Propagation delays induced in GPS signals by dry air, water vapor, hydrometeors, and other particulates," *Journal of Geophysical Research*, vol. 104, no. D8, pp. 9663–9670, Apr. 27, 1999.
45. S. Karapantazis and F.-N. Pavlidou, "Broadband communications via high-altitude platforms: a survey," *IEEE Communications Surveys & Tutorials*, vol. 7, no. 1, pp. 2–31, First Quarter 2005.
46. J. Mitola and G. M. Jr., "Cognitive radio: making software radios more personal," *IEEE Personal Communications*, vol. 6, no. 4, pp. 13–18, Aug. 1999.
47. J. Mitola III, "Cognitive radio an integrated agent architecture for software defined radio," Ph.D. dissertation, KTH Royal Institute of Technology, Stockholm, Sweden, May 8, 2000. [Online]. Available: http://www.it.kth.se/~jmitola/Mitola_Dissertation8_Integrated.pdf
48. S. Yarkan and H. Arslan, "On extracting the higher order radio channel statistics through location aided services for cognitive radio," in *Proceedings of 50th*

Annual IEEE Global Communications Conference (IEEE GLOBECOM 2007), 2007, invited Paper.

49. M. Bimpas, N. Paraskevopoulos, K. Nikellis, D. Economou, and N. Uzunoglu, "Development of A Three Band RADAR System for Detecting Trapped Alive Humans Under Building Ruins," *Progress in Electromagnetics Research, PIER 49*, pp. 161–188, 2004.
50. D. Good, "On why the blind leading the blind is a good idea," in *Proc. Int. Conf. Humanizing the Information Age*, Aizu-Wakamatsu City, Japan, Aug. 1997, pp. 173.
51. N. Shankar, "Pricing for spectrum usage in cognitive radios," in *Proc. IEEE Consumer Commun. and Networking Conf. (CCNC)*, Las Vegas, NV, USA, Jan. 2006, pp. 579–582.
52. C. Kloeck, H. Jaekel, and F. Jondral, "Dynamic and local combined pricing, allocation and billing system with cognitive radios," in *Proc. IEEE Int. Symp. New Frontiers in Dynamic Spectrum Access Networks, (DySPAN)*, Baltimore, MD, USA, Nov. 2005, pp. 73–81.

14

Cross-Layer Adaptation and Optimization for Cognitive Radio

Hüseyin Arslan[1] and Serhan Yarkan[2]

[1] University of South Florida, Tampa, FL, USA, arslan@eng.usf.edu
[2] University of South Florida, Tampa, FL, USA, syarkan@eng.usf.edu

14.1 Introduction

When we look at the evolution of wireless communication systems, as in most systems, two major developments are most prominent: 1) addition of new features and 2) improvement of already existing capabilities. The first development arises from the fact that communication through a wireless medium makes life easier. Every development brings several new features and these new features contribute positively to this fact. These developments can be seen clearly in the evolution of cell phones. In the past, cell phones were used only for transmitting voice and short text-based messages. Currently, there are cell phones in which an operating system runs and several multimedia applications are available.

The second type of development solely originates from the fact that every physical concept is finite. Therefore, according to the principal of parsimony,[1] the newer systems restrain themselves from wasting resources. Using resources adequately under dynamically changing conditions introduces the notion of adaptation and optimization.

Although systems evolve in terms of 1) and 2), particularly in communication systems, it can be observed that the fundamental design architecture, which is known as "layered architecture," still remains the same. Despite the inefficiency of contemporary communication systems, they still accomplish their tasks. Nevertheless, it can be seen that the aforementioned evolutionary developments are approaching a saturation point for contemporary communication systems. The fundamental design architecture inherently hinders

[1] *Entia non sunt multiplicanda, præter necessitatem* [1]. The famous statement, which is also known as "Occam's Razor" and believed that phrased by William of Ockham, which means "Entities should not be multiplied unnecessarily" [2]. A direct consequence of this statement is that for two systems which accomplish the same task, the one that accomplishes the objective in lesser amount of effort, element, unit, etc., is preferable to the other.

H. Arslan (Ed.), Cognitive Radio, Software Defined Radio, and Adaptive Wireless Systems, 421–452.

applicability of some of the new developments in the world of communications. This fact was seen by researchers, but they have only patched the flaws rather than applying radical changes. However, with the emergence of cognitive radio, which was coined by Joseph Mitola III [3], the perception of adaptation and optimization of wireless communication systems gained new dimensions and perspectives. The emergence of cognitive radio (and cognitive engine) is a promising solution for the barrier which arises from the flaws of the fundamental design architecture.

In this chapter, we search for an answer to the question how can a global (or multilayer) adaptation and optimization be established for cognitive radio? In order to be able to provide a concrete answer, first we will briefly review the fundamental design architecture while providing the reasons and relevant efforts of migration from the traditional architecture to cross-layer design. Next, we will outline cross-layer architecture and the place of adaptation along with optimization for the past and contemporary wireless communication systems. Subsequently, we will provide the essentials for an overall adaptation and optimization process in terms of cross-layer architecture. Later, we will introduce three cross-layer application examples depending on the relationships between layers as adjacent layer interaction, nonadjacent layer interaction, and composite interaction. Then, we will investigate the optimization problems from a formal perspective to be able to gain some insight into cross-layer optimization for cognitive radio. We will also address multi-objective optimization problems (MOPs) and relevant solutions which are going to run on cognitive engine. In addition, we will summarize the challenges related to cross-layer adaptation and optimization for cognitive radio. As a final remark, we will extend individual cross-layer adaptation and optimization problems to a network in which there are multiple individuals.

14.2 Why We Need Cross-Layer Design, Adaptation, and Optimization

14.2.1 Traditional Layered Design and Its Evolution

Traditional protocol stack has been designed for dealing with complicated problems by breaking them into smaller parts. It consists of layers whose definitions and tasks are defined explicitly and independently. In other words, each layer is isolated from the others except for providing output to and getting input from adjacent layers [4]. According to the direction of the flow upward/downward, each layer conducts its own task by taking inputs from the layer below/above and conveys the outputs obtained to above/below. This architecture has several advantages. First, defining the tasks explicitly provides modularity, which means simplicity in the design. Second, explicit definitions facilitate the standardization process. Therefore, several vendors can produce various types of products by following the explicit abstractions, and

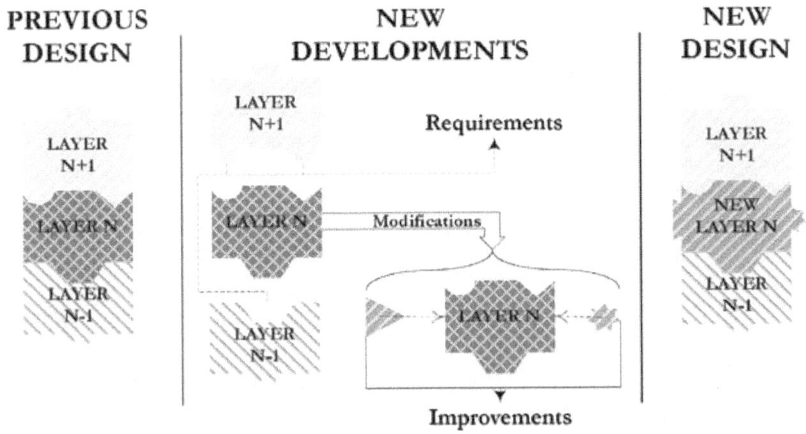

Fig. 14.1. The only consideration in designing a new layer for traditional architecture is that the new layer design should meet the requirements of the adjacent layers. The requirements of layer $N+1$ are denoted as a round and a triangle, whereas that of layer $N-1$ is represented with a half–hexagon.

at the same layer level, any of them can communicate with the others without having any problem. Finally, modularity that comes with independence allows any type of alteration at any layer level as long as the input/output requirements of the adjacent layers are met, which gives rise to the concept of expandability of the layers as illustrated in Figure 14.1.

Even though layered structure overcomes many problems successfully, it has been realized that the stringent architecture creates some problems such as asynchrony and inefficiencies. A brief list of the major advantages and disadvantages of the traditional layered architecture and their effects is given in Table 14.1. At the beginning, the disadvantages have been tried to alleviate by increasing the amount of information flow among the layers [5]. Afterwards, with the emergence of new high-speed communication networks, the major handicap of the traditional design, which is the obligation of an algorithmic ordering, has been emphasized more. As in algorithms, in the traditional architecture, a process which resides in the next adjacent layer cannot be executed unless a process defined in a layer is completed. In early '90s, the obligated ordering concept, which causes long communication delays and low throughput for wired networks, has been tried to overcome by inter-layer optimization [6]. Inter-layer optimization re-organizes the processes for layers in such a way that some of the processes that do not interfere with each other can be integrated and executed in parallel rather than serial.[2] This approach is extended to a new protocol scheme called Horizontally Oriented Protocol

[2] At this point, it must be stated that even though some of the processes are integrated, inter-layer optimization still preserves the strictly isolated structure of layers [6].

Table 14.1. The major advantages and disadvantages of the traditional layered architecture and their effects.

+/−	Explanation	Effect
Advantages		
Modularity	Each layer can be designed independent of others	Simpler design
Standardization	Design only requires to have the knowledge of explicit definitions and abstractions	Interoperability
Expandability	Layers can be updated, altered, or expanded "independently"	Individual flexibility
Disadvantages		
Ordering	Execution of any process in any layer has to be after the execution of previous processes in former layers	• Inefficiency • Latency
Interaction	Due to strict isolation, information cannot cross other layers	• Unawareness • Redundant processes • Sub-optimal performance
Adaptation	In wireless communications, rapid channel variations cannot be responded immediately	• Decrease in capacity • Sub-optimal performance
Topologies	Some of the network topologies need flexible layer architecture	Inefficiency

Structure (HOPS). In HOPS, the building stone of the architecture is defined as "function," instead of "layer." Thus, the functions that do not need to wait for others' outcome can be binded and executed in parallel, which is the reason why the scheme is called as "horizontally oriented" [7].

Migration from strict layered architecture to a more flexible interactive one has another very strong motivation: *wireless* networks. Because of the different nature of wireless communications, numerous concepts defined in wired networks need careful re-consideration or even modification, so does the protocol stack. Peculiar to wireless communications, due to small-scale fading, wireless channel conditions may change drastically in a very short duration of time [8]. Therefore, in order to take advantage of the durations in which the channel is identified as "good," a flexible design is essential [9]. Large-scale channel variations contribute to the necessity of flexible architecture as well [10]. Interference and time-varying capacity property due to multipath, relative mobility, and shadowing are other very crucial parameters that affect the wireless networks [8, 11]. Apart from those, new transmission schemes for wireless communications such as relay networks [12] may not be established via a strictly isolated layered architecture [9] and might require a different design.

Especially in wireless networks, *cross-layer approaches* emerged by increasing the amount of information provided to the adjacent layers. As an immediate example, packet losses in wireless networks can be considered. In wireless networks, packet losses can occur because of bad wireless channel conditions, congestion, or some other reasons. However, operating in transport layer, the Transmission Control Protocol (TCP) cannot comprehend the reason behind these losses. Therefore, it assumes that packet losses solely depend on congestion [13]. Thus, any loss detected is going to be handled[3] in terms of congestion even though the actual reason might be different. In order to compensate for this flaw, the link reliability is also tried to be improved as much as possible. Hence, in case of a loss, wireless link is eliminated from other possible causes, so TCP can handle the issue by considering other factors [13]. In fact, this approach solely is not enough to optimize the problem because of the drop of throughput. However, if the information about the reason of the loss can be obtained from the layers, the system can only focus on alleviating the actual reason without considering the other possibilities [10], which improves the performance.

14.3 Cross-Layer Design, Adaptation, and Optimization

Up to this point, we have observed that under some circumstances, strictly layered architecture performs inefficiently. As discussed in Section 14.2.1, in order to overcome this major problem, increasing the amount of information flow between layers and re-organizing the processes according to their dependency on each other have been proposed. All these efforts lead to a novel concept called "cross-layer architecture." Hence, *cross-layer design* can be defined generally as follows: "Any kind of innovation on the traditional structure that blurs, changes, or even removes the boundaries between layers."

In the literature, there are numerous types of cross-layer designs in the frame of the definition stated above. Some of the designs only allow the information to flow upward and/or downward direction [15], whereas some of them are based on coupling of some of the layers [16] or merging some adjacent layers [10]. Even though these innovations are considered as solutions for some

[3] TCP uses slow start, congestion avoidance, fast retransmit, and fast recovery algorithms together to avoid and handle the congestion. The transmission is initiated with "slow start," which is based on a gradual increase of sending rate of the segments. This gradual increase is kept until a congestion is detected. In case of a congestion, TCP employs a special algorithm which slows down the sending rate of segments. When the congestion is cleared, TCP employs the slow start again and tries to attain maximum throughput based on the gradual increase by avoiding congestion [14].

problems, they come at the expense of different problems such as more complicated designs compared to the traditional one. Besides, blurring or removing completely the boundaries between layers in the traditional architecture causes the tasks defined explicitly for each layer to spread into other layers and become others' problems as well. In other words, violating the independence of layers introduces additional dimensions to the tasks of other layers. Consequently, optimization of the tasks is converted from a narrow (single layer) domain to a broader (multi-layer) domain.

Having a cross-layer architecture with optimization is not going to be sufficient for ultimate system design goal. The missing link in the chain is adaptation [8, 17, 18]. In Section 14.2.1, it is outlined that the status information of a wireless communication system needs to travel among the layers because of changing wireless channel conditions, network load, and Quality of Service (QoS). Allowing the status information to travel among the layers is a starting point to complete the chain [17, 18]. Therefore, in the subsequent sections, the infrastructure for merging the concepts of cross-layer, adaptation, and cognitive radio will be discussed.

14.3.1 Cognitive Radio, Cross-Layer Design, and Adaptation

Cognitive radio is a radio that can sense, be aware of, learn, and adapt to the surrounding environment according to its inner and outer stimuli. These properties of cognitive radio take their places in the cognition cycle [19]. Overall cognition cycle can be seen as an instance of Artificial Intelligence (AI), since it encompasses observing, learning, reasoning, and adaptation.

Adaptation itself is a complex problem in the cognition cycle, because cognitive radio needs to take into account several input sources at the same time including its own past observations as a result of learning property. For instance, during its adaptation, cognitive radio needs to consider several requirements simultaneously such as user and application preferences, its own capabilities such as battery status, environmental conditions such as the availability of spectrum and propagation characteristics, and so forth. A compromise point, which can be regarded also as optimization, is tried to attain between these requirements. Note that some of the requirements fall into the tasks of specific layers in the traditional design. More explicitly, cognitive radio needs to consider QoS requirements, physical medium options as in traditional architectures beside some additional constraints such as battery consumption and past experiences. Therefore, one can conclude that, cognitive radio needs an overall adaptation that covers multiple layers with the aid of optimization.

Currently, there is no architecture that can meet all the aforementioned requirements of cognitive radio simultaneously. A fundamental reason behind that is the absence of any sort of controller and coordinator governing the overall adaptation process. The obligation of the presence of a controller and coordinator for complete adaptivity can be explained by an analogy. Since cognitive radio has AI capabilities, it is adequate to consider the most intelligent systems on the Earth: humans. Humans have aural, olfactory, tactile, taste, and visual sensors. These sensors help humans to perceive the surrounding environment. Humans are aware of themselves with the aid of inner sensors called as nerves. Humans are also equipped with very complex structures called "organs" to carry out vital operations. Each organ in the human body is physically isolated from the others such as the heart, the kidneys, and the liver. Interestingly, as in the layered architecture, some of the organs operate in algorithmic order such as digestive system. In the human digestive system, the intestines should wait for the stomach to operate. At the end, there is another structure that controls and coordinates every single organ in the human body: the brain. The brain is aware of both inner and outer world of the body via the nervous system and sensors. It gathers all the information from inner and outer world, processes and compares it with its past knowledge, chooses the best (or, in engineering terminology, optimum) decision, acts on, and observes the consequences for future usage. This procedure that the brain follows highly resembles the cognition cycle.

The human body analogy stresses that, a special structure, which has the capability of both controlling and coordinating, is essential in order to obtain a complete adaptive architecture. In cognitive radio domain, the counterpart of this special structure is known as "cognitive engine." Even though currently there is no formal definition of cognitive engine, it is agreed that cognitive engine is responsible for the overall adaptation and optimization process. However, a question arises automatically by introducing cognitive engine to the cross-layer design: What kind of an architecture should be adopted to include both cross-layer design and cognitive radio? Now, we seek for appropriate approaches, if possible, an answer to this question.

14.3.2 Cognitive Engine and Cross-Layer Architecture Design

As stated in Section 14.3.1, when a completely adaptive system is considered, the control and coordination of already defined layers have to be organized. The initial step is to establish the flow of information between each layer regardless of the levels of layers. In the earliest attempts toward cross-layer design, several layers were connected to each other in bi-directional way [17]. With the help of combinatorics, for an n-layered architecture ($n \geq 2$, $n \in \mathbb{Z}$), the number of single-direction flow must be defined is given by $\binom{n}{r}$ and $r = 2$.[4] Considering that the flow has two directions (upward and downward) as in [17] and there are already information paths between adjacent layers, the total number of new information paths to be defined becomes:

$$\mathcal{R} = (n-1)(n-2). \tag{14.1}$$

As can be seen in (14.1), a linear increase in the number of layers causes a quadratic increase in the number of new paths to be defined, which should

[4] This representation is known as binomial coefficient and given by $(n!)/((n-r)!\,r!)$.

be avoided. Besides, solely $\mathcal{R}$ paths are not going to be enough to attain a complete adaptation since pre-defined layers are not capable of converging to a complete adaptation. In addition, cognitive radio not only consists of pre-defined layers, but also includes several other sensors that need to be in connection with the adaptation process. This emphasizes that cognitive engine needs to form a sort of interface between the layers and sensors. In Section 14.3.1, when the cognition cycle was introduced, one of the abilities of the cognitive radio, learning from its past experiences, has been brought forward. Learning from past observations includes memory related processes. Therefore, apart from layers and sensors, cognitive engine needs to interact with different parts of the hardware. Finally, considering the evolution of cognitive radio, cognitive engine can be updated and modified.

By considering the aforementioned aspects, it is reasonable to think cognitive engine as a new layer that has connections with each layer, sensors, and hardware. This assumption, as a side-effect, also removes the quadratic behavior defined in (14.1) and allows one to facilitate the already defined layered architecture to some extent by converting (14.1) into a linear form.

One of the important characteristics of this new architecture is that it should preserve the previous achievements related to cross-layer organization and architecture. More explicitly, cognitive engine must take advantage of the present architecture rather than changing it entirely. The major impact of the cognitive engine on the cross-layer design is to remove the distance between layers on the edges. Consequently, a cognitive engine that is considered as a separate layer can be placed between layered architecture and several other peripherals such as memory and sensors. This is illustrated in Figure 14.2.

In a contextual model as shown in Figure 14.2, the operation of cognitive engine is extremely important. In this architecture, cognitive engine is "attached" to an already existing structure. Thus, the presence of cognitive engine should not create any problem to the layered architecture, since current layered architecture can already handle various issues very efficiently. What cognitive engine introduces is to spread the adaptation among all the layers, which cannot be achieved through traditional layered architecture. On the other hand, cognitive engine forms an interface for the available information coming from peripherals (such as memory and additional sensors) to improve the performance and adaptivity. Hence, it can be said that cognitive engine intervenes only when it is needed.[5] According to the situation, cognitive engine can take over the optimization process because of the necessity

[5] Actually, this behavior of the cognitive engine has also several counterparts in the human body analogy, which is discussed earlier. One of them is known as "involuntary (stereotyped) reflex actions." A person who touches a hot stove immediately pulls his/her hand back without "thinking." This response to the stimuli is handled by a mechanism called "reflex arc." Even though the brain is not involved with the first stage of the action (pulling the hand immediately), later on, the information "pain" is sent to the brain and the brain relates the actions and learns.

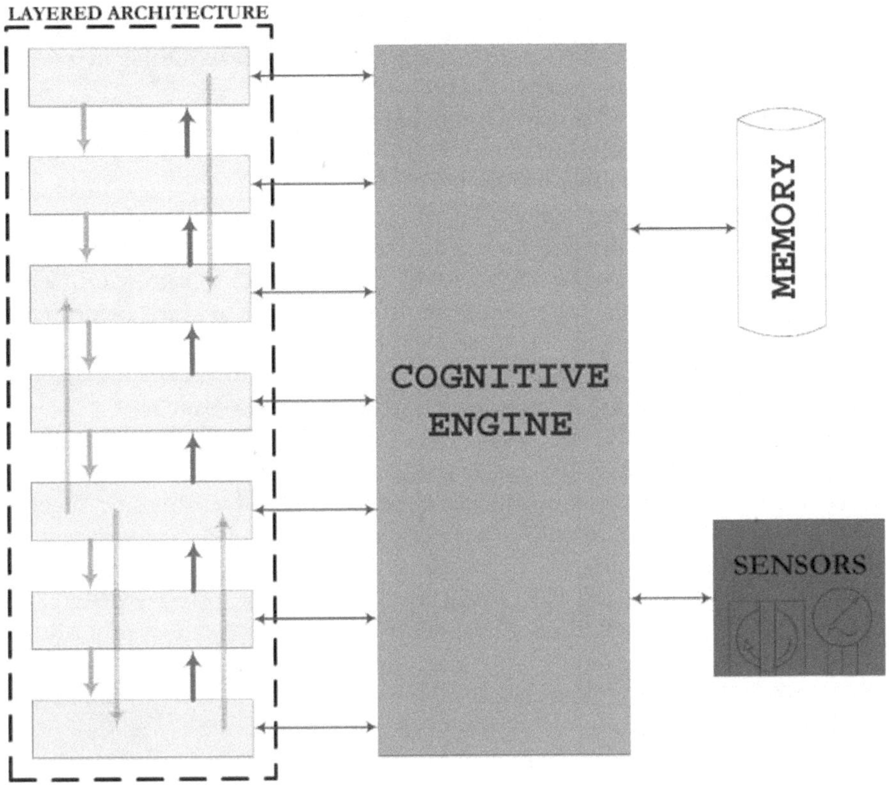

Fig. 14.2. A contextual model for cognitive radio. Cross-layer architecture and peripherals such as memory and sensors are all inter-connected through cognitive engine.

of the contribution of other layers. Similarly, cognitive engine takes action to by-pass some of the layers for the sake of optimization and/or speed or battery requirements.

Some illustrative examples – which do not contain any structure such as cognitive engine – for cross-layer adaptation and optimization can be presented for the sake of further comprehension before going deep into cognitive engine.

Illustrative Examples of Cross-Layer Adaptation and Optimization

Wireless channel possesses various characteristics. Multipath phenomenon introduces spreading in time, whereas due to the Doppler effect, the signal spreads in frequency domain. In addition to multipath propagation and Doppler spread, the transmission bandwidth is of great importance in understanding the characteristics of the wireless channel [8].[6] These concepts are the most prominent factors affecting the small-scale fading in wireless channels, which can cause erroneous data reception.

In order to be able to achieve a reliable communication over fading channels, channel coding is used to detect and correct possible errors. The essence of channel coding is to introduce redundancy into the data to be sent in order to restore it at the receiver side. There are several channel coding schemes such as block codes, convolutional codes, and turbo codes. One of the most important parameters in the channel coding is the coding rate. Coding rate is defined as $\mathbf{R}_c = k/n$, where k denotes the number of bits before channel coding and n represents the number of bits after encoding operation. Thus, $n-k$ is the total redundancy which is a measure of spectral inefficiency due to coding process [8]. Consequently, $\mathbf{R}_c = 1$ means no redundancy at all, in other words 100% efficiency. It is important to remember that even though $\mathbf{R}_c = 1$ promises 100% spectral efficiency, in case of an error, the data may not be recovered appropriately, which requires re-transmission meaning low data rate. We can exemplify channel coding with contemporary communication technologies such as Global Service for Mobile (GSM) and Global Packet Radio Service (GPRS). GSM uses 0.5 code rate for speech data [20], whereas GPRS uses four different code rates between 0.5–1 [21], according to the pre-defined channel quality schemes. Measuring the link quality consecutively allows GPRS to switch between different code rates to establish high throughput [21].

It is also possible to get higher data rates by preserving the spectral efficiency. This is achieved by switching to higher order modulations in case of having a good link [8]. In Enhanced GPRS (EGPRS), two different modulations are available with different coding rates [8,21]. After link measurements, the communication system adjusts itself with the aid of Adaptive Modulation and Coding (AMC). If the link quality is good, then EGPRS can switch from lower order modulation (Gaussian Minimum Shift Keying [GMSK]) to a higher order modulation (8–Phase Shift Keying [PSK]) with reducing the channel coding power $((\mathbf{R}_c \rightarrow 1) \equiv (n \rightarrow k))$ [22].

EGPRS also provides a different cross-layer application which includes again the collaboration of data link and physical layer. Instead of AMC, this time, a technique, which is known as "incremental redundancy" (a method in Type–II Hybrid Automatic Repeat Request (ARQ)[7]), is used [21]. In incremental redundancy, first, the data is sent through the channel with a weak coding power ($\mathbf{R}_c$). If no error occurs, a high bit rate is achieved. Unless there is an erroneous reception, which is going to be notified by an ARQ scheme

[6] Note that if the transmission bandwidth is less than the coherence bandwidth of the channel, the effects of signal spreading can be neglected.

[7] It is named as Type-II, because it stores the erroneous packet whereas Type-I discards it. The other method of Type-II Hybrid ARQ is known as "chase combining."

Table 14.2. Separation theorem and JSCC.

Traditional architecture	Cross-layer architecture
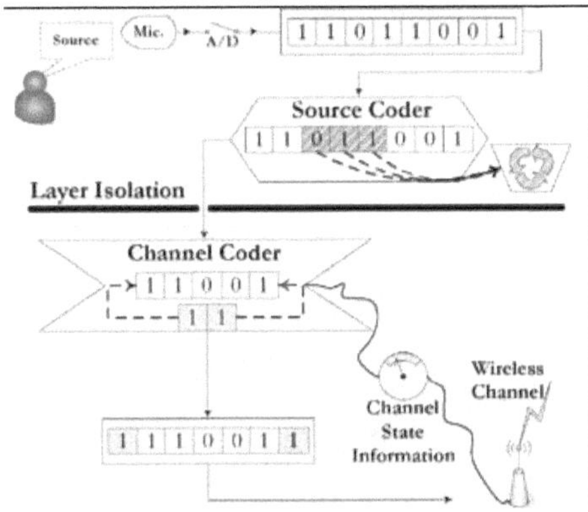	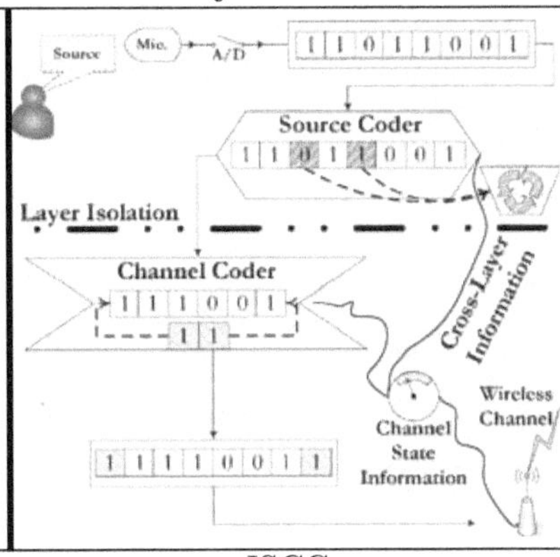
Separation theorem [34]	JSCC

to the transmitter side, the coding power is never increased. In case of an erroneous packet reception, the coding power is increased step by step until the reception becomes error-free [23].[8]

Up to this point, some applications that facilitate the collaboration of the layers close to each other have been outlined. There are also other applications that can use cross-layer approach for the layers residing at the edge of the traditional architecture such as the collaboration of application and physical layer. A very well-known example of these sorts of applications is known as Joint Source–Channel Coding (JSCC).

In multimedia applications the notion of perceptual quality is of vital importance. Therefore, the main purpose of the multimedia transmission can be defined as to obtain the best perceptual quality. However, when the transmission is carried out over wireless channels, numerous constraints (such as fading, shadowing, interference, and so on) that affect the perceptual quality must be considered as well (see Table 14.2). Especially video transport through wireless channels is one of the prominent applications for JSCC. A JSCC that is aware of the wireless channel[9] improves the performance significantly [25, and references therein]. The main purpose of any type of communications is to make sure that the information that is intended to be sent arrives at the receiver side. Otherwise, by definition, there is no communications, which means that there is no need to think further. Therefore, a link must exist between transmitter and receiver. In terms of JSCC for video transportation, we cannot talk anything about the source coder unless a link exists. If there is a link, then

[8] The working principle of incremental redundancy method can be considered as the reciprocal of slow start method for TCP, which was introduced before.

[9] Note that some JSCC approaches do not assume the presence of any sort of channel knowledge [24].

both source and channel coder get into the picture. In order to protect the video information from link degradations, channel coding is applied. However, this comes at the expense of reducing the efficiency of the bandwidth. If the video information is not protected, it may not arrive or may arrive at the receiver side but with unrecoverable errors, which causes the video to lose its intelligibility. Source encoder removes redundancies from the video and prepares the encoded scheme in an error-resilient way [26]. Before the delivery, the information is packed as video frames. At this point, bit rate of each frame is controlled by the rate controller according to the Channel State Information (CSI), which is provided by lower layers such as physical layer. At the final stage, channel coder adds redundancies depending on the link qualifications. Hence, in JSCC, the problem reduces to the optimization or allocation of the total bit rate between source and channel coding operation [27].[10]

Cross-layer design (and therefore, cross-layer optimization) manifests itself in a more complicated way for wireless ad hoc networks compared to the previous examples. Lack of communication infrastructure brings several other issues into the picture. Unlike the systems that have infrastructure, in ad hoc networks, each node should consider very challenging tasks in several layers such as routing in network layer due to the dynamic topology; scheduling for wireless channel access in Medium Access Control (MAC) layer; and power control in physical layer [29]. Security can also be added on top of these considerations [30]. Optimizing each layer one by one (as in traditional architecture approach) may end up with high network throughput, but this gives rise to several other considerations such as unfair transmission rates for some of the nodes in the network [31]. Therefore, the overall optimization must include throughput and resource utilization, congestion control [32, 33], scheduling [33, and references therein], and efficient routing [31], which are established in different layers in traditional architecture.

After solidifying the cross-layer concept with several examples including interaction of closer layers, distant layers, and multiple layers, it is appropriate to introduce some of the adaptation parameters that are being used in contemporary communication systems.

14.3.3 Some of the Adaptation Parameters That Are Popularly Used in Contemporary Communication Systems

Although the emergence of cognitive radio emphasizes the term "adaptivity" stronger than the previous communication technologies, we must note that the evolution of wireless communications has already been going toward adaptivity. It is not hard to see this reality when the progression of wireless communications in time is reviewed. When we look at the whole wireless

[10] As stated in [27], this sort of approach of JSCC is limited only to the source and channel encoders. The actual optimization includes more detailed investigation such as power considerations [28].

communication history, we can see that the early technologies or standards use fixed schemes in the system design such as fixed resource allocation, fixed frequency assignment, or fixed average signal quality for the receiver designs, and so forth [8]. Fixed scheme provides a simple design architecture, but it also comes at the expense of sub-optimum performance because of similar reasons discussed about traditional and cross-layer design architecture, in Section 14.2.1. This trade-off has been realized and tried to overcome by flexing the design, which was led by adaptation.

As stated in Section 14.3.1, there are numerous examples of adaptive wireless systems that have already been used. Especially recent standards such as WiMAX include many adaptation capabilities. However, these recent efforts as well as the previous adaptation methodologies focus on individual layers and look at the problem from a narrower perspective compared to cognitive radio. This is not surprising, because global adaptation requires perfect knowledge about all the parameters in every level including the relationships between them. But, cognitive radio, by its very definition, aims global adaptation. In global adaptation, some of the parameters conflict with each other for specific optimization criteria. Therefore, cognitive radio must be aware of what to change for adaptation and how those changes affect the system.

There are adaptation forms that serve to attain the same goal in wireless communication systems. One of the very well-known examples of these sorts of adaptation methods is to maintain Bit-Error-Rate (BER) at a certain level (constant BER). In many adaptive wireless communication systems, maintaining the desired BER level is established by increasing the power level or applying Forward-Error-Correction (FEC) techniques [8]. Increasing the power level has several side-effects such as faster battery consumption, increase in interference, and so forth. Similarly, applying FEC techniques reduces the efficiency of the use of the bandwidth. In this case, exploiting the options of attaining the same goal requires the evaluation of the side-effects of each path.

Typically, when upper layer requirements are also taken into account as in multimedia transmission, the global adaptation encompasses several constraints farther such as delay, perceptual quality, and so on. At this point, cognitive radio chooses one available option that takes it to the global optimum, if possible. However, introducing more constraints into the optimization process increases the probability of conflict between constraints. When an application that requires both high data rate and a constant BER is considered, applying adaptive modulation will cause the two goals to conflict, because maintaining BER under a desired level is possible with reducing the order of the modulation. This automatically reduces the data rate, under the assumption that the other limitations are constant. Conversely, under the same conditions, a high-data rate communication requires higher order modulation, which increases BER.

As we see, introducing even one constraint complicates the problem. Thus, cognitive radio needs to consider the trade-offs mentioned above comprehensively, since there are many parameters to be adjusted. Table 14.3 provides

Table 14.3. Some of the writable parameters for adaptive wireless communication systems.

Layer	Parameters
RF	Antenna powers Dynamic range Pre-distortion parameter Pre-equalization parameter
Physical layer	Transmit power Digital modulation order Carrier frequency Operation bandwidth Processing gain Duty cycle Waveform Pulse shaping filter type FFT size (for OFDM) Cyclic prefix size (for OFDM)
Data link layer	Channel coding rate Channel coding type Packet size Packet type Data rate Interleaving depth Channel/Slot allocation Carrier allocation (in multi-carrier systems) MAC scheduling algorithm Handover (Handoff) Number of slots
Network	Routing algorithm/metric Clustering parameters Network scheduling algorithm
Transport	Congestion control parameters Rate control parameters
Upper	Communication modes (simplex, duplex, etc.) Source coding Encryption Service personalization

some of the currently used popular adaptation parameters with respect to the layers. In the frame of global optimization, cognitive radio needs to consider these and many others which will appear jointly in the future.

14.4 Cross-Layer Optimization

Although cognitive engine conceptually looks like the missing part of overall cross-layer adaptation and optimization, in the implementation stage, the real challenge is to construct the formal methods that are going to run on cognitive engine. Currently, there is no unified mathematical model that can handle each and every one of the capabilities mentioned above. Consequently, cognitive engine needs to have –at least until a unified model appears–several mathematical models to cope with different aspects of the cognition cycle.

Fortunately, there are very successful individual formal models that can operate in particular domains of the cognition cycle such as learning, reasoning, multiobjective optimization, and so on. Specifically, in this section, we are going to investigate how a MOP can be handled via available formal models. Before getting into the details of multiobjective optimization concept, it is appropriate to introduce the optimization problem in a general context.

14.4.1 Optimization Problems

No matter how complex the optimization problems are, the main goal of all of them is the same: "to find the best solution among available set of solutions under limited resources." In order to be able to visualize this definition, we can consider a very well-known example called as "0–1 knapsack problem." According to the story, a hiker wants to put several items (such as cans of food, bed roll, and so on) into his bag, but he cannot carry more than 70 lb [35].[11] He wants to find "the best" combination of the items which weight as close as possible to 70 lb according to the relative value of each item determined by himself. For instance, he may think of cans of food as more valuable than bed roll, because food is essential for survival even though its price is less than that of the bed roll. More explicitly, in 0–1 knapsack problem, the limited source is the bag (or the hiker) that cannot carry more than 70 lb, whereas the best solution corresponds to the combination of the hiker's favorite items that weight as close as possible to 70 lb without exceeding it.

Establishing a 'reasonable' solution requires the optimization problems to have a formal model. In a general optimization problem, the formal model relies on defining the following three items:

- **Variables:** They comprise the essence of the problem via the mathematical relations between each other.
- **Objective Function:** It represents the concept that is going to be optimized. It can be univariate or multivariate depending on the structure of the problem. The purpose of the problem corresponds to obtaining either maximum or minimum value of this function.

[11] The problem is called as 0–1 knapsack problem, because the hiker either chooses an item (which is represented by "1") or leaves it (which is represented by "0").

- **Constraints:**[12] As the name implies, these are the limitations by which the objective function is going to be optimized. Along with the domain of the objective function, it defines the feasibility region, which means that any probable solution outside this region is going to be ignored.

Having the items listed above on our hands, the statement of the optimization problem can be written as follows:

$$\begin{aligned} &\text{Find } \mathbf{x}^* \text{ which} \\ &\text{minimizes } f(\mathbf{x}) \\ &\text{subject to } c_i(\mathbf{x}) \leq 0, \quad i = 1, 2, \ldots, r \\ &\text{with } m_j(\mathbf{x}) = 0, \quad j = 1, 2, \ldots, h, \end{aligned} \tag{14.2}$$

where $\mathbf{x}^* = [x_1, x_2, \ldots, x_n]^T$, $(\cdot)^T$ denotes the transpose operation, f represents the objective function, c_i and m_j denotes the constraints.[13]

If one wanted to apply (14.2) to 0–1 knapsack problem, the formal model would be as follows:

$$\begin{aligned} &\text{Maximize } f = \sum_{i=1}^{n} b_i x_i \\ &\text{subject to } \sum_{i=1}^{n} a_i x_i \leq 70 \\ &\text{with } x_i \in \{0, 1\}, \end{aligned} \tag{14.3}$$

where b_i denotes the relative value of the i-th item according to the hiker, a_i is the weight of the i-th item.

Even though (14.2) gives the formal statement of a general optimization problem, there are numerous types of different optimization problems. Since there is no unified and comprehensive method available yet as stated at the beginning of Section 14.4, classification of the problem is extremely important, because the approach (or the strategy) for obtaining the solution depends on the particular class (or category). Therefore, we can briefly glance at classifications of optimization problems.

14.4.2 Classifications of Optimization Problems

Generally, the classification of optimization problems is divided into three coarse categories as follows:

1. existence of constraints,
2. structure of the variables,
3. equation types of objective function and/or constraints.

Let us now briefly introduce each classification item.

[12] Note that, some types of optimization problems may not require a set of constraints [36].

[13] In the literature, sometimes m_j is omitted.

Existence of Constraints

This sort of classification of optimization problems has already been introduced previously as constrained and unconstrained optimization problems. Even though there are numerous types of unconstrained optimization problems (and a significant amount techniques devoted), in cognitive radio domain, most of the adaptations require at least one or two constraints such as battery level, channel state, and so on. Therefore, most of the time the optimization problems related to cognitive radio fall into the category of constrained.

Structure of the Variables

Structure of the variables determines the domain of the problem in which it is going to be investigated. There are several structure categories such as continuous–discrete, deterministic–stochastic, and so on. In cognitive radio domain, for instance, in the optimization problem which includes rate controller as mentioned in Section 14.3.2, the number of bits in a frame is of discrete type. Conversely, channel state information based on Received Signal Strength Indicator (RSSI) is of both continuous and stochastic type.

Apart from the categorization above, a different categorization for this topic is also possible. A problem can be considered as combinatorial or variational depending on the cardinality of the set of variables. In combinatorial category [37], the solution set is finite.[14] In variational category, basically, the solution set is infinite [38]. Especially combinatorial optimization problems are very important for cross-layer adaptation, as mentioned in Section 14.3.2 while discussing AMC.

Equation Types of Objective Function and/or Constraints

Formal aspects of the objective function and/or constraints are extremely important to treat an optimization problem. This type of classification is slightly different from the others, since particular categories define particular mathematical tools. The following categories are generally referred in the literature according to the types of the equations within the problem: linear, quadratic, polynomial, non-linear, and sparse.

The classification of optimization problems introduced above provides just an idea about the concept. Apart from the classification above, there can be defined several other categories such as "number of variables in the objective function." The formal and detailed classification of the optimization problems and relevant approaches are out of the scope of this chapter. Interested readers, who want to gain more information about the classification, may refer to [39–41].

[14] Knapsack problem is a combinatorial optimization problem, since the hiker can establish his favorite combination of items over a finite set.

Cognitive Radio and Optimization Problems

Two fundamental concerns behind the conceptual model of cognitive radio are (i) to provide backward compatibility and at the same time (ii) to improve an already working architecture. Since the layered architecture is going to be kept to some extent and this approach is going to be combined with overall adaptation, we automatically face to a different type of major classification in optimization problems, which is known as MOPs. Before investigating MOPs, it is worth mentioning another approach in the literature, for cross-layer architecture, which comprises a mid-step between single objective function and multiobjective function.

This approach considers the use of the interaction of multiple layers to optimize only one of the objectives. The main idea behind this approach is to take advantage of the multi-modal structure of the current communication standards with the aid of combinatorial optimization. Recall that in EGPRS, the number of possible modulation schemes is limited to two, as stated in Section 14.3.2. This means that the contribution of the physical layer to the problem in terms of modulation is a set which has only two elements. Similarly, in a IEEE 802.11a system, the modulation set has four different modulation options, which are Binary PSK (BPSK), Quadrature PSK (QPSK), 16-Quadrature Amplitude Modulation (QAM), and 64-QAM. As in modulation, channel coding can be treated in the same way. If this approach is followed for every possible layer, at the end, a comprehensive set which is composed of elements formed by the Cartesian product of every possible parameter set across the layers is obtained. In other words, the Cartesian product of every possible parameter set forms the solution set in which "the best" is going to be sought for, as illustrated in Figure 14.3. Although the Cartesian product can form a set that has a large cardinality, it is still finite [42]. Of course, this reasoning gives rise to an obvious question: Who is responsible for the optimization? This question can be answered in two ways: (I) without cognitive engine and (II) with cognitive engine. For (I), there are several approaches [42]:

- *Bottom-up approach:* The lower layers try to save the upper layer from the losses. This approach cannot provide an overall optimization, since it is going to fail in multimedia applications.
- *Application-centric approach:* In contrast to bottom-up approach, this approach gives the priority to the application layer to have the control on the optimization process. This approach cannot support the overall optimization either, because the response time of the application layer to sudden changes in the lower layers (especially in the channel) is not sufficient.
- *MAC-centric approach:* After seeing that pushing the responsibility towards edges comes at the expense of sub-optimality, this approach tries to keep the control of the optimization process around the center of

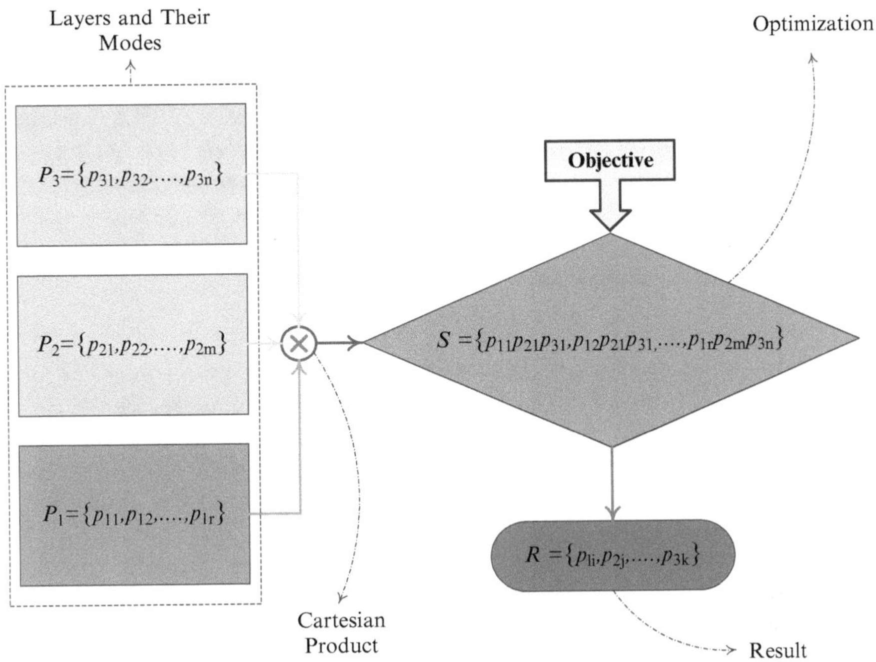

Fig. 14.3. The mid-step between single and multiobjective optimization. The design still has one objective, but, the variables are coming from different layers, which means that the type of the variables are not unique, as introduced in classification of optimization problems.

the stack. It gathers the information from the upper and lower layers and decides upon its own criteria. The major drawback occurs in JSCC.[15]

For (II), the answer is easy: "cognitive engine." Since we have not provided a substantial answer to the optimization in terms of cognitive engine, one may ask how does cognitive radio know which combination in the solution set provides the optimum solution? As being aware of its inner and outer world, cognitive engine defines the constraints according to its inner and outer environment. In one scenario, cognitive engine may realize that it is running out of battery. This automatically affects the selection of the best combination and forces cognitive engine to find a power efficient one. Conversely, as soon as the device understands that it is plugged into the electric outlet, cognitive engine immediately drops the constraint of power efficiency and defines another constraint accordingly.

[15] In [42], there is another approach proposed for (I), which is called "integrated approach." Interestingly, the key point of this approach is stated as learning and classification techniques [43].

Combinatorial approach that takes advantage of multi-modal standards draws us near our ultimate propose, which is to get solutions for MOP. However, we have to be aware of that MOPs are of different formalization from that in (14.2), since the number of objective functions is more than one. Nevertheless, (14.2) can be modified to reflect the multiobjective structure while maintaining the general frame of the optimization concept as follows:[16]

$$\begin{aligned} \text{Minimize } \mathbf{F}(\mathbf{x}) &= [\mathbf{F}_1(\mathbf{x}), \mathbf{F}_2(\mathbf{x}), \ldots, \mathbf{F}_n(\mathbf{x})]^T, \\ \text{subject to } c_i(\mathbf{x}) &\leq 0, \\ g_j(\mathbf{x}) &= 0. \end{aligned} \tag{14.4}$$

In multi-dimensional cases, the solution for optimization problem becomes harder.[17] Considering the ultimate boundary, which is time, the challenge becomes clearer. Recalling from the previous sections, in cognitive radio, it is desired to have a complete adaptivity across all the layers in a short period of time since the wireless channel is highly dynamic. Apart from that, formalizing the requirements of each layer depending on stochastic events (such as channel conditions) is extremely difficult. We will investigate the challenges of cross-layer optimization in detail later. First, we outline the statement of the problem formally.

Multiobjective Optimization Problems and Related Approaches

As their name implies, MOPs occur in a single design having multiple objectives which usually contend against each other. If MOPs are also considered in terms of the definition given at the beginning of Section 14.4.1, it is seen that the challenging part is the allocation of resources between contenders.

We have seen some examples explicitly referring to MOPs such as the relation between channel coding power and spectral efficiency. Recall that, it is impossible to have 100% spectral efficiency and maximum data protection simultaneously. When MOPs are considered, aiming maximum (or minimum) generally loses its meaning. We say "generally," because, in typical MOPs, it is extremely difficult to find a solution that can maximize (or minimize) each individual objectives simultaneously. Instead, the term solution corresponds to a set which is composed of some alternatives representing the trade-offs between objective functions. However, there may be some extreme cases that

[16] Note that, mathematically, maximization of any function f is equivalent to minimizing $-f$. Therefore, for the sake of brevity, every optimization problem can be defined only through minimization and vice versa.

[17] Again, the concept "harder" can also be defined formally. As a special form of the problem of sum-of-subsets, knapsack problem is classified as Non-deterministic Polynomial-time Complete (**NP**-Complete) in connection with the decision problem (*Entscheidungsproblem*) [44] which is believed that coined by David Hilbert. For the relations between optimization problem and computational complexity, [45] can be referred. For proofs and further discussion please see [46, 47].

lead to a solution rather than a set of alternatives [48]. These extreme cases are known as utopia [49]. Then, that solution which satisfies the maximization (or minimization) of each individual objective simultaneously is called optimal.

Ignoring extreme cases for MOPs and focusing on typical ones, we want to consider what to do when the decision time comes. At that time, an element has to be drawn from the set of alternatives for a decision to act on. At this point, a new notion called *preference* gets into the picture. It is important to note that the notion of preference distinguishes MOPs from Global Optimization Problems (GOPs). GOPs search for a single *solution*, whereas MOPs are based on getting the best *compromise* between multiple objectives in a set of alternatives. Hence, for MOPs, trade-off approach is adopted rather than a search.[18] According to the preference of the decision-maker, the notion of optimality turns into another concept called *efficiency*. Efficiency is the assessment of an element in the set of alternatives, which is chosen by decision-maker according to his preference. There are several definitions of assessment of the efficiency of an element in the set of alternatives. Before explaining the relationship between efficiency and its assessment, we need to introduce the notion of *dominance*. In MOPs, dominance expresses the preference level of the elements to each other. If one of the elements in the set, say x_2, is less preferred to another element, say x_1 (because x_1 provides better values for each individual objective function simultaneously), then, it is said that x_1 dominates x_2. Now, putting all the things together, we can assess the efficiency of the choice of the decision-maker. Edgeworth–Pareto optimal (or Pareto optimal [50], efficient solution, nondominated solution, a noninferior, or functional efficient solution [51]) is one of the most important assessment definitions of the efficiency of the element of interest. Informally, Edgeworth–Pareto[19] optimality can be defined as follows:

Definition 1 (Edgeworth–Pareto optimality) *In the set of alternatives (trade-offs or nondominated set), if there is no other element that can dominate the element chosen from the set, the element chosen from the set is called Edgeworth–Pareto optimal.*

In order to be able to solidify the concepts mentioned up until now, it is appropriate to examine Figure 14.4. Figure 14.4 illustrates a simple, two-objective function optimization problem which has a convex solution set or feasibility region. The horizontal and vertical axes denote the objective function 1 and 2, respectively. The solution set is represented with diagonally

---

[18] In the literature, "preferences" belong to decision-maker. Decision-maker is the one who is responsible for the final decision [48]. Note that, even in mid-step example introduced in Section 14.4.2 (see also Figure 14.3), the *responsibility* must be taken by someone.

[19] In the literature, Pareto optimal has a vaster usage than Edgeworth–Pareto optimal. Edgeworth is the one who proposed the correspondent term of optimum for multiobjective optimization problem, whereas Pareto is the one who generalized it. For further historical discussions, please see [52,53].

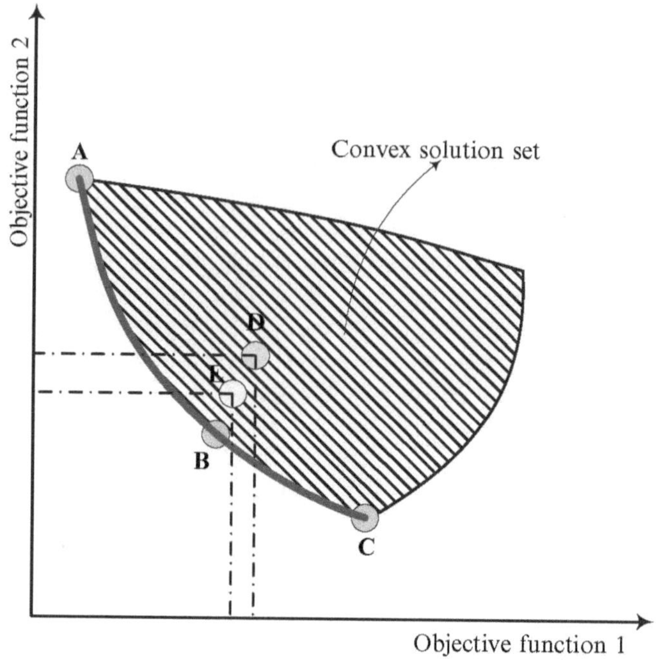

Fig. 14.4. Convex solution set, efficiency, nondominated set, and Edgeworth–Pareto optimality are represented. The circles labeled as "**A**, **B**, **C**, **D**, and **E**" denote points inside the set.

shaded, convex shape. Recall that this region is determined by the constraints peculiar to the problem. The decision-maker is allowed to choose any point within the convex solution set including its border. As stated above, the assessment is going to be established by the notion of dominance. In Figure 14.4, point **D** is being dominated by point **E**, because point **E** provides lower values for both objective functions 1 and 2, simultaneously [46,48]. This fact can be seen through the area bounded by two dashed lines originating from point **D** to the axes. Similarly, the relationship of the dominance between point **E** and point **B**; due to the property of transitivity, the relationship between point **D** and point **B** can be seen as well. Conversely, the curve $\widehat{\mathbf{ABC}}$ denotes the nondominated set, since no element can be found in the convex solution set which can provide lower values for both objective functions 1 and 2, simultaneously. Thus, depending on the preference of the decision-maker, any of the elements laying on $\widehat{\mathbf{ABC}}$ is considered as Edgeworth–Pareto optimal or efficient.

As a final remark, it must be stated that the mathematical definitions of efficiency, inefficiency, Pareto optimality, and dominance have slight differences between each other. Since these are out of the scope of this section, we will not discuss them here. However, the readers who are interested in may refer to [46, 48–51, and references therein].

Related Approaches

As mentioned previously, MOPs are complicated structures by their nature. Number of objective functions and vast variety of the constraints are two prominent factors that increase their complexities. Besides, there are some newer concepts that are not in single objective optimization problems such as decision-maker and preferences. These complexity factors and newer concepts together lead to have a different classification from the one introduced in Section 14.4.2.

MOPs are categorized as follows depending on when the preferences of the decision-maker involve with the algorithm [49, 51]:

1. ***No articulation of preferences:*** In this sub-category, the algorithm totally ignores the decision-maker before and during its run. However, after the algorithm ends, it strives to provide the whole feasible set to the decision-maker.
2. ***A priori:*** This approach assumes that the algorithm to be used has the knowledge of the preference of the decision-maker before it runs.
3. ***Progressive (interactive):*** In this one, unlike 2, there is no chronological order between the preference of the decision-maker and the initiation of the relevant algorithm. The decision-maker can provide its feedback during the operation of the algorithm. The algorithm provides a candidate solution to the decision-maker and waits for the response of the decision-maker. If the candidate solution is not accepted by the decision-maker, then algorithm strives to find a better candidate until it is accepted.[20] A further sub-classification of this category is also possible in terms of how the preferences expressed during the procedure as follows [55]:
   - target values,
   - ranking of alternatives or objectives,
   - other than above.
4. ***A posteriori:*** Unlike 1, these sorts of algorithms have the capability of narrowing down the solution space to Pareto set. The essence of these sorts of algorithms can be summarized as "generate-first-choose-later."

As discussed before, due to lack of a unified method, we need to categorize the problems so that we can apply specific tools to specific classes. Table 14.4 tabulates some of the methods available.

If the list of taxonomy above is examined in detail, it can be observed that the involvement of the decision-maker with the process needs formalization

[20] An application of such an interactive optimization procedure and the relevant flow chart is presented in [54].

Table 14.4. Multiobjective optimization problems, their classifications, and relevant solution methods.

Sub-classes	Solution methods
No preferences	Global criterion Achievment function Compromise function Objective sum Minimax method Nash arbitration Objective product Ideal distance minimization Maximal effectiveness principle
A priori preferences	Global criterion method Weighted sum Lexicographic method Weighted Tchebycheff Exponential weight criterion Weighted product method Goal programming Bounded objective function method Physical programming Multiobjective decomposition
Interactive preferences	Hierarchical decomposition method STEM method Multiobjective graph theory Method of constraints Parameter space investigation method Random search method Vector-relaxation method Interactive ϵ-grid method Method of local improvements Pareto boundary maps method
A posteriori preferences	Physical programming Normal boundary intersection method Normal constraint method Dynamic multiobjective programming Reachable set method Piecewise linear approximation method Genetic algorithms

as well. Then, how can we put the preferences of the decision-maker into the mathematical model we developed? The answer to this question, again, comes from the roots of the theory, which is economics. The function which represents the preferences of the decision-maker is called preference function or utility function.[21] With the aid of utility function, the formal model can be applied for the models which require decision-maker's feedback.

According to the taxonomy of MOPs and Table 14.4, we see that for "no articulation of preferences," the decision-maker is being dictated by the algorithm, which means that decision-maker has no control on the process [49, 51, 55]. When we consider cognitive engine, it looks a bit controversial, because whatever the solution method is chosen, it is going to run on cognitive engine. This problem arises from the fact that proposed solution methods are implemented on a different platform such as computer. Conversely, for cognitive radio, the problem, the solution, and even the decision-maker are all in the same platform. This implication proves that cognitive engine must have different units which govern the separated processes such as running the algorithm, evaluating the preferences, and so on.[22]

14.4.3 Challenges for Cross-Layer Optimization

Before integrating cognitive engine into the traditional architecture, we have to pay more attention to several aspects of cross-layer design and optimization. First and foremost, MOPs are naturally challenging due to the number of objectives involved with the problem. It is clear that the less the number of objectives, the less complex the problem. In addition, since we want to maintain the layered architecture to some extent, we must be aware of that each layer has its own design criteria. Besides, we seek for a formal model that encompasses all the layers, variables, constraints, and even objectives which are defined on different domains [18]. For instance, minimum bit-error-rate and no congestion are two separate objectives for physical and network layer, respectively. Thus, formalizing these different objectives together and putting them onto a common mathematical platform are very challenging problems. Furthermore, in spite of the fact that some solutions for MOPs have really fast convergence rates, for wireless devices that operates in very dynamic environments, the delay requirements are very tight. Another dimension of delay bottleneck challenges us in sharing the information through layers [9,18], sensors, and other peripherals of cognitive radio. In order for the solution to work appropriately, all the information should be available to cognitive engine before the optimization process starts. As a direct consequence of number of

---

21 In economics, the concept of "utility" (or satisfaction) comes from the basic consumer–entrepreneur relationship. In a system that includes both consumer and entrepreneur, the purpose is to attain maximum utility (or satisfaction) for the consumer and maximum profit for the entrepreneur [56]. Morover, including/removing uncertainty (or risk) into/from the system causes this function to be named utility function/value function, respectively [48]. Nevertheless, the term preference function can be used for both risk and risk-free systems.

22 Interestingly, current computers are designed based on von Neumann architecture, which has separate control, calculation, memory, input and output units [57]. Furthermore, this design operates as Universal Turing Machine.

objectives and delay requirements, solution for MOPs themselves may need an optimization since it consumes resources of the device. In other words, calculating the optimum allocation of a resource consumes the other system resources as well. Finally, it is known that cross-layer design can cause loops due to some interactions of layers bringing forward another challenging notion to be considered, which is known as stability [58].

At this point, we need to address another concern about cross-layer design and optimization. Before the concept of cognitive radio and therefore cognitive engine were introduced, one of the biggest challenges for cross-layer architecture had been that: "who has control? [18]"[23] With the emergence of cognitive radio, we can say that this question becomes obsolete.

Throughout Section 14.4.2, we expressed that the concept of decision-maker lies on the heart of MOPs along with preferences. We can say the same thing for cognitive engine in cognitive radio architecture. However, we also state that some of the problems have already been solved by the traditional architecture very efficiently. Therefore, in cognitive radio architecture, it is very reasonable to avoid MOPs unless they are needed. In other words, unless the intervention of cognitive radio is inevitable, cognitive engine does not need to get involved with the optimization process.

14.5 Further Notes

Up to this section, we outlined the methods to solve optimization problems for cognitive engine. However, it is worth mentioning that there are two very important accessories, which assist cognitive engine in solving the optimization problems: game theory and neural networks.

Our main focus was the optimization of the cross-layer architecture for a single device up to this point. One can suspect that optimizing one device in a network may not end up with a fully optimized network. Actually, this can easily be seen by using the same notions of multiobjective optimization discussed at the beginning of Section 14.4.2. In a network in which each node is trying to optimize its own objectives, a conflict between each node for several resources is highly probable. This phenomenon manifests itself in ad hoc networks very clearly [32]. Therefore, individual optimizations may need to be established for the sake of optimization of the network. Of course, these sorts of optimizations are of different type compared to individual optimizations, since there are two nested systems to be optimized: individual optimization and network optimization. Meanwhile, as in MOPs, we need to refrain ourselves from using the term "optimization" under those circumstances. Instead, the term "common satisfaction" may be much more appropriate. Then, the crucial question becomes: how do we attain a common satisfaction? In order to answer this question, we are going to review game theory briefly.

---

[23] Recall that, this question is investigated in Section 14.4.2 in a different form.

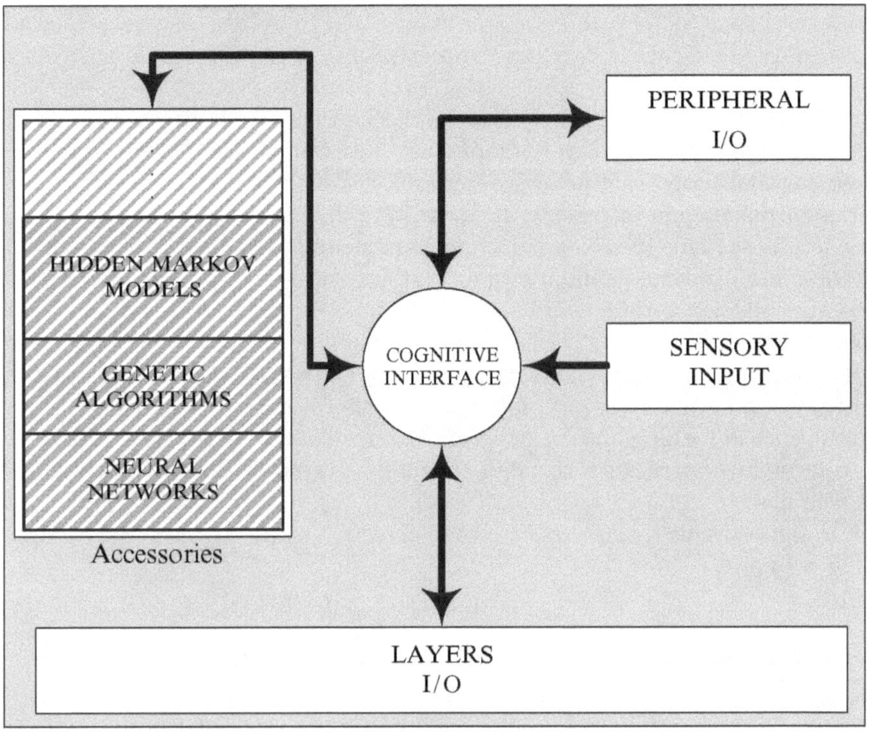

Fig. 14.5. A basic cognitive engine structure. Note that the construction highly resembles von Neumann architecture [57] because of the simplicity in the design simplicity.

Foundation of game theory is based on economics. The main idea behind the game theory is that, as defined in [56], the quantification of the problem of the rational behavior. The theory is constructed first for an isolated economical system, which is inspired from Robinson Crusoe.[24] Then, by introducing several individuals more, a social economy concept is introduced. Each individual acts rationally to attain its own satisfaction. Each individual develops a strategy depending on the system. Then, as in MOPs, each strategy is defined by a utility function which is tried to be reached to a *static equilibrium* rather than an "optimum" [56].

With this new perspective of rational behavior, many engineering problems in which there are several participants can be formalized and treated. In the

[24] As a historical note, it must be stated that the idea of rational behavior of an isolated human first appears with Abu Bakr Ibn Tufail (known as Abubacer in Latin) who was the author of "Hay bin Yakzan" (Hayy Ibn Yaqzan), the predecessor of Robinson Crusoe. In fact, the name of the hero of the story comes from another story with the same name written previously by a Turkish philosopher İbni Sina [59], who is known as Avicenna in Latin.

literature, related to wireless communications, there are also several studies that apply game theory to solve complex problems such as random access and power control [60].

Similar to game theory, neural networks must be addressed as another assisting tool for cross-layer adaptation and optimization. Stemming from their parallel processing power, neural networks are one of the essential tools for cognitive engine to cope with several problems. Inspired by the operation of nervous systems of biological organisms, neural networks have the capability of generalization, familiarity recognition, categorization, error-correction, and time sequence retention [61]. As can be seen, such properties can assist cognitive engine in many different areas including cross-layer adaptation and optimization. There are several applications related to amalgamation of neural networks and cross-layer design [62,63]. Furthermore, we begin to see that the intelligence governing and/or assisting the protocol process is being embedded into cross-layer structure through specialized layers called "cognitive layer," as well [64].

In light of this discussion, a basic cognitive engine structure can be as in Figure 14.5.

14.6 Conclusion

In this chapter, we tried to elucidate the entangled relationship between cognitive radio and cross-layer architecture. The rise of cognitive radio implies an inherent cross-layer design by its very definition including self- and environmental-awareness and adaptation. In parallel, the efforts to relax the traditional architecture point out a new intelligent structure that can provide a complete adaptation. The missing link in the chain of amalgamation of cognitive radio and cross-layer design is cognitive engine. Various work has been carried out to establish this unification. It is not hard to see that there are very challenging issues including both formal and design aspects of the merge. However, no matter how hard the problems are, the innovations that come together with cognitive radio tempt the research community to make the dream come true.

References

1. W. M. Thorburn, "The Myth of Occam's Razor," *Mind*, vol. 27, no. 107, pp. 345–353, July 1918.
2. R. K. Standish, "Why Occam's Razor," *Foundations of Physics Letters*, vol. 17, no. 3, pp. 255–266, June 2004.
3. J. Mitola and G. Q. Maguire Jr., "Cognitive radio: making software radios more personal," *IEEE Personal Communications*, vol. 6, no. 4, pp. 13–18, Aug. 1999.
4. J. D. Day and H. Zimmermann, "The OSI reference model," in *Proceedings of IEEE*, vol. 71, no. 12, Dec. 1983, pp. 1334–1340.

5. G. H. Cooper, "An argument for soft layering of protocols," Massachusetts Institute of Technology, Cambridge, MA, U.S.A, Tech. Rep., May 1983.
6. D. D. Clark and D. L. Tennenhouse, "Architectural considerations for a new generation of protocols," *SIGCOMM Computer Communication Review*, vol. 20, no. 4, pp. 200–208, Sept. 1990.
7. Z. J. Haas, "A protocol structure for high-speed communication over broadband ISDN," *IEEE Network*, vol. 5, no. 1, pp. 64–70, Jan. 1991.
8. H. Arslan, "Adaptation techniques and the enabling parameter estimation algorithms for wireless communication systems," *Signal Processing for Mobile Communications Handbook, M. Ibnkahla, Ed. Boca Raton, FL: CRC.* CRC Press, 2004, pp. 28-1 – 28-26.
9. V. Srivastava and M. Motani, "Cross-layer design: a survey and the road ahead," *IEEE Communications Magazine*, vol. 43, no. 12, pp. 112–119, Dec. 2005.
10. S. Shakkottai, T. S. Rappaport, and P. C. Karlsson, "Cross-layer design for wireless networks," *IEEE Communications Magazine*, vol. 41, no. 10, pp. 74–80, Oct. 2003.
11. X. Lin, N. B. Shroff, and R. Srikant, "A tutorial on cross-layer optimization in wireless networks," *IEEE Journal on Selected Areas in Communications*, vol. 24, no. 8, pp. 1452–1463, Aug. 2006.
12. B. Zhao and M. C. Valenti, "Practical relay networks: A generalization of hybrid–ARQ," *IEEE Journal on Selected Areas in Communications*, vol. 23, no. 1, pp. 7–18, Jan. 2005.
13. H. Balakrishnan, V. N. Padmanabhan, S. Seshan, and R. H. Katz, "A comparison of mechanisms for improving TCP performance over wireless links," *IEEE/ACM Transactions Network*, vol. 5, no. 6, pp. 756–769, Dec. 1997.
14. W. Stevens, "TCP slow start, congestion avoidance, fast retransmit, and fast recovery algorithms," *Request for Command – 2001, RFC2001*, Jan. 1997.
15. V. T. Raisinghani and S. Iyer, "Cross-layer design optimizations in wireless protocol stacks," *Computer Communications*, vol. 27, no. 8, pp. 720–724, May 2004.
16. G. Xylomenos and G. C. Polyzos, "Internet protocol performance over networks with wireless links," *IEEE Network*, vol. 13, no. 4, pp. 55–63, July/Aug. 1999.
17. Z. J. Haas, "Design methodologies for adaptive and multimedia networks," *IEEE Communications Magazine*, vol. 39, no. 11, pp. 106–107, Nov. 2001.
18. T. S. Rappaport, A. Annamalai, R. M. Buehrera, and W. H. Tranter, "Wireless communications: past events and a future perspective," *IEEE Communications Magazine*, vol. 40, no. 5, pp. 148–161, May 2002.
19. J. Mitola III, "Cognitive radio an integrated agent architecture for software defined radio," Ph.D. dissertation, KTH Royal Institute of Technology, Stockholm, Sweden, May 8, 2000.
20. K. Järvinen, J. Vainio, P. Kapanen, T. Honkanen, P. Haavisto, R. Salami, C. Laflamme, and J.-P. Adoul, "GSM enhanced full rate speech codec," in *Proc. IEEE International Conference on Acoustics, Speech, and Signal Processing, ICASSP-97*, vol. 2, Munich, Germany, Apr. 21–24, 1997, pp. 771–774.
21. A. Furuskär, S. Mazur, F. Müller, and H. Olofsson, "EDGE: enhanced data rates for GSM and TDMA/136 evolution," *IEEE Personal Communications [see also IEEE Wireless Communications]*, vol. 6, no. 3, pp. 56–66, June 1999.
22. S. Nanda, K. Balachandran, and S. Kumar, "Adaptation techniques in wireless packet data services," *IEEE Communications Magazine*, vol. 38, no. 1, pp. 54–64, Jan. 2000.

23. S. Kallel, "Analysis of a Type II hybrid ARQ scheme with code combining," *IEEE Transactions on Communications*, vol. 38, no. 8, pp. 1133–1137, Aug. 1990.
24. T. R. Fischer and M. W. Marcellin, "Joint trellis coded quantization/ modulation," *IEEE Transactions on Communications*, vol. 39, no. 2, pp. 172–176, Feb. 1991.
25. R. van Dyck and D. J. Miller, "Transport of wireless video using separate, concatenated, and joint source-channel coding," *Proceedings of the IEEE*, vol. 87, no. 10, pp. 1734–1750, Oct. 1999.
26. Y. Wang, S. Wenger, J. Wen, and A. K. Katsaggelos, "Error resilient video coding techniques," *IEEE Signal Processing Magazine*, vol. 17, no. 4, pp. 61–82, July 2000.
27. A. Ortega and K. Ramchandran, "Rate-distortion methods for image and video compression," *IEEE Signal Processing Magazine*, vol. 15, no. 6, pp. 23–50, Nov. 1998.
28. Y. Eisenberg, C. E. Luna, T. N. Pappas, R. Berry, and A. K. Katsaggelos, "Joint source coding and transmission power management for energy efficient wireless video communications," *IEEE Transactions on Circuits and Systems for Video Technology*, vol. 12, no. 6, pp. 411–424, June 2002.
29. P. Gupta and P. R. Kumar, "The capacity of wireless networks," *IEEE Transactions on Information Theory*, vol. 46, no. 2, pp. 388–404, Mar. 2000.
30. C. K. Toh, "Maximum battery life routing to support ubiquitous mobile computing in wireless ad hoc networks," *IEEE Communications Magazine*, vol. 39, no. 6, pp. 138–147, June 2001.
31. M. Johansson and L. Xiao, "Cross-layer optimization of wireless networks using nonlinear column generation," *IEEE Transactions on Wireless Communications*, vol. 5, no. 2, pp. 435–445, Feb. 2006.
32. M. Chiang, "Balancing transport and physical layers in wireless multihop networks: jointly optimal congestion control and power control," *IEEE Journal on Selected Areas in Communications*, vol. 23, no. 1, pp. 104–116, Jan. 2005.
33. A. Eryilmaz and R. Srikant, "Fair resource allocation in wireless networks using queue-length-based scheduling and congestion control," in *Proc. 24th Annual Joint Conference of the IEEE Computer and Communications Societies (INFOCOM 2005)*, vol. 3, Miami, Florida, U.S.A., Mar. 13–17, 2005, pp. 1794–1803.
34. C. E. Shannon, "The mathematical theory of communication," *The Bell System Technical Journal*, vol. 27, pp. 379–423; 623–656, July, October, 1948.
35. G. B. Dantzig, "Discrete-variable extremum problems," *Operations Research*, vol. 5, no. 2, pp. 266–277, Apr. 1957.
36. J. Nocedal, "Theory of algorithms for unconstrained optimization," *Acta Numerica*, vol. 1, pp. 199–242, 1992.
37. M. Bruglieri, F. Maffioli, and M. Ehrgott, "Cardinality constrained minimum cut problems: complexity and algorithms," *Discrete Applied Mathematics*, vol. 137, no. 3, pp. 311–341, Mar. 2004.
38. A. A. Amini, T. E. Weymouth, and R. C. Jain, "Using dynamic programming for solving variational problems in vision," *IEEE Transactions on Pattern Analysis and Machine Intelligence*, vol. 12, no. 9, pp. 855–867, Sept. 1990.
39. R. Horst and P. M. Pardalos, *Handbook of Global Optimization.* Boston, U.S.A.: Kluwer Academic Publishers, 1995.
40. Y. Collette and P. Siarry, *Multiobjective Optimization: Principles and Case Studies.* Berlin, Germany: Springer, 2003.

41. G. E. Evans, "An overview of techniques for solving multiobjective mathematical programs," *Management Science*, vol. 30, no. 11, pp. 1268–1282, Nov. 1984.
42. M. van Der Schaar and S. Shankar, "Cross-layer wireless multimedia transmission: challenges, principles and new paradigms," *IEEE Wireless Communications [see also IEEE Personal Communications]*, vol. 12, no. 4, pp. 50–58, Aug. 2005.
43. M. van Der Schaar and M. Tekalp, "Network and content-adaptive cross-layer optimization for wireless multimedia communication by learning," *IEEE International Symposium on Circuits and Systems*, 2005.
44. A. M. Turing, "On computable numbers, with an application to the Entscheidungsproblem," *Proceedings of the London Mathematical Society*, vol. 42, no. 2, pp. 230–265, 1936.
45. C. Papadimitriou and M. Yannakakis, "Optimization, approximation, and complexity classes," in *STOC '88: Proceedings of the twentieth annual ACM symposium on Theory of computing.* New York, NY, U.S.A: ACM Press, 1988, pp. 229–234.
46. M. Ehrgott, *Multicriteria Optimization*, 2nd ed. Berlin, Germany: Springer, 2005.
47. P. Serafini, "Some considerations about computational complexity for multi objective combinatorial problems," in *International Conference on Vector Optimization – Lecture Notes in Economics and Mathematical Systems [294]*. New York, NY, U.S.A: Springer, Aug. 4–7, 1986, pp. 222–232.
48. Y. Sawaragi, H. Nakayama, and T. Tanino, *Theory of Multiobjective Optimization*, ser. Monographs and Textbooks [Volume 176]. Academic Press, Inc., Orlando, FL, 1985.
49. R. T. Marler and J. S. Arora, "Survey of multi-objective optimization methods for engineering," *Structural and Multidisciplinary Optimization*, vol. 26, no. 6, pp. 369–395, Apr. 2004.
50. V. Pareto, *Manual of Political Economy [translated by A. Schwier]*. Edited by A. Schwier and A. Page, reprinted by Augustus M. Kelley: New York, NY, 1971.
51. U. M. Diwekar, *Introduction to Applied Optimization*, ser. Applied Optimization [Volume 80]. Norwell, Massachusetts, U.S.A.: Kluwer Academic Publishers, 2003.
52. W. Stadler, "A survey of multicriteria optimization or the vector maximum problem, part I: 1776–1960," *Journal of Optimization Theory and Applications*, vol. 29, no. 1, pp. 1–52, Sept. 1979.
53. W. Stadler, "Inititators of multicriteria optimization," in *International Conference on Vector Optimization – Lecture Notes in Economics and Mathematical Systems [294]*. New York, NY, U.S.A: Springer, Aug. 4–7, 1986, pp. 229–234.
54. J. P. Hämäläinen, K. Miettinen, P. Tarvainen, and J. Toivanen, "Interactive solution approach to a multiobjective optimization problem in a paper machine headbox design," *Journal of Optimization Theory and Applications*, vol. 116, no. 2, pp. 265–281, Feb. 2003.
55. E. R. Lieberman, "Soviet multi–objective mathematical programming methods: an overview," *Management Science*, vol. 37, no. 9, pp. 1147–1165, Sept. 1991.
56. J. von Neumann and O. Morgenstern, *Theory of Games and Economic Behavior*, 60th-Anniversary ed. Princeton, New Jersey, U.S.A.: Princeton University Press, 2004.

57. J. von Neumann, "First draft of a report on the EDVAC," Moore School of Electrical Enginnering, University of Pennsylvania, Tech. Rep. W-670ORD-4926, June 30, 1945.
58. V. Kawadia and P. R. Kumar, "A cautionary perspective on cross-layer design," *IEEE Wireless Communications [see also IEEE Personal Communications]*, vol. 12, no. 1, pp. 3–11, Feb. 2005.
59. A. M. Sayili, "Was Ibni Sina an Iranian or a Turk?" *Isis*, vol. 31, no. 1, pp. 8–24, Nov. 1939.
60. A. B. MacKenzie and S. B. Wicker, "Game theory and the design of self-configuring, adaptive wireless networks," *IEEE Communications Magazine*, vol. 39, pp. 126–131, Nov. 2001.
61. J. J. Hopfield, "Neural networks and physical systems with emergent collective computational abilities," *Proceedings of the National Academy of Sciences*, vol. 79, no. 8, pp. 2554–2558, Apr. 15, 1982.
62. C. Wang, P. C. Lin, and T. Lin, "A cross-layer adaptation scheme for improving IEEE 802.11e QoS by Learning," IEEE Transactions *on Neural Networks*, vol. 17, no. 6, pp. 1661–1665, Nov. 2006.
63. P. C. Lin, C. Wang, and T. N. Lin, "A context–aware approach for multimedia performance optimization using neural networks in wireless LAN environments," in *Proc. IEEE International Conference on Multimedia & Expo (ICME '06)*, Toronto, Ontario, Canada, July 9–12, 2006, pp. 1177–1180.
64. T. Ahmed, A. Mehaoua, R. Boutaba, and Y. Iraqi, "Adaptive packet video streaming over IP networks: a cross-layer approach," *IEEE Journal on Selected Areas in Communications*, vol. 23, no. 2, pp. 385–401, Feb. 2005.

Index

B

C

M

N

Q

R

SPRINGER NATURE

GPSR Compliance

The European Union's (EU) General Product Safety Regulation (GPSR) is a set of rules that requires consumer products to be safe and our obligations to ensure this.

If you have any concerns about our products, you can contact us on ProductSafety@springernature.com

In case Publisher is established outside the EU, the EU authorized representative is:

Springer Nature Customer Service Center GmbH
Europaplatz 3
69115 Heidelberg, Germany

Kolibri 360 GmbH
Ferdinand-Jühlke-Straße 7
99095 Erfurt, Deutschland
produktsicherheit@kolibri360.de